5th International Conference on Mechanical Instrumentation and Automation (ICMIA 2024)

Journal of Physics: Conference Series Volume 2815

Wuhan, China
19-21 April 2024

ISBN: 979-8-3313-0288-7
ISSN: 1742-6588

Printed from e-media with permission by:

Curran Associates, Inc.
57 Morehouse Lane
Red Hook, NY 12571

Some format issues inherent in the e-media version may also appear in this print version.

Institute of Physics
Dirac House, Temple Back
Bristol BS1 6BE UK

Phone: 44 1 17 929 7481
Fax: 44 1 17 920 0979

techtracking@iop.org

Additional copies of this publication are available from:

Curran Associates, Inc.
57 Morehouse Lane
Red Hook, NY 12571 USA
Phone: 845-758-0400
Fax: 845-758-2633
Email: curran@proceedings.com
Web: www.proceedings.com

TABLE OF CONTENTS

Journal of Physics: Conference Series **2815** (2024) 011001 doi:10.1088/1742-6596/2815/1/011001

Preface

In the midst of the rapidly evolving landscape of technological advancements, the field of mechanical instrumentation and automation has emerged as a pivotal cornerstone in driving industrial progress and scientific exploration. It is with this recognition that we gathered once again for the 2024 5th International Conference on Mechanical Instrumentation and Automation (ICMIA 2024), a forum that aims to showcase the latest research achievements, foster collaborations, and pave the way for innovative solutions in this dynamic domain.

The ICMIA 2024, held in Wuhan, China, from April 19th to 21st, 2024 via hybrid form, serves as a gathering place for experts, scholars, researchers, and practitioners from around the world. It provides a unique platform to share insights, discuss challenges, and explore opportunities in the field of mechanical instrumentation and automation. This conference serves not only as a showcase for the latest research but also as a catalyst for cross-disciplinary collaborations and knowledge exchange.

The keynote speakers at ICMIA 2024 brought with them a wealth of experience and expertise. Their presentations, focusing on cutting-edge research and industry applications, provided valuable insights into the current state of the art and future directions of mechanical instrumentation and automation. Among them, Prof. Ning Sun (Nankai University, China) made a speech on Intelligent Control for Complex Underactuated Cranes with Applications. He first shared some of his team's recent results on dynamics analysis, motion planning, and intelligent control of different crane systems, and then discussed some of their extended and related researches on robotic systems with similar dynamic characteristics. We are grateful to these esteemed speakers for sharing their knowledge and insights.

The collected papers in this volume represent the culmination of rigorous research and innovative ideas presented at ICMIA 2024. These papers span a wide range of topics, including but not limited to Mechanical and Instrument Design, Instrument Science and Technology, Electronic Design Automation, Fault Diagnosis and Maintenance Theory, Automatic Detection and Instrumentation, Intelligent Instrument System, and their applications in various industrial sectors. The authors, representing a diverse array of backgrounds and institutions, have contributed their valuable research to this volume, reflecting the global reach and impact of the conference.

The selection process for these papers was rigorous and thorough. All submitted papers were subjected to peer review by experts in the field, ensuring the highest standards of quality and relevance. The accepted papers were then carefully curated and compiled into this volume, providing a comprehensive overview of the latest advancements in mechanical instrumentation and automation.

We are proud to present this collection of papers as a testament to the excellence and vibrancy of the research community in this field. We hope that this volume will serve as a valuable resource for researchers, practitioners, and students alike, inspiring new ideas, fostering collaborations, and propelling the field of mechanical instrumentation and automation forward.

The Committee of ICMIA 2024

Journal of Physics: Conference Series **2815** (2024) 011001 doi:10.1088/1742-6596/2815/1/011001

Committee Member

Conference Chair
Prof. Hamid Reza Karimi, Politecnico di Milano, Milan, Italy
Prof. Ning Sun, Nankai University, China
Prof. Richard (Chunhui) Yang, Western Sydney University (WSU), Australia

Publication Chair
Prof. Shaojian Zhang, Nanchang University, China
Prof. Maria Pia Fanti, Polytechnic of Bari, Italy

Technical Program Committee Chair
Prof. Shubin Yan, Zhejiang University of Water Resources and Electric Power, China

Organizing Chair
Assistant Researcher Qingxiang Wu, Nankai University, China
Lecturer Tong Yang, Nankai University, China

Technical Program Committee Member
Prof. Menghua Zhang, University of Jinan, China
Prof. Yinan Wu, Nankai University, China
Prof. Zeashan Hameed Khan, Air University, Pakistan
Prof. Fushuan Wen, Zhejiang University, China
Prof. Huajun Dong, Dalian Jiaotong University, China
Prof. Yiming Zhang, Fuzhou University, China
Prof. Intan Zaurah Mat Darus, Universiti Teknologi Malaysia, Malaysia
Prof. Ian McAndrew, Capitol Technology University
Prof. Miadreza Shafie-khah., University of Vaasa, Finland
Prof. Xianbin Wang, Western University, Canada
Prof. Ramash Kumar K., University of Madras, India
Prof. Angelo Cangelosi, University of Manchester, UK
Prof. Himani Goyal Sharma, CHANDIGARH UNIVERSITY, India
Prof. Ozgur Koray, Biruni University, Turkey
Prof. Jelena Misic, Ryerson University, Canada
Prof. Surjeet Dalal, Teerthanker Mahaveer University, India
Prof. GOH HUI HWANG, Guangxi University, China
Assoc. Prof. He Chen, Hebei University of Technology, China
Assoc. Prof. Hongze Wang, Shanghai Jiao Tong University, China
Assoc. Prof. Dr.M.Balakrishnan, M.Kumarasamy College of Engineering, India
Assoc. Prof. Taza, City University of Science and Information Technology, Pakistan
Assoc. Prof. Valerian Croitorescu, University POLITEHNICA of Bucharest, Romania
Assoc. Prof. SAIDA (Boukhris) GTIFA, UNIVERSITY OF CARTHAGE, Tunisia
Assoc. Prof. Noor Zaman Jhanjhi, Taylors University, Malaysia
Assoc. Prof. Antti Rasila, Guangdong Technion – Israel Institute of Technology, China
Assoc. Prof. Pinial Khan, Sindh Agriculture University Tando Jam Pakistan, Pakistan
Assoc. Prof. Jinsong Li, Dalian University of Technology, China

Organizing Committee Member
Prof. Hamid Bentarzi, University of M'hamed Bougara Boumerdes, Algeria
Prof. Dalin Zhang, Aalborg University, Denmark
Prof. Zaifa Zhou, Southeast University, China
Prof. Shuming Yang, Xi'an Jiaotong University, China
Prof. Behnam Mohammadi-ivatloo, University of Tabriz, Iran
Prof. Minoru Sasaki, Gifu University, Japan
Prof. Katarina Monkova, Technical University of Kosice, Slovakia
Prof. Intan Zaurah Mat Darus, Universiti Teknologi Malaysia, Malaysia
Prof. Mazhar Shaikh, Nizwa College of Technology, Oman

Journal of Physics: Conference Series **2815** (2024) 011001 doi:10.1088/1742-6596/2815/1/011001

Journal of Physics: Conference Series **2815** (2024) 011002 doi:10.1088/1742-6596/2815/1/011002

Peer Review Statement

All papers published in this volume have been reviewed through processes administered by the Editors. Reviews were conducted by expert referees to the professional and scientific standards expected of a proceedings journal published by IOP Publishing.

- **Type of peer review:** Single Anonymous
- **Conference submission management system:** Morressier
- **Number of submissions received:** 98
- **Number of submissions sent for review:** 86
- **Number of submissions accepted:** 56
- **Acceptance Rate (Submissions Accepted / Submissions Received × 100):** 57.1
- **Average number of reviews per paper:** 2
- **Total number of reviewers involved:** 10
- **Contact person for queries:**
  Name: Xuexia Ye
  Email: xx.ye@keoaeic.org
  Affiliation: AEIC Academic Exchange Information Centre

Journal of Physics: Conference Series **2815** (2024) 012001
doi:10.1088/1742-6596/2815/1/012001

Influence of tunnel cone angle and lip deflection on the performance of coaxial twin propeller

Kaichao Du [1,*], Yongkang Shi [2] and Ke Li [3]

[1] Xinjiang University, Urumqi City, Xinjiang Uygur, 830001, China
[2] Xinjiang University, Urumqi City, Xinjiang Uygur, 830001, China
[3] Xinjiang University, Urumqi City, Xinjiang Uygur, 830001, China
[1,2,3] Intelligent Manufacturing Modern Industry College, Xinjiang University, Urumqi, Xinjiang Uygur, 830001, China

[*] Corresponding author's e-mail: 3357846894@qq.com

Abstract: To investigate the influence of the culvert cone angle and the angle of the mouth on the aerodynamic performance of the culvert coaxial twin rotor and its mechanism, the quasi-steady numerical simulation of the flow field of the culvert coaxial twin rotor under a hovering state was carried out based on the Reynolds mean Navier-Stokes (RANS) equation and the multiple reference coordinate system method (MRF). The influence of cone angle from -6° to 9° and lip deflection angle from -20° to 30° on the aerodynamic performance of coaxial double-rotors with ducted passages is analyzed. The results show that the contractile culvert can generate a large gain lift force, which is conducive to improving the aerodynamic efficiency of the system. The inward deflection of the lip obstructs the flow of air into the culvert, which will worsen the local aerodynamic environment and lead to serious air separation, increasing the power consumption of the system. However, the structure of the outward deflection of the lip (open culvert) better complies with the flow trend of air, helps to reduce aerodynamic interference, and improves the additional lift of the culvert, which is beneficial to improving the aerodynamic performance of the system.

1. Introduction

In recent years, the low-altitude economy as a strategic emerging industry ushered in new development opportunities. Electric vertical take-off and landing vehicle (eVTOL), the main product in the low-altitude field, is widely used in many fields such as smart city, logistics and transportation, emergency rescue, aerial survey, and environmental protection. As a power or lift device, the ducted propeller has advantages such as high propulsion efficiency, compact structure layout, low aerodynamic noise, and good safety [1], and is widely used in VTOL aircraft. Researchers at home and abroad have done a lot of research on duct propellers. Taylor [2] conducted an in-depth study on the influence of structural design parameters such as culvert height, paddle disk position, lip shape, and culvert cone angle on the aerodynamic performance of the culvert propeller system through wind tunnel experiments for the first time and described in detail the influence of parameter changes on aerodynamic efficiency. Bento et al. [3] used steady and unsteady numerical simulation methods to study the aerodynamic characteristics of circular and square trenches and the interaction between propellers and culverts. They found that airflow separation would occur at the corner of square trenches, resulting in strong separation vortices, which reduced the aerodynamic efficiency of the ducted propeller system. Some domestic scholars have also studied the factors affecting the performance of channeled propellers. Xu and Ye [4] used the unsteady numerical simulation method based on the unstructured, dynamic nested mesh to conduct a comparative study and analysis of the aerodynamic characteristics of the ducted propeller and the isolated propeller. The results showed that the existence of the ducted propeller could not only improve the flow characteristics around the tip and reduce the tip loss but also, the negative pressure zone at the lip of the ducted tunnel could provide some extra lift for the ducted tunnel. Hu et al. [5] combined the slip grid method to conduct unsteady numerical simulation research on the flow field and aerodynamic performance of ducted fans with different tip clearance and relative positions between the paddle disc and the culvert. Based on the quasi-steady numerical simulation method of the RANS equation and multiple reference coordinate systems, Han et al. [6] analyzed and summarized the influence of design variables such as culvert length to diameter ratio and lip deflection angle on thrust characteristics and flow mechanism, but there was a lack of description of the influence of culvert shape changes on flow characteristics. Wang [7] used free wake and surface vortex methods to study the effects of different pitch angles, culvert clearance, culvert expansion angle, and lip shape on the performance of culvert propellers and drew some regular conclusions.

In this paper, the aerodynamic calculation model of the ducted coaxial twin propeller is established, and the influence of different ducted cone angles and lip deflection on the performance and flow mechanism of the ducted coaxial twin propeller are numerically simulated and analyzed.

2. Computational models and grid generation

To ensure consistent clearance between the upper and lower propeller wing tips and the inner wall of the culvert, the culvert model in this paper is obtained by using Clark-Y flat-convex airfoil rotation of 360°, and the propeller model is obtained by reverse reconstruction after three-dimensional scanning. The coaxial twin-propeller adopts a positive and negative layout, with the upper propeller turning forward and the lower propeller reversing. The rotation speed is set to 6000r/min, and the basic model and related parameter values are shown in Figure 1 and Table 1.

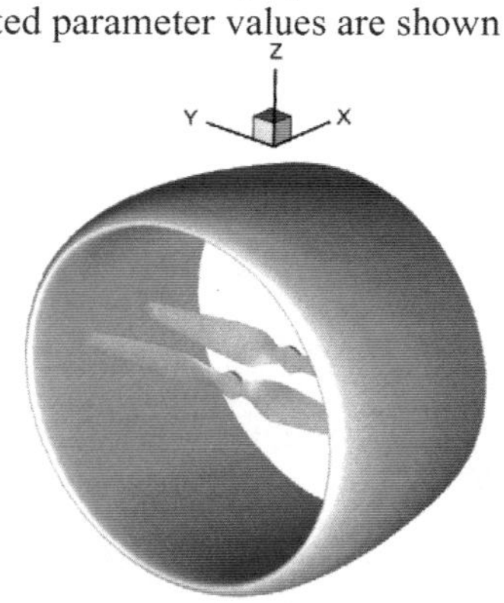

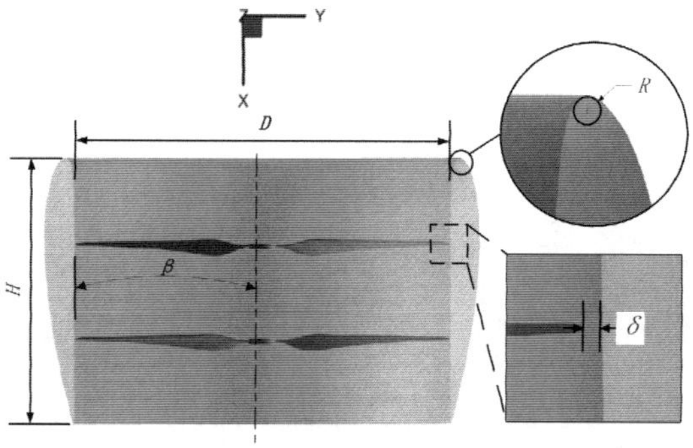

Figure 1 Basic model of coaxial twin rotor with duct.

Table 1 Basic Model Geometric Parameters.

NAME	ARGUMENT
Duct height (H)	250mm
Disc Diameter (15 inches)	381mm
Culvert clearance (δ)	4mm
Inside diameter of culvert (D)	389mm
Cone angle (β)	0°
Lip radius (R)	3mm

As the local curvature of the model varies greatly, the meshing strategy adopts unstructured meshes for discretization and adopts the best mesh elements in each part of the calculation region. The calculation region is filled with a combined grid consisting of a layered boundary layer grid near the culvert wall, a pure polyhedron grid in the transition area, and a hexahedron core grid in the core area. This results in a volume grid with common nodes, which enhances the overall quality of the grid and reduces the total number of grids and solution time. Figure 2 depicts this combined grid. The hybrid mesh, composed of polyhedral and hexahedral mesh, has the advantages of being of high quality, low quantity, and high precision. It is more suitable for the external aerodynamic numerical simulation analysis of complex aircraft.

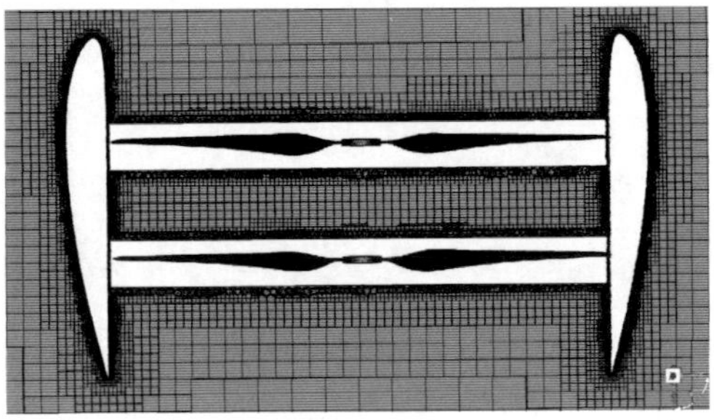

Figure 2 High-quality Poly Hexcore grid based on Mosaic technology.

To better capture the flow field details, mesh encryption was carried out on the front and back edges of the culvert, the front and back edges of the propeller, and the gap between the blade and the culvert. The culvert and blade surface mesh are shown in Figures 3-4. The total grid size is 11 million, of which 4 million are static, and 7 million are dynamic.

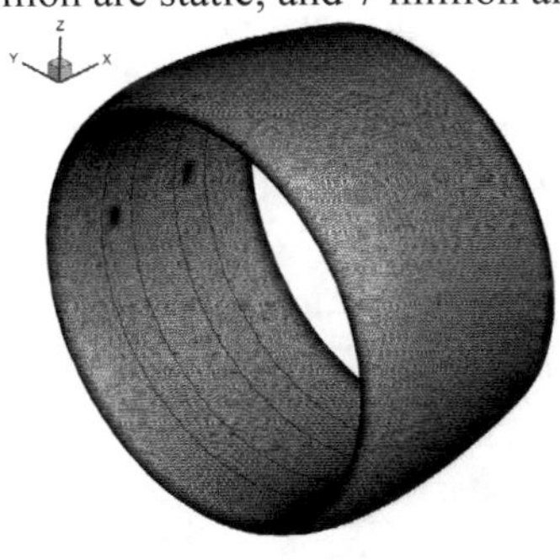

Figure 3 Surface Grid of Culvert.

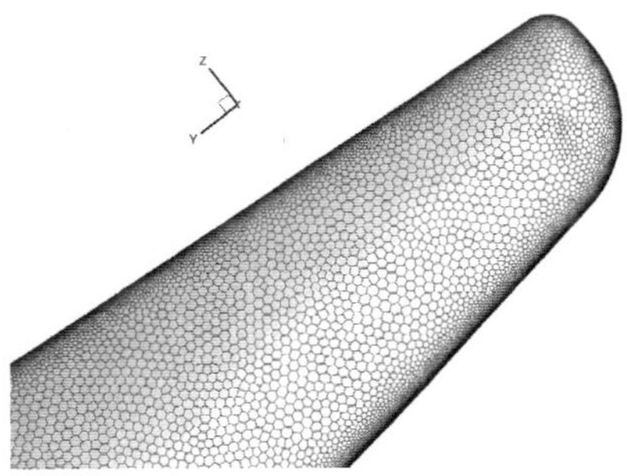

Figure 4 Grid of rotor surface.

3. Numerical solution method

This paper uses the RANS equation and multiple reference coordinate system method. The method of multiple reference coordinates is a steady-state solution with good stability and easy convergence, which essentially solves the quasi-steady flow of the flow field at a certain moment. The quasi-steady method can not only retain the main characteristics of the flow field but also has higher computational efficiency than unsteady simulation, so it is a widely used calculation method [8].

When calculating using the MRF model, the computing domain is divided into three subdomains. The rotating domain includes two rotating domains and an external stationary domain. The rotating domain rotates with the propeller contained within it, but this rotation defines the motion of the rotating component through the rotating coordinate system. There is no relative motion between the two adjacent moving regions, and the grid used for calculation is still fixed. The rotating coordinate system solves the fluid control equation in the rotating domain, and the other regions outside the rotating domain are the static domain. The static coordinate system solves the fluid control equation in the static domain. The interface is set at the contact point between the two subdomains, and the flow field information at the interface is exchanged through data interpolation between the dynamic and static domains and mathematical transformation between the coordinate systems. The numerical simulation of a multi-region flow field with rotating motion is realized. The Reynolds mean NS equation is the governing equation of the mean variable of the flow field, and the compressible RANS equation can be written as a Cartesian tensor with a summation convention.

Continuity equation (conservation of mass):

$$\frac{\partial \rho}{\partial t}+\frac{\partial}{\partial x_i}(\rho u_i)=0 \tag{1}$$

Momentum equation (conservation of energy):

$$\frac{\partial}{\partial t}(\rho u_i)+\frac{\partial}{\partial x_j}(\rho u_j u_i) = -\frac{\partial p}{\partial x_i}+\frac{\partial \sigma_{ij}}{\partial x_j}+\frac{\partial}{\partial x_j}\left(-\rho \overline{u_i' u_j'}\right) \tag{2}$$

The $\mathbf{u_i}$ Reynolds mean velocity component, which represents the omitted mean symbol, $\boldsymbol{\rho}$ is density, $\mathbf{p}$ is pressure, $\mathbf{u_i'}$ is pulsation velocity, and $\boldsymbol{\sigma_{ij}}$ is stress tensor component [9]. The K-$\boldsymbol{\omega}$SST (shear stress transport) two-equation model is used to solve the momentum model. This model combines the advantages of K-$\boldsymbol{\omega}$ calculation in the near wall region and K-Epsilon model calculation in the far field and considers the transport process of turbulent shear stress in the definition of turbulent viscosity, which can accurately predict the flow field characteristics dominated by the inverse pressure gradient and is suitable for calculating the development of separated air leaving the surface in space [10]. The velocity inlet boundary condition is set at the entrance of the calculation domain, the pressure outlet boundary condition is set at the exit, the non-slip wall boundary condition is set for the culvert and blade, and the interface boundary condition is set at the interface between the moving domain and the static domain. As shown in Figure 5, the calculation domain is a cylinder region with a diameter of 15D and a height of 30D.

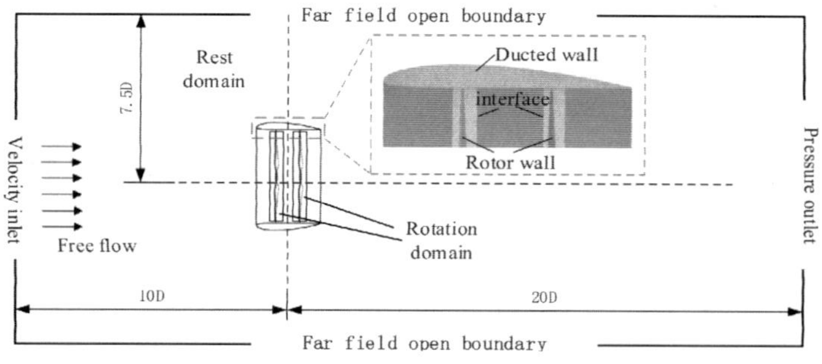

Figure 5 Schematic diagram of boundary condition setting.

4. Calculation results and analysis

4.1. Analysis of the influence of cone angle on aerodynamic performance

Proper duct cone angle can improve the aerodynamic characteristics of the ducted propeller. In this section, the influence of different cone angles on the performance of the channeled propeller system is analyzed. Figure 6 and Figure 7 show the variation trend of the tension, total tension, and power load of each part of the system with the cone angle of the culvert, respectively. Power load is the ratio of the total tension of the system to the required power, which is used to characterize the aerodynamic efficiency of the system. The higher the power load, the higher the aerodynamic efficiency of the system. The power P is used to represent the energy consumption of the system, and its definition is shown in Equation (3). The higher the power is, the higher the system energy consumption is, and the lower the system energy consumption is. The power load Q is used to represent the efficiency of the system, and its definition is shown in Equation (4). The greater the power load is, the higher the aerodynamic efficiency of the system is.

$$p = \frac{\pi\ (M_1 + M_2)\ n}{60} \qquad (3)$$

where P is propeller power, unit W; M is the propeller torque, unit Nm; n is the propeller speed, expressed in r/min.

$$Q = \frac{T}{P} \qquad (4)$$

where Q is the power load in s/m, and T is the total force of the system, expressed in N.

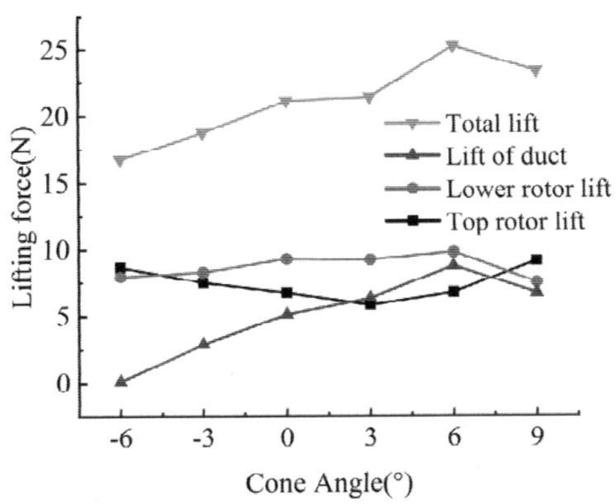

Figure 6 Lift forces of various components.

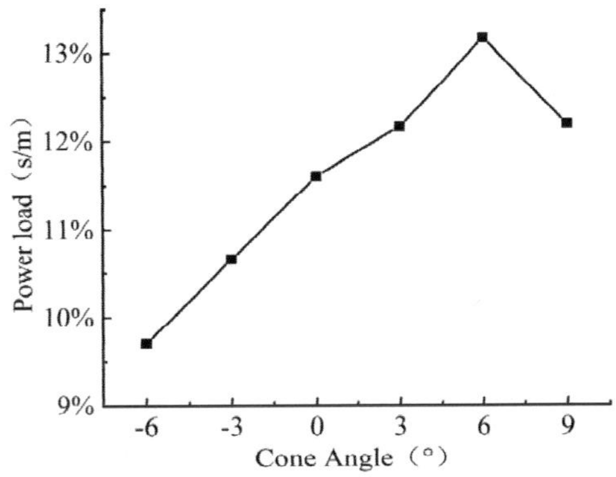

Figure 7 System aerodynamic efficiency curve under different culvert cone angles.

It can be seen that with the increase of the duct cone angle, the total pull and power load of the system increase first and then decrease; the pull of the upper propeller decreases first and then increases, and the pull of the lower propeller increases first and then decreases, mainly because the increase of the cone angle causes the gap between the upper propeller and the inner wall of the duct to increase continuously, while the gap between the lower propeller and the inner wall of the duct to decrease continuously. According to the research conclusion proposed by Deng et al. [1], the increase in culvert clearance leads to an increase in inter-propeller vortex strength, induced drag, and decreased propeller aerodynamic performance. However, when the clearance continues to increase beyond a certain range, the unloading effect of the inner wall of the culvert on the propeller is weakened, and the propeller lift is increased. The tension generated by the culvert increases with the increase of the cone angle. Figure 6 shows the changing trend of the tension of the culvert with the cone angle. It can be seen that with the increase of the cone angle within a certain range, the lifting effect of the culvert gradually strengthens. Due to the increase of the cone angle, the entrance area of the culvert continues to increase, the outlet area continues to shrink, and the shape of the culvert gradually changes from the expansion to the contraction type. According to the fluid continuity theorem and Bernoulli theorem, the axial velocity in the culvert speeds up, the airflow through the culvert increases, and the additional tension of the culvert increases. At 6°, the additional lift of the culvert body accounts for 34.7% of the total lift of the system, and the increase effect is significant. However, the total lift decreases when the cone angle continues to increase to 9°, which is because the excessive cone angle leads to narrowing of the outlet of the culvert, and the slip flow behind the paddle accumulates at the outlet of the culvert and cannot flow out smoothly. The blocking effect of the culvert on the wake begins to appear, and the waste resistance in the downwash flow field of the inner wall of the culvert also increases. The change in the cone angle of the duct directly affects the airflow flow at the entrance and exit of the duct [11]. Figure 8 shows the axial velocity cloud image and lip flow in symmetrical sections with cone angles of -6°, 0° and 6°, respectively. It can be seen that the velocity distribution in most regions of the wash flow field at the cone angle of -6° is in the range of 4.7m/s-16.5m/s, and the axial velocity gradient is large, resulting in uneven flow field velocity distribution. At a 6° cone angle, the velocity distribution is between 5.8m/s-18.8m/s, the velocity gradient of the flow field is small and evenly distributed, and the velocity decreases roughly linearly along the axis. The accumulation effect of the culvert on the downwash flow field is obvious, indicating that the contractile culvert is conducive to improving the aerodynamic performance of the culvert. This is due to the airflow around the lip and the airflow acceleration effect caused by the contraction of the culvert flow path, resulting in the reduction of the pressure on the inner surface of the culvert and forming a low-pressure area near the entrance of the culvert. The negative pressure at the entrance of the culvert provides part of the additional lift for the system. It is worth noting that when the culvert cone angle is -6°, the gain lift generated by the culvert body only accounts for 0.7% of the total lift of the system, and the culvert hardly generates lift at this time

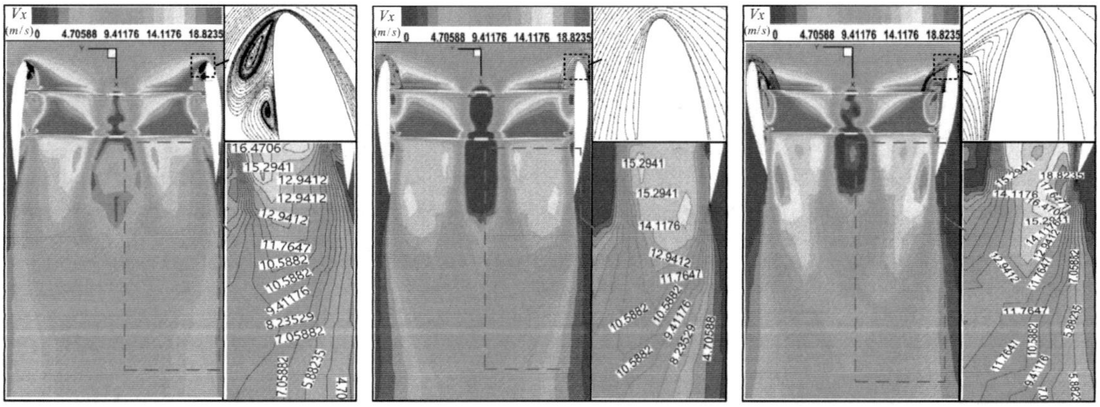

Figure 8 Axial velocity cloud map and lip flow at symmetric cross-sections with different cone angles Z=0.

At a -6° cone angle, airflow separation occurs at the lip area, resulting in eddy current and energy dissipation. Compared with 0° and 6° cone angle, the flow velocity around the lip area of the duct at -6° cone angle is larger, and the pressure is smaller, so the additional lift of the duct is smaller, which is the main reason for the sharp reduction of the additional pull of the duct at -6° cone angle. Figure 9 shows the pressure distribution on the culvert surface of section Z=0 when the cone angle is -6° and 6°. When the cone angle is -6°, the pressure on the culvert surface is mainly generated by the pressure difference between the negative pressure zone of the inner wall abc section and the outer wall de section. However, due to the downward positive pressure generated in the latter half of the inner wall of the culvert, part of the culvert tension is lost. The pressure on the surface of the culvert is mainly generated by the difference in pressure between the negative pressure zone and the inner wall in the dabc section of the outer wall, and there is no positive pressure zone. The pressure curve is integrated in the X direction to represent the additional tension of the culvert. The pressure of the 6° cone angle is greater than that of the -6° cone angle, which verifies the above analysis.

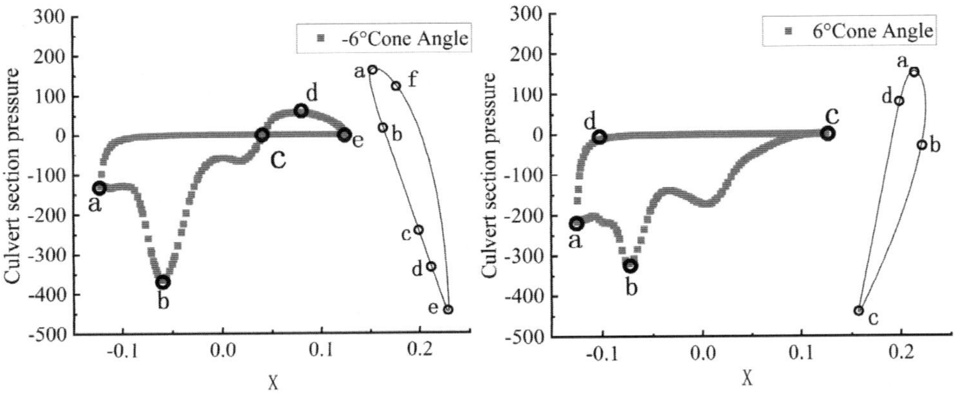

Figure 9 -6° and 6° pressure distribution of the section of the culvert with a cone angle.

The presence of the cone angle changes the area of the outlet and the entrance of the culvert, thus changing the airflow characteristics at the entrance and the pressure distribution on the surface of the culvert. If the cone angle is too small, the flow distortion at the lip will result in serious air separation and failing lift gain in the culvert. In summary, the increase of duct cone angle has a beneficial effect on aerodynamic efficiency. Still, the value of the cone angle should not be too large or too small and should be within a reasonable range. The area ratio (the ratio of the impeller disk area to the outlet area of the culvert) is an important parameter for improving the static thrust of the culvert, generally ranging from 1-1.25 [11]. In the research range, the system performance of the duct twin propeller of this configuration is the best when the cone angle is 6°and the area ratio is 1.25.

4.2. Analysis of the influence of lip deflection on aerodynamic performance

Lip shape is an important geometrical parameter that affects the flow characteristics and additional lift of the culvert, as well as the cause of the tension generated by the culvert. This section analyzes the influence of different lip deflection angles on the aerodynamic performance of the system and each component and its influence mechanism. As shown in Figure 10, the lip deflection is defined by the 1/5th chord length of the basic airfoil of the culvert. The leading edge is partially deflected -20°, -10°, 5°, 15°, 25°, and 30°, so that the inward lip deflection (closed culvert) is negative and the outward lip deflection (open culvert) is positive.

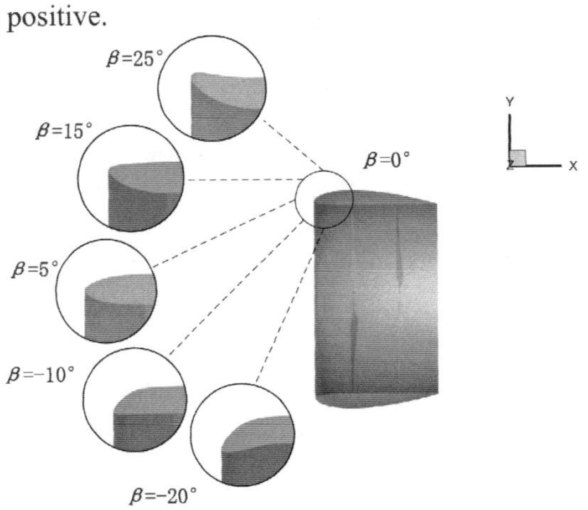

Figure 10 Schematic diagram of culvert lip deflection.

Figures 11-12 show the influence trend of lip deflection angle change on the total tension of the system and the tension and power load of each component. It can be seen that with the increase of lip deflection angle, the culvert tension first increases and then remains unchanged, while the aerodynamic efficiency of the system first increases and then slowly decreases. Overall, the efficiency of the internal lip deflection culvert system is low. The outward deflection has a beneficial effect on the aerodynamic efficiency and total lift of the system.

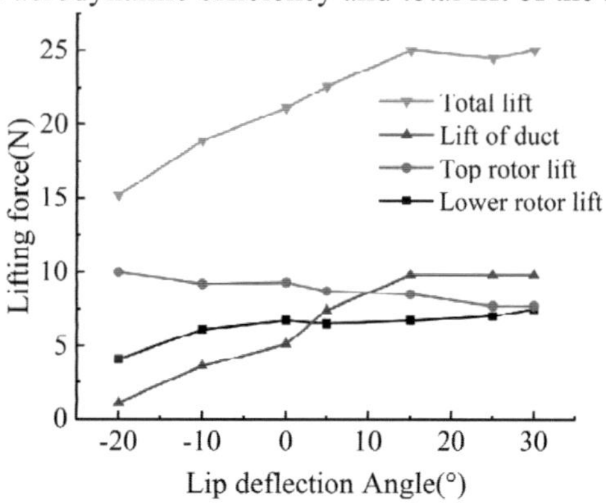

Figure 11 Lift forces of various components.

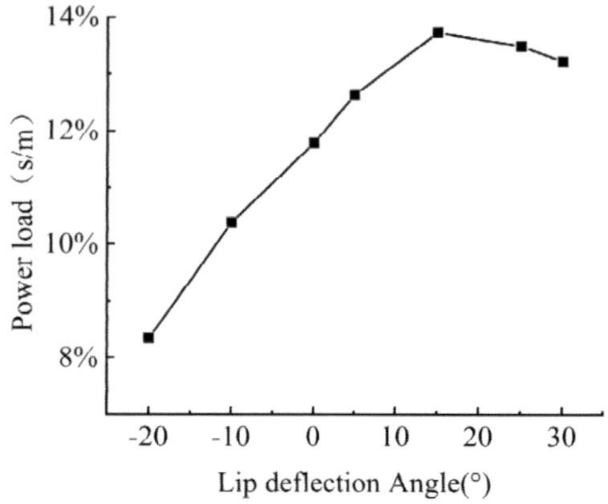

Figure 12 System aerodynamic efficiency curve under different lip deflection angles.

Figure 13 and Figure 14 are the flow diagrams inside the culvert when the lip deflection is 25° and -20°. It can be seen that the inward deflection of the lip of the culvert blocks the normal flow of air into the culvert, and serious flow separation occurs at the deflected part. The separated flow exhibits strong, unsteady characteristics, resulting in aerodynamic interference between the separated air and the upper propeller. A large interference resistance is generated, which reduces the lift force generated by the upper propeller. The inward deflection structure of the lip not only slows down the inflow speed but also increases the pressure difference resistance of the upper and lower surfaces of the lip. It also changes the flow state around the lip. It disturbs the flow around the lip, resulting in a lower additional lift force of the culvert and lower aerodynamic efficiency of the system. The outward deflection of the lip of the open structure better conforms to the flow trend of the inlet airflow. This shape can make the airflow under the suction effect of the propeller smoothly into the inside of the passageway without causing too much resistance or interference, help to maintain a more stable aerodynamic environment conducive to the normal operation of the upper propeller, and increase the local angle of attack of the upper propeller. The outward deflection of the lip increases the inlet area of the culvert, thereby increasing the airflow into the culvert. At the same time, the outward deflection also increases the flow radius around the lip area, so the airflow is accelerated at the lip area, and the area and peak value of the negative pressure zone inside the lip wall are increased, thus increasing the additional lift of the culvert. The lip deflection has little effect on the total lift of the upper and lower propellers.

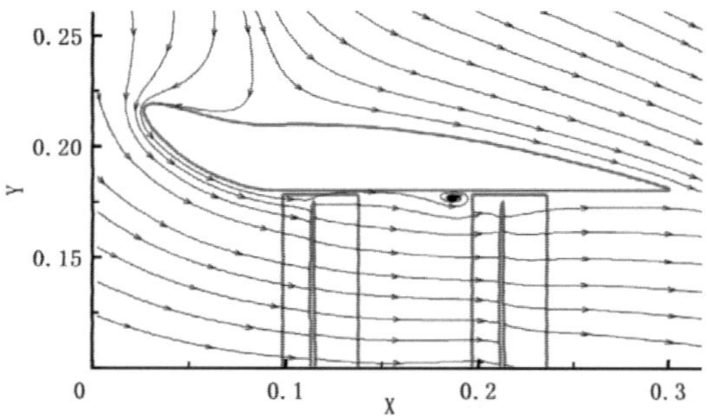

Figure 13 Flow diagram inside the culvert with lip deflection 25°.

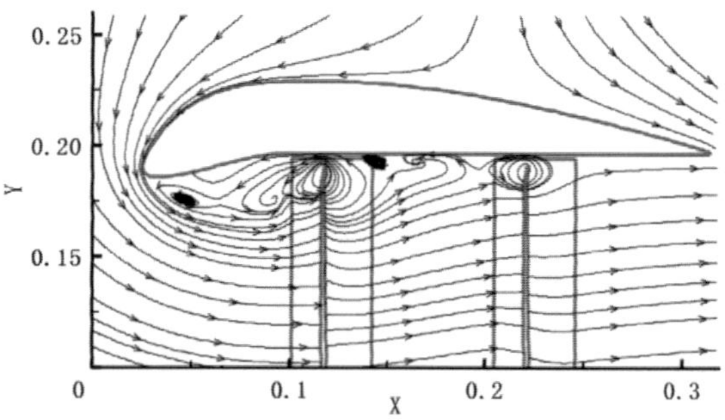

Figure 14 Flow diagram inside the culvert with lip deflection -20°.

5. Conclusion

1) Increasing the culvert cone angle can improve the inlet flow environment and increase the effective lift area of the culvert, thus strengthening the overall lifting effect of the culvert propeller system. However, with the increase of the cone angle, the blocking effect of the culvert on the downstream slipstream of the blade will also increase. Therefore, in the design stage, selecting the best cone angle value is necessary.

(2) The outward deflection of the lip can make the inlet flow smoother, improve the flow state inside the lip and duct, improve the quality of the flow field, reduce the possibility of air diversion and stall, and improve the efficiency and thrust of the system.

(3) According to the conclusions of this paper, it can be inferred that the aerodynamic performance of the ducted propeller system is closely related to the flow state of the airflow in the axial flow state, and the shape characteristics of the duct have a very significant impact on the flow around and inside the duct. In the subsequent research, the geometric shape of the duct can be appropriately optimized to obtain an optimal aerodynamic shape.

References

[1] DENG Yang-ping, MI Bai-gang, ZHAN Hao, et al. Ground test and numerical simulation on ground effect of ducted propeller system [J]. Journal of Northwestern Polytechnical University, 2020, 38(5): 1038-1046. J. Clerk Maxwell, A Treatise on Electricity and Magnetism, 3rd ed., vol. 2. Oxford: Clarendon, 1892, pp.68–73. [CrossRef]

[2] TAYLOR RT. Experimental investigation of the effects of some shroud design variables on the static thrust characteristics of a small-scale shrouded propeller submerged in a wing: NACA TN-4126 [R]. Washington, D.C.: NACA, 1958. K. Elissa, "Title of paper if known," unpublished. [CrossRef]

[3] BENTO H F, DE VRIES R, VELDHUIS L L. Aerodynamic performance and interaction effects of circular and square ducted propeller: AIAA-2020-1029 [R]. Reston: AIAA, 2020. [CrossRef]

[4] Xu He-yong, YE Zheng-yin. Numerical simulation comparison of aerodynamic characteristics between ducted propellers and isolated propellers [J]. Journal of Aerodynamics, 2011, 26 (12): 2820-5. [CrossRef]

[5] Hu Rui, Cao Chen-kai, Zhao Guo-qing, et al. Study on the influence of duct fan layout parameters on its aerodynamic characteristics [J]. Flight Mechanics, 2023, 41 (03): 7-12. [CrossRef]

[6] Han Kai, Bai Jun-qiang, Qiu Ya-song, et al. The influence of design variables on ducted propellers and their flow mechanism [J]. Journal of Aeronautics, 2022, 43 (07): 167-83. [CrossRef]

[7] Wang Hai-peng. Analysis of aerodynamic characteristics of ducted propellers in axial flow [D]; Nanjing University of Aeronautics and Astronautics, 2007. [CrossRef]

[8] SILVA P A SF, TSOUTSANIS P, ANTONIADIS A F. Simple multiple reference frame for high-order solution of hovering rotors with and without ground effect [J]. Aerospace Science and Technology, 2021, 111: 106518. [CrossRef]

[9] QIU Ya-son, BAI Jun-qiang, LI Ya-lin, et al. Study on influence of complex geometry details on the aerodynamic performance of high-lift system [J]. Acta Aeronautica et Astronautica Sinica, 2012, 33(3): 421-429. [CrossRef]

[10] Xu Jia, Fan Ning-jun, Zhao Shu. Research on the aerodynamic characteristics of the culvert body of a culvert aircraft [J]. Journal of Missile and Guidance, 2009, 29 (04): 174-8. [CrossRef]

[11] Long Li-min. Overall design and research of ducted fan unmanned aerial vehicles [D]. Shanghai: Shanghai University, 2008 [CrossRef]

Vibration characteristics analysis and structural optimization of bio-organic fertilizer mixer based on finite element method

Hualong He[1], Mamtimin Geni[1,*] and Gulbahar Tohti[1]

[1]Xinjiang University, Urumqi 830049, China

[*]Corresponding author's e-mail: 3057929116@qq.com

Abstract: In order to solve the waste produced by grape planting and cattle and sheep breeding in the Turpan area, this paper puts forward an organic liquid fertilizer mixer, and analyzes the vibration characteristics of the main structure-mixing barrel and auger. The finite element model of the mixer structure is established by Ansys, and the modal analysis is carried out in combination with the actual application. The analysis structure shows that the minimum natural frequency of the mixer barrel is 51.993 Hz, the minimum natural frequency of the auger is 12.601 Hz, in which the natural frequency of the auger is less than its working frequency of 48.5 Hz, and there is resonance risk. To improve this problem, the structure of the mixing auger is optimized, the length of the auger is reduced, and the pipe thickness and diameter are increased. The improved modal analysis shows that the minimum natural frequency of the auger is 50.371 Hz, which avoids the working frequency and will not cause resonance.

1. Introduction

In recent years, the planting area of crops and the number of livestock raised in Xinjiang have continued to increase, which has promoted the vigorous development of planting and animal husbandry and provided a rich source of raw materials for the production of organic fertilizer [1]. Xinjiang has performed well in the application of agricultural organic fertilizer, its application amount is large, and its proportion in fertilizer structure is appropriate, which provides a solid raw material base and sales market for the industrial development of organic fertilizer [2]. Taking Turpan as an example, the scale of grape planting and cattle and sheep breeding is huge, but the problems of grape pruning and the treatment of cattle and sheep manure have become difficult problems that restrict the sustainable development of local agriculture. Therefore, the application of organic liquid fertilizer mixers is particularly important in this area, which provides an effective way to solve these problems [5].

In the production process of modern bio-organic liquid fertilizer, the stirring process plays a vital role [6]. At present, most bio-organic liquid fertilizer mixers adopt the vertical cylindrical mixing barrel design, and the core mixing mechanism is usually the auger [3]. In the process of mixing, the organic liquid fertilizer forms a circular flow under the guidance of the auger, and generates friction with the auger, mixing barrel, etc. The auger shears the organic liquid fertilizer to realize the full mixing of the solid phase and the liquid phase in the organic liquid fertilizer, which is beneficial to the later fermentation process of the organic liquid fertilizer.

Therefore, in order to solve the waste produced in grape planting and cattle and sheep breeding in the Turpan area, this paper puts forward an organic liquid fertilizer mixer and analyzes the vibration characteristics of the main structure according to the vibration problem of the mixer and the practical application[11].

Journal of Physics: Conference Series **2815** (2024) 012002 doi:10.1088/1742-6596/2815/1/012002

2. Overall structural design

The main structure of the bio-organic liquid fertilizer mixer is shown in Figure 1. The mixer is mainly composed of the auger, transmission device, and mixing barrel. The transmission device is a planetary gear, which provides power through a motor and transmits power to the auger [8][9][10]. When the planet carrier revolves around the main shaft of the mixing barrel, it rotates around the main shaft of the planet carrier at the same time. The auger revolves around the main axis of the planet carrier and rotates around the main axis of the auger. By using different transmission ratios, augers are arranged in different radius ranges, and different planetary carriers have different rotational speeds, so that it takes less time to evenly mix solid and liquid phases in organic liquid fertilizer.

1. Overall drawing of organic liquid fertilizer mixer 2. Assembly drawing of auger and transmission system 3. Planetary mechanism diagram

Figure 1 Design drawing of the main mixing mechanism of bio-organic liquid fertilizer mixer.

3. Finite element analysis of shell

The bio-organic liquid fertilizer mixer will produce an obvious vibration phenomenon in the power transmission of the motor. This vibration not only brings noise interference to the production personnel but also may damage the structure of the mixer, thus affecting its normal operation. Under long-term vibration, the key parts of the mixer are easily damaged [7], thus shortening its service life. Therefore, it is of great theoretical and practical significance to make an accurate finite element analysis of the mixing barrel of the bio-organic liquid fertilizer mixer.

As an advanced structural analysis method, the core idea of the finite element method is to divide a complex structure into several small units and to reveal its mechanical characteristics more accurately by simplifying the internal deformation and stress of each small unit. These small units are connected with each other through nodes to form a complete analysis model. In finite element analysis, solving the stiffness matrix is one of the key steps. By solving the stiffness matrix of the whole structure, the stiffness matrix of each small element can be further obtained, and then the mechanical response and vibration characteristics of the whole structure can be deeply understood.

Through finite element analysis, we can effectively analyze the natural frequency of the bio-organic liquid fertilizer mixer [4] to avoid overlapping with the working frequency, thus achieving the purpose of vibration reduction and noise reduction. In addition, the finite element method can also be used to optimize the structural design of the mixer and improve its stability and durability. Therefore, it is of great significance to analyze the mixing barrel of feed mixer by finite element method for improving the performance of equipment, prolonging the service life, and promoting the sustainable development of agriculture.

The following is the process of finite element analysis.

1) Establishing the entity model of the required solution;
2) Discretizing the region of the required solution;
3) Selecting an interpolation function or a displacement function;
4) Establishing the stiffness matrix of the infinitesimal element;
5) Combining the micro-cell stiffness matrix with a whole stiffness matrix;
6) Introducing boundary conditions and loads;
7) Solving and analyzing the data.

doi:10.1088/1742-6596/2815/1/012002

A simplified model of the mixing barrel of the bio-organic liquid fertilizer mixer is established by SolidWorks, and it is imported into ANSYS Workbench software. The simplified model of the mixing barrel and auger of the bio-organic liquid fertilizer mixer is shown in Figure 2. The mixing barrel and the auger are regarded as an assembly respectively, and are divided by tetrahedral grid. The result of grid division is shown in Figure 3. When the bio-organic liquid fertilizer mixer works, the bottom is fixed on the shelf, so a fixed constraint is added to the bottom of the mixing barrel, and the upper plane of the auger is fixed on the gear and bearing, as shown in Figure 4. Then, the first six natural frequencies of the mixing barrel and auger of the bio-organic liquid fertilizer mixer are solved. The first six natural frequencies of the mixing barrel and auger are shown in Table 1 and Table 2, respectively.

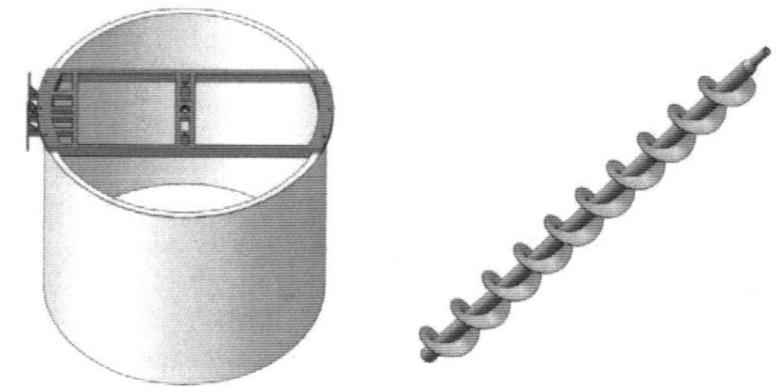

1. Simplified model of mixing barrel 2. Simplified model of auger

Figure 2 Simplified model of main components of bio-organic liquid fertilizer mixer.

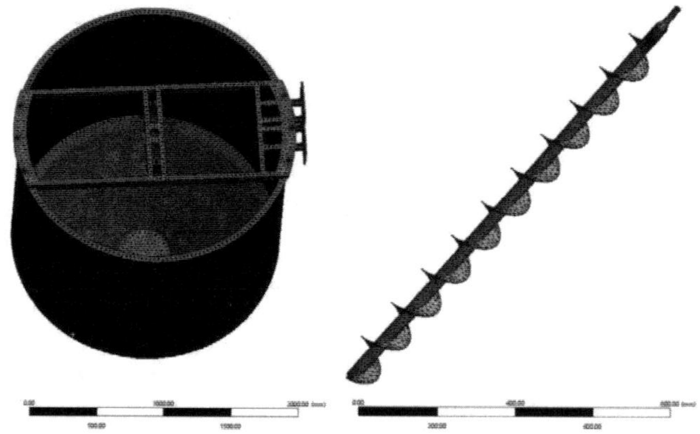

1. Grid division of mixing barrel 2. Grid division of auger

Figure 3 Grid division of main components of bio-organic liquid fertilizer mixer.

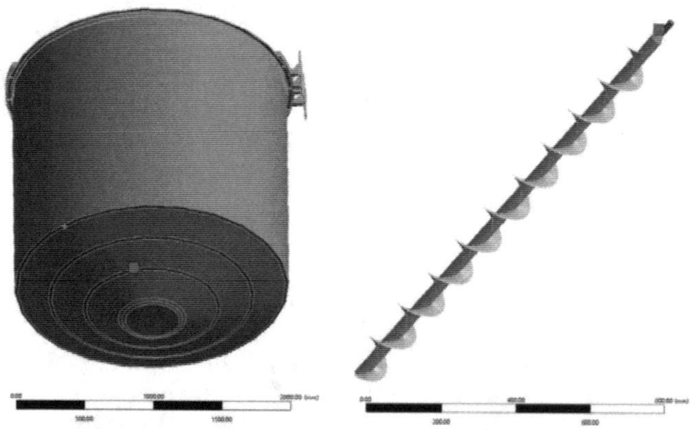

1. Boundary condition of mixing barrel 2. Boundary condition of auger

Figure 4 Boundary conditions of main components of bio-organic liquid fertilizer mixer.

Table 1 First Six Natural Frequencies of Mixing Barrel.

Order	1	2	3	4	5	6
Modal /Hz	51.993	57.071	73.101	87.738	90.696	96.603

Table 2 First Six Natural Frequencies of Auger.

Order	1	2	3	4	5	6
Modal /Hz	12.601	12.622	80.555	80.697	227.02	227.0

The cloud images of the first six vibration modes of the mixing barrel and auger of the bio-organic liquid fertilizer mixer are shown in Figure 5:

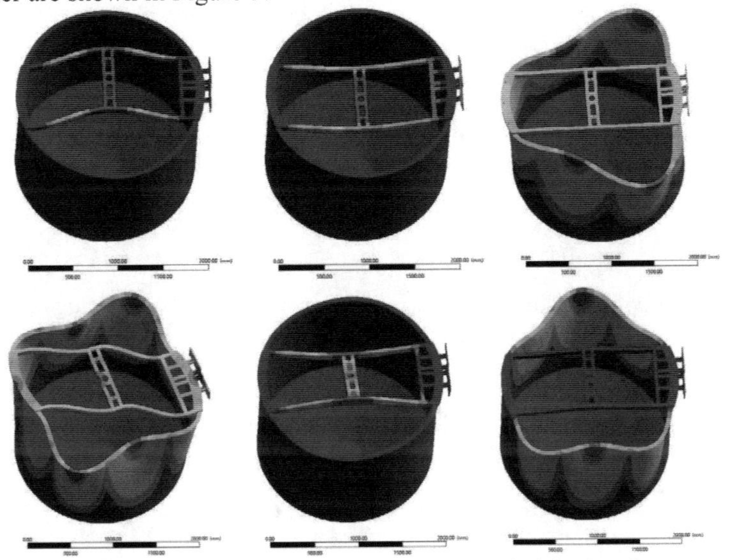

1. Cloud image of sixth-order vibration mode of mixing barrel

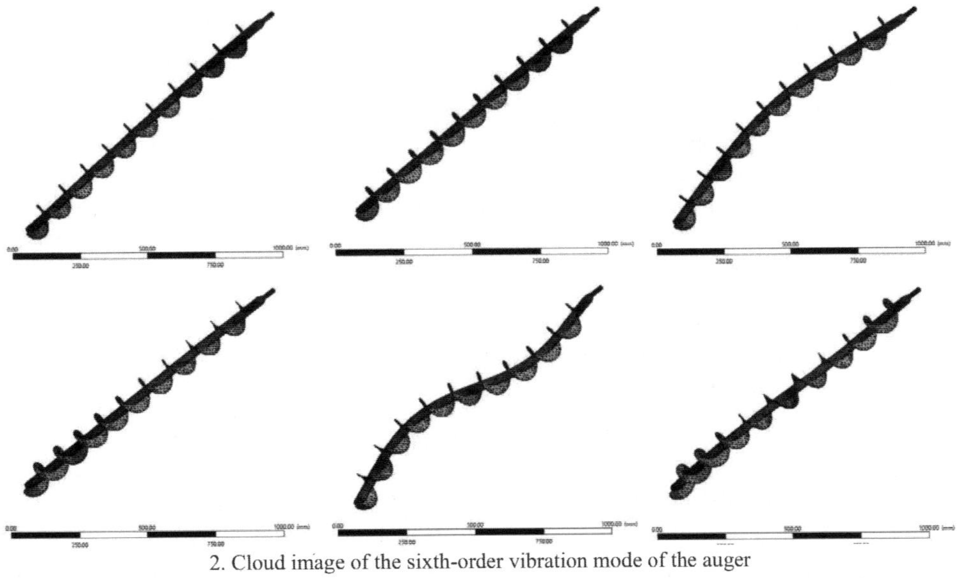

2. Cloud image of the sixth-order vibration mode of the auger

Figure 5 Cloud pictures of the first six vibration modes of the main components of bio-organic liquid fertilizer mixer.

It is preliminarily selected that the driving motor speed of the organic liquid fertilizer mixer is 2910 rpm, and its working vibration frequency is 48.5 Hz through calculation. The reducer of the organic liquid fertilizer mixer is selected as a single-head worm gear reducer, and its working frequency is the same as that of the motor. According to the previous calculation results, its working frequency is inevitable, and the auger in the organic liquid fertilizer mixer will inevitably produce resonance. Therefore, on the basis of previous work, the stiffness and stability of the auger in the organic liquid fertilizer mixer are improved and optimized by increasing the thickness, decreasing the length, and increasing the diameter of the auger. The improved model is shown in Figure 6:

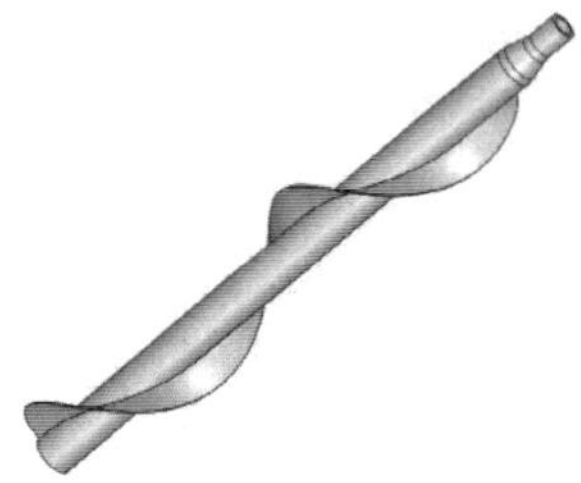

Figure 6 Design of auger of improved organic liquid fertilizer mixer.

The same modal analysis is carried out on the improved mixing barrel, and the results are shown in Table 3. As can be seen from Table 3, the first natural frequency of the auger of the organic liquid fertilizer mixer is 50.371 Hz, which effectively avoids its working frequency of 48.5 Hz and reduces the noise generated by resonance.

Table 3 The First Six Natural Frequencies of the Improved Mixing Barrel.

Order	1	2	3	4	5	6
Modal /*Hz*	50.371	50.5	318.86	319.7	383.52	640.6

4. Conclusion

When the organic liquid fertilizer mixer works normally, the vibration of the motor, the meshing of the gears in the reducer, and the vibration of the main shaft of the auger every time it rotates and stirs will produce noise. Therefore, these vibrations should be considered in the design analysis, and the machine should avoid these vibration frequencies.

Through the modal analysis of the auger of the organic liquid fertilizer mixer, it can be found that increasing the thickness of the auger, decreasing the length of the auger, and increasing the diameter of the auger can effectively enhance the stiffness of the elongated structure, so that its natural frequency can effectively avoid its working frequency, thus achieving the effects of avoiding resonance, reducing noise and improving stiffness.

Acknowledgments

The research was supported by the National Natural Science Foundation of China (Project No. 12162031) and the State Key Laboratory for Manufacturing Systems Engineering of Xi'an Jiaotong University (Project No. sklms2022022).

References

[1] Lu Wenyu and He Zhongwei. Development Status, Problems and Countermeasures of Organic Fertilizer Industry in China [J]. Science and Technology and Industry, 2022, 22 (09): 258-262.
[2] Wu Jian, Geng Yin, Zhou Yanjun, et al. Research status and development ideas of organic fertilizer applicator [J]. Agricultural Engineering, 2022,12 (04): 19-22.doi: 10.19998/j.cnki.2095-1795.2022.04

[3] Zhang Xiaowei, Cao Haifeng and Wang Deming. Research and design of organic fertilizer horizontal mixer [J]. Use and maintenance of agricultural machinery, 2022, (02): 6-10.doi: 10.14031/j.cnki.njwx.2022.02.002.

[4] Liu Jun and Ou Kehuo. Simulation analysis of planetary concrete mixing equipment based on EDEM [J]. Construction Machinery, 2020, (11): 62-65+71. DOI: 10.14189/j.cnki.cm1981. 2020.11. 006.

[5] Yang Liuhua, Wang Hongjiang, Wu Aixiang, et al. Present situation and development trend of whole tailings paste mixing technology [J]. Metal Mine, 2016, (07): 34-41.

[6] Yang Guijie. Analysis of the present situation and development prospect of organic fertilizer in China [J]. Friends of Farmers Getting Rich, 2013, (12): 84.

[7] Shen Jijun. Research status and development prospect of concrete vibration mixing technology [J]. Construction Machinery Technology and Management, 2009, 22 (12): 112-115.

[8] Feng Zhongxu, Wang Weizhong, Yao Yunshi, et al. Study on the reasonable rotational speed of mixer [J]. china journal of highway and transport, 2006, (02): 116-120. DOI: 10.19721/j.cnki.1001-7372.2006.02.0206.

[9] Hao G . Modal and fatigue characteristics analysis of key components of tower crane [J]. Vibroengineering Procedia, 2024, 54 72-77.

[10] Xinghao C . Modal analysis and vibration response test of tank body [J]. Vibroengineering Procedia, 2024, 54 90-95.

[11] Yan L ,XingKun Z ,WenHua L . Subsea jumper damage detection based on fractal analysis and modal characteristics [J]. Brodogradnja : An International Journal of Naval Architecture and Ocean Engineering for Research and Development, 2024, 75 (2): 1-17.

Development of a novel PCB electric field sensor with high-resolution

Jinyang Li [1], Yi-Horng Lai [2,*] and Lijie Qu [3]

[1] Pen-Tung Sah Institute of Micro-Nano Science and Technology, Xiamen University, Xiamen, Fujian, 361005, China
[2] School of Mechanical and Electrical Engineering & Automation, Xiamen University Tan Kah Kee College, Zhangzhou, Fujian, 363105, China
[3] School of Civil Engineering and Architecture, Wuyi University, Wuyishan City, Fujian, 354300, China

[*] Corresponding author's e-mail: lai81@xujc.com

Abstract: Electric field sensors with high resolution and high sensitivity have an important engineering application. Developing a compact, cost-effective Electric Field Sensor (EFS) for aerospace applications, this study introduces a novel use of PCB parallel board probes, enhancing affordability and simplicity. A uniquely designed three-stage adaptive signal amplification module quickly adjusts amplification factors, ensuring high sensitivity and resolution even in complex electric field environments. Due to the non-linear response characteristics of the PCB probes in complex electric field environments, the GTEM cells and the high voltage parallel plate method are used to simulate the complex electric field scenarios. In the weak electric field experiment, the proposed EFS achieved a high resolution of 10 mV/m. In the strong electric field experiment, the linearity of fitting between the proposed EFS and commercial NBM-550EFS reached 0.9999. Finally, based on RBFNN, the overall measurement model of EFS achieves the dynamic measurement range of 0-1 kV/m.

1. Introduction

Electric field sensors (EFS) have important practical significance in modern engineering applications. For example, monitoring the electrical field is critical information for power equipment inspection. In addition, for the aerospace field, aviation electrical equipment is more susceptible to the impact of pulse electrical fields caused by surrounding discharge in the working environment. Designing high-resolution and high-sensitivity EFS is an important issue [1-3].

Based on MEMS technology, the EFS has the advantage of being small [4, 5]. Compared with traditional silicon micro-machined sensors, the manufacturing process of flexible polyimide materials reduces costs and exhibits higher durability and shock resistance [6]. The non-contact EFS that can measure sudden discharge has been proposed in [7]. The sensor board is made of PCB FR4 material and integrates multiple functions such as sensing, collection, timing, recognition, data storage, and transmission. Suo et al. [8] utilized the induced charge principle of parallel plates to detect the electric field. The proposed hexahedral structure design solves the problem of nonlinear measurement caused by the non-perpendicularity between the parallel plate probe and the electric field source.

The EFS based on parallel PCB boards is affected by factors such as PCB size and the fixed insulation dielectric coefficient between electrode plates. The performance of low-frequency response and high resolution is poor. If the EFS can work in an adaptive amplify mode, detection efficiency in weak electrical fields can be improved. Moreover, EFS can adapt to different measurement scenarios.

Amidst existing EFS technologies, this paper introduces a novel approach by employing a PCB-based parallel plate probe that combines low cost with high efficiency, starkly contrasting traditional, more complex systems. Furthermore, we propose an innovative three-stage adaptive signal amplification algorithm, which dynamically adjusts to signal strength variations, a significant advancement over static models.

Two verification experiments were designed to investigate the effectiveness of the proposed EFS. First, the Gigahertz Transverse Electromagnetic (GTEM) cell produces a microelectric field environment in the weak electric field scenario. On the other hand, in the strong electric field scenario, the high-voltage parallel plate method is utilized to evaluate the performance of EFS. Furthermore, the proposed EFS uses RBF neural networks to construct an overall electric field measurement model for the two kinds of measurement data in different electric field scenarios. Finally, the experimental data are aligned and compared with the commercial standard NBM-550, EHP50 sensor (Narda Ltd. Germany).

2. Principle of EFS measurement with parallel plate capacitors

The one-dimensional parallel plate capacitor is placed in a uniform electrical field $\overrightarrow{E_1}$, and the inductive electrical field inside the capacitor is $\overrightarrow{E_2}$. Without considering the edge effects, the charges on parallel plate capacitors' upper and lower surfaces exhibit a uniform distribution. We assume that the surface charge density of the sensor's upper electrode plate is ρ_1, the induced charge density on the lower surface is $\rho_2 = -\rho_1$. The dielectric coefficient of the medium between the plates is ε_2, and the dielectric constant of the medium outside the plates is ε_1, respectively. The distance between the upper and lower electrode plates of the capacitor is d_{pcb}. We take a cylindrical Gaussian surface S_1 and S_2 on top and bottom of the electrode plate. The areas of Gaussian surfaces are ΔS_1 and ΔS_2 respectively. The schema of the induced electrical field between parallel plate sensors is shown in Figure 1.

Applying the Gaussian theorem, the electrical flux passing through a closed surface is proportional to the amount of charge enclosed by the closed surface. The induced charge of the electrode plate is related to the strength of the external electric field as follows:

$$\oiint_{S_1} \overrightarrow{E_1} \cdot \mathrm{dS} = -E_1 \Delta S_1 = \frac{\rho_1 \Delta S_1}{\varepsilon_1} \tag{1}$$

$$\oiint_{S_2} \overrightarrow{E_2} \cdot \mathrm{dS} = -E_2 \Delta S_2 = \frac{\rho_2 \Delta S_2}{\varepsilon_2} \tag{2}$$

Due to $\rho_2 = -\rho_1$, we have:

$$\overrightarrow{E_2} = \frac{\varepsilon_1}{\varepsilon_2} \overrightarrow{E_1} \tag{3}$$

The fluctuation in the external electrical field can cause a variation of induced charge Q on the electrode plate. Moreover, the voltage U is induced at the parallel plate capacitor. When the dielectric coefficient and electrode distance of the parallel plate sensor are fixed, the induced voltage of the sensor is directly proportional to the external electrical field strength.

$$U = \frac{Q}{C} = \frac{\varepsilon_2 \overrightarrow{E_2} \Delta S_2}{C} \propto \overrightarrow{E_1} \tag{4}$$

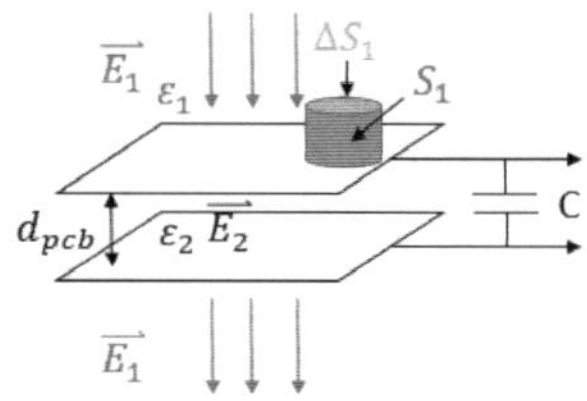

Figure 1. Schema of the induced electrical field.

3. Hardware design of EFS

The hardware assembly of the proposed system includes power management, adaptive signal preprocessing, ADC, DSP, and data transmission modules. The power management module provides the necessary power and reference voltage for other modules; the signal preprocessing module can adaptively amplify the gain of signals under the control of DSP and filter the noise of signals. After completing the DSP module, the acquired data is transmitted to the host computer through RS485. The overall framework is shown in Figure 2.

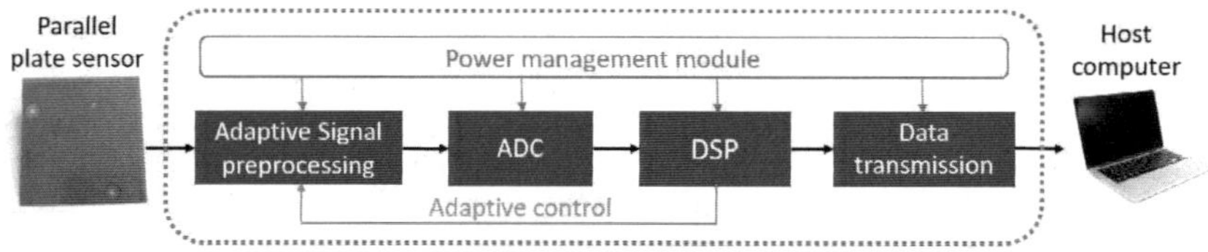

Figure 2. The overall framework of EFS.

3.1. Design of EFS probe

According to the Chapter 2 principle of EFS measurement, the design of sensor probes mainly considers the dielectric coefficient ε_2 of the PCB and the distance between the two electrode plates d_{pcb}. This study uses the PCB boards with specification FR4 as the sensor probe. The FR4 board is made of copper foil metal plate with an interlayer of epoxy resin and fiberglass cloth. By checking the datasheet of the PCB, the dielectric coefficient ε_2 of the FR4 is 4.4 under the operating frequency range of 25-120 Hz. Furthermore, according to the simulation results, the distance between parallel plates is confirmed to be 0.8 mm within the EFS measurement range of 0.1-1000 V/m. The hardware of the PCB parallel plate probe of EFS is shown in Figure 3.

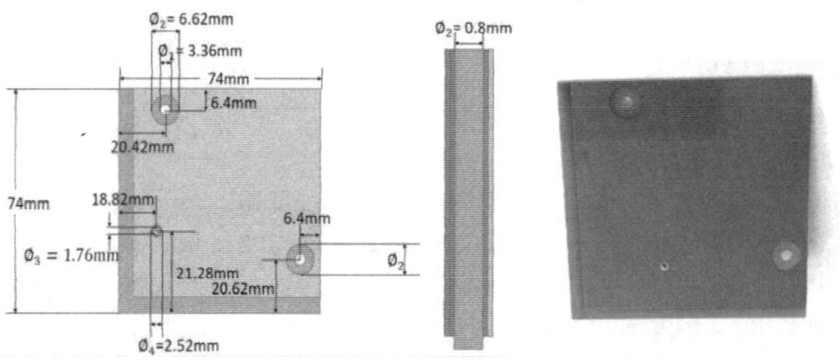

Figure 3. The design of the EFS probe.

3.2. Design of adaptive signal preprocessing module

The adaptive signal preprocessing module integrates filtering and amplification functions. It employs second-order Butterworth high-pass and low-pass filters connected in series to create a band-pass filter with a range of 20-120 Hz. Due to the micro inductive voltage collected by the parallel plate probe (i.e., μV scale), the adaptive signal amplification module adopts a 3-stage amplification design. The first stage features an LTC6241 instrument amplifier with a fixed 5x gain. The second stage includes a programmable gain module comprising an ADG781 analog switch and an operational amplifier (OP amp), allowing gain adjustments of 1x, 5x, 10x, 20x, and 40x through DSP-controlled resistance changes. The final stage uses an AD8331 voltage-controlled gain amplifier, with gain dynamically regulated by an LTC1663 chip via DSP, achieving up to 10, 000 times signal enhancement.

3.3. Flowchart of the adaptive algorithm

The proposed system features an adaptive gain control algorithm to accurately and swiftly measure electrical field values in complex environments. This algorithm operates on a dynamic feedback process, setting high and low ADC thresholds and a default measurement period of 1 second with a sampling rate of 1000 Hz. If the maximum signal during a measurement period falls below the ADC's low threshold, indicating no valid signal, the DSP's adaptive algorithm adjusts the analog switches in the second stage and the gain voltage in the third stage to incrementally amplify the total gain by ten times each period, up to the system's maximum amplification limit. Conversely, if the signal exceeds the high threshold, the gain is reduced by ten times each period until the minimum gain level is achieved, ensuring optimal signal fidelity. The flowchart of the adaptive gain algorithm is shown in Figure 4.

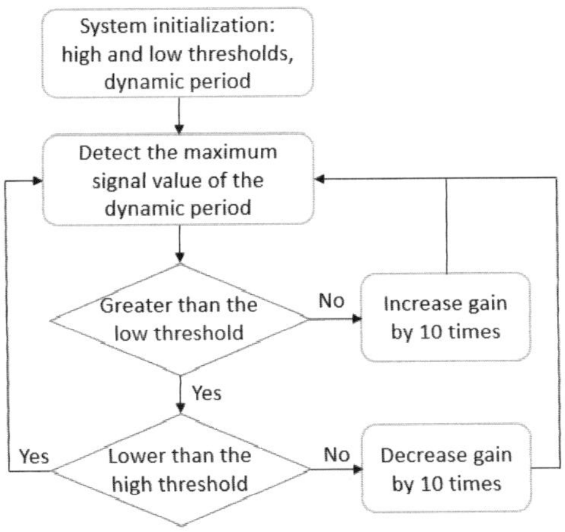

Figure 4. The flowchart of the adaptive gain algorithm.

3.4. RBFNN algorithm

Due to the inability to achieve a completely linear relationship in the complex electric field data measured by the PCB sensor probe, the Radial Basic Function Neuro network (RBFNN) is designed in the DSP module. The RBFNN has a strong approximation ability for nonlinear continuous functions [9]. The proposed measurement model based on RBFNN can establish a nonlinear correspondence between sensor probe voltage and electric field strength.

The RBFNN algorithm consists of three layers: the input layer, the hidden layer with a nonlinear activation function, and the output layer. Each input node corresponds to an element in the input layer. Each hidden node consists of a radial activation function, which includes local perception nodes. The Gaussian distribution function is used as an activation function in this study. The RBFNN parameters can be trained using the backpropagation (BP) algorithm. The output y of RBFNN can be represented as:

$$y = \sum_{i=1}^{n} \omega_i \varphi(\|x - u_i\|^2) \tag{5}$$

where x is input, ω_i is the weighting value, and φ is an activation function.

4. Experiment of EFS

For the parallel plate probes, the induced voltage of the probe is usually at the μV level in a weak electric field environment. On the other hand, the PCB-based EFS does not always have a linear relationship in the overall measurement range. To investigate the effect of the proposed EFS, the complex electric field scenarios are designed to simulate weak and strong electric field environments.

Firstly, we evaluate that the motherboard of EFS can detect the μV level signals. Secondly, the GTEM cell and high voltage parallel plate method will simulate the complex electric field environments and verify the consistency between the proposed EFS and commercial NBM-550 EFS.

4.1. Verification of minimum measurable voltage of EFS

The experimental protocol for verifying the minimum measurable voltage of EFS is as follows. Three series resistors are used to produce the aligned voltage of the signal generator. By adjusting the ratio between the middle resistance and the resistance at both ends, we calibrate the aligned voltage level as 3 μV. We check whether the measured voltage values in the 3-axis direction match the aligned voltage values (i.e., 3 μV). The experiment result is shown in Figure 5. From the GUI interface of the host computer, the measured voltage values in the 3-axis direction almost match the aligned voltage value.

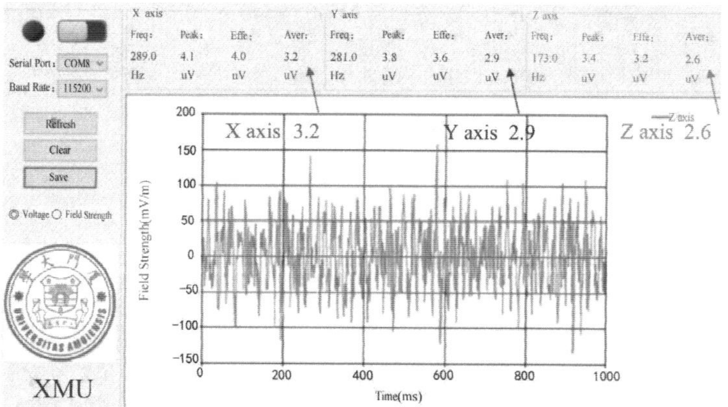

Figure 5. Verification of minimum measurable voltage of EFS.

4.2. Consistency comparison in complex electric field environments

The experimental scenario for complex electric field environments is as follows:

1. The proposed EFS uses a single probe in complex electric field environments.
2. The commercial EFS (NBM-550 and the EHP-50D) are the reference EFS. We analyze the linear relationship between the data of the proposed system and that of the reference sensor in the same testing condition.
3. Experimental protocol of weak electric field (below 20 V/m) scenario: First, we place the EHP-50D in the GTEM cell. Different intensity signals of 70 Hz are applied to the GTEM cell to simulate weak electric field environments. We record the applied voltage of the GTEM cell based on the electric field value measured by NBM-550. We replace the proposed EFS probe in the GTEM cell. We record the measured signal value of the proposed system at the same location and intensity, as shown in Figure 6(a).
4. Experimental protocol for the strong electric field (above 20 V/m) scenario: We apply a voltage of different intensities (70 Hz) sequentially between two parallel plates to simulate a strong uniform electric field environment. We set the proposed probe and the EHP-50D probe at the same height of

IOP Publishing
doi:10.1088/1742-6596/2815/1/012003

the uniform electric field. We record the signal strength measured by the proposed system and NBM-550 simultaneously (Figure 6(b)).

5. Due to the characteristics of PCB technology, the data measured in two kinds of electric field environments cannot achieve a completely linear relationship. The proposed system uses RBFNN to construct an overall measurement model of EFS.

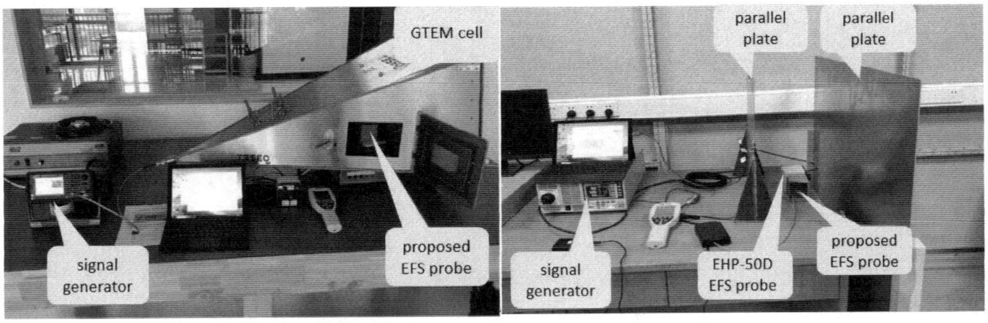

Figure 6. Experiment with (a) a weak electric field scenario and (b) a strong electric field scenario.

The experimental result of the weak electric field scenario is shown in Figure 7(a). The maximum resolution of the proposed EFS can achieve 10 mV/m. However, the linearity of the measurement results for commercial NBM-550 EFS and the proposed EFS is not high. On the other hand, the experimental result of the strong electric field scenario is shown in Figure 7(b). The linearity of fitting for commercial NBM-550 EFS and the proposed EFS is very high ($R^2 = 0.9999$). The dynamic measurement range of the proposed EFS is 0-1 kV/m.

We train RBFNN using two kinds of data of electric field scenarios simultaneously. The overall measurement model of the proposed EFS based on RBFNN is shown in Figure 8.

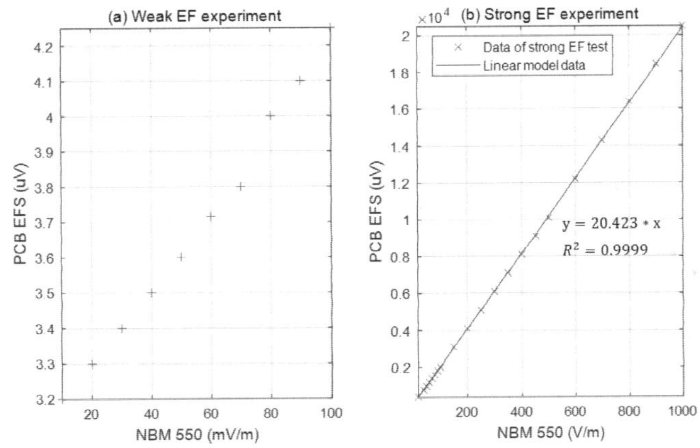

Figure 7. Experiment results of (a) weak electric field and (b) strong electric field.

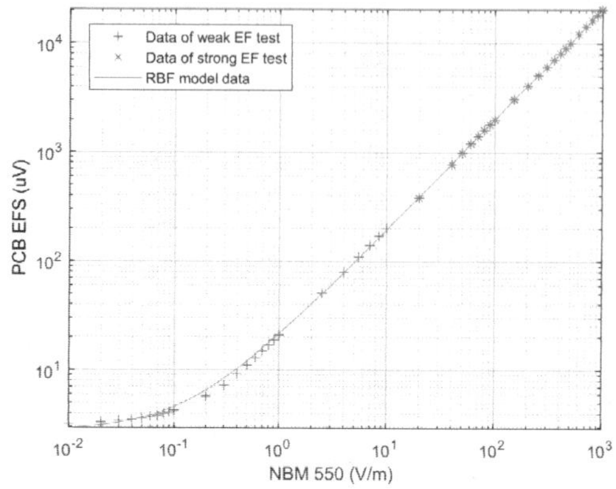

Figure 8. The overall measurement model of the proposed EFS based on RBFNN.

After analyzing the experimental data, the measurement range, resolution, and sensitivity of the proposed Electric Field Sensor (EFS) meet the industry standard design specifications. Utilizing the adaptive signal amplification design and the Radial Basis Function Neural Network (RBFNN) measurement model, the resolution and sensitivity of the proposed EFS have proven to be superior to those reported in previous literature. This enhancement suggests a significant advancement in the capabilities of electric field sensors, contributing new methodologies to the field.

5. Conclusions.

This study introduces a new Electric Field Sensor (EFS) that effectively combines high resolution and sensitivity, employing a cost-effective PCB probe design that simplifies manufacturing while maintaining quality. The sensor incorporates a novel three-stage adaptive signal amplification algorithm, which dynamically adjusts to environmental electric field changes, ensuring a precise resolution of 0.01 V/m and high sensitivity up to 4.5 V (kV/m). Additionally, enhanced by a Radial Basis Function Neural Network (RBFNN), our measurement model allows the EFS to perform reliably in weak and strong electric field conditions. This approach enhances the flexibility and applicability of the sensor in aerospace applications and offers new insights and methodologies for future research in the field.

Acknowledgments

This work is partially supported by (No. 2023J011045). The authors also gratefully acknowledge the helpful comments and suggestions of the reviewers, which have improved the presentation.

References

[1] Z. Han, F. Xue, J. Hu and J. He, (2021) Micro Electric Field Sensors: Principles and Applications, IEEE Industrial Electronics Magazine, vol. 15, no. 4, pp. 35-42. [CrossRef]

[2] Z. Han, F. Xue, G. Yang, Z. Yu, J. Hu and J. He, (2021) Micro-Cantilever Capacitive Sensor for High-Resolution Measurement of Electric Fields, IEEE Sensors Journal, vol. 21, no. 4, pp. 4317-4324. [CrossRef]

[3] Z. Chu, C. Peng, R. Ren, B. Ling, Z. Zhang, H. Lei and S. Xia, (2018) A High Sensitivity Electric Field Microsensor Based on Torsional Resonance. Sensors, pp. 18(1): 286. [CrossRef]

[4] C. Tao, (2013) MEM electric field sensor using force deflection with capacitance interrogation. IEEE Power & Energy Society General Meeting. pp. 1-4. [CrossRef]

[5] X. Wen, P. Yang, Z. Chu, C. Peng, Y. Liu and S. Wu, (2021) Toward Atmospheric Electricity Research: A Low-Cost, Highly Sensitive and Robust Balloon-Borne Electric Field Sounding Sensor. IEEE Sensors Journal, vol.21, no.12, pp. 13405-13416. [CrossRef]

[6] T. Chen, B. Hill, S. Isik, C. Shafai and L. Shafai, (2021) DC electric field sensor based on polyimide substrate. IEEE Sensors. pp. 1-4. [CrossRef]
[7] Y. Wang, S. Gu, Y. Fang, Y. Xu, Y. Chen and P. Li (2020) Compact electric field change meter and its application in lightning detection and fault analysis for power grids. IEEE International Conference on High Voltage Engineering and Application. pp. 1-4. [CrossRef]
[8] C. Suo, R. Wei, W. Zhang, and Y. Li (2021) Research on the Three-Dimensional Power Frequency Electric Field Measurement System. Journal of Sensors, vol. 2021, pp. 8859022. [CrossRef]
[9] J. Perng, G. Chen, S. Hsieh, (2014) Optimal PID Controller Design Based on PSO-RBFNN for Wind Turbine Systems. Energies vol. 7, pp. 191-209. [CrossRef]

Journal of Physics: Conference Series **2815** (2024) 012004 doi:10.1088/1742-6596/2815/1/012004

Characterization of underwater manipulator based on the fluid-structure interaction under pulsating flow

Yongqi Li [1], Xia Liu [1] and Derong Duan [1,*]

[1] School of Mechanical Engineering, University of Jinan, Jinan, 250022, China;

[*]Corresponding author's e-mail: me_duandr@ujn.edu.cn

Abstract: Underwater manipulators are taking on an increasingly significant function in marine biological fishing operations. Under the action of the ocean flow field, vortex-induced vibration will be induced, which will cause fatigue damage or even fracture of the underwater manipulator. Based on the Ansys fluid-solid coupling software, this paper simulates the vortex-induced vibration response characterizations of underwater manipulators under pulsating flow conditions. The pulsation flow at the entrance is realized by changing the velocity amplitude and pulsation frequency. The influence laws of pulsating frequency and amplitude on the C_L, C_D, and wake vortex shedding are investigated. The results showed that the C_L and C_D of the underwater manipulator increased by 133.33 % and 39.18 %, respectively, with an increase in the pulsation parameter. Additionally, with the increase of pulsating frequency, the banded vortex behind the underwater manipulator is gradually obvious, and the energy in the vortex is also increasing.

1. Introduction

Underwater manipulator is essential equipment for marine resource collection. Its vortex-induced vibration (VIV) phenomenon occurs under the influence of the incoming flow, which leads to the reduction of the positioning accuracy of the structure [1, 2]. Therefore, it is very important to conduct a fluid-solid coupling analysis of marine engineering equipment under pulsating flow [3]. Li et al. [4] experimentally studied the performance of the VIV response of the tube in an oscillating flow. They found that the width of the spectrum in the oscillating flow became wider, and more frequencies were excited. Zhao et al. [5], Yuan et al. [6], and Fu et al. [7] researched the response characterizations of VIV under pulsating flow. The intermittent changes in the vibration frequency of the cylinder were all observed.

The analysis of VIV response characteristics of underwater manipulators under pulsating flow is very important for establishing a precise positioning underwater manipulator model. Due to the universality and versatility of Ansys, this paper uses the Ansys Workbench platform. It is based on the two-way fluid-solid coupling calculation method to study the impacts of pulsating frequency (f) and pulsating amplitude (A) on the lift coefficient (C_L), drag coefficient (C_D), and wake vortex shedding of the underwater manipulator.

2. Numerical method research

2.1. Mathematical models and analysis method

This work uses the underwater manipulator model installed on the fixed base. The physical model is shown in Figure 1. When the underwater manipulator is in operation, the upper arm (Arm 1) is always in the vertical direction and remains fixed, while the lower arm (Arm 2) rotates in the range of 0°-360° (rotation angle is β). This work mainly analyses the lift-drag coefficient and vortex shedding of the lower arm when the underwater manipulator rotates 135°.

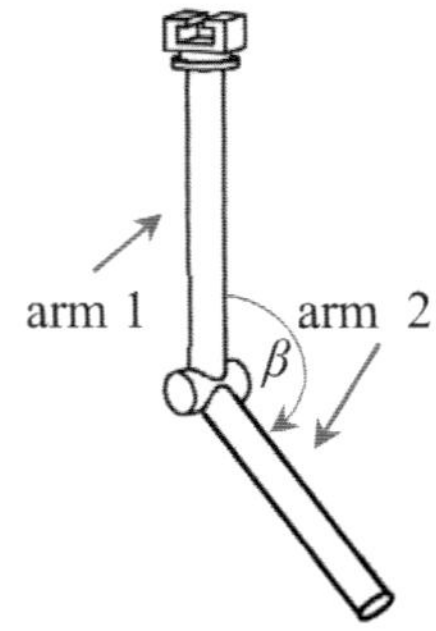

Figure 1. Calculation domain diagram.

Figure 2 illustrates the computational area of the underwater manipulator. The diameter of the manipulator model is D. The model is 10D from the inlet boundary and 24D from the outlet boundary. The width of the fluid domain is 20D, and the height of the fluid domain is 34D. The length of the Arm 1 and Arm 2 are both 12D.

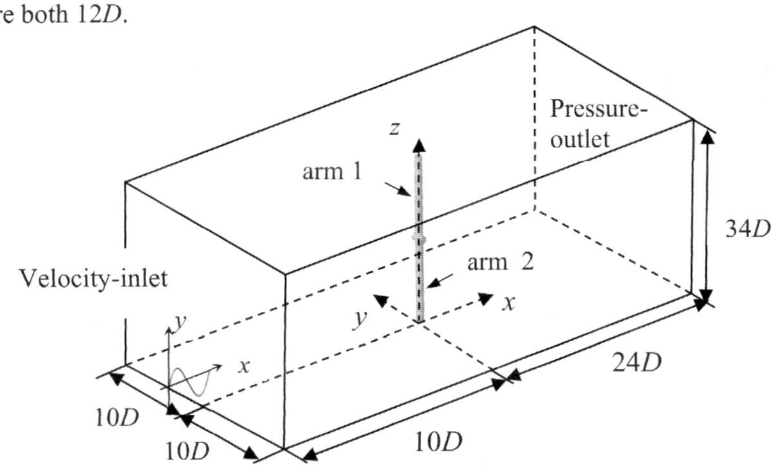

Figure 2. Calculation domain diagram.

This paper simplifies the ocean current into a sinusoidal pulsating flow based on the hydrological environment of the Shandong Peninsula to better study the VIV characteristics of the underwater manipulator under actual sea conditions. The UDF function loads the pulsating velocity function, and the velocity is set as a function of time. The calculation formula is shown in Equation (1):

$$u(t) = u_m\left(1 + A\sin(2\pi ft)\right) \tag{1}$$

where v(t) is the inlet pulsating flow velocity, m/s; t is the flow time, s; f is the pulsating frequency, Hz; A is the dimensionless pulsating amplitude.

The outlet was the boundary of the pressure outlet, given static pressure and appropriate reflux conditions. The pulsating flow varies with the fluctuating frequency and the dimensionless fluctuating amplitude. The pulsating frequency and the dimensionless pulsating amplitude are determined by the wave conditions of the Yellow Sea and the Bohai Sea in the Shandong Peninsula. Table 1 shows seven different pulsation conditions.

Table 1. Basic parameters of pulsating flow.

Parameters	Value		
A	0.2	0.3	0.4
f/Hz	0.2	0.3	0.4

2.2. Grid segmentation and independence verification

Taking the vertical attitude model as an example, the model of the underwater manipulator is meshed. The overall three-dimensional meshing results are shown in Figure 3. The meshing scale in the axial direction is relatively rough, and the meshing in the two-dimensional plane is more refined, as shown in Figure 4. Among them, the red part is the unstructured grid area, and the rest of the area is the structured grid.

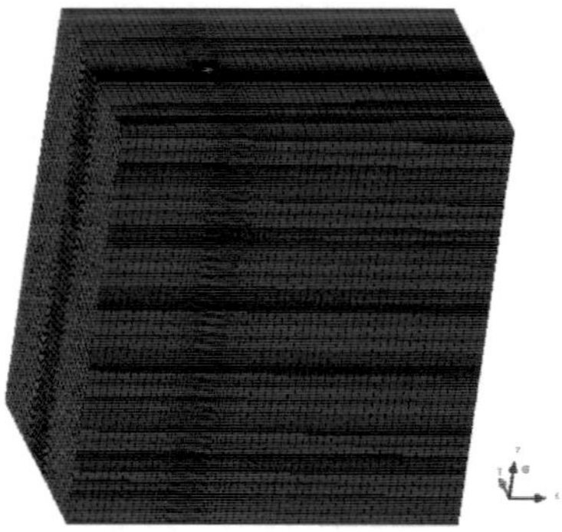

Figure 3. Three-dimensional schematic diagram of mixed grid division.

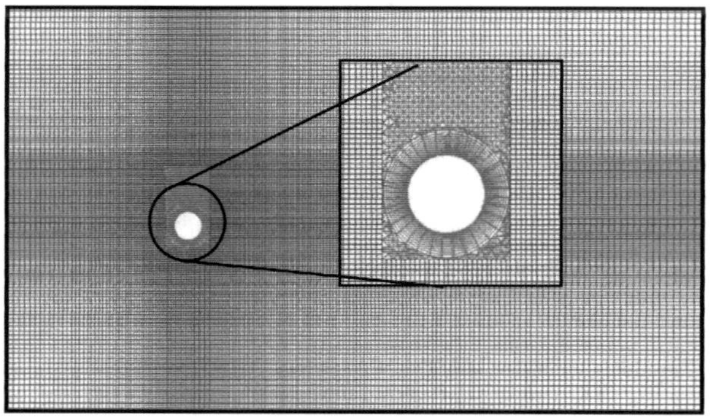

Figure 4. Mixed grid division horizontal cross-section.

The mesh independence verification is carried out to ensure that the grid is independent of the final computational flow results. Taking the underwater manipulator in the vertical state as an example, the working conditions of u = 0.2 m/s, f = 0, and A = 0 are selected. The C_L and the C_D obtained under different mesh numbers are compared to observe whether the number of meshes affects the final calculation results. Table 2 shows the results of mesh independence verification. It is clear from Table 2 that both the C_L and the C_D change very little as the number of meshes changes. After the vibration of the manipulator is stable, the maximum errors of the C_D and the C_L are 7.2 % and 6.1 %, respectively. Therefore, selecting the second set of grids can meet the calculation requirements and obtain accurate calculation results.

Table 2. Grid independence verification results.

Mesh	Elements	$\overline{C_D}$	$C_{L,\ rms}$
1	931533	0.83	0.49
2	1238612	0.84	0.51
3	1532891	0.89	0.52

3. Results and discussion

3.1 Effect of pulsation frequency on fluid-solid coupling

Figure 5 shows the trend of the C_L and the C_D of Arm 2 in the A2 attitude. Figure 5(a) shows the C_L obtained for Arm 2 at f = 0.2, 0.3, and 0.4 Hz, respectively. When the pulsation frequency is 0.2Hz, the $C_{L,\ max}$ is stable near 0.06, and the periodicity is also poor. As the pulsation frequency increases to 0.3Hz, the maximum lift coefficient increases to 0.11, and the periodicity is gradually obvious. When the f= 0.4 Hz, the $C_{L,\ max}$ = 0.19 is about 3.12 times that at f=0.2 Hz. The pulsation frequency increases the pulsatility in the flow field, causing the lift coefficient to increase with the pulsation frequency [8]. In addition, the periodicity of the lift coefficient is progressively more pronounced with increasing pulsation frequency. From the results in Figure 5(b), the $C_{D,\ max}$ = 0.74 when the f= 0.2 Hz. When the pulsation frequency is increased to 0.4Hz, the $C_{D,\ max}$ is increased by 9.46%. The lower the pulsation frequency is, the lower its $\overline{C_D}$ is, but the larger the fluctuation period is.

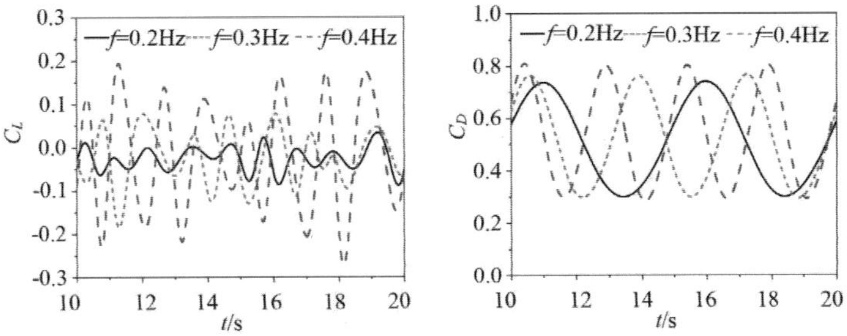

(a) C_L (b) C_D

Figure 5. The C_L and the C_D at different f.

To further explore the effect of the f on the flow field behind Arm 2, the vortex shedding at different f was comparatively analyzed. Figure 6 shows the vorticity cloud diagram of Arm 2 at the same time under three groups of pulsation frequencies. From Figure 6(a), as can be seen, when f = 0.2Hz, the trailing wake is formed behind the bottom end of Arm 2. With increased pulsation frequency, the banded vortex behind the underwater manipulator gradually increases. It is speculated that the reason for this phenomenon is the increase of the f, which increases the pulsation of the flow field, resulting in an increase in vorticity.

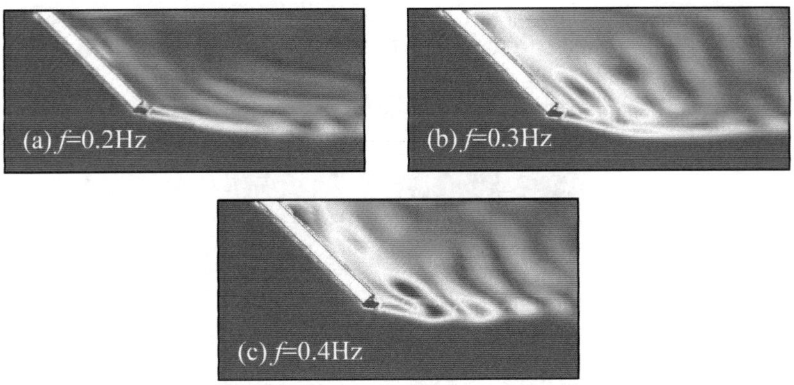

Figure 6. The vorticity cloud diagram at different f.

3.2 Effect of pulsation amplitude on fluid-solid coupling

This part first analyses the C_L and the C_D under three groups of pulsating amplitude (A = 0.2, 0.3, 0.4). Figure 7 shows the C_L and the C_D under different pulsation amplitudes. It is observed from Figure 7 that the values of the C_L and the C_D tend to increase significantly as the pulsation amplitude increases. When A = 0.2, then $C_{L,\ max}$ = 0.06 and $C_{D,\ max}$ = 0.74. Compared with the two-dimensional simulation results, the lift and drag coefficients are reduced to a certain extent. This is because the 3D effect reduces the movement of the Arm 2 to a certain extent. When the pulsation amplitude is 0.4, the $C_{L,\ max}$ and the $C_{D,\ max}$ increase to 0.14 and 1.03, respectively, 133.33 % and 39.18 % higher than those at A=0.2.

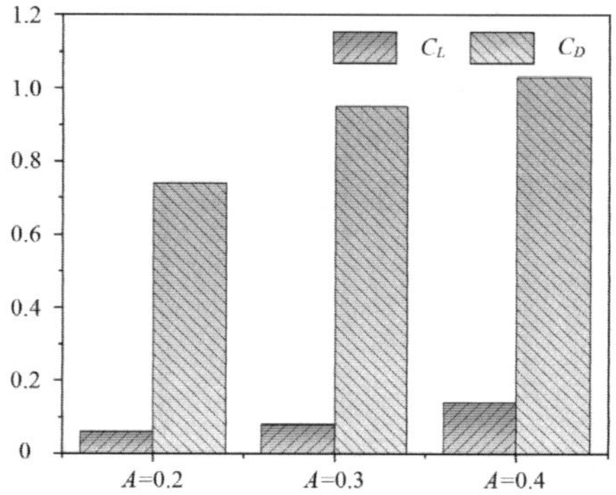

Figure 7. The value of the $C_{L,\ max}$ and the $C_{D,\ max}$.

Figure 8 shows the Arm 2 half-profile vorticity cloud diagram when the A= 0.2, 0.3, and 0.4, respectively. In Figure 8 (a), when the pulsation amplitude is 0.2, the vortex is centralized in the middle and lower part of the Arm 2 backside, and the upper vortex scale is small. When the pulsating amplitude increases to 0.3, a weak in-line vorticity is formed in the wake region at the bottom of the manipulator, gradually decreasing with the increase in height. When the pulsation amplitude increases to 0.4, the incoming flow velocity increases, the coupling effect between the fluid and the manipulator is enhanced, the wake area behind it further increases, and the vortex power is greater. The larger response amplitude leads to a wider wake.

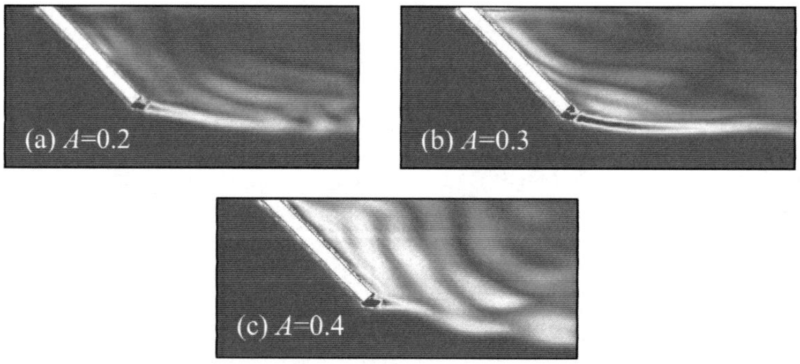

Figure 8. The vorticity cloud diagram at different pulsation amplitude.

4. Conclusion

The fluid-structure coupling of the underwater manipulator under pulsating flow was numerically simulated. The VIV responses of the underwater manipulator under different working conditions were obtained by changing the f and the A. The conclusions are as follows:

(1) With the increase of pulsation parameters, the C_L and the C_D of Arm 2 increased significantly. The pulsation amplitude has a greater influence on the underwater manipulator. When the pulsation amplitude increases from 0.2 to 0.4, the C_L and the C_D increase by 133.33 % and 39.18 %, respectively.

(2) With the increase of the pulsating frequency, the wake vortex area behind the underwater manipulator is further increased, and the vortex gradually appears at the upper end of Arm 2.

(3) The underwater manipulator will be affected by different flow rates at different times. The fluid force and uniform flow of the underwater manipulator are more complicated, which will ultimately complicate the vibration response characteristics of the underwater manipulator in pulsating flow.

In this paper, it can be found that the variation of pulsation parameters further increases the energy consumption of the underwater manipulator by analyzing the CL, CD, and vortex shedding of the underwater manipulator under pulsating flow. Finally, in conjunction with the findings of this paper, the response characteristics of the underwater manipulator help to optimize and improve the hydrodynamic model, which provides a preliminary basic study for the stable operation of the underwater manipulator in the marine environment.

Acknowledgments
This project is supported by the National Natural Science Foundation of China (No. 51905211) and A Project of the "20 Regulations for New Universities" funding program of Jinan [No. 202228116].

References

[1] Wang, J. S., Fan, D., Lin, K. (2020). A review on flow-induced vibration of offshore circular cylinders. Journal of Hydrodynamics, 32(3), 415-440.
[2] Ali, U., Islam, M., Janajreh, I., Fatt, Y., Alam, M. M. (2021). Flow-induced vibrations of single and multiple heated circular cylinders: A review. Energies, 14(24), 8496.
[3] Deng, D., Zhao, W., Wan, D. (2021). Numerical study of vortex-induced vibration of a flexible cylinder with large aspect ratios in oscillatory flows. Ocean Engineering, 238, 109730.
[4] Li, X., Yuan, Y., Duan, Z., Xue, H., Tang, W. (2023). Experimental investigation on vortex-induced vibration of a flexible pipe in combined uniform and oscillatory flow. Ocean Engineering.
[5] Zhao, M., Pearcey, T., Cheng, L., Xiang, Y. (2017). Three-Dimensional Numerical Simulations of Vortex-Induced Vibrations of a Circular Cylinder in Oscillatory Flow. Journal of Waterway Port Coastal and Ocean Engineering-asce, 143, 04017007.
[6] Yuan, Y., Xue, H Tang, W. (2018). Numerical analysis of Vortex-Induced Vibration for flexible risers under steady and oscillatory flows. Ocean Engineering, 148, 548-562.
[7] Fu, B., Zou, L., Wan, D. (2018). Numerical study of vortex-induced vibrations of a flexible cylinder in an oscillatory flow. Journal of Fluids and Structures, 77, 170-181.
[8] Liang C, Papadakis G. Large eddy simulation of pulsating flow over a circular cylinder at subcritical Reynolds number [J]. Computers & fluids, 2007, 36(2): 299-312

Experimental research on the flow flied of the nacelle and inlet of a civil aircraft with wind-mounted engines at low speed

Renyu Hu, Dongyun Zhang and Haoran Sun *

Shanghai Aircraft Design and Research Institute, Shanghai 201210, China

* Corresponding author: sunhaoran@comac.cc

Abstract: Modern civil transport aircraft with wing-mounted engines are exposed to the upwash induced by the wing, resulting in the local angle of attack at the nacelle leading edge, notably greater than that of aircraft. Large local angles of attack can be combined with different engine airflow conditions, causing adversary aerodynamic phenomena to occur at different locations. Total and static pressures obtained in a wind tunnel test using a full-span aircraft model and simulating required engine airflow with an ejector indicate that during the around with landing configuration, the high suction peak will form on the bottom of the inlet with local air flow being supersonic and could lead to inlet distortion. In contrast, the engine that fails during takeoff climb will have a significant suction peak form on the top of the nacelle and might experience external separation and additional drag. Test results also demonstrate that well-designed nacelles can postpone these adversary phenomena until the aircraft stalls first.

1. Introduction

It is almost a common practice for modern civil transport aircraft to have engines wing-mounted, which places the engines in a relatively clean environment in front of the wing and thus eliminates inlet distortion issues caused by the ingestion of wakes of wing or fuselage with a rear-mounted configuration [1, 2]. In the case of wing-mounted engines, the presence of an air-frame affects the flow field of the nacelle leading edge and inlet through increased local angle of attack at the highlight plane, a plane formed by the leading edge of the nacelle, resulting from the up-wash induced by the wing. The greater the lift generated by the wing is, the more significant this up-wash in front of the wing becomes. When high-lift devices are fully deployed at low speeds, the local angle of attack at the engine highlight plane could exceed the aircraft angle of attack by more than 5 degrees.

In addition to the local angle of attack, the flow flies off the nacelle leading edge, and the inlet is affected by engine airflow. Mass flow ratio (MFR) is defined by Equation (1) as the ratio of the area of the captured stream tube at the free stream (A0) to the area of the highlight plane (A_1) [3]. With fixed highlight plane area and given flight speed and attitude, *MFR* is only the function of engine airflow. During the takeoff and go-around phase, with engine power set at TOGA (Take Off and Go Around) rating, A_0 will be greater than A_1 and thus MFR>1 because of higher engine airflow [4]. As shown in Figure 1, part of the airflow within the captured stream tube has to flow over the leading edge of the nacelle inwards to enter the inlet, generating a suction peak on the inner side of the leading edge (on the surface of the inlet). With the engine running with lower airflow like that associated with idle rating, A_0 will be less than A_1 and thus MFR<1 [5], which means part of the airflow outside the captured stream tube has to pass over the leading outwards before flowing down the fan cowl, having a suction peak form on the external side of the leading edge.

$$MFR = \frac{A_0}{A_1} \tag{1}$$

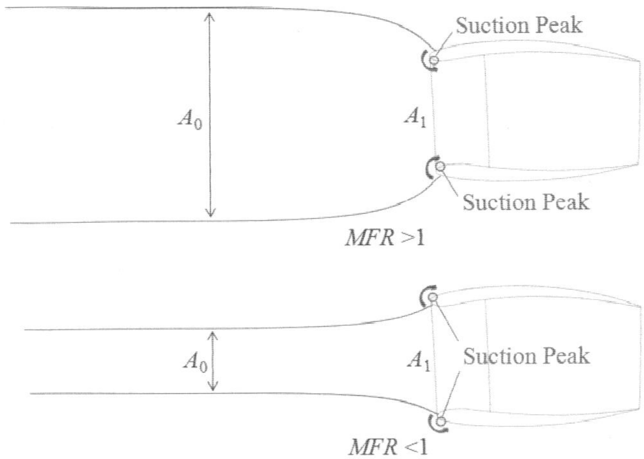

Figure 1. The flow field is around the nacelle leading edge at different *MFR*s.

The local angle of attack and mass flow ratio will generate suction peaks in the different areas of the nacelle leading edge in different civil aircraft operation scenarios. Figure 2 illustrates the locations of critical areas on the nacelle for different angles of attack (AOA) and airflow combinations where excessive suction peaks might form. These suction peaks could further develop into adversary aerodynamic phenomena like separation and impact the safety of aircraft and engine operation [6]. In some go-around cases, a large angle of attack is combined with high engine airflow to result in internal separation in the keel area and inlet distortion. In contrast, during the climb with one engine inoperative, external separation may occur in the crown area of the failed engine, generating additional drag and further compromising the climb performance and obstacle-clearing capability of the aircraft. One of the goals to achieve in nacelle and inlet design for civil transport aircraft with engines wing-mounted is to avoid separations in those scenarios [7, 8], and wind tunnel tests dedicated to analyzing flow field in these critical areas would be an effective means of design validation [9-13].

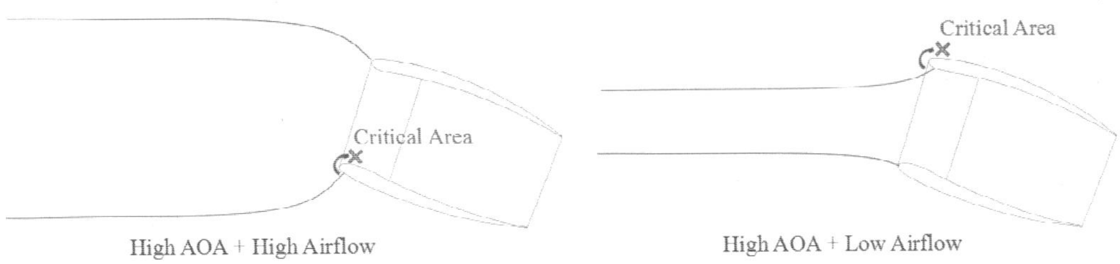

Figure 2. Critical Areas for different flight conditions.

2. Test method

To study the flow field of the nacelle leading edge and inlet of an engine wing-mounted civil aircraft during takeoff and landing and to evaluate the proposed nacelle and inlet design, a wind tunnel test was conducted in a commercial low-speed wind tunnel facility with a test section dimension of 3m, using a full-span aircraft model with slat and flap parts capable of simulating both takeoff and landing configuration. As shown in Figure 3, the model was installed in the test section through a ventral support system, which can change the model's attitude to simulate the angle of attack effect.

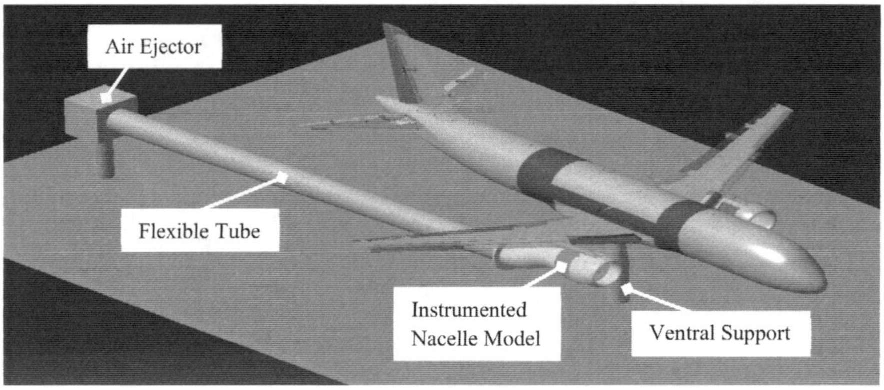

Figure 3. Installation of aircraft model in a wind tunnel.

To correctly simulate engine airflow, especially those high mass flow conditions associated with takeoff and go-around engine power rating, an ejector was used in the test, which used a small amount of high energy air supplied externally to induce a large airflow coming into the nacelle [14]. Engine power conditions simulated during the test ranged from idle to TOGA. Three mass flow conditions were simulated during the test: "low mass flow" for idle power rating, "high mass flow" for takeoff and go-around power rating, and a "medium mass flow" condition in between. The ejector was installed downstream from the aircraft model close to the exit of the test section and connected to the starboard nacelle through a flexible tube. To minimize the impact of the presence of the tube on the flow field around the wing, an S-shape duct was added to the rear end of the nacelle to keep the flexible tube as far away from the wing as possible. It was demonstrated by numerical simulation that the alteration to the flow field by the tube was acceptable.

3. Data collecting and processing

To analyze the flow field quantitatively, total pressures inside the inlet and static pressures around the leading edge of the starboard nacelle were measured in the test. As illustrated in Figure 4, 8 total pressure rakes had been equally distributed in a circumferential direction inside the inlet, with the angle between two adjacent rakes being 45°. Each rake had five total pressure ports on it. The radial distribution of these pressure ports followed the principle of equal area, which means that the measurement area was divided into five rings of the same area, and a set of 8 pressure ports were distributed along a circumference that equally divided each ring. Static pressure taps were provided on the crown section (0° position) and the keel section (180° position). Due to limited space, there were only five crown pressure taps and four keel pressure taps. As shown in Figure 5, tap C1 was on the external side of the leading edge, and tap C2 and tap K1 were right on the leading edge, with the rest of the static pressure taps located inside the inlet. Static and total pressures measured during the test were used to calculate the pressure coefficient (C_p) and total pressure recovery coefficients, respectively. The total pressure recovery coefficient indicates the ratio of measured total pressure at the rake to the total pressure of the free stream [15].

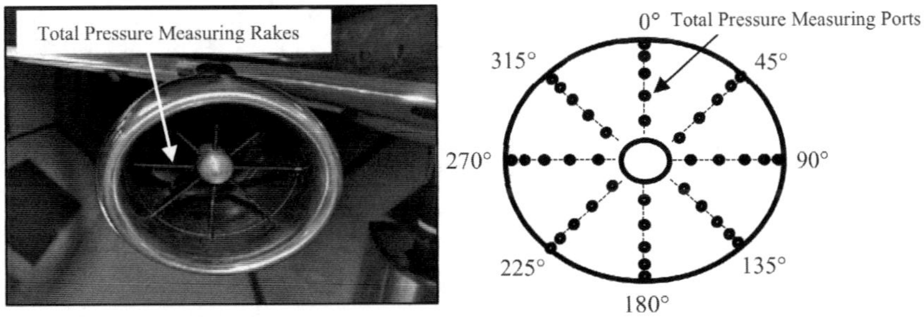

Figure 4. Arrangement of total pressure measuring ports inside the inlet.

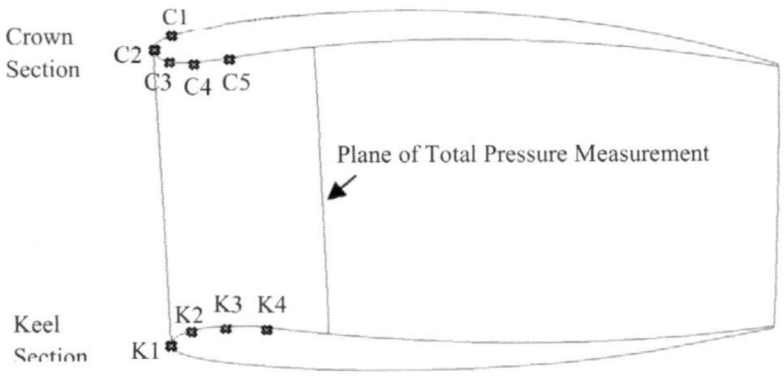

Figure 5. Arrangement of static pressure taps on nacelle leading edge and inlet.

4. Data analysis

4.1. Analysis of the flow field inside the inlet

Figure 6 and Figure 7 present the variation of static pressures with the nacelle airflow and the aircraft angle of attack, respectively, on the leading edges of both the crown section and the keel section of the nacelle. α_{max} refers to the angle of attack at which the maximum lift coefficient was observed in a previous force measurement test with the same aircraft model. As shown in Figure 6, with the increase in air entering the nacelle per second, the air moving along the surface of the inlet accelerates rapidly, generating a suction peak immediately downstream of the leading edge. As the air mass flow level increases from low to medium, the pressure distribution pattern at the crown position (Figure 6 left) appears to be "flipped over", which suggests that the stagnation point of the incoming flow in that area moves from the inner side of the leading edge to the external side as *MFR* increases, virtually changing the local angle of attack from positive to negative. In the keel area, engine airflow and local angle of attack affect the local flow field around the leading edge similarly and might combine to produce a much higher suction peak (Figure 6 right) than that in the crown area. As indicated by the pressure distribution variation in Figure 7, increasing the aircraft angle of attack will augment the leading-edge suction peak inside the inlet at the keel position, just as what could be expected with a wing section. It is noted that the aircraft angle of attack affects the local angle of attack of the keel section both directly through the altered orientation of the free stream and indirectly through the shift in the lift of the wing, thus induced up-wash. Calculation based on the data obtained during the test shows that with the slats and flaps deployed for landing and an aircraft angle of attack of α_{max}+10°, the local airflow speed at tap K2 on the bottom of the inlet had exceeded the speed of sound and could have impacted the flow field downstream adversely. The plots of total pressure recovery coefficients in the plane of measurement at the exit of the inlet in Figure 8, however, demonstrate that even with a test Reynolds number considerably lower than that of flight condition, no significant deterioration of the flow field was observed up to an angle of attack of α_{max}+10°. Local supersonic flow on the inlet surface has only reduced downstream total pressure slightly, and the minor decrease in total pressure recovery coefficients shown in the upper part of the plots in Figure 8 might result from the wake of the spinner (center body of the inlet).

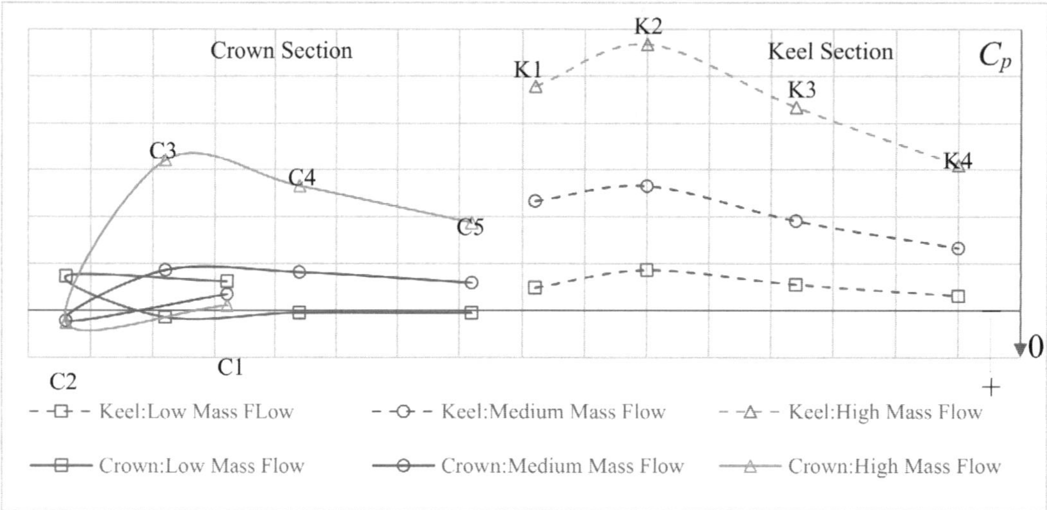

Figure 6. Effect of airflow on static pressure on crown and keel sections (landing configuration, α_{max}).

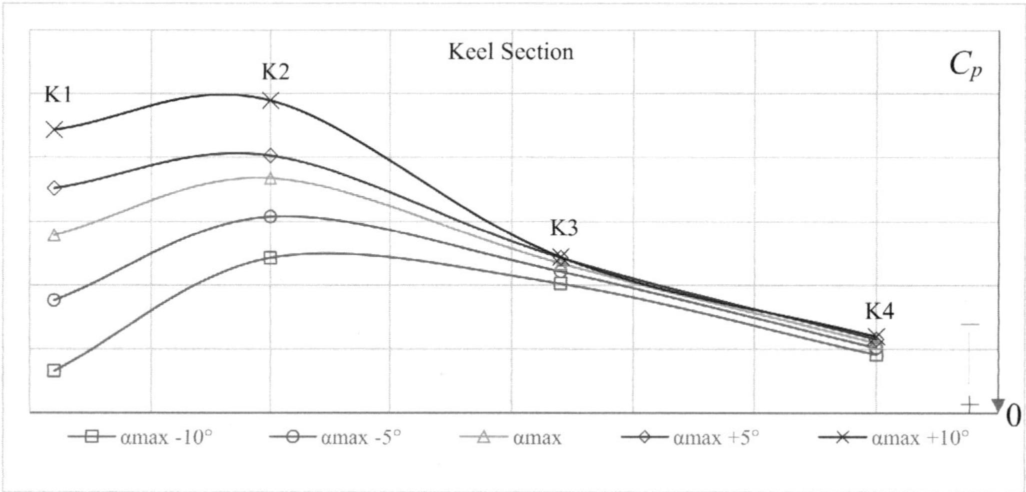

Figure 7. Effect of aircraft angle of attack on static pressure on keel section (landing configuration, high mass flow).

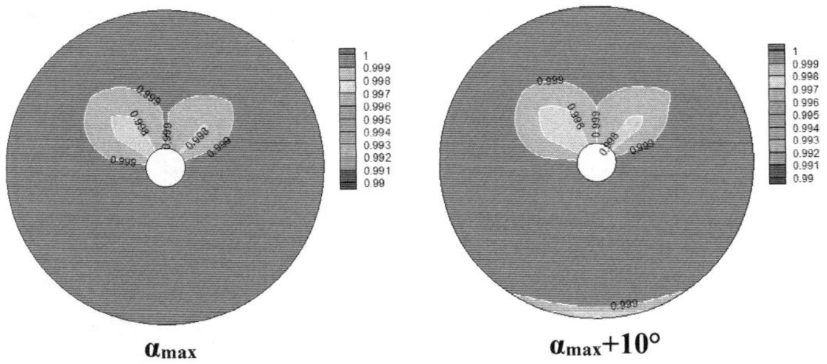

Figure 8. Total pressure recovery coefficients in the plane of measurement of the inlet (landing configuration, high mass flow).

4.2. Analysis of flow field around nacelle leading edge at crown position

Figure 9 presents the variation of static pressures on the crown section with aircraft angle of attack for takeoff configuration and idle airflow, which indicates that under the combined effect of flow pattern at low engine airflow and local angle of attack, a suction peak forms on the external side of the nacelle leading edge and continues to develop until separation occurs. Pressure data show that airflow at the top of the nacelle might have separated at an angle of attack between α_{max}+5°and α_{max}+10°, resulting in the visible collapse of the suction peak at α_{max}+10°. Although nacelle external separation in case of engine failure is usually studied in tests with a test Reynolds number high enough to facilitate evaluation and confirmation to be done for real flight conditions, it could still be concluded here that for the current nacelle design, the risk of having separations on the nacelle external surface and associated additional drag when the engine fails at normal climbing attitude is relatively low.

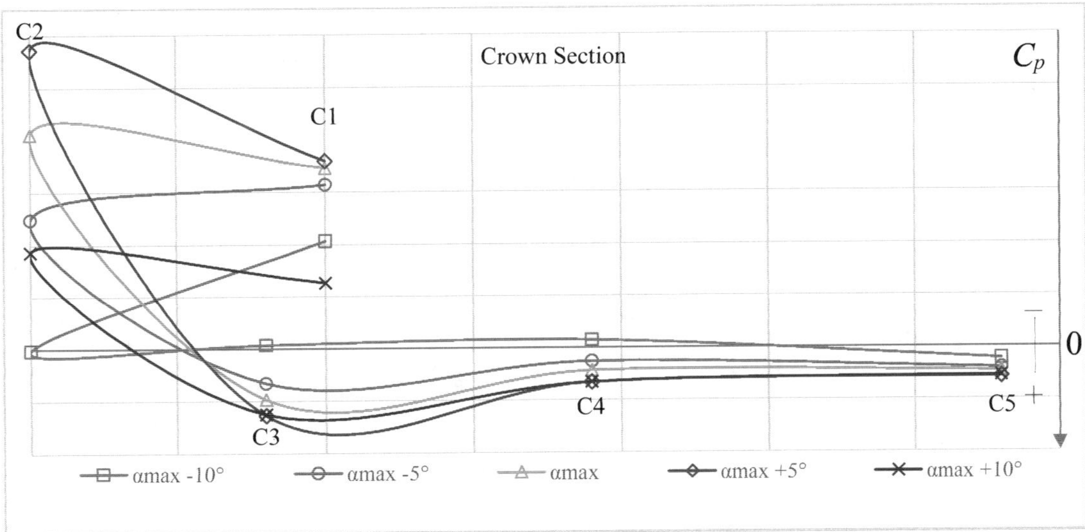

Figure 9. Effect of aircraft angle of attack on static pressure on crown section (takeoff configuration, high mass flow).

5. Conclusion

Engines mounted under the wing are subjected to the up-wash induced by the wing, causing the local angle of attack at the leading edge of the nacelle to be notably greater than the aircraft angle of attack. In real flight operations, a large local angle of attack would be combined with different engine conditions to have high suction peak form in the different areas of the nacelle and inlet. This could lead to separation and impact engine operation. A low-speed wind tunnel was performed using a full-span model to study the flow field of a proposed nacelle design for a wing-mounted engine, in which various engine airflow was simulated with an air ejector and pressures on the leading edge and inside the inlet of the nacelle was measured. The following are what was observed during the test or could be concluded from the results:

- Engine airflow combined with the angle of attack strongly affects the flow field of the nacelle leading edge and inlet regarding stagnation point position, the location and magnitude of the suction peak, and pressure distribution pattern.
- Air flow on the bottom of the inlet accelerates more significantly than on the top during a go-around with a landing configuration where both maximum airflow and maximum local angle of attack are expected and can go supersonic with the angle of attack large enough. This tendency suggests that internal separation from excessive air flow acceleration at the keel position could directly cause inlet distortion for wing-mounted engines if the Nacelle and inlet design do not check the airflow acceleration.
- It has been observed during test runs that there is a risk of the external part of the engine separating during ascent with one engine not working. This is due to a significant suction peak that occurs on the external side of the nacelle leading edge at the crown position, even when the airflow is set as low as idle power. This suction peak can lead to flow separation at an excessive aircraft angle of attack.
- With a well-designed nacelle, it is possible to limit the negative impact of the acceleration of airflow around the leading edge and inside the inlet to a reasonable level. This ensures that the separation of the nacelle from the aircraft's external structure will not happen until the aircraft stalls first, thus guaranteeing the aircraft's performance and safety during normal operating conditions.

Acknowledgments

The authors gratefully acknowledge the support from the Shanghai Sailing Program (Grant No. 22YF1459700) and the Natural Science Foundation of Shanghai (Grant No. 23ZR1478900).

References

[1] Zhang D, Hu R, Wang Z and Xu H 2024 Design and verification of thrust reversers efflux- pattern for wing-mounted layout civil aircraft based on CFD *J. Xihua Univ. (Natural Science Edition)* **43** 1-9

[2] Chen G 2021 Research on aerodynamic optimization of underwing long ducted nacelle *Aircraft Design* **41** 16-20

[3] Obert E 2009 *Aerodynamic design of transport aircraft* (Amsterdam: IOS) p480

[4] Hu R, Zhang D, and Shi Y 2014 Design of through-flow nacelle for low-speed wind tunnel testing of civil aircraft *Civil Aircraft Design and Research* **115** 7-9

[5] Haftmann B, Debbeler F and Gielen H 1988 Takeoff drag prediction for airbus A300-600 and A310 compared with flight test results *J. Aircraft* **25** 1088-96

[6] Wang L, Liu K, Chen Y, Si J, Zhou S and Shao Q 2021 Wind tunnel test and numerical study on civil aircraft inlet performance under low speed and high angle of attack condition *J. Propulsion Technology* **42** 1235-44

[7] Silva V, Lundbladh A, Petit O, and Xisto C 2022 Multipoint aerodynamic design of ultra-short nacelles for ultra-high bypass engines *J. Propul. Power.* **10** 541-58

[8] Andrew T Lehnig and Rahm S 1991 Inlet/fan cowl design for increased angle-of-attack requirements of transport aircraft - methodology and test results *AIAA/ASME/SAE/ASEE 27th Joint Conf. (Sacramento, CA)* (Washington: AIAA) AIAA-91-2481

[9] Hu B, Huang Y, Zhang R, Zeng W, and Gu Y 2019 Reverse thrust calibration test technology of turbofan nacelle simulator *J. Aerospace Power* **34** 2056-62

Journal of Physics: Conference Series **2815** (2024) 012005 doi:10.1088/1742-6596/2815/1/012005

[10] Zhong Y, Ma H, Li J, and Guo J 2020 Technological progress of ground test measurement of aeroengine inlet total pressure distortion *Aeroengine* **46** 62-77
[11] Chen H, Liu L and Wu C 2017 Inlet and jet test techniques for large scale model in 8m× 6m Low Speed Wind Tunnel *ACTA AERODYNAMICA SINICA* **35** 875-8
[12] Li Q and Gui F 2019 Design and experimental development overview of foreign TBCC inlet Technology *Gas Tubine Experiment and Research* **32** 58-62
[13] Burgsmuller W, Castan C, Hoheisel H and Kooi J 1995 Preparation and use of TPS-technique for low speed investigations on transport aircraft *Int. Forum on Turbine Power Simulation (Amsterdam)* (Amsterdam: DNW) pp 34-42
[14] WU C, Xu B, Zhang W, Tang W and Li Y 2020 Design and optimization of columniform distributed ejector for low speed wind tunnel inlet test *J. Aerospace Power* **35** 2482-8
[15] Cao G, Li B, Liang S and Tan H 2020 Numerical study on aerodynamic performance of CK drone aircraft air inlet in maneuvering flight *Transactions of Nanjing University of Aeronatutics and Astronautices* **37** 739-49

Research on butterfly coil EMAT sound field characteristics

Chaoqun Wang [1], Jian Ma [1,*], Luyao Wang [1], Xue Bai [1] and Jianwei Chen [1]

[1] Laser Institute, Qilu University of Technology (Shandong Academy of Sciences), 250104, Jinan, China

[*] Corresponding author's E-mail: mj@sdlaser.cn

Abstract: As a main coil structure of an electromagnetic ultrasonic transducer (EMAT), the butterfly coil can excite transverse waves under the action of the vertical magnetic field, which is widely used in the thickness measurement of metal plates and the detection of internal buried defects. This paper discusses the directivity of the butterfly coil in the acoustic field to facilitate the design and application of the electromagnetic ultrasonic transducer and obtain the energy distribution state of the radiated acoustic field. The acoustic field directivity of butterfly coil EMAT is discussed in this paper. Firstly, the ultrasonic excitation principle of the butterfly coil EMAT is analyzed, and the radiated acoustic field of the butterfly coil is theoretically calculated by equating the main flap of the butterfly coil to a rectangular wave source. The experimental system of acoustic field directivity measurement was constructed to measure the directivity data of the radiated acoustic field of the ultrasonic shear wave. The acoustic field was analyzed for different working frequencies and different polarization directions of the relative semicircular specimen. The experimental data matched the theoretical calculations well, and this paper provides a reference basis for the engineering application of this transducer.

1. Introduction

Non-destructive testing tech advances, drawing broad industry attention to ultrasonic testing. Electromagnetic ultrasonic testing, a key method, relies on electromagnetic induction to generate ultrasonic waves [1]. Compared to traditional piezoelectric methods, electromagnetic ultrasound detection technology boasts faster detection speed and eliminates the need for a coupling agent [2]. Compared to laser ultrasonic detection technology, which relies on pulsed and continuous lasers for non-contact detection, it is influenced by the surface roughness of the workpiece due to its reliance on optical methods to detect ultrasonic vibrations on the material surface [3]. However, electromagnetic ultrasonic detection is widely applied in aerospace, railway transport, steel casting, and more due to its simplicity and no-pretreatment requirement [4]. Dixon et al. [5] presented a high-frequency, non-contact electromagnetic ultrasonic transducer applied successfully in aluminum plate thickness measurement. Petcher et al. [6] devised a lightweight, easily adjustable electromagnetic ultrasonic sensor structure. Mounting on a train enables non-contact scanning of the rail surface.

An electromagnetic ultrasonic inspection system comprises an EMAT and a tested workpiece. Various types of ultrasonic waves are used by adjusting the coil type and permanent magnet structure of the EMAT. Each parameter influences the radiated acoustic field. Currently, the most common EMAT coils include spiral, foldback, runway, and butterfly coils [7]. Permanent magnets are mainly classified into unipolar and horseshoe shapes, which provide spatial magnetic fields distributed in different orientations [8]. The EMAT can generate ultrasonic body waves within the specimen and surface waves on its surface. When ultrasonic waves encounter reflective interfaces or defects during propagation, an impedance mismatch occurs, leading to the reflection of ultrasonic waves [9]. EMAT operates on this principle, enabling thickness measurement, defect detection, and imaging applications. However, in practical applications, it is necessary to conduct an in-depth study of the ultrasonic radiation sound field excited by EMAT to obtain more accurate detection results.

In recent years, many scholars have conducted relevant studies on the radiated acoustic field of EMAT. Wu et al. [10] investigated the shear-wave acoustic field in the direction of EMAT-excited horizontal polarisation. They verified it experimentally, and the experimental results were consistent with the theoretical predictions. Xie et al. [11] simulated and experimentally investigated the radiated acoustic field of a curved coil EMAT. Wang et al. [12] measured the acoustic field directivity of SH waves by creating a three-dimensional finite element model of SH-guided waves generated by a periodic permanent magnet (PPM) EMAT.

The butterfly coil EMAT consists of a butterfly coil and a unipolar-type permanent magnet. The magnetic field direction is perpendicular to the coil plane. A high-frequency alternating current is applied to the butterfly coil, which induces vertical incident longitudinal and shear waves inside the object to be measured under the influence of the biased magnetic field. However, studies on the directivity of the radiated acoustic field of butterfly coil electromagnetic ultrasonic probes are limited.

In summary, electromagnetic ultrasonic testing has been rapidly developing and widely used as a non-contact detection method in recent years. An in-depth study of the radiated acoustic field's directivity from the butterfly coil EMAT can offer a theoretical foundation for EMAT's engineering applications.

2. Theory

2.1. EMAT physical structure and detection principles

The EMAT consists of a permanent magnet, an excitation coil, and an object to be measured. The permanent magnet and the coil generate a biased static and dynamic magnetic field. Under the joint action of the biased magnetic field and the excitation coil, the surface of the object to be measured generates Lorentz force and magnetostrictive force. For non-ferromagnetic materials, the mechanism of action is the Lorentz force; for ferromagnetic materials, the mechanism of action includes the above two. In this study, aluminum is used as the object to be measured to study the acoustic field characteristics of the shear wave generated by the butterfly coil EMAT. The structure of EMAT and its working principle are shown in Figure 1. The butterfly coil has axisymmetric characteristics, and it is divided into the main flap in the middle and the side flaps on both sides, with the width of the main flap as 2a and the effective length of the coil as 2b. In practice, the non-working area is shielded to deprive it of eddy current excitation on the surface of the workpiece. The square magnet is positioned above the butterfly coil's working area, with the magnet's center projection onto the workpiece surface along the z-axis serving as the origin, denoted as O.

EMAT's transducer process relies on electromagnetic coupling, generating forces in various directions on the specimen's surface to produce ultrasonic waves. The biased magnetic field and the current direction in the excitation coil determine the vibration direction of the ultrasonic signal [13]. A high-frequency AC passes through the butterfly coil, creating an alternating magnetic field. Following the law of current induction, eddy currents form on the specimen's surface at the same frequency but in the opposite direction from the current. The paranoid magnetic field induces Lorentz force within the skin-convergent layer on the specimen's surface. This force causes horizontal vibration of the mass, propagating in the depth direction perpendicular to the specimen surface [14]. Therefore, when the specimen surface experiences shear force, the plasma point's vibration is perpendicular to the ultrasonic wave's propagation, exciting a shear wave propagating along the depth direction [15]. Polarisation refers to the phenomenon of the vibration vector of the shear wave deviating from certain directions. When the current in the main valve is perpendicular to the semicircular plane, the polarisation direction of the shear wave is parallel to the semicircular plane, such as the polarisation direction is perpendicular to the X-axis in Figure 1(a); after rotating the probe by 90°, when the direction of the current in the main valve is parallel to the semicircular plane, the polarisation direction of the shear wave is perpendicular to the semicircular plane, such as the polarisation direction is perpendicular to the Y-axis in Figure 1(b).

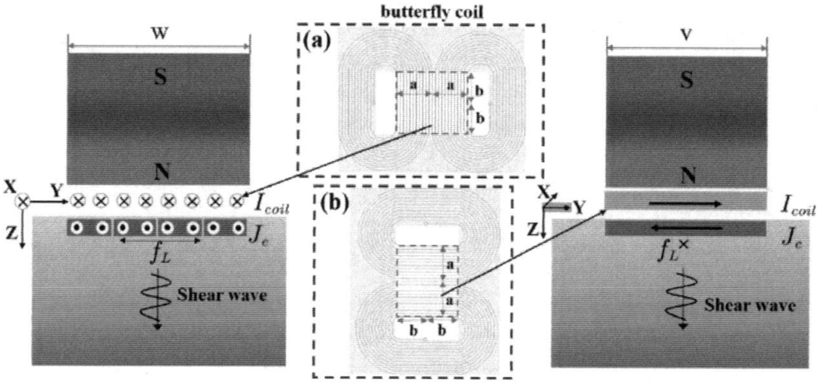

Figure 1. Structure and principle of EMAT with butterfly coil.

2.2. Directivity of the butterfly coil EMAT

As shown in Figure 2, the shear stresses are uniformly distributed in a finite rectangular region τ_{zx} , and the position of the particle Q relative to the sound source in a uniform elastic half-space is represented by the coordinates $(R_0, \theta_0, \varphi_0)$. When the surface shear stress area cell ds tends to infinity, the shear wave displacement of the particle Q in the far field of a point source can be obtained by integrating over the entire converging skin depth [16].

$$u_{\theta\_\tau} = 4abA_{\tau 0}e^{-ik_tR}\frac{1-\exp(-\sqrt{\pi f\mu\sigma\sigma})}{R\sqrt{\pi f\mu\sigma}}f(\theta)\frac{\sin(k_ta\sin\theta\cos\varphi)}{k_ta\sin\theta\cos\varphi}\frac{\sin(k_tb\sin\theta\sin\varphi)}{k_tb\sin\theta\sin\varphi} \quad (1)$$

where $f(\theta)=\dfrac{\cos\varphi\cos\theta\left(1\text{-}2\sin^2\theta\right)}{\left(1\text{-}2\sin^2\theta\right)^2+4\sin^2\theta\cos\theta\sqrt{\left(\frac{v_t}{v_l}\right)^2-\sin^2\theta}}$, k_t are the shear wavenumber, θ is the azimuthal angle at a point, v_t and are the wave speeds of shear and longitudinal waves in the medium, and a and b are the half-lengths and half-widths of the rectangular wave source. According to Equation (1), the directivity function of the transverse wave sound field is shown in Equation (2) when the main flap of the butterfly coil is under the action of the rectangular magnet.

$$d(\theta)=\frac{u_{\theta\_\tau}}{\max\left(u_{\theta\_\tau}\right)}=f(\theta)\frac{\sin(k_ta\sin\theta\cos\varphi)}{k_ta\sin\theta\cos\varphi}\frac{\sin(k_tb\sin\theta\sin\varphi)}{k_tb\sin\theta\sin\varphi} \quad (2)$$

In the XOZ plane, the directivity pattern is shown in Figure 2(b), which shows that the energy of the transverse acoustic field is more concentrated and reaches its maximum in the z-axis.

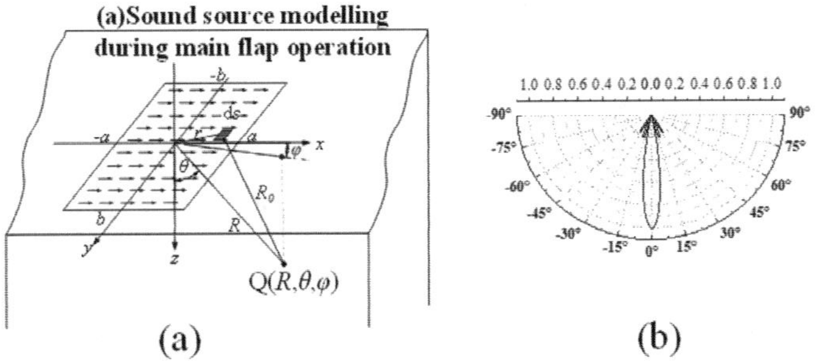

Figure 2. Sound source model and directivity pattern in the XOZ plane for coil main valve operation.

3. Experiment

3.1. Experimental system

The experimental equipment includes the host computer, RITEC RAM-5000 SNAP nonlinear high-energy ultrasonic test system, butterfly coil EMAT, and oscilloscope. The butterfly coil used in the experiment has a main flap width 2a of 11 mm and a length of 2b of 10 mm and uses a sector-shaped non-ferromagnetic material with a thickness of 60 mm and a radius of 120 mm as the ultrasonic acoustic field carrier and selects the model of 6061 Aluminium. The constructed radiation sound field detection platform is shown in Figure 3.

Figure 3. Diagram of the experimental platform.

3.2. Experimental program

The operating frequency and polarisation direction influence the electromagnetic ultrasonic transducer's directivity. This study explores butterfly coil directivity at various frequencies by adjusting the transducer's operating frequency. Additionally, the investigation includes butterfly coil directivity under different polarisation directions. In Figure 4, the transducer acoustic field measurement adopts a "one transmitter, one receiver" mode, with the excitation end fixed at the center of the semicircular aluminum block. The receiving end uses the same transducer as the excitation end, acting as the receiving probe to capture signals from all angles of the semicircular circumference.

To measure the radiated acoustic field from the butterfly coil EMAT at various operating frequencies 1 M, 1.5 M, and 2 M, it is crucial to ensure the excitation and receiving probes share the same polarisation direction on the same horizontal plane during the experiment. Additionally, to examine the acoustic field characteristics under different polarisation directions, the two probes must be rotated by 90°, followed by similarly measuring the acoustic field's directivity in each polarisation direction.

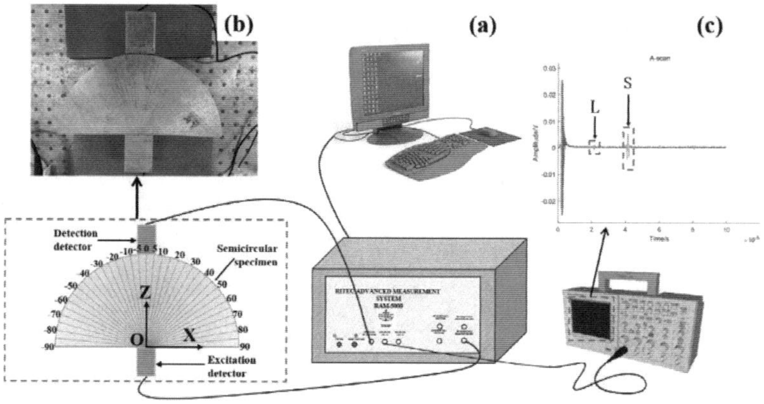

Figure 4. Diagram of the experimental model of acoustic field directionality. (a) The experimental equipment; (b) The semicircular aluminum specimen, and the excitation and reception EMATs employed in the experiment.; (c) The A-scan data acquired by an oscilloscope.

3.3. Experimental program

A broadband bandpass filter was implemented on the RITEC control page to minimize noise interference and ensure stable experimental data. Calculating the shear wave's speed in the aluminum plate from the A-scan waveform of approximately 3100 m/s, band-pass filters with different operating frequencies were designed to refine the original A-scan signal. The filters target the shear and longitudinal signals while eliminating initial impulse and electrical noise signals. Post-filtering, the shear wave effect signal was isolated, and peak-to-peak values were measured as the radiated sound field amplitude at that angle. The data were then presented as a two-dimensional polar plot to illustrate the radiated sound field across the entire sector plane.

4. Analysis of results

4.1. Radiated sound field versus operating frequency

To evaluate the effect of EMAT on the directivity of the sound field at different frequencies, the sound field of the same butterfly coil EMAT was measured at 1 M, 1.5 M, and 2 M frequencies and compared with its theoretical directivity. For the convenience of observation, Figure 5(a)-(c) show the theoretically calculated and experimentally measured directivity curves. From the figure, it can be seen that the theoretical calculation results are basically in line with the experimental results in the sound field concentration range of 15 degrees from the center, and there is still a certain deviation between the experimental results and the theory due to experimental errors and large sampling intervals and other factors.

To precisely compare shear wave energy trends at different frequencies and angles, Figure 6 presents four A-scan signals with good directivity at the 2 M frequency. Maximum and peak-to-peak shear wave data at each angle are detailed in Table 1. Comparing (a) and (b), the shear wave at the centroid position is clear, with a relatively higher amplitude of 11.1 mV. The maximum shear wave value in (b) is 5.7 mV, 5.4 mV lower than in (a). As the detection point moves from the centroid position, the shear wave amplitude decreases gradually, with a smaller change in the longitudinal wave amplitude. Observing Figure 6(c) and (d), the shear wave amplitude notably decreases. At the 10° detection position, the maximum shear wave value drops to 3.8 mV; at the 15° position, the peak shear wave value is only 4 mV.

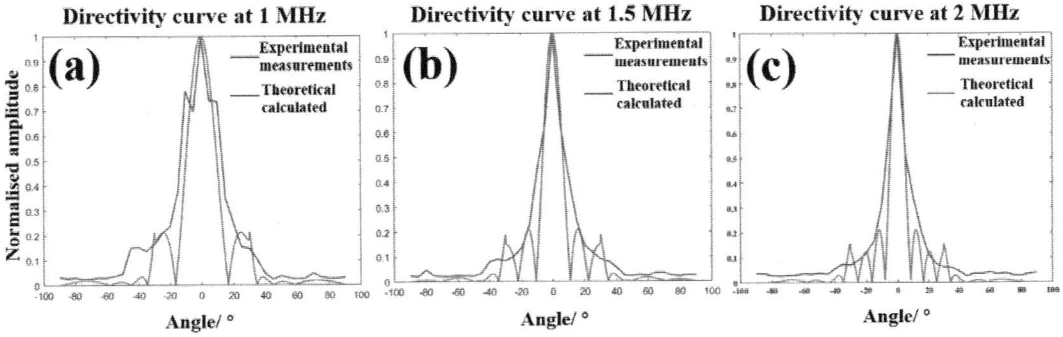

Figure 5. Directivity patterns at different frequencies. The blue and red curves are experimentally measured and theoretically calculated pointing curves, respectively.

4.2. Radiated sound field versus direction of polarization

Measurements were taken under two polarisation directions to examine acoustic field directivity with diverse polarization directions. Comparisons were made using the radiated acoustic field at 1 M and 2 M frequencies, as depicted in Figure 7. There is no normalization in the plotting process to compare the shear wave amplitudes under the two polarisation directions. In Figure 7, with polarization parallel to the semicircular plane, the sound field concentrates at 0°, maintaining high energy amplitude but exhibiting poor directivity within -20° to 20°. With polarization perpendicular to the semicircular plane, the amplitude rapidly decreases from 0° to 20°, reaching a low point around -10°. The amplitudes in the 0° direction show minimal difference between the two polarization directions.

Table 1. The maximum and peak-to-peak data of the shear wave at each angle.

	0°	5°	10°	15°
Maximum values (mv)	11.1	5.7	3.8	2.3
Peak-to-peak ratio (mv)	22.1	10.2	7.5	4.0

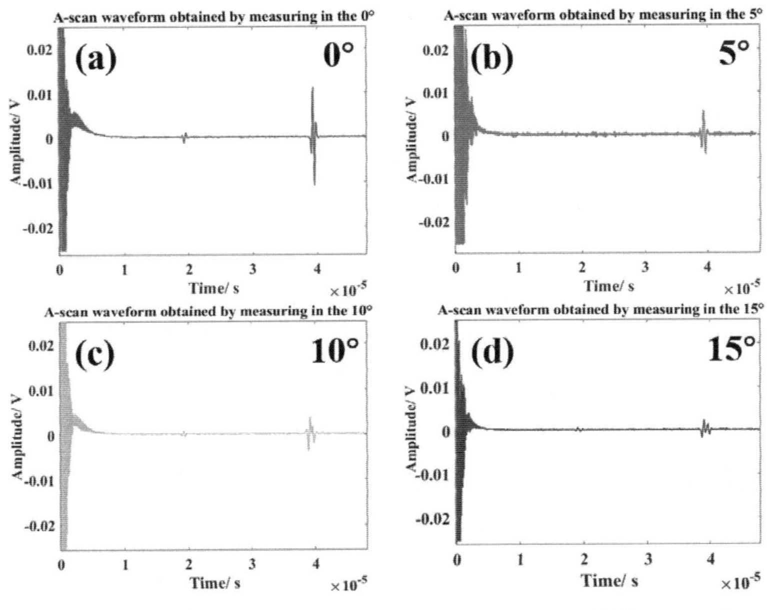

Figure 6. A-scan waveforms were obtained from measurements at different angles. (a-c) shows the filtered A-scan waveforms at 0°, 5°, 10° and 15° positions, respectively.

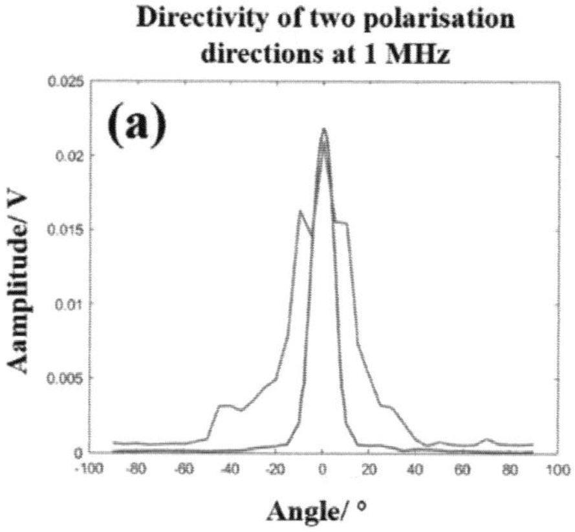

Figure 7. Sound field directivity patterns in two polarisation directions. The blue and red curves are the directivity curves with the polarisation direction perpendicular to the X and Y axes.

5. Conclusion

This paper establishes an experimental system for measuring the radiation acoustic field of butterfly coil EMAT. When sound field characteristics at various frequencies are investigated, the results demonstrate that increased frequency concentrates transverse wave radiation energy, improving sound field directivity. Experimental measurements align with theoretical calculations. Comparing results with perpendicular and parallel polarization directions to the semicircular plane reveals better directivity and concentrated energy within -10°~10° when the polarization direction is perpendicular to the Y-axis. Changes in polarization direction have minimal impact on shear wave acoustic field amplitude. The observed directivity of the butterfly coil EMAT facilitates future applications in plate thickness measurement and defect localization.

Acknowledgment

The work was supported in part by the Natural Science Foundation of Shandong Province, China (Project No. ZR2023ME073); the National Natural Science Foundation of China (Project No. 51805304); the Education Department of Shandong Province, China (Project No. 2022KJ130); and Qilu University of Technology (Shandong Academy of Sciences), China (Projects No. 2023PY009, 2021JC02008, and 2022GH005).

References

[1] Liu H, Liu T, Li Y, et al. Uniaxial stress in-situ measurement using EMAT shear and longitudinal waves: Transducer design and experiments [J]. Applied Acoustics, 2021, 175: 107781.
[2] He M, Shi W, Lu C, et al. The application of the pulse compression technique in metal materials crack detection with LF-EMATs [J]. Nondestructive Testing and Evaluation, 2023, 38(1): 45-66.
[3] BAI Xue, MA Jian, XU Wanwei, et al. Research on online testing technology for metal additive manufacturing based on laser ultrasonic [J]. Aeronautical Manufacturing Technology, 2022, 65(20): 70–82.
[4] Amosov, Oleg S., Svetlana G. Amosova, and Ilya O. Iochkov. "Defects detection and recognition in aviation riveted joints by using ultrasonic echo signals of non-destructive testing." IFAC-PapersOnLine 54.1 (2021): 484-489.
[5] Dixon, S., C. Edwards, and S. B. Palmer. "High accuracy non-contact ultrasonic thickness gauging of aluminium sheet using electromagnetic acoustic transducers." Ultrasonics 39.6 (2001): 445-453.

Journal of Physics: Conference Series **2815** (2024) 012006 doi:10.1088/1742-6596/2815/1/012006

[6] Petcher, P. A., M. D. G. Potter, and S. Dixon. "A new electromagnetic acoustic transducer (EMAT) design for operation on rail." Ndt & E International 65 (2014): 1-7
[7] Zhou J, Jia H, Zheng Y, et al. Research on racetrack coil EMAT FE model and sound field characteristics [J]. Journal of Mechanical Engineering. Switzerland, 2020, 56: 9-19.
[8] Sun C Z, Sinclair A, Filleter T. Influence of magnetostriction induced by the periodic permanent magnet electromagnetic acoustic transducer (PPM EMAT) on steel [J]. Sensors, 2021, 21(22): 7700.
[9] Patey S J, Corcoran J P. Physics of ultrasound [J]. Anaesthesia & Intensive Care Medicine, 2021, 22(1): 58-63.
[10] Di W, Ming-Xuan L, Xiao-Min W. Shear wave field radiated by an electromagnetic acoustic transducer [J]. Chinese Physics Letters, 2006, 23(12): 3294.
[11] Xie Y, Liu Z, Yin L, et al. Directivity analysis of meander-line-coil EMATs with a wholly analytical method [J]. Ultrasonics, 2017, 73: 262-270.
[12] Wang S J, Zao-Jun M A, Yong-Qian L I, et al. 3-D Simulation for the Acoustic Field Directivity of SH-Wave EMATs [C]//2017 Far East NDT New Technology & Application Forum (FENDT). IEEE, 2017: 187-191.
[13] Wang P, Zhang Y, Yao E, et al. Method of measuring the mechanical properties of ferromagnetic materials based on magnetostrictive EMAT characteristic parameters [J]. Measurement, 2021, 168: 108187.
[14] McMillan R, Hampson R, Tabatabaeipour M, et al. Design and Manufacture of an Optimised Side-Shifted PPM EMAT Array for Use in Mobile Robotic Localisation [J]. Sensors, 2023, 23(4): 2012.
[15] Joshi D, Mehta D S. Recent Trends and Diversity in Ultrasonics [M]//Handbook of Metrology and Applications. Singapore: Springer Nature Singapore, 2023: 1-18.
[16] Tang X M, Zhu Z, Toksöz M N. Radiation patterns of compressional and shear transducers at the surface of an elastic half-space [J]. The Journal of the Acoustical Society of America, 1994, 95(1): 71-76.

Journal of Physics: Conference Series **2815** (2024) 012007 doi:10.1088/1742-6596/2815/1/012007

Analysis and structural optimization of tree-type dynamic pressure groove mechanical seal performance

Zechen Wu[1, *]

[1]College of Automation, Chengdu University of Information Technology, Chengdu, Sichuan, 610225, China

[*]Corresponding author's e-mail: wu.zechen@outlook.com

Abstract. This study investigates the influence of the order and groove ratio of a tree-shaped groove on the sealing performance of a mechanical seal system using numerical simulations. A three-dimensional geometric model of the tree-shaped groove is established using SOLIDWORKS, and the mesh is divided in ANSYS Meshing before being imported into Fluent software for analysis. The single-factor experimental method is employed to evaluate the impact of different parameters on the opening force and leakage rate by varying one parameter while keeping others constant. The simulation results reveal that in the low-speed range, the tree-shaped groove with a groove ratio of 3:4:5 exhibits a larger opening force. Conversely, in the high-speed range, the three-stage tree-shaped groove with a groove ratio of 1:1:1 demonstrates a lower leakage rate and a larger opening force. The findings highlight the significant impact of the order and groove ratio of the tree-shaped groove on the sealing performance of the non-contact mechanical seal. The three-stage tree-shaped groove outperforms the second and fourth stages in terms of mechanical sealing performance, particularly in regard to the opening force. This study provides valuable insights into optimizing the design of tree-shaped grooves in mechanical seal systems to enhance their overall performance and efficiency.

1. Introduction

As is known to all, mechanical seals are divided into contact mechanical seals and non-contact mechanical seals. For the former, researchers mainly focus on the friction characteristics parameters, the influence of surface topography and its characterization, and the end face material matching [1]. Due to the issue of surface wear and tear, once the leakage rate reaches a certain level, the system can be considered a failure. This not only results in economic losses but also leads to the waste of related energy and even causes safety incidents such as combustion and explosion [2]. In order to avoid this situation, people have subsequently developed non-contact mechanical sealing systems. According to the direction of rotation, non-contact mechanical seals are divided into unidirectional rotation and bidirectional rotation. The asymmetry of the groove structure makes the unidirectional rotating non-contact mechanical seal only work when it is rotating forward, and it cannot establish the dynamic pressure of isolating the dynamic and static ring rigid solid contact when it is rotating in the opposite direction [3]. This situation has led people to pay more attention to the research of bidirectional rotating non-contact mechanical seals. In bidirectional rotating non-contact mechanical seals, the tree-shaped groove can establish an isolated fluid film on the end face of the dynamic and static ring to form a stable opening force during both forward and reverse rotation, thus attracting widespread attention from many scholars.

John Crane Company was the first to develop a bidirectional rotating tree-shaped groove end face structure for dry gas seals in 1992 [4]. In 2001, Wang successfully applied for a related patent [5]. In 2012, Ma et al. conducted research on the dynamic pressure characteristics of the gas film of the bidirectional rotating tree-shaped groove seal end face and concluded that the bidirectional rotating tree-shaped groove dry gas seal can produce a significant dynamic pressure effect, forming a stable and uniform pressure distribution of the gas film, which is conducive to the opening of the seal end face [6]. In 2016, Wang et al. performed numerical simulation calculations on the tree-shaped groove non-contact mechanical seal using Fluent software, and based on the numerical simulation results, they provided the optimal selection range for the number of grooves, groove depth, width, and angle of the tree-shaped groove [7]. Despite this, the related research on the structural parameters of the tree-shaped groove is still not perfect. Based on this, this article will further explore the selection of related parameters, such as the order of the tree-shaped groove and its groove circumference ratio, and improve the research on the related structural parameters of the tree-shaped groove. In assessing the effectiveness of the mechanical seal, this essay chooses the opening force and leakage rate as the assessment metrics, with the aim that the opening force should be maximized and the leakage rate minimized. The opening force is viewed as a crucial element influencing the degradation of mechanical seals, as it is the force capable of parting the sealing surfaces. Hence, it is one of the assessment standards; the leakage rate is a primary parameter in gauging the quality of seal performance and ensuring minimal leakage is a fundamental aspect of the design of mechanical seals. Thus, it is also selected as an assessment metric [8].

In summary, this article mainly studies the influence of the geometric parameters of the tree-shaped dynamic pressure groove in the mechanical seal system on its opening force and leakage rate. First, the three-dimensional geometric leakage model of the tree-shaped groove is established through SOLIDWORKS 3D modeling software, and it is imported into ANSYS Meshing for grid division processing. Then, the successfully divided grid is imported into Fluent software for numerical simulation, and the optimal choice of the order of the tree-shaped groove and its groove circumference ratio is given based on the numerical simulation results. The research results obtained have laid a theoretical foundation for the tree-shaped groove mechanical seal system.

2. Physical model

As shown in Figure 1, it is the structural design diagram of the non-contact mechanical seal. It mainly pumps the high-pressure blocking gas on the outer diameter side into the seal end face through the dynamic ring fluid dynamic pressure groove under the action of viscous shear force, causing the gas film pressure from the outer diameter to the groove diameter to gradually increase. The gas film pressure from the groove diameter to the inner diameter gradually decreases. Because the increase in end face film pressure makes the opening force formed greater than the closing force acting on the seal ring, it forces the two end faces that remain in contact in a stationary state to separate and be in a stable non-contact state. The groove type of dynamic pressure groove can be divided into many types, such as spiral groove, straight groove, T-type groove [9-11], etc. This article mainly explores the impact of different orders of the tree-shaped groove and its groove circumference ratio on its sealing performance. As shown in Figures 2 to 4, they are the gas film flow fields of the second-order tree-shaped groove, the third-order tree-shaped groove, and the fourth-order tree-shaped groove, respectively. Considering that the ratio of the gas film thickness size to the radial size reaches four orders of magnitude, in order to facilitate the observation of its corresponding details, the gas film thickness size is magnified 1000 times [12].

In order to facilitate the exploration of the impact of different orders of the tree-shaped groove and its groove circumference ratio on its sealing performance, the remaining geometric parameters are defined as shown in Table 1.

Table 1. Geometric parameters.

Geometric parameters	Numerical value
Inside radius r (mm)	25
Outside radius R (mm)	40

Number of grooves N_g	12
Depth of grooves D_g (μm)	5
Included angle θ (°)	30
breadth b (mm)	2

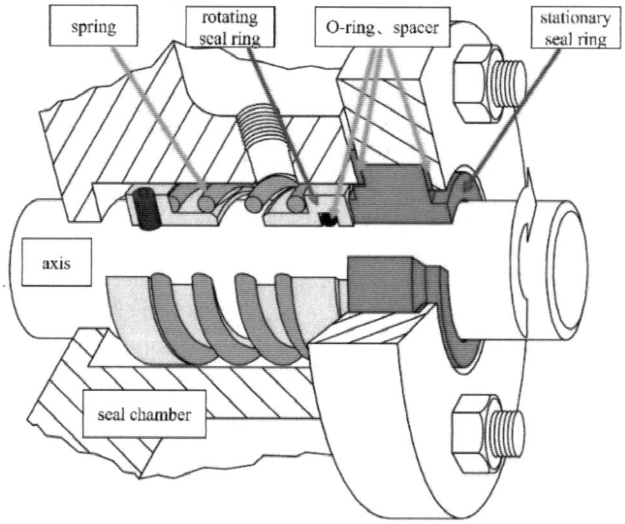

Figure 1. Schematic diagram of non-contact mechanical seal.

Figure 2. Second-order tree-shaped groove.

Figure 3. Third-order tree-shaped groove.

Figure 4. Fourth-order tree-shaped groove.

3. Numerical model

3.1 Theoretical model of non-contact mechanical seal system

3.1.1 Assumptions. Based on the basic theory of fluid mechanics and considering the structure of the mechanical seal system itself, the following assumptions are made when analyzing the steady-state flow field of the gas film between the dynamic and static ring end faces [13-14]:

(1) The gas between the sealing end faces is an isothermal, equal-viscosity ideal gas;

(2) The gas flow between the sealing end faces is a continuous medium flow;

(3) The gas in the sealing gas film moves in laminar flow, belongs to a Newtonian fluid, and complies with Newton's law of viscosity [14-15];

(4) The impact of the roughness of the sealing end face on the gas flow is ignored [16].

3.1.2 Mathematical model. Based on the above assumptions, we can obtain the dimensionless Reynolds equation for calculating the pressure of the end-face gas film [17]:

$$\frac{1}{R^2}\frac{\partial}{\partial\theta}\left(PH^3\frac{\partial P}{\partial\theta}\right)+\frac{1}{R^2}\frac{\partial}{\partial R}\left(RPH^3\frac{\partial P}{\partial R}\right)=\Lambda\frac{\partial(PH)}{\partial\theta} \tag{1}$$

Each dimensionless parameter is defined as follows:

$$R=\frac{r}{r_i},\ P=\frac{p}{p_i},\ H=\frac{h}{h_o},\ \Lambda=\frac{6\mu\omega r_i^2}{\left(p_i h_o^2\right)} \tag{2}$$

where:

r_i represents the inner radius of the sealed end face;

p represents the air film pressure between the end faces;

p_i represents the pressure on the inner diameter side;

h represents the air film thickness at any point on the end face;

h_o represents the film thickness of the air film in the non-slotted area between the end faces;

μ represents the viscosity of the gas;

ω represents the rotational angular velocity;

Λ represents the dimensionless compression number.

3.1.3 Selection of operating parameters. This paper uses the single-factor experimental method in numerical simulation; that is, one-factor changes, and other factors under consideration remain unchanged. Table 2 gives the operating parameters given in the numerical simulation process of this paper.

Table 2. Operating parameters.

Operating parameters	Numerical value
Inlet pressure P_{in} (MPa)	0.1013
Outlet pressure P_{out} (MPa)	0.5
Revolving speed N (r/min)	3000
Viscosity μ ($Pa \bullet s$)	1.8×10^{-5}
Film thickness h (μm)	2

3.2 Simulation analysis of the tree-shaped slot mechanical sealing system
When conducting Fluent simulation in this paper, the calculation model used is the laminar flow model [18-21]. The way of establishing the geometric model, setting the boundary conditions, and condition assumptions refer to the setting methods of other types of slot mechanical seals [22].

3.2.1 Grid division and solution settings. This paper uses ANSYS Meshing for grid division and takes into account that as the number of grids continues to increase, the results of the simulation values gradually converge, that is, grid independence [23]. After testing, the final number of grid divisions is determined to be 883, 616. The divided model is shown in Figure 5.

Figure 5. Grid division.

3.2.2 Grid division and solution settings. The mandatory boundary conditions for solving Equation (1) are [24]: On the inner diameter exit surface, there is $p = p_i$ (atmospheric pressure); on the outer diameter inlet surface, there is $p = p_o$ (medium pressure).

The periodic boundary condition is:

$$p\left(\theta + 2\pi / N_g\right) = p(\theta) \tag{3}$$

3.2.3 Setting of convergence condition. In Fluent, this paper chooses the SIMPLEC method and improves the calculation accuracy through the Second Order Upwind. In order to ensure that the residual curve converges as quickly as possible during the simulation process, the pressure and momentum in the under-relaxation factor are set to 0.5, and the absolute accuracy of model convergence is set to 1×10^{-6}.

Journal of Physics: Conference Series **2815** (2024) 012007 doi:10.1088/1742-6596/2815/1/012007

3.2.4 Verification of simulation methods. To verify the correctness of the grid partitioning method and numerical simulation in this paper, the numerical results obtained are compared with the numerical results in [7], as shown in Figure 6. It can be seen that as the membrane thickness increases, the opening force corresponding to its mechanical seal also decreases accordingly, which is consistent with the trend shown in reference [7], thus proving the correctness of the simulation method.

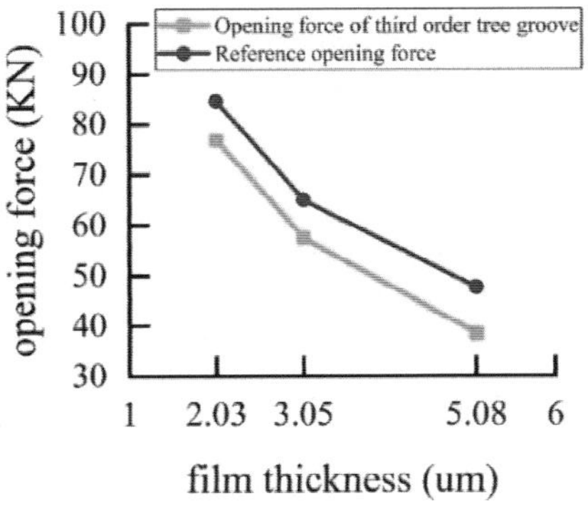

Figure 6. Comparison of the simulated values of the opening force corresponding to different membrane thicknesses with the simulated values in the reference literature.

4. Results and discussion

According to the previously set geometric parameters and operations, the corresponding results can be obtained through simulation with Fluent software.

4.1 Fluid film pressure distribution

The prominent feature of the tree-shaped groove is that there are more high-pressure areas [25-26], as shown in Figure 7 for the fluid film pressure distribution diagram. As can be seen from the figure, the third-order tree-shaped grooves all generate pressure peaks at the root of the groove. The closer the groove root is to the high-pressure side, the greater the pressure peak will be.

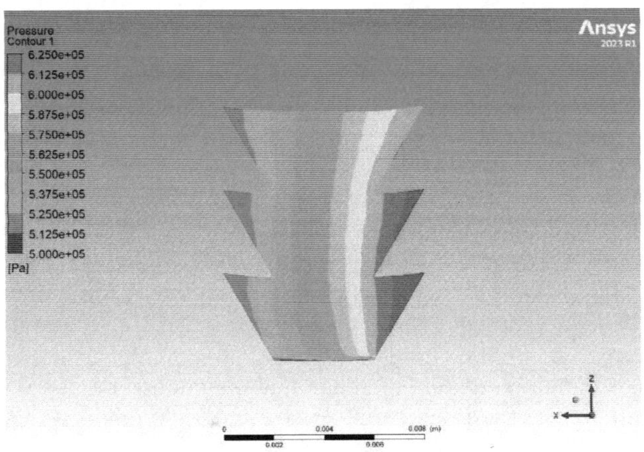

Figure 7. Fluid film pressure distribution diagram.

4.2 The impact of geometric parameters on sealing performance

4.2.1 Order of tree-shaped grooves. The correlation between the revolving speed and the opening force and leakage rate of tree-shaped grooves of different orders is shown in Figure 8.

From the correlation diagram of revolving speed and opening force, it can be seen that the opening force of the third-order tree-shaped groove is almost always greater than that of the second and fourth-order tree-shaped grooves within the corresponding speed range, and it has a consistent change trend with the second-order tree-shaped groove. Although the fourth-order tree-shaped groove is also superior to the second-order tree-shaped groove, considering the cost and difficulty of manufacturing [22], its opening force is not as large as the third-order tree-shaped groove. Therefore, in the subsequent manufacturing process, the third-order tree-shaped groove can be used to improve the overall opening force of the mechanical seal.

From the correlation diagram of revolving speed and leakage rate, it can be seen that the leakage rate of the third-order tree-shaped groove is slightly higher than that of the second and fourth-order tree-shaped grooves except in the speed range of 8~10 Kr/min. However, in the rest of the range, its leakage rate has always been lower than that of the second-order tree-shaped groove.

Comparing the obtained correlation between revolving speed opening force, and leakage rate with the study in [27], it is not difficult to find that the performance of the third-order tree-shaped groove mechanical seal is similar to that of other non-contact mechanical seal systems. As the revolving speed increases, its opening force will gradually increase, while the leakage rate will first decrease to a certain extent and then increase with the increase of revolving speed.

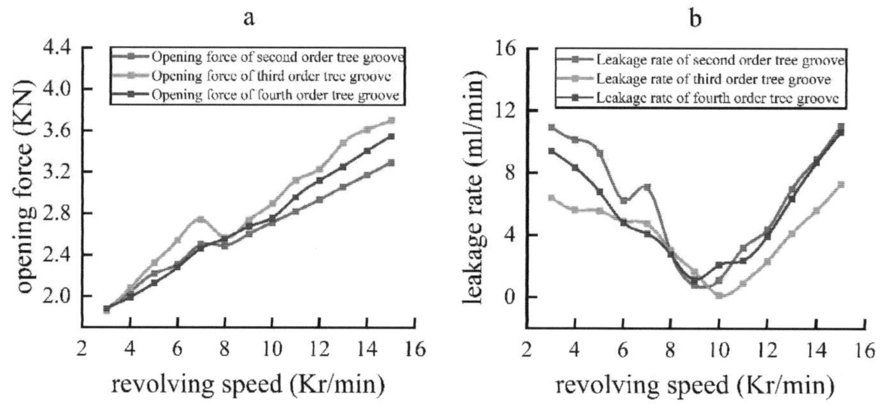

Figure 8. The correlation between the revolving speed and the opening force and leakage rate of tree-shaped grooves of different orders.
(a) The correlation between the revolving speed and opening force of tree-shaped grooves of different orders;
(b) The correlation between the revolving speed and leakage rate of tree-shaped grooves of different orders.

4.2.2 Groove circumference ratio of tree-shaped groove. The correlation between the revolving speed and the opening force and leakage rate of tree-shaped grooves with different groove circumference ratios is shown in Figure 9.

From the correlation diagram of revolving speed and opening force, it can be seen that when the groove circumference ratio of the third-order tree-shaped groove is 3:4:5, its opening force is larger in the low-speed range. When the revolving speed is about 5~7 Kr/min, the opening force of the third-order tree-shaped groove with a groove circumference ratio of 1:1:1 is larger. It is not difficult to find that as the revolving speed continues to increase, the opening force corresponding to the tree-shaped grooves with different groove circumference ratios does not differ much. Considering the manufacturing cost and the difficulty of processing, it is recommended to use the third-order tree-shaped groove with a groove circumference ratio of 1:1:1 at higher speeds.

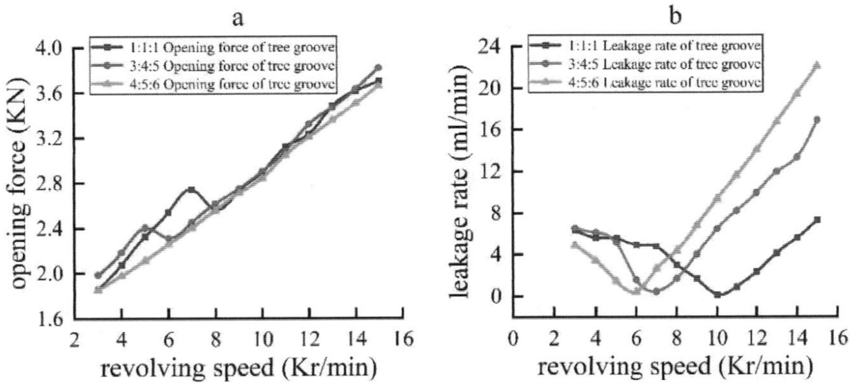

Figure 9. The correlation between the revolving speed and the opening force and leakage rate of tree-shaped grooves with different groove ratios.
(a) The correlation between the revolving speed and opening force of different slot ratios of tree-shaped slots;
(b) The correlation between the revolving speed and leakage rate of different slot ratios of tree-shaped slots.

5. Conclusion

Through the numerical analysis of the non-contact mechanical seal of the tree-shaped groove, the results show that the order of the tree-shaped groove and its groove circumference ratio have a certain impact on the sealing performance of the tree-shaped groove. In the order of the tree-shaped groove, the sealing performance of the tree-shaped groove is not monotonically related to its order, but in the third order, it has the best opening force and leakage rate. In the groove circumference ratio of the tree-shaped groove, if the overall mechanical seal system is in a lower speed range, the tree-shaped groove with a groove circumference ratio of 3:4:5 can obtain a larger opening force. If the overall mechanical seal system is in a higher speed range, the tree-shaped groove with a groove circumference ratio of 1:1:1 can obtain a lower leakage rate and a larger opening force.

References

[1] Haomuming, Zhou Rui, Sun Pengtao, et al. Experimental Study on Friction Characteristics of Contact Mechanical Seal during Start-up and Shut-down Process[J]. Lubrication and Sealing, 2023, 48(09):41-46.
[2] Haomuming, Song Yong, Wang Shunji, et al. Study on Steady-State Friction Characteristics of Contact Mechanical Seals[J]. Petroleum Machinery, 2023, 51(04):1-7.
[3] Wang Yan, Sun Jianjun, Ma Chenbo, et al. Progress in the study of bidirectional rotating non-contact mechanical seal technology[J]. Fluid Machinery, 2013, 41(03):34-40.
[4] Goldswain I M. Mechanical face seals [P]. European Patent: 0499370 A1, 1992.
[5] Wang Yuming. Spiral groove end face seal device capable of bidirectional rotation[P]. China: ZL01102213. 2, 2001.
[6] Ma Yuanyuan, Peng Xudong, Bai Shaoxian, et al. Dynamic pressure characteristics of bidirectional rotating tree-shaped groove seal end face air film[J]. Fluid Machinery, 2012, 40(01):13-16+5.
[7] Wang Yan, Yu Xuemei, Lu Long, et al. Simulation calculation of tree-shaped groove non-contact mechanical seal based on Fluent[J]. Fluid Machinery, 2016, 44(10):19-24.
[8] Jianlei Wang, Hiroshi Kadokawa, Weigang Zhao, et al. Research on Structural Design and Optimization of End Face Groove Type for Hydrodynamic and Hydrostatic Mechanical Seals. Journal of Mechanical Engineering, 2021, 57(09): 108-117.

[9] Wang Zhi, Wu Yuxiao, Han Xu, et al. Performance Analysis of Supercritical Carbon Dioxide Combined Dry Gas Seal[J]. Journal of Power Engineering, 2023, 43(09):1143-1150+1189.
[10] Li Zhentao, Pang Minmin, Wei Huijuan, et al. Influence of groove parameters on the sealing performance of oblique straight groove liquid film seal[J]. Lubrication and Seal, 2024, 49(03):7-14.
[11] Xiong Zhongfen, Liu Meihong, Wei Qiufa, et al. Nonlinear regression analysis of structural parameters of T-type groove cylindrical air film seal under steady state performance[J]. Lubrication and Seal, 2023, 48(12):62-67+116.
[12] Mo Longgang, Ding Xuexing, Yan Ruqi, et al. Analysis of Steady-State Performance of Tree-like Groove Dry Gas Seal[J]. Lubrication and Sealing, 2022, 47(12):68-74.
[13] Hu Qiong, Zhu Minjun, Wang Yan, et al. Performance Analysis of Fir Tree-shaped Groove Dry Gas Seal with Radial Ordered Micro-structure[J]. Engineering Science and Technology, 2020, 52(1):153-160.
[14] Wang Jingmo, Ding Xuexing, Wang Shipeng, et al. Study on the Performance of Micro-texture Structure at the Bottom of T-type Groove in Dry Gas Seal[J]. Lubrication and Sealing, 2023, 48(9):99-107.
[15] Shang Hao, Chen Yuan, Li Xiaolu, et al. Study on the Impact of Nonlinear Effects under Film Thickness Disturbance on the Performance of Dry Gas Seals[J]. Journal of Chemical Industry, 2021, 72(4):2213-2222.
[16] Zhang Weizheng, Zhao Jijun, Zhang Xianzhong, et al. Fluid-Solid Coupling Analysis of New Spiral Groove Dry Gas Seal[J]. Lubrication and Sealing, 2022, 47(09):1-10.
[17] Wang Yufei, Ding Xuexing, Ma Gaofeng, et al. Influence of Derived Spiral Groove on the Steady-State Characteristics of Micro-film in Supercritical Carbon Dioxide Dry Gas Seal[J]. Lubrication and Sealing, 2022, 47(08):90-99.
[18] Wang Yue, Zou Longqing, Li Wei, et al. Study on the Influence of Rheological Characteristics of Dry Gas Seal Air Film on Friction Characteristics[J]. Machine Tool & Hydraulics, 2023, 51(16): 178-183.
[19] Ding Junhua, Yu Shurong, Wang Shipeng, et al. Simulation of Flow Field and Sealing Performance Test of Ultra-High Speed Dry Gas Seal under Multiple Effects[J]. Journal of Chemical Industry, 2023, 74(5): 2088-2099.
[20] Song Pengyun, Fu Chaobo, Li Zhiyong. Study on the Sealing Performance of Spiral Groove Dry Gas Seal in Static State[J]. Lubrication and Sealing, 2021, 46(3):4-10.
[21] Geng Jun, Mutalif Ahmed, Yin Runsheng. Numerical Calculation of Sealing Performance of Regular Polygonal Micro-pore Structured End Face [J]. Mechanical Design and Manufacturing, 2024, (02):83-87+93.
[22] Jiang Ziwei, Chen Qing. Study on the Performance and Structural Optimization of Spiral Groove Dry Gas Seal[J]. Journal of Jilin Institute of Chemical Technology, 2021, 38(1):13-17.
[23] Wang Jingmo. Dynamic Analysis and Micro-texture Optimization Design of T-slot in Dry Gas Seal [D]. Lanzhou University of Technology, 2023.
[24] Wang Heshun, Liu Xiaoming, Zhang Chening, et al. Comparative Study on the Characteristics of Double-Tip Groove Dry Gas Seals [J]. Journal of Tribology, 2021, 41(6): 974-982.
[25] Wang Yuming, Liu Wei, Liu Ying. Current status and prospect of basic research on non-contact mechanical seal[J]. Hydraulic Pneumatic & Seals, 2011, 31(02):29-33.
[26] Mao Wenyuan, Song Pengyun, Deng Qiangguo, et al. Progress in the study of mechanical seal fluid dynamic pressure groove processing technology[J]. Lubrication and Seal, 2017, 42(11):124-136.
[27] Jia Qian, Liu Chenyang, Li Sihan, et al. Study on the Working Characteristics of Spiral Groove Mechanical Seal under Gas-Liquid Two Phases [J]. Fluid Machinery, 2023, 51(4): 31-37, 43.

Journal of Physics: Conference Series **2815** (2024) 012008 doi:10.1088/1742-6596/2815/1/012008

Simulation and optimization of fuze circuit housing structural strength under a high overload environment

Qi Wu[1,2], Lide Yan[2], Hang Zhang[3], Lei Hou[2], Pengfei Liang[2], Yaoyun Hao[2], Xiaochang Duan[2], Yubo Xue[2] and Jinshan Ding[1,*]

[1] School of Electronic Engineering, Xidian University, Xi'an 710071, China
[2] Xi'an Institute of Electromechanical Information Technology, Xi'an 710089, China
[3] System Design Institute, Hubei Aerospace Technology Academy, Wuhan 430040, China

[*] Corresponding author's e-mail: ding@xidian.edu.cn

Abstract. With the continuous improvement of modern military equipment protection systems, the demand for ammunition to penetrate hard targets has become increasingly apparent. The projectile is subjected to a high overload environment during launching and penetrating hard targets. Fuze as the detonation control system of the projectile, is prone to significant deformation or even damage to its circuit housing structure in a high overload environment. The dynamic response of fuze circuit housing under high overload conditions is analyzed, and the housing structure is further optimized according to the simulation results to ensure that the fuze circuit housing can meet the strength requirements under a high overload environment.

1. Introduction

After entering the 21st century, with the continuous development of military science and technology in various countries, intelligence and information technology have become a mainstream trend in modern war. The application of cutting-edge technology makes the complexity of modern weapon systems continue to increase, so the update and iteration speed of weapon equipment is also accelerating, which requires that the overall cycle of development, production, and testing must be shortened under the premise of ensuring quality. For the development and production stage, we can increase efficiency by increasing manpower and material resources. For the test part, due to the limitations of test range number and field conditions, it is often difficult to obtain final test data for too long, resulting in a significant extension of the development cycle and an increase in development costs [1].

In order to solve the above technical bottlenecks, digital design, numerical simulation, and other technologies came into being. The application of data-driven visual simulation in the field of ammunition engineering is an effective technical way to solve the problem of long test cycles, high development costs, and actual combat damage capability evaluation [2,3]. By simulating products through digital models, defects in design can be found as early as possible, and the waste of resources caused by the direct use of physical models can be effectively avoided. Compared with traditional ammunition design technology, digital design technology can effectively shorten ammunition research and development cycle, but also can reduce the development process, simplify the manufacturing process, and improve ammunition design accuracy.

Recently there have been many cases of digital simulation applied to ammunition engineering. Yin et al. [4] used joint simulation and optimization of Isight and Abaqus to model and analyze the stress distribution of the ballistic test circuit module during the artillery firing process. Liang et al. [5] used the finite element method and LS-DYNA software to calculate structural buffers in reducing the effectiveness of missile-borne equipment. Yu et al. [6] analyzed the stress of the missile-borne module circuit by using Hypermesh and Abaqus software for the problem that missile-borne components of small-caliber weapons are prone to failure under high impact. Zhao et al. [7] analyzed the failure behavior of a missile-borne chip under high overload impact, and the failure was caused by an impact fracture of the internal movable structure of the packaged chip. Lyu et al. [8] analyzed the failure mechanism under three kinds of impact attitudes with the help of finite element software LS-DYNA for the impact direction on the reliability of encapsulated devices. Cao et al. [9] investigated the coupled dynamic characteristics between impact load and fuze in the process of projectile penetrating the hard target and revealed the change rule of fuze by impact load.

According to a previous study, ammunition will have a large speed change in a short time during the launch and penetration process, and vibration and shock are the main factors causing the failure of electronic equipment, so missile-borne equipment needs to have high overload mechanical characteristics [10-12]. Since it is usually difficult to use vibration isolation measures for the whole machine, it is necessary to protect the key parts to avoid the failure of ammunition in the operation process [13-15]. Missile-borne circuit housing is used to protect important components, and once the surface is fractured, the components will be directly exposed to the external environment, so it is necessary to have an intuitive understanding of force characteristics and failure mechanisms [16]. This article takes an innovative perspective to analyze the overload dynamic response of circuit housing in ammunition penetration fuze, and proposes targeted improvements for the structural part, which is beneficial for the development and optimization of kinetic energy destructive munition, and has important significance for the development of national defense.

2. Demonstrate program design

Based on the above, it can be seen that the internal stability of a missile-borne body is the core factor affecting its destructive effect [17]. Under an impact environment, missile-borne structures will inevitably be deformed, and excessive deformation will seriously affect the overall performance index, and may even lead to premature explosion or dud. For the inside of the missile-borne body, circuit housing once damaged will make some of the key components lose physical protection. When stress generated by collision propagates to the circuit board or chip interior, it will cause the dislodgement of component pins on the circuit board, breakage of gold wires inside the chip, power failure of energy storage components and other failure modes, which will lead to the whole fuze failure [18], and hidden dangers brought by the above reasons should not be ignored. In order to analyze the stability situation, this paper takes missile-borne circuit housing as the object, conducts kinetic failure studies, and makes structural improvements based on the deformation results as a way to enhance its survivability, overload strength, and overall reliability.

Specifically, the impact response of circuit housing mainly depends on impact conditions, geometry, and material. For impact conditions, we expect a higher load-bearing capacity to be better. For geometry, we first take the current common shell shape shown in Figure 1 as an example to demonstrate, and then improve the structure according to simulation results. For material, we take the common shell materials superhard aluminum alloy and titanium alloy as an example to analyze, as shown in Table 1.

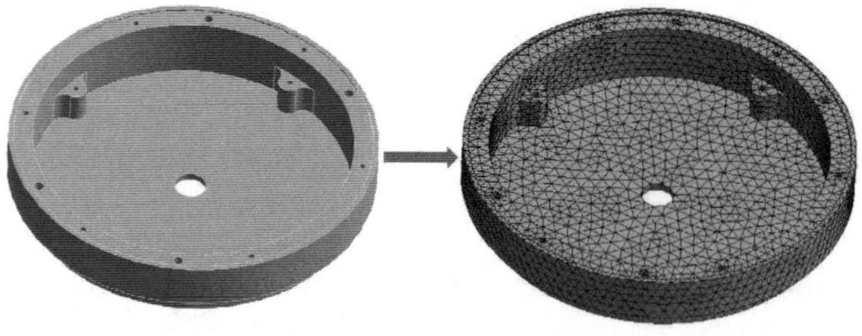

Figure 1. Finite element model of circuit housing.

During the launching process, the upper surface of the circuit housing is subjected to extremely normal tensile loads with peak durations of milliseconds, and the tensile stresses may cause the upper surface to fracture. The above conclusion is consistent with the first strength theory description, so we assume in the next simulation that two of the three principal stresses in impact overload are zero, and the maximum principal stress is the tensile stress applied on its upper surface. When surface tensile stress reaches the strength limit of material in unidirectional tension, the housing will fracture. According to this method, the load limit can be derived as an analytical basis for judging the strength of circuit housing.

Table 1. Basic material parameters.

Materials	ρ (g·cm^{-3})	E (GPa)	σ_b (MPa)	μ
Superhard aluminum alloy	2.79	71.8	524	0.30
Titanium alloy	4.42	112.3	1400	0.34

3. Scheme simulation demonstration

In order to demonstrate the rationality and feasibility of this simulation, we first set up a mathematical model of the traditional circuit housing and applied a dynamically adjusted normal tension on the housing surface to simulate the overload phenomenon in the launching environment of a missile-borne body. This model uses the Johnson-Cook model to describe material state, which is widely used in numerical simulation of penetration and impact. The Johnson-Cook model describes the Von-Mises stress as shown in equation (1).

$$\sigma = (A+B\varepsilon^n)(1+C\ln\varepsilon_1)(1-T^m) \tag{1}$$

where A is yield stress, B is the hardening modulus, C is strain rate hardening coefficient, m is thermal softening coefficient, n is strain hardening coefficient, ε is equivalent plastic strain, ε_1 is relative equivalent plastic strain rate, and T is homology temperature.

The material fracture can be derived from the cumulative damage rule as shown in equation (2).

$$D=\sum\frac{\Delta\varepsilon}{(D_1+D_2\exp D_3\dfrac{p}{\sigma_e})(1+D_4\ln\varepsilon_1)(1+D_5T)} \tag{2}$$

where $\Delta\varepsilon$ is the increment of effective plastic strain when the load is applied, D_1~ D_5 is the constant, p is pressure, and σ_e is Von-Mises equivalent stress, when $D=1$ failure occurs and the material breaks.

When the external force acting on the circuit housing is too large, the material at the danger point will be brittle along the section where the maximum tensile stress is located, and we need to observe at what amplitude the circuit housing will fracture. For the above schemes, this paper analyzes them from the perspective of aluminum alloy and titanium alloy housing respectively.

3.1. Aluminum alloy housing

In this paper, the aircraft aluminum alloy 7075 is taken as an example, which has the characteristics of high strength, easy processing, low density, and excellent mechanical properties, and has a wide range of applications in the aerospace field. For the aluminum alloy housing model, after several rounds of testing, we screened out the two results that are closest to the strength limit of the housing: 12.1 MPa (about 20408 g overload) and 13.3 MPa (about 22449 g overload) normal tensile stress, as shown in Figure 2 and Figure 3.

The results show that the upper surface has obvious deformation under the condition of 12.1 MPa tensile stress. The maximum deformation is 14.12 mm and the maximum stress is 523 MPa, but no obvious fracture occurs. Under the loading condition of 13.3 MPa, the upper surface has completely fallen off, and the maximum stress is 533 MPa. It can be seen that the circuit housing will break when it reaches the material strength limit, and the corresponding tensile stress is within the range of 12.1

MPa~13.3 MPa. The housing has certain strength characteristics, but for the high-overload environment during projectile launch and penetration, the strength result has a failure risk.

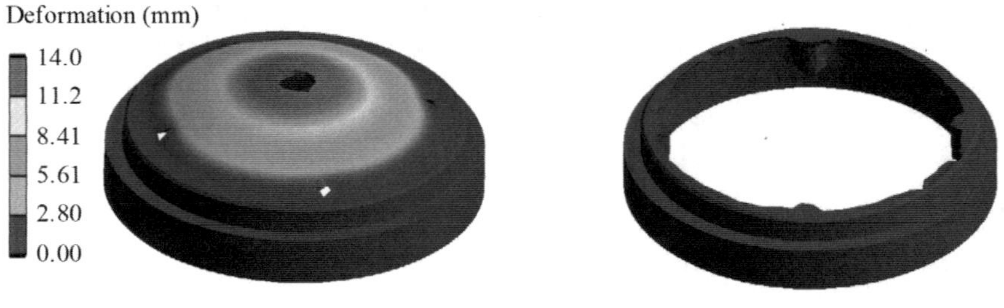

Figure 2. Deformation cloud diagram of aluminum alloy housing.

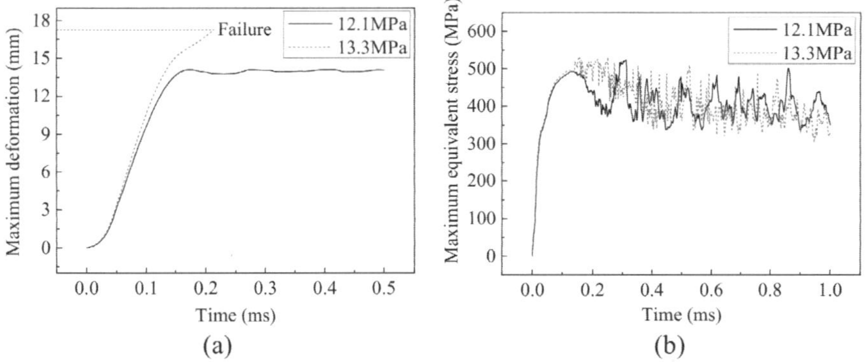

Figure 3. Numerical results of aluminum alloy housing.

3.2. Titanium alloy housing

The above analysis shows that the traditional circuit housing has risks in the missile-borne body launch environment, so we try to solve this problem by replacing a more robust material. In the aerospace, aviation, medical, and weapons industries, in order to effectively overcome the impact of high overload, people usually replace the material of key components with high-strength titanium alloys. Titanium alloy has higher strength, good corrosion resistance, and high heat resistance, and is mainly used in the production of aircraft engines, rockets, missiles, and high-speed aircraft structural parts. Although the processing cost of titanium alloy is relatively high, its ultra-high strength characteristics make it more suitable for the application requirements of missile-borne materials.

Titanium alloy will be selected as the housing material in this simulation. After multiple rounds of testing, we selected the two results that are closest to the strength limit: 30.3 MPa (about 51020 g overload) and 31.5 MPa (about 53061 g overload) normal tensile stress among all the results. The simulation results are shown in Figure 4 and Figure 5.

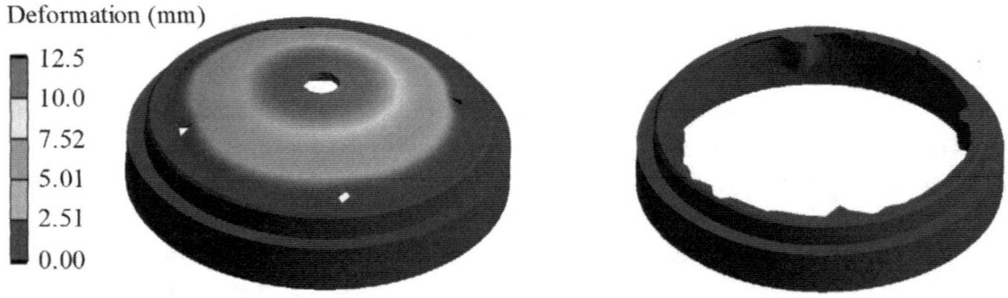

Figure 4. Deformation cloud diagram of titanium alloy housing.

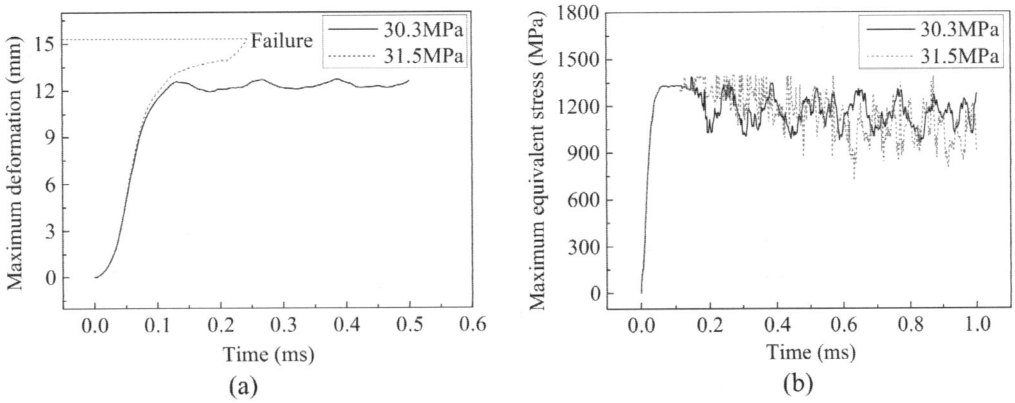

Figure 5. Numerical results of titanium alloy housing.

The results show that the circuit housing has large deformation under 30.3 MPa load, the maximum deformation is 12.70 mm, and the maximum stress is 1386 MPa, but no obvious fracture occurs. Under the loading condition of 31.5 MPa, the upper surface has completely broken, and the maximum stress is 1400 MPa. It can be seen that the circuit housing will break after reaching the strength limit, and the corresponding tensile stress is within the range of 30.3 MPa~31.5 MPa, which has greater strength improvement compared with the aluminum alloy housing.

From the perspective of the traditional scheme, there is the possibility of the circuit housing falling off in a harsh environment. Although the risk can be effectively reduced by changing housing material, the improvement of material strength comes at the cost of increasing processing difficulty. Therefore, we are expected to optimize and improve the housing strength from the perspective of structural design to ensure that the existing cost is basically unchanged. The following contents will be developed from this perspective.

4. Improvement scheme demonstration

In traditional scheme designs, because the circuit housing part of the product usually does not have a large dynamic response, people usually ignore its strength check. However, the high overload characteristics of ammunition make this technical defect more and more prominent, so we expect to strengthen the circuit housing structure to further strengthen its anti-load ability.

Specifically, the first thing that can be thought of is to improve housing thickness, because the impact load usually occurs on the upper surface, so its thickness will be a key factor in determining overall structural strength. However, the adjustment of thickness is relatively limited, because it will compress the occupied space of other parts in the missile-borne body. After comprehensive coordination, the upper surface thickness is increased from 2.4 mm to 4 mm. On the other hand, in traditional schemes, the connection part of circuit housing is often used in the form of a square edge, which is easy to cause the appearance of stress concentration from a mechanical perspective, and is not conducive to the survival of high-strength loads. Therefore, we added a 5 mm rounded corner design to the surface connection position, and the improved model is shown in Figure 6.

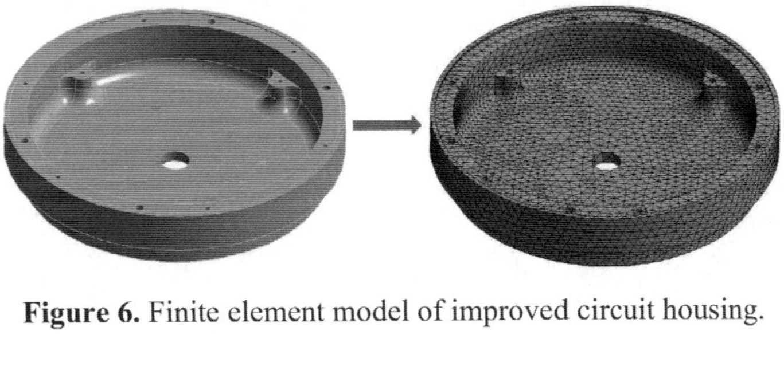

Figure 6. Finite element model of improved circuit housing.

4.1. Improved aluminum alloy housing

After structural improvement according to the above method, aluminum alloy is selected again for simulation analysis. Among all the results, we screened out the two results closest to the strength limit: 23.0 MPa (about 38776 g overload) and 24.2 MPa (about 40816 g overload) normal tensile stress, and the results are shown in Figure 7 and Figure 8.

Figure 7. Deformation cloud diagram of improved aluminum alloy housing.

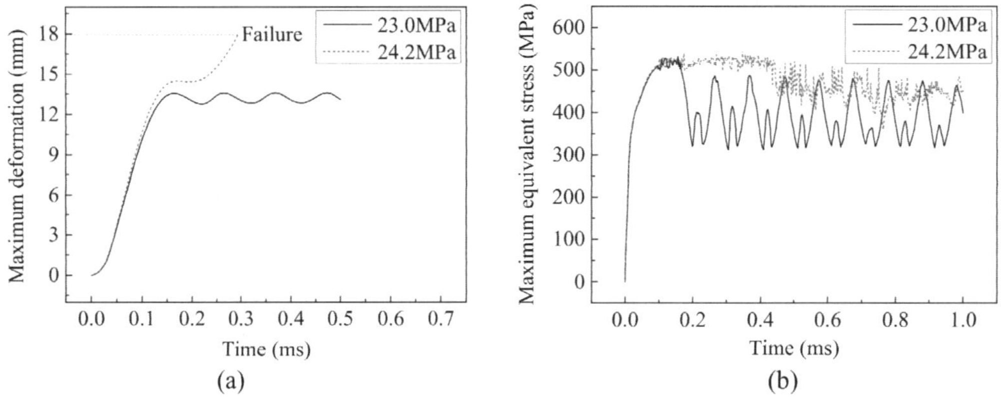

Figure 8. Numerical results of improved aluminum alloy housing.

From the above results, it can be concluded that the improved housing deforms 13.59 mm under an overload stress of 23.0 MPa, and the maximum stress is 532 MPa, while obvious fracture occurs under the overload stress of 24.2 MPa, with a maximum stress of 545 MPa. It can be seen that the strength limit has been slightly improved, and the corresponding tensile stress is within the range of 23.0 MPa~24.2 MPa. Compared from the perspective of bearing capacity, the structural improvement has improved the bearing capacity by about 85.8%, and it has been able to meet the reliability requirements under most overload conditions in practical applications.

4.2. Improved titanium alloy housing

In order to further compare the improvement effect, we analyzed the titanium alloy housing in order to expect a higher load-bearing limit. After several rounds of tests, we screened out the two results closest to the strength limit: 61.8 MPa (about 104082 g overload) and 63.0 MPa (about 106122 g overload) normal tensile stress, as shown in Figure 9 and Figure 10.

From the above results, it can be concluded that the housing will deform 12.59 mm under a tensile stress of 61.8 MPa, the maximum stress is 1390 MPa, while the upper surface will completely fall off under a tensile stress of 63.0 MPa, and the maximum stress is 1407 MPa. It can be seen that the strength limit has been slightly improved, and the corresponding tensile stress is in the range of 61.8 MPa~63.0 MPa. Comparing the results before and after the improvement, it can be seen that the bearing capacity has been improved by about 101.9%, which can be used in more severe environments.

Figure 9. Deformation cloud diagram of improved titanium alloy housing.

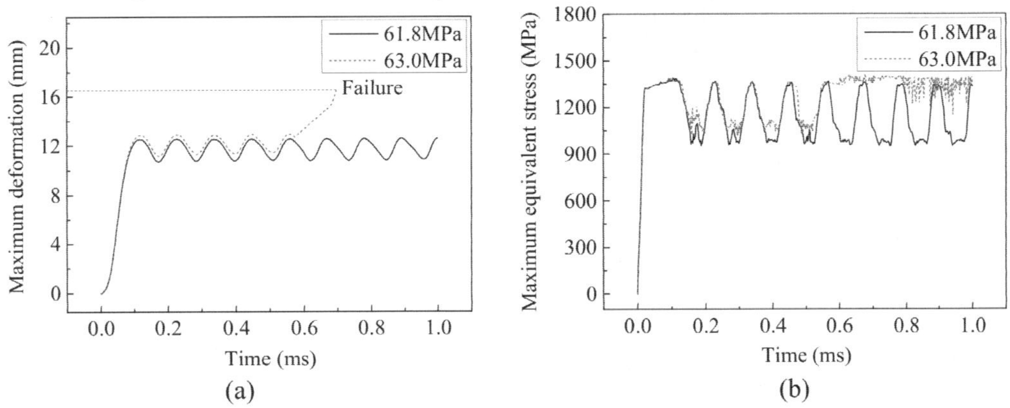

Figure 10. Numerical results of improved titanium alloy housing.

Based on the above analysis, the improvement effect of this structure is remarkable, and the comparison results are shown in Figure 11. We optimized the structural details of missile-borne circuit housing on the premise of ensuring no major changes in the existing process, so as to ensure that no adverse factors such as fracture, fall off and failure of the missile-borne body occur during launch or penetration.

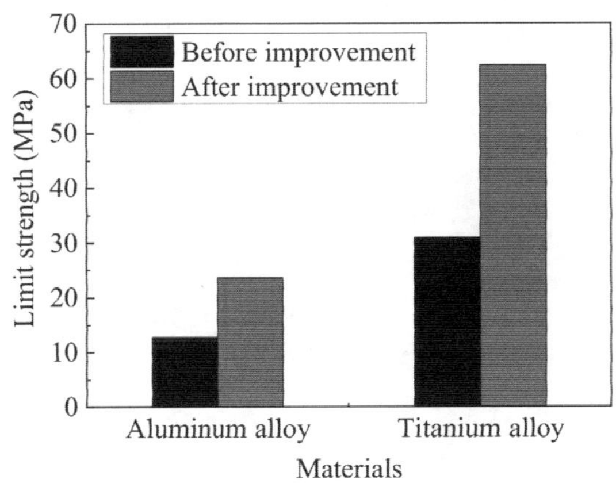

Figure 11. Comparison results of limit strength for the improved scheme.

5. Conclusion

In response to the high overload environment experienced by the missile-borne body during launch and penetration, this paper takes missile-borne circuit housing as the research object, uses the numerical simulation method to analyze load deformation under aluminum alloy and titanium alloy materials, and improves the housing structure so that it can be applied to the more severe overload environment. The analysis results in this paper can provide a reference for relevant research. In future work, we will carry out overload analysis under various conditions for a complete ammunition system, and optimize the key components to ensure safe survival and reliable ignition of ammunition under an overload environment.

References

[1] Wang Y H 2022 Design of digital simulation platform for guided munitions *North University of China*

[2] Mancini A, Frontoni E, Zingaretti P 2015 Development of a low-cost Unmanned Surface Vehicle for digital survey *ECMR* pp 1-6

[3] Morsi I, Zaghloul M S, Essam N 2010 Future voyage data recorder based on multi-sensors and human machine interface for marine accident *ICCAS* pp 1635-1638

[4] Yin W H, Yang G L, Ge J L, Li J K and Su A 2016 Design and Optimization of Cushioning and Protective Structure for a Circuit Module of Projectile-borne Testing Equipment *Journal of Gun Launch & Control* vol 37 chapter 3 pp 56–60

[5] Liang Z Y, Deng D Z, Wu H C, Zhang Z T, Shi W C, Wang Y Q and Zhao W J 2016 Research on Aluminum Foam and Thin-walled Metal Tube Composite Structure to Reduce Onboard Equipment Overload *Acta Armamentarii* vol 37 chapter S2 pp 1–7

[6] Yu Y F, Li Z X and Wu Z L 2019 Analysis on Stress for Circuit Module Under Launching High Overload Environment in Small Caliber Smart Bullet *Ordnance Industry Automation* vol 38 chapter 7 pp 92–96

[7] Zhao X L, Ma T H and Fan J B 2013 Failure analysis of common missile-borne chip in high g shock *Chinese Journal of Scientific Instrument* vol 34 chapter 10 pp 2358–2364

[8] Lyu M T and He H 2022 Impact Reliability Analysis of 3D Stacked Packaging under High Acceleration Loads *Navigation and Control* vol 21 chapter Z1 pp 181–191+165

[9] Cao J, Zhang H and Wang X F 2015 Optimization of isolated protection for the hard-target penetration fuze *Journal of Vibration and Shock* vol 34 chapter 24 pp 192–196

[10] Liu P 2023 Design and Dynamic Properties Analysis of a Shell-Carried Antenna under Mechanical Environment *Environmental Technology* vol 41 chapter 6 pp 24–29

[11] Qian L Z, Jiang B A and Guo J H 2023 Anti-High-Overload Design Method of Information Projectile *Acta Armamentarii* vol 44 chapter 5

[12] Wang X G and Hu S G 2022 Research on structure optimization and reliability of positionable missile-borne flight recorder *Journal of Ordnance Equipment Engineering* vol 43 chapter 7 pp 134–139

[13] Zhao W W, Niu F Y, Zuo X R, Wu X and Wang Z Y 2022 Study on Fracture Behavior of Low Alloy Wear-resistant Steel *Hot Working Technology* vol 51 chapter 22 pp 30–34

[14] Tian F, Chen H J and Jin H J 2022 Analysis on Cracking Failure of Sail Plate *Hot Working Technology* vol 51 chapter 19 pp 157–159

[15] Wang Y, Pei D X, Zhang Y, Shen D W and Zhao X L 2022 Research on anti-high-overload of projectile bottom pressure test system for small caliber gun *Metrology and measurement technology* vol 42 chapter 2 pp 85–90

[16] Xu W Z and Yang R 2010 Numerical Study on Deceleration Character of Projectile and Charge with Different Geometrical Properties *Journal of Projectiles, Rockets, Missiles, and Guidance* vol 30 chapter 2 pp 97–100

[17] Liu G L, Jiao G L, Ma Y, Hu J and Shi W J 2023 Study on Safety Test Method of Ammunition *Modern Defence Technology* vol 51 chapter 1 pp 107–118

[18] Liu W B 2021 Analysis of Impact Failure Mechanism and Protection Methods for Hard Target Penetration Fuze Circuit *Nanjing University of Science and Technology*

Modeling and motion simulation study of a three-dimensional sewing device

Shaohua Xu[1,*], Daxin Xin[2], Ming Yang[3] and Ying Zhang[4]

[1] Xi'an Technological University, School of Electronic Information Engineering, Xi'an, Shaanxi, 710021, China
[2] Xi'an Technological University, School of Electronic Information Engineering, Xi'an, Shaanxi, 710021, China
[3] Shaanxi New Rayon Technology Co. Shaanxi New Rayon Technology Co. Xi'an, Shaanxi, 710021, China
[4] Xi'an Technological University, School of Electronic Information Engineering, Xi'an, Shaanxi, 710021, China

[*]Corresponding author's e-mail: appleatz@foxmail.com

Abstract. In response to the sewing demands of steering wheels and their unique shapes, a multi-robot arm cooperative three-dimensional sewing device was designed to achieve efficient and precise sewing operations. Utilizing SolidWorks, a collaborative three-dimensional sewing device involving multiple robotic arms was engineered, and its mechanism model was established in ADAMS software for motion simulation analysis to evaluate the device's performance and motion characteristics. This three-dimensional sewing device is capable of fulfilling the sewing requirements of special curved surface items such as steering wheels, enhancing production efficiency and saving labor. Taking the constructed physical prototype as the research subject, motion experiments simulating the sewing process were conducted. Combined with the simulation results, curves, and three-dimensional motion effects, the feasibility of the design was successfully verified, providing a foundation for subsequent sewing trajectory planning and motion control of the device.

1. Introduction

Due to changes in production methods and the increase in labor costs, manual sewing work has gradually decreased while the demand for clothing and decorations continues to rise. To meet market demands and improve production efficiency, sewing equipment is continuously transitioning towards full automation and intelligence, leading to the emergence of various sewing robots [1]-[2]. The continuous increase in labor costs has also driven the development of industrial robots, forcing some enterprises to choose industrial robots for production [3]. Industrial robots are widely used in various industries today, with the automotive manufacturing industry having the highest demand for industrial robots [4].

Ningbo Cixing Co., Ltd., in collaboration with the 716th Research Institute of CSSC, has successively overcome the sewing mechanism and corresponding specialized machines for 3D three-dimensional sewing needs.Kudo M introduces a new automatic sewing system that handles fabric on the workbench during sewing as if it were human work[5].

The steering wheel is a special item to be sewn, requiring the tightening of its leather cover edges. However, due to the unique shape of the steering wheel, which cannot be laid flat, traditional sewing machines cannot complete the sewing action, necessitating the use of three-dimensional sewing devices for the task [6]. Three-dimensional sewing devices, integrating industrial robot technology on the basis of intelligent sewing equipment, have the significant advantage of achieving spatial freeform surface sewing, solving three-dimensional problems that flat sewing cannot address [7].

China is only a large sewing machine production country, not a sewing machine production powerhouse, it must be seen that China and Japan, Germany and other developed countries in the industry on the huge gap [8].

Research on three-dimensional sewing devices for steering wheels is relatively novel and more innovative for practical applications, and such research has not yet been conducted domestically or internationally.

2. Ease of design of the three-dimensional sewing device

2.1. Requirement analysis

In the paper, a design was established for both the sewing robotic arm and the lifting robotic arm of the three-dimensional sewing device. The lifting robotic arm is composed of the main body, pitch platform, rotation platform, moving arm one, and moving arm two, among other components. Similarly, the sewing robotic arm consists of parts such as the pitch platform, rotation platform, moving arm one, and moving arm two. The construction of three-dimensional models was facilitated using SolidWorks software. Figures 1 illustrate the designed three-dimensional models of the three-dimensional sewing device.

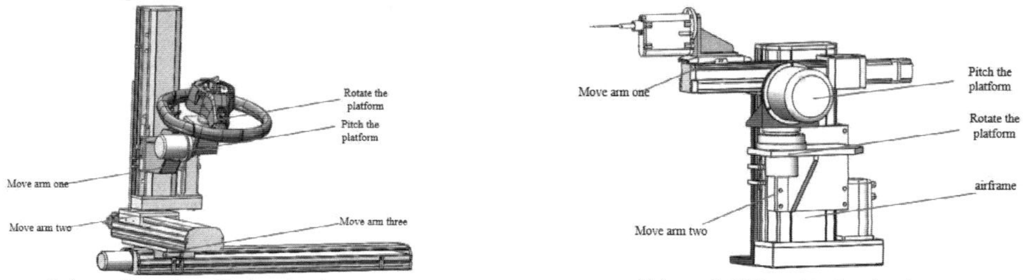

(a) Sewing Robotic Arm. (b) Lifting Robotic Arm.

Figure 1. The three-dimensional model diagram of a comprehensive sewing equipment.

2.2. Modeling and workspace analysis of three-dimensional sewing equipment

During the entire sewing process, the dual robotic arms within the three-dimensional sewing equipment collaborate with each other to accomplish the sewing task. By analyzing the motion patterns of the robots and integrating the theory of robot kinematics modeling, a kinematic analysis of the robots is conducted. This analysis leads to the establishment of the robots' kinematic model, which connects the robot joint space with the end-effector task space, resulting in the determination of the positional and velocity relationships between them [9].

In Table 1, the working ranges and operational workspaces of the supporting robotic arm with five degrees of freedom and the four-degree-of-freedom sewing robotic arm are both presented.

Table 1. Robotic arm motion range.

Coordinate Name	Moving arm one (mm)	Moving arm two (mm)	Moving arm Three (mm)	Pitch Platform (°)	Rotation platform (°)
Sewing Robotic Arm	-150~150	-125~125	-350~350	-30~30	0~360
Lifting Robotic Arm	-100~100	-50~50	/	-30~30	-30~30

2.3. Model simplification and parameter configuration.

A three-dimensional solid model of the sewing equipment is created using SolidWorks software. The model is then imported into ADAMS for editing. Due to the large number of components in the model, in order to increase computational speed, some simplifications can be made to the model before importing. If the entire assembly is imported, the processing will be more complicated. Non-interacting or non-essential components can be merged into one part and then imported separately[10]. The ADAMS virtual prototype technology is closer to the real situation in a general environment, and it is very useful for the analysis and research of the more complex motion characteristics and end poses of robots when encountering difficulties[11].

The two mechanical arm devices in the three-dimensional sewing equipment have large mechanical systems. Directly importing the dissimplified 3D model into ADAMS would increase the time required for prototype construction due to several components. Therefore, the overall equipment model is simplified to ensure the efficiency of the three-dimensional modeling and simulation of the mechanical arms, making it more convenient and clear to add constraints, drives, and motion parameters.

The simplification primarily considers the following requirements: the motion of the mechanical arms is a rigid body motion, ignoring the issue of elastic deformation, and certain components and features with minimal impact on the motion of the mechanical arms are omitted. The end effector and actuator housing, as well as some motors and fixed housings, are merged into a single entity using Boolean operations. Based on the specific sewing process, the additions are detailed in Table 2.

Table 2. Constraint-driven relationships among components of a sewing machine arm.

Component A	Component B	Constraint type	Constraint name
Needle base	Ground	Fixed pair	JOINT_1
rotating arm	Needle base	Moving pair	JOINT_2
Needle holder interface	Rotating arm	Moving pair	JOINT_3

The arm is operated with the same simplification method and constraint approach for adding operations, as shown in Figure 2 (a) and (b).

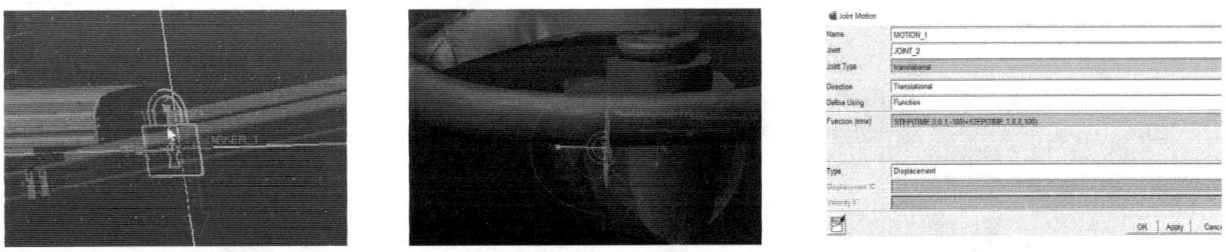

(a) Fixed pair. (b) Revolute pair. (c) Setting parameters for STEP.

Figure 2. The process of adding constraints and setting drive parameters for the robotic arm.

When adding drive functions, it is necessary to first set the units of the model, with time, length, and angle units set as s, mm, and rad, respectively. The drive functions of the robotic arm joints are set according to the form of Equation (1), where time is the independent variable of the function, t is the initial time value, h_0 is the initial value of the function, t_1 is the final time value, and h_1 is the final value of the function. The STEP function is a step function constructed using a cubic polynomial, which can be used as a design process function[12].

$$\text{STEP}\ (x, t, h_0, t_1, h_1) \quad (1)$$

The drive functions are separately added to each joint to control the sewing equipment. This paper implements the control of the sewing robot arm using a sequential process, with each joint sequentially activated to locate the needle entry point and then return to the initial state. The drive functions for some joint drives are set as follows: STEP (..), and the configuration process is depicted in Figure 2(c).

Slider drive function: STEP (time, 0, 0, 1, -100) + STEP (time, 1, 0, 2, 100).

Rotational joint drive function: STEP (time, 0, 0, 1, 30D) + STEP (time, 1, 0, 2, -30D)

Similarly, the lifting arm is controlled in a sequential manner, with each joint sequentially activated to locate the corresponding sewing position before hoisting the steering wheel to a fixed state. The drive functions for some joint drives are set as follows:

Move arm one: STEP (time, 0, 0, 1, -80) + STEP (time, 1, 0, 2, -20).

Rotational joint drive function: STEP (time, 0, 0, 1,-30D) + STEP (time, 1, 0, 2, 30D).

The simulation lasts 2 seconds with 500 steps for the sewing arm model and 4 seconds with 500 steps for the lifting robotic arm model. Both simulations start with the arms in their natural state and involve moving to specific points before returning to an adjusted state. The lifting arm and sewing arm designed in this paper are driven according to their constraint conditions, completing the construction of the virtual prototype as shown in Figure 3.

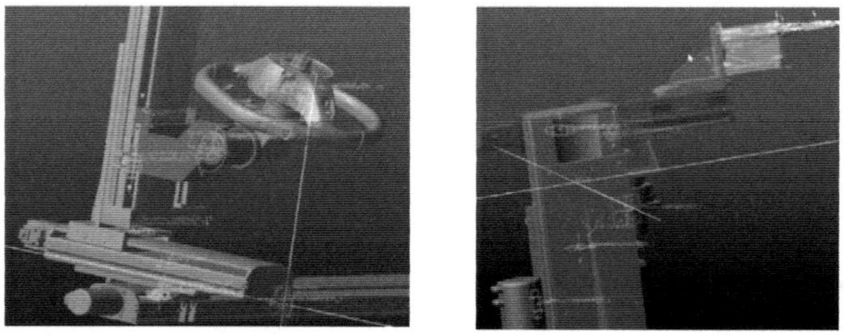

(a) Lifting Robotic Arm. (b) Sewing Robotic Arm.

Figure 3. Equipment Motion Range.

2.4. Motion process of three-dimensional sewing equipment

When designing three-dimensional sewing equipment, it is necessary to comprehensively consider safety, sewing effectiveness, and the characteristics and motion properties of the steering wheel. The driving functions of all the above-mentioned joints are operated, and the virtual prototype model of the sewing machine arm is simulated. Due to the natural primitive state of the mechanical arm import, the simulated motion reaches the sewing point and then returns to the initial state.

The simulated results show that in action simulated in Figure 4(a), the rotary joint deviates outward by 15°, and the pitch joint deviates upward by 10°, forming a certain working space. Then, simulate rotating inward by a certain angle and start descending, and eventually, the mechanical arm returns to its initial position.

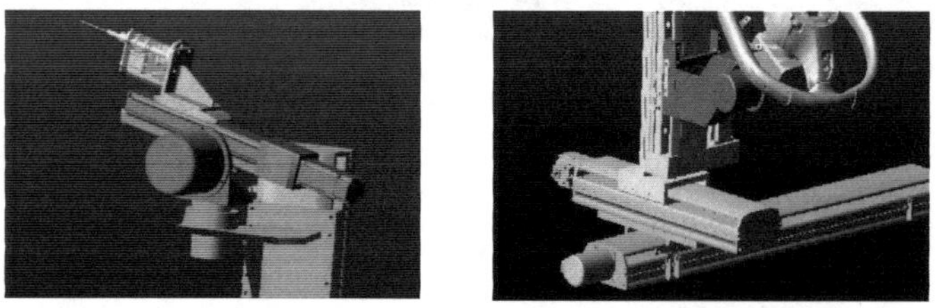

(a) External Deflection. (b) Move to the right.

Figure 4. Simulation process of the robotic arm.

Running the drive functions for all the above-mentioned joints, simulating the virtual prototype model of the lifting arm, and as the arm is imported into its natural original state, the simulation starts with the posture reaching a fixed point and then returning to the fixed adjusted final state.

The simulation results, as shown in Figure 4(b), simulate the following actions: Axis one starts moving to the right, axis two starts moving forward, axis three starts lifting to a specified height, gradually reaching fixed motion positions. The end effector steering wheel begins rotating by 20°. Finally, the arm lifts the steering wheel to a fixed position. From the above motion simulation, it can be observed that the structure of the three-dimensional sewing equipment designed in this study can basically meet the requirements of sewing operations.

3. Simulation results and analysis

The post-processing module of ADAMS has multiple curve analysis and processing functions, which can analyze parameters such as velocity, acceleration, force, and displacement of the robotic arm during motion.

By observing the position changes of the center point of mass in different directions, as shown in Figure 5, it can be noted that the maximum distance traveled occurs between 0.75s and 1.5s, representing the motion path of the end needle. The displacement curve is smooth, indicating a high controllability of the sewing robotic arm. When the robotic arm finds the sewing point and returns to the initial state, the total displacement is 0. During this process, the speed first increases and then decreases, with the overall device executing according to the set parameters, eventually reducing the speed to 0. When the device is in operation, it needs to enter the initial preparation state. After finishing sewing, the three-dimensional sewing equipment must return to its original position to ensure safety for the next use, preventing collisions between devices and potential damage caused by worker-machine collisions during startup. Therefore, the designed actions essentially meet the sewing task requirements.

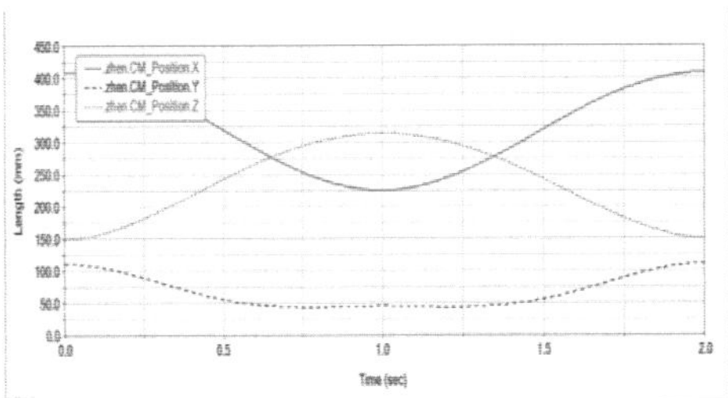

Figure 5. Curve of End Effector Center Position Variation.

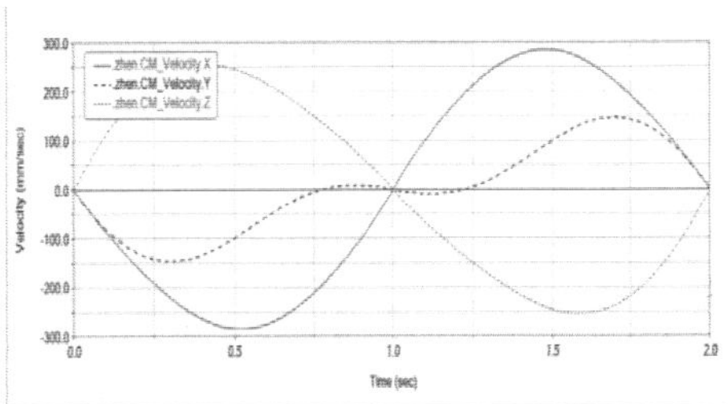

Figure 6. Graph of End Effector Center Velocity Variation.

Figure 6 shows that during the simulation time, the device rotates according to the set angles, with smooth velocity curves for the joints and stable motion speeds. The initial velocity is 0 at the starting position, and after completing the motion process and reaching the final position, the velocity returns to 0, consistent with the laws of motion.

Figure 7 shows the speed variation of the joints in the lifting arm section. During the simulation time, the equipment is driven according to the set function, and the joints start to accelerate 1 second after the start, then slowly decelerate after reaching the fixed lifting point, and the speed returns to 0 upon reaching the lifting position.

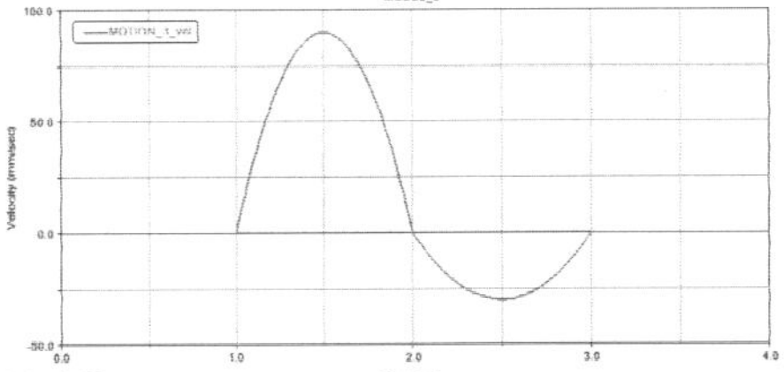

Figure 7. Joint Velocity Variation Curve Graph.

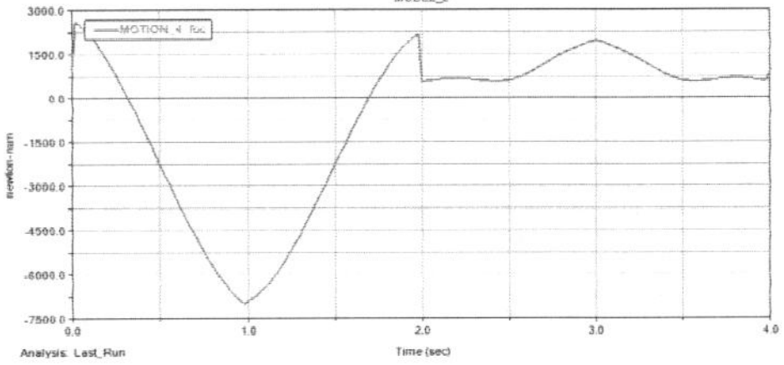

Figure 8. Joint Drive Force Variation Curve.

Journal of Physics: Conference Series **2815** (2024) 012009 doi:10.1088/1742-6596/2815/1/012009

In Figure 8, starting from 2 seconds, the end effector of the lifting arm continuously adjusts the direction of the wheel's initial position by rotating. The driving force of the joints shows significant fluctuations, but the overall force is relatively stable. There are no major impact forces on individual joints, and they eventually reach the fixed lifting position smoothly. This ensures that the sewing machine arm can reach the fixed stopping point of the lifting arm.

The simulation results indicate that the design structure is reasonable and conducive to the sewing work of the inner and outer peripheral covers of the steering wheel. The application of virtual prototype technology under the dynamics model lays the foundation for further motion control, providing a reference basis for trajectory planning of the three-dimensional sewing equipment.

4. In conclusion

The establishment of a three-dimensional solid model of the three-dimensional sewing equipment, the kinematic and dynamic modeling, and the analysis of the working space ensure the application's safety distance. Then, the model is simplified using ADAMS software to simulate the steering wheel sewing process, and motion planning for the two robotic arms is conducted to achieve the sewing effect of the mechanical equipment. In the three-dimensional sewing equipment, the main components and joints smoothly transition during the sewing process, with good continuity in the velocity, torque, and acceleration curves without significant abrupt changes. This ensures the stable realization of the drive systems of various platforms in the three-dimensional sewing equipment, thereby meeting the sewing requirements during the equipment's motion.

References

[1] Ding Min. Opportunities and Challenges Brought by Sewing Robots[J]. China Textile Inspection, 2018, No.510(02):124-125.

[2] Li Haitao. Trends in Sewing Equipment Technology Development from CISMA2015[J]. Textile Machinery, 2015, No.270(11):16-17.

[3] Rong Wei. Current Situation and Development Trend of Industrial Robots and Intelligent Manufacturing[J]. Times Farm Machinery, 2017, (10):46+49.

[4] Zhang Kai. Analysis of the Technological Development and Application of Industrial Robots[J]. Scientific Technology Innovation, 2019, (02):165-166.

[5] Kudo M, Nasu Y, Mitobe K, et al. Multi-arm robot control system for manipulation of flexible materials in sewing operation[J]. Mechatronics, 2000, 10(3): 371-402.

[6] Wang Chuhan, Xin Daxin, Yang Ming, Liu Yangyang. Modeling and Workspace Analysis of a Sewing Dual Manipulator[J]. Modern Manufacturing Technology and Equipment, 2023, 59(02):47-52.

[7] Liu Guojiang. Research on Key Technologies of Three-dimensional Sewing[D]. Xi'an University of Technology, 2019(08).

[8] Brandstetter,P, Kreeck,T. Korbel,P. Non-model,based sensorless vector controlof Permanent magnet synchronous motor.IEEE International Symposium onindustrial Electronics, 2008,618~623

[9] Jiang Yimu. Research on Visual Servo Control Based on Robot Dynamics Model[D]. Harbin Institute of Technology, 2021(03)

[10] Fan, W. Design and Analysis of the Mechanical Hand Structure of Banana Picking Robot [D]. Guangxi University, 2018 (06).

[11] Liang Xiaobo, Jiang Gang. Three-Dimensional Spatial Kinematic Analysis of Three-Link Mechanical Arm[J]. Mechanical Design and Manufacturing, 2013, No.274(12):67-69+72.

[12] Liu Zhe. Analysis and Motion Simulation of the Seven Degrees of Freedom Polishing Robot Manipulator. Liaoning University of Technology, 2015 (06).

Journal of Physics: Conference Series **2815** (2024) 012010 doi:10.1088/1742-6596/2815/1/012010

Optimization of electrical control equipment interface layout with cognitive load based on pso algorithm

Ting Zhang[1, 2], Ruizhi Liu[2,*] and Liang Sun[2]

[1]Department of Art and Design, Shanghai Dianji University, Shanghai 200240, China

[2]School of Mechanical Engineering, Shanghai Dianji University, Shanghai 201306, China

[*]Corresponding author's e-mail: liuruizhi9918@163.com

Abstract. Addressing the complexity of current impregnation machine electrical control interfaces, this study proposes optimization methods for their layout design. It incorporates factors affecting cognitive load during interface interaction, formulating three principles: highlighting essential elements, operation sequence, and element relevance. These principles guide the construction of a mathematical model for the optimization of electrical equipment control interface layouts. The Particle Swarm Optimization algorithm is applied to solve the model's objective function, producing optimized layouts, which are then validated through eye-tracking experiments. The results demonstrate the method's effectiveness in optimizing control interface layouts, aligning with operators' intuitive logic.

1. Introduction

In human-machine interaction interfaces, users are required to process a vast amount of information for various operations. If the interface is designed to be overly complex or chaotic, it increases the cognitive load on users, making it difficult for them to understand and operate[1]. Electrical control equipment on the market is often custom-designed for specific site work environments and individual operating habits[2]. Given the significant potential for development in human-machine interaction interfaces, research in this field is extensive both domestically and internationally. In all areas of industrial design, researchers have been continuously exploring more effective and rational design solutions[3,4].

Sun developed a mathematical model for workshop facility layout and employed a genetic algorithm to solve the function, resulting in an optimized layout of workshop facilities [5]. Zhou established an optimization model using a multi-objective genetic algorithm, solving for the Pareto solution set to obtain multiple optimal solutions for different optimization objectives [6]. Shen and his team utilized the Simulated Annealing (SA) algorithm to optimize the element layout of electronic components, reducing the acceleration of component pins and simultaneously improving the layout environment of the circuit board interface [7]. In summary, compared to traditional problem-solving, which requires constructing a model relative to the problem itself, computational intelligence algorithms do not require the establishment of an exact model of the problem, significantly simplifying the complexity of the issue [8,9].

This paper addresses the deficiencies in the design of existing electrical control cabinet interaction interfaces. By integrating principles of human-machine interface layout based on cognitive load, it employs particle swarm optimization to scientifically enhance the arrangement of interface elements and the overall interface, thereby improving usability and user experience.

2. Interface layout optimization method integrating cognitive load

The design of the interface layout integrates cognitive load, which is a user-centric design method that emphasizes the cognitive load of users. The aim is to lighten the cognitive load pressure of users by optimizing the interface layout, thus improving the user experience and satisfaction[10]. That is, by taking into account factors such as the user's cognitive level, task complexity, and the amount of information, the interface layout is optimized to provide the most suitable form of interface presentation. This section organizes the approach to reduce cognitive load during the interaction interface operation process and summarizes the three aspects of interface optimization layout principles for the electrical control cabinet mentioned in the text.

2.1. Principle of highlighting important elements

The "Z" pattern is a concept in human-computer interaction and interface design that primarily observes users' reading habits and visual search paths during web browsing. It is widely used to enhance user experience and information retrieval efficiency. The "Z" pattern leverages users' visual tendencies by strategically placing the most critical elements in the top left area, aligning with the initial focal point of vision. This optimizes interaction speed and reduces cognitive load. The center serves as a focal point for important content, facilitating in-depth exploration. The path concludes in the bottom right, reserving space for secondary information or controls.

2.2. Principle of operation sequence

The principle of operation sequence aims to gradually guide users to complete the entire workflow according to tasks to provide a more intuitive and efficient user experience. This principle follows the rules of visual cognition, where elements should be laid out in order from left to right, from top to bottom, to reduce the likelihood of confusion and erroneous operations. Moreover, this principle guides users from start to end through orderly layout and interface design, allowing for progressive next steps and reducing learning costs and feelings of disorientation.

2.3. Principle of element relevance

The principle of element relevance involves grouping and classifying relevant features and information, forming a logical association, so users can quickly find the content they need, optimizing user experience and operational efficiency. This principle combines the Principle of Key Element Highlighting and the Principle of Operation Order to achieve the organization and layout of elements and functions on the interface. In conjunction with the content of this study, the operation interface is divided into a display area and an operation area, as shown in Figure 1.

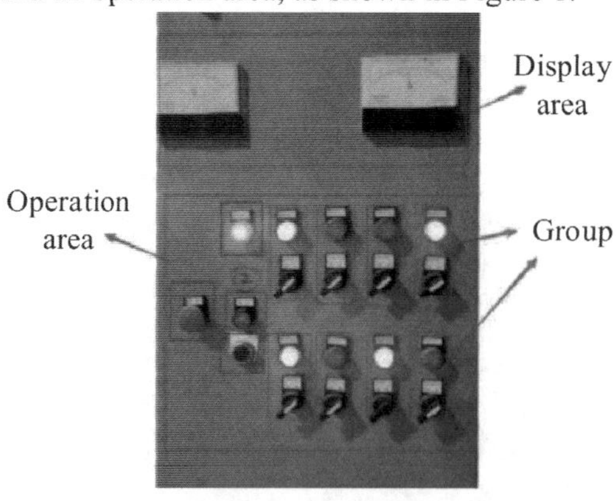

Figure 1. Operation interface area division.

3. Mathematical model for electrical control interface optimization

3.1. Layout boundary constraints

In multi-objective optimization, layout boundary constraints confine element positions and sizes within a predetermined area. Optimizing spacing can prevent improper space utilization and user burden while also avoiding erroneous operations and accidental activations. The layout must avoid overlaps and extremes to ensure the control panel is compact, efficient, and user-friendly. According to the above layout scope limits and element spacing displays, the formulas for satisfying spacing constraints and boundary constraints can be obtained:

$$\begin{cases} |M_x - N_x| - \dfrac{(M_i + N_i)}{2} \geq \Delta x_{\min} \\ |M_y - N_y| - \dfrac{(M_j + N_j)}{2} \geq \Delta y_{\min} \end{cases} \tag{1}$$

where M_i and M_j represent the length and width of the element M ; $\Delta x_{\min}$ and $\Delta y_{\min}$ represent the minimum spacing between elements M and N in the direction x and the direction y .

3.2. Objective function

In the study of optimizing the layout of electrical control equipment interfaces, the application of expert evaluation is particularly important. Such evaluations are not only based on professional knowledge but also integrate the experts' practical experience, offering solid theoretical support and practical guidance for interface layout design. This section employs the expert evaluation method to analyze the three principles involved in the previous section.

In accordance with the principle of highlighting important elements, the significance of the elements to be laid out is evaluated and rated by experts. The expression of importance data is as follows:

$$I = [I_{ij}] = \begin{bmatrix} I_{11} & I_{12} & \cdots & I_{1h} \\ I_{21} & I_{22} & \cdots & I_{2h} \\ \cdots & \cdots & \ddots & \cdots \\ I_{n1} & I_{n2} & \cdots & I_{nh} \end{bmatrix} \tag{2}$$

where I_{ij} is the importance evaluation of element ci to be laid out by expert j. The element importance evaluation with the introduction of expert weight values is expressed as $I_i = I\,W$, where W is the expert weight matrix.

On the other hand, the operational frequency of each element profoundly reflects its importance within the device interface. Through the various task operations carried out during the device's working process, an operational frequency matrix F is constructed and represented as follows:

$$F = [f_{ij}] = \begin{bmatrix} f_{11} & f_{12} & \cdots & f_{1d} \\ f_{21} & f_{22} & \cdots & f_{2d} \\ \cdots & \cdots & \ddots & \cdots \\ f_{n1} & f_{n2} & \cdots & f_{nd} \end{bmatrix} \tag{3}$$

where f_{ij} represents the operation frequency of element ci in the process of task j . $v_g = [v_{g1}, v_{g2}, v_{g3} \ldots v_{gr}]^T$ indicates the operation weight of the element, and the weighted operation frequency of the element ci is $G_i = v_g f_{ij}$. Combining the principle of element importance and the principle of operating frequency described above, the mathematical model is described as follows:

$$Q_i = I_i + G_i \tag{4}$$

Regarding the principle of operation sequence, the operational order of elements to be laid out is obtained by organizing the sequence of tasks executed by the equipment under different operational tasks, denoted as c1-c14, respectively. Elements not utilized during the operation are assigned a value of 0, the first element requiring operation is assigned a value of 1, and so on in a deferred sequence. The operation sequence matrix is obtained through the actual operational tasks of the experimental subjects:

$$E = [E_{ij}] = \begin{bmatrix} E_{11} & E_{12} & \cdots & E_{1d} \\ E_{21} & E_{22} & \cdots & E_{2d} \\ \cdots & \cdots & \ddots & \cdots \\ E_{n1} & E_{n2} & \cdots & E_{nd} \end{bmatrix} \tag{5}$$

where E_{ij} represents the operation sequence assignment of element ci in task j. The mathematical model of the operation sequence of elements is obtained:

$$S_i = \sum_{j=1}^{d} \left[\frac{E_{ij}}{\sum_{i=1}^{n} C_{ij} \bullet w_{li}} \right] \tag{6}$$

where n represents the total number of elements awaiting layout, d represents different tasks of the device during the working process, and the research object w_{li} is a fixed value, representing the operation sequence weight of the element ci.

Regarding the principle of element relevance, the closeness between elements can be represented by the distance between the elements. Referring to the relevant equipment data, the correlation between each element is evaluated, represented by 0-1, with the degree of correlation ranging from weak to strong. After obtaining the four internal evaluation results through group discussion, the data is preliminarily processed and organized, and the evaluation result is evaluated for reliability with experts and operators. The obtained correlation matrix expression is as follows:

$$R = [R_{ij}] = \begin{bmatrix} R_{11} & R_{12} & \cdots & R_{1n} \\ R_{21} & R_{22} & \cdots & R_{2n} \\ \cdots & \cdots & \ddots & \cdots \\ R_{n1} & R_{n2} & \cdots & R_{nn} \end{bmatrix} \tag{7}$$

where R_{ij} represents the degree of correlation between elements ci and cj in the elements to be arranged, and n is the total number of elements to be arranged. Therefore, the mathematical model of the correlation between elements is obtained:

$$R_i = \sum_{j+1}^{n} \left(\frac{R_{ij}}{d_{ij}} \right) \tag{8}$$

where d_{ij} represents the distance between the center points of element i and element j.

Under the constraints, the maximum extent of obtaining the minimum value of the objective function is achieved. The mathematical expression is represented as follows:

$$\begin{cases} H(i)_{best} = \max\left[\sum_{i}^{n}(w_{li}S_i + w_{2i}Q_i + R_i)\right] \\ w_{li} = 1 - \dfrac{1}{n} \\ w_{2i} = 1 - \dfrac{d_{ia}}{d_{a\max}} \end{cases} \tag{9}$$

where w_{li} and w_{2i} are control coefficients, d_{ia} is the distance from the center point of element ci to the optimal layout position point a, and $d_{a\max}$ is the maximum distance from the element position point to the optimal layout point a.

3.3.Implementation process based on PSO

1. Initialization: Initialize the parameters, with the particle swarm size as N and the initial position randomly generated. "ger" indicates the maximum number of iterations. In this study, the swarm exists in a two-dimensional space, and the speed K is randomly generated within a fixed range. The learning factors are c_1 and c_2.

2. Fitness computation: Compute the fitness of particles in the swarm. The mathematical formula is as follows:

$$F_i(K) = f(x_{i1}(K), x_{i2}(K) \ldots x_{ij}(K)) \tag{10}$$

where $F_i(K)$ indicates the fitness of particle i in the Kth generation; $x_{ij}(K)$ represents the d-dimensional component of the Kth generation particle i; f represents the fitness function.

3. Update by comparing with the particle's own historical best: We compare the historical best fitness value $k^{best}(K+1)$ with $F_i(K)$ and update with the most optimal fitness value. The mathematical formula is as follows:

$$p_i^{best}(K+1) = \begin{cases} p_i^{best}(K), \; p_i^{best}(K) > F_i(K) \\ F_i(K), F_i(K) > p_i^{best}(K) \end{cases} \tag{11}$$

4. Update by comparing with the swarm's historical best fitness: The optimal value of the new generation swarm is $k^{best}(K+1)$. Compared with the historical best fitness value, we take the optimal value for updating. The mathematical formula is as follows:

$$k^{best}(K+1) = \begin{cases} \max(p^{best}(K+1)), \max(p^{best}(K+1)) > k^{best}(K+1) \\ k^{best}(K+1), k^{best}(K+1) > \max(p^{best}(K+1)) \end{cases} \tag{12}$$

where $\max(p^{best}(K+1))$ represents the historical optimal of all particles in the $K+1$ generation.

5. Update particle velocity v_i*:* The particle adjusts its own speed and positional information based on the p_i^{bst} and k^{best} values obtained from the above steps *3* and *4*. The adjustment mathematical formula is as follows:

$$\begin{cases} v_i = wv_i + \left[c_1 \cdot (x_{p_i^{bset}} - x_i) + c_2 \cdot (x_{k^{best}} - x_i) \right] \cdot rand() \\ x_i(K+1) = x_i(K) + v_i \end{cases} \tag{13}$$

where w represents the inertia weight; $rand()$ represents a random number uniformly distributed in the interval [0,1]; $x_{p_i^{bset}}$ represents the optimal solution coordinates of particle i; $x_{k^{best}}$ represents the optimal solution coordinates of the population.

6. Algorithm termination: The algorithm will terminate when the set number of iterations is reached or when the optimal solution is obtained. The algorithm flow is shown in Figure 2.

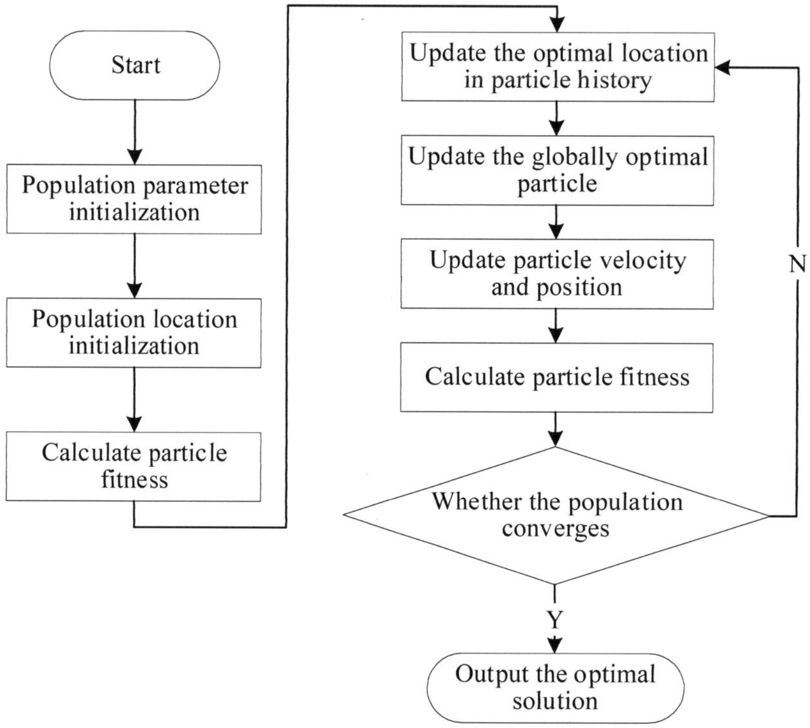

Figure 2. Flow chart of PSO.

4. Experimental results and analysis

In the human-machine interaction interface of the electrical control cabinet discussed in this paper, based on the preliminary statistical opinions on element grouping from the operators, the page divides the circuit breaker (rotary switch in the diagram) and the corresponding signal lights above it into a rectangle using the minimum bounding box, named after the knob. All other elements are abstracted as rectangles of various sizes, with element coding as shown in Figure 3.

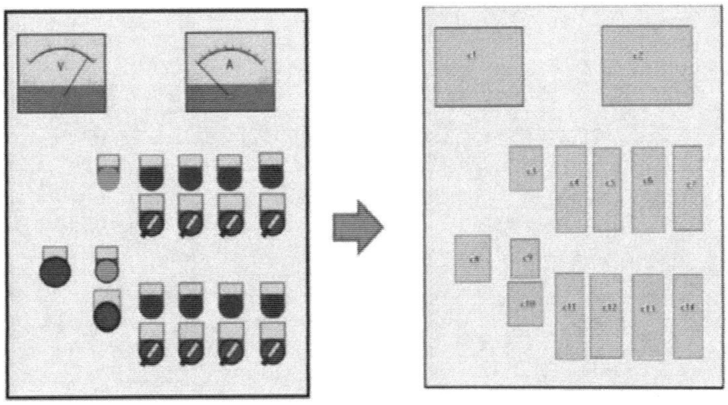

Figure 3. Element encoding.

In order to ensure the compactness and efficiency of the control panel, it is crucial to prevent erroneous operations and unnecessary actions by the user, thus enhancing user experience and interface usability. Special attention is given to avoid elements being too close or too far apart to prevent accidental touches or confusion. Reference is made for the spacing between elements. The corresponding initial information of element encoding in the coordinate system is illustrated in Table 1.

Table 1. Electrical control equipment interface element information.

Elements	Length (mm)	Width (mm)	Center point coordinates
c1	95.00	95.00	(289.50, 401.50)
c2	95.00	95.00	(428.70, 401.5)
c3	32.00	27.00	(242.00, 253.00)
c4	92.00	25.00	(298.00, 199.00)
c5	92.00	25.00	(353.00, 199.00)
c6	92.00	25.00	(408.00, 199.00)
c7	92.00	25.00	(263.00,199.00)
c8	40.00	35.00	(176.00, 145.00)
c9	32.00	27.00	(242.00, 149.00)
c10	35.00	32.00	(242.00, 106.00)
c11	92.00	25.00	(298.00, 77.00)
c12	92.00	25.00	(353.00, 77.00)
c13	92.00	25.00	(408.00, 77.00)
c14	92.00	25.00	(463.00, 77.00)

By investigating the various tasks of the electrical equipment in actual work, the weighted operation frequency weight and operation sequence weight were obtained. After repeatedly referencing sources and consulting experts, the correlation weight was ultimately determined.

Based on the four aspects of significance, operation frequency, operation sequence, and correlation data obtained earlier, the PSO algorithm is used for computations, solved using MATLAB. The parameter settings of the PSO algorithm include a population size (N) of 80, a dimension of 2, a maximum iteration number of 500, an inertia factor set at 0.6, and both individual and social learning factors set at 0.8. Figure 4 details the optimization process based on the PSO algorithm, with the final solution of the objective function value being 4.743. The optimized interface element coordinate data is shown in Table 2. The optimized interface is shown in Figure 5.

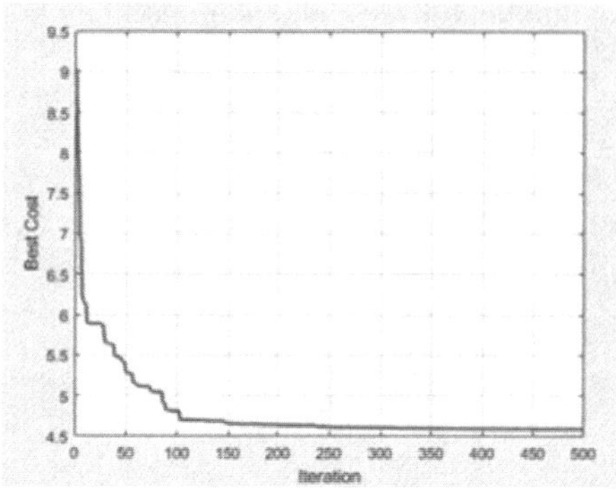

Figure 4. The curve of fitness with the number of iterations.

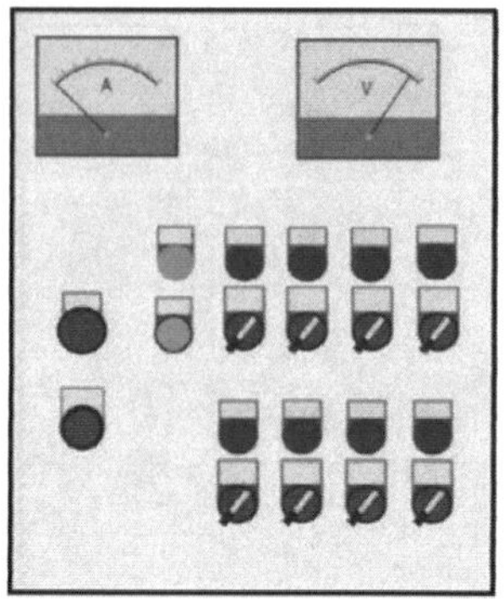

Figure 5. Interface optimization.

Table 2. Optimize interface element coordinates.

Elements	Center point coordinates	Elements	Center point coordinates
c1	(428.7, 401.5)	c8	(192.5, 215.5)
c2	(296.5, 401.5)	c9	(242.7, 208.0)
c3	(242.0, 265.0)	c10	(242.0, 253.0)
c4	(298.0, 211.0)	c11	(298.0, 89.5)
c5	(353.0, 211.0)	c12	(353.0, 89.5)
c6	(408.0, 211.0)	c13	(408.0, 89.5)
c7	(463.0, 211.0)	c14	(463.0, 89.5)

Eye-tracking technology evaluates the cognitive load on different interfaces by tracking eye movement metrics such as users' eye movement trajectories and fixation points, thereby validating and optimizing the effectiveness of the interfaces. This experiment involved a total of 20 participants, who were randomly divided into two groups, A and B, with each group consisting of 10 individuals. The participants were university students aged between 21-25 years old. They had normal or corrected-to-normal vision, no color blindness, and were not highly myopic. All participants were novices and had no prior experience with operating electrical control cabinets. Groups A and B each conducted eye-tracking experiments on the optimized interface and the original operating interface, respectively. The data obtained is presented in Table 3.

Table 3. Eye-tracking experiment data.

Description		N	Average value	Standard deviation
Number of fixation points	A	10	69.60	24.690
	B	10	90.20	44.0877
Average fixation duration	A	10	1.68	0.1992
	B	10	2.86	0.6717
First fixation duration	A	10	0.111	0.0570
	B	10	0.127	0.0184

The mean data indicates that the number of fixation points collected from the optimized interface is less than that of the original interface. Compared to the original interface, the optimized interface also exhibits a shorter average fixation duration and initial fixation time in the test results. This suggests that the optimized interface imposes a lower cognitive load on users during testing, meaning that the user experience surpasses that of the original interface. The eye movement trajectory diagrams before and after the interface optimization are shown in Figure 6 and Figure 7.

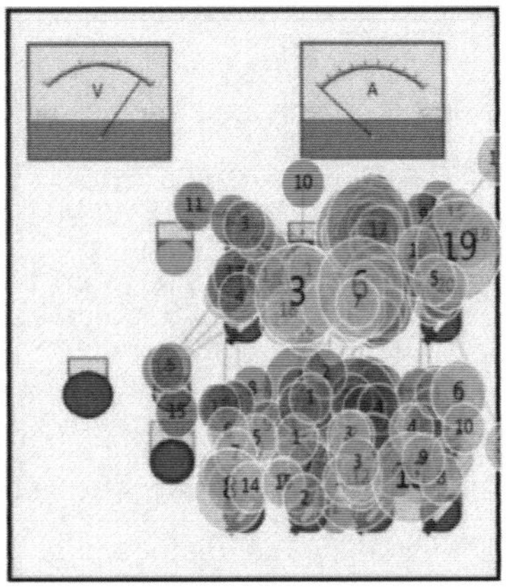

Figure 6. Original interface eye movement trajectory diagram.

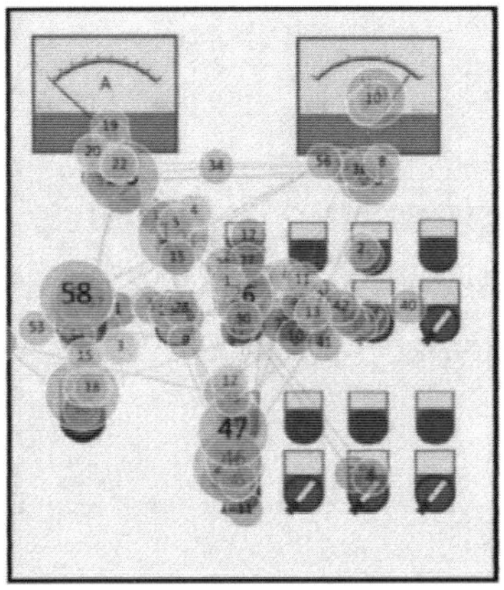

Figure 7. Optimized interface eye movement trajectory diagram.

5. Conclusions

This paper summarizes the design principles for the cognitive load on electrical operation interfaces, decomposes complex multi-objective layout problems, analyzes judgment matrices to establish mathematical models, and effectively optimizes the interface layout of electrical control equipment using particle swarm optimization algorithms. Eye-tracking experiments were designed to simulate the operation of equipment, and experimental data objectively verified the optimization results. This provides a more scientific interactive interface for the optimization design of the electrical control interface of the impregnation machine.

References

[1] Jafari R A, Choobineh A, Barahmand N, et al. Research Outputs in Ergonomics and Human Factors Engineering: A bibliometric and co-word analysis of content and contributions[J]. International journal of occupational safety and ergonomics, 2021,4(28): 41-42.

[2] Bao H, Fang W, Guo B, et al. Adaptive Human–Computer Interface Design for Supervision Task Based on User Attention and System State[J]. International Journal of Human-Computer Interaction, 2023: 1-13.

[3] Ma G R, He J X, Chen C H, et al. Trigger motion and interface optimization of an eye-controlled human-computer interaction system based on voluntary eye blinks[J]. Human-Computer Interaction, 2023: 1-31.

[4] Daniels T. Goldsmith T. Basu A. Overview of a Robust Human Factors Engineering Process[J]. Biomedical instrumentation & technology,2021,1(55): 16-20.

[5] Sun H R and Feng G H. Application of genetic algorithm in plant layout optimization[J]. New Technology & New Products of China, 2022(21): 133-135.

[6] Zhou Y, Ma C Y and Shi T H. Optimization design of magnetic core layout for inductive powering transfer based on genetic algorithm[J]. Electric Machines and Control,2022,26(12): 108-115.

[7] Shen B, Pan Z W and Lin H, et al. Optimization design of component layout based on simulated annealing algorithm[J]. Missiles and Space Vehicles, 2019(06):17-21.

[8] Zheng Y. Optimization of computer programming based on mathematical models of artificial intelligence algorithms[J]. Computers and Electrical Engineering, 2023, 110: 108834.

[9] D'Antonio G, Saja A, Ascheri A, et al. An integrated mathematical model for the optimization of hybrid product-process layouts[J]. Journal of Manufacturing Systems, 2018, 46: 179-192.

[10] Salim F, Mostafa E, Mustafa H, et al. Anxiety is associated with extraneous cognitive load during teaching using high-fidelity clinical simulation.[J]. Medical education online,2021,26(1).

The supporting structure strength analysis of indexing stamping machine for manufacturing thin-walled inverting steel tubes

Zijun Zhu [1], Shenghang Cai [1], Gaofeng Si [2], Jiaxin Yu [1], Hangtian Zhu [1] and Yuqing Zheng [1,*]

[1] School of Engineering, Huzhou University, Huzhou, Zhejiang Province, China, 313000
[2] Zhejiang MINGHE Steel Pipe Co., Ltd., Huzhou, Zhejiang Province, China, 313299

[*] Corresponding author's e-mail: zhengyuqing@zjhu.edu.cn

Abstract: Due to the occurrence of cracks in thin-walled inverting steel tubes during the inverting process leading to fractures, thereby failing to achieve the desired energy absorption effect, they were often manufactured by interstitial stamping process with the manual rotation of thin-walled tubes to reduce stress concentration points, which is not conducive to mass production. Therefore it proposes a new design scheme of the indexing stamping machine for mass production of thin-walled inverting tubes. And the tube supporting structure of the indexing stamping machine will undergo shear stress and axial stress during manufacturing operation. To ensure that the machine can operate properly, ANSYS software is employed to conduct strength and stability analysis for the supporting structure. The calculation results indicate that the main supporting structure can meet the strength requirements of the thin-walled inverting tube forming machine. These study findings provide some references for optimizing the structure design for indexing stamping machines.

1. Introduction

In current researches on tube forming, Zheng Yuqing [1, 5] established the optimal radius of curvature for tube forming dies, and analyzed the stress situation of the tube during the forming process. Shi Xiaokuan [2] conducted a comparative study on the cost of tube forming dies, identifying different processing scenarios for various types of forming dies. Through experiments and simulations determine the energy absorption characteristics of inverting tubes and developed an accurate calculation method for calculating the initiation load of tube energy absorption[3, 9-11]. Gao Qiang[4] improves the energy absorption performance of thin-walled roll over tube by referring to the foam filling structure and introducing gradient configuration. However, traditional tube forming techniques are still used in inverting tube forming. Traditional inverting tube forming techniques utilize axial pressure extrusion, which requires high plastic deformation capabilities of the material. This technique is prone to cracking during the inverting process, leading to fractures during usage and compromising the energy absorption effectiveness. Additionally, manual rotation of thin-walled tubes is required to adjust stress positions during the inverting process, which is not conducive to mass production. To address the aforementioned issues with traditional inverting tube forming techniques, this paper proposes a thin-walled inverting tube indexing stamping machine. Its main feature lies in replacing one-time extrusion plastic forming

Journal of Physics: Conference Series **2815** (2024) 012011 doi:10.1088/1742-6596/2815/1/012011

with a combination of rolling and striking operations. This multi-stage plastic deformation method avoids local tearing failure caused by excessive plastic deformation in specific areas, effectively improving the success rate of inverting tube manufacturing. Furthermore, it is suitable for plastic forging processing of other types of thin-walled workpieces (thickness 0.5-3.5mm). Workpieces processed by swing rolling exhibit rational metal material fiber flow and high forming efficiency. Additionally, swing rolling plasticity enables batch production of thin-walled inverting tubes. The power source of this thin-walled inverting tube indexing stamping machine utilizes a three-phase asynchronous motor, which drives the punching mechanism and workpiece rotating mechanism. By utilizing the motion characteristics of the crankshaft and pulleys, the curled end of the thin-walled tube is completely rolled over to achieve the desired swing rolling forming effect.

2. Composition of machine and working principle

2.1. The structural composition of the thin-walled inverting tube indexing stamping machine

The structural composition of the thin-walled inverting tube indexing stamping machine can be divided into four main mechanisms: the punching mechanism, the workpiece rotation mechanism, the supporting mechanism, and the resetting mechanism. The punching mechanism consists of a crankshaft and a spline coupling. The workpiece rotation mechanism comprises a rack, gears, worm, worm gear, and pulleys. The supporting mechanism is composed of a supporting body, a lifting plate, and two trapezoidal threaded rods. The resetting mechanism consists of a worm, a worm gear, and trapezoidal threaded rods. A simplified model is shown in Figure 1.

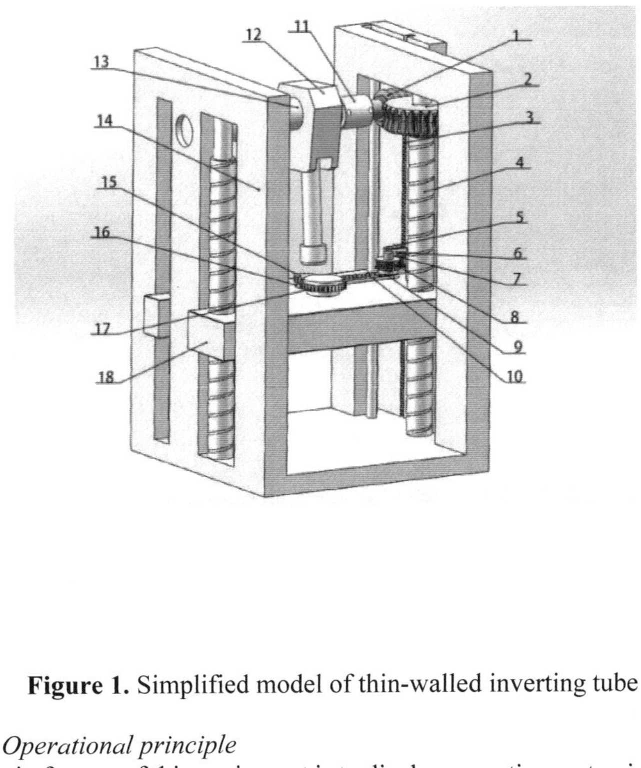

Structural	
1	worm A
2	worm gear
3	rack
4	screw
5	gear A
6	gear bracket
7	worm B
8	worm gear
9	pulley A
10	belt
11	spline
12	punch
13	crankshaf
14	Support
15	gear B
16	pulley B
17	gear C
18	lifting

Figure 1. Simplified model of thin-walled inverting tube indexing stamping machine.

2.2. Operational principle

The main feature of this equipment is to displace one-time extrusion plastic forming with a combination of swing rolling and striking operations. Compared with one-time extrusion plastic forming, this multi-stage plastic deformation method allows for more reasonable flow of metal material fibers in the workpiece, thus improving forming efficiency. The processing of the workpiece can be divided into the following three steps:

1) Stamping process: The motor drives Part 1 (worm A), and then Part 1 (worm A) drives Part 13 (crankshaft) through Part 11 (spline coupling), causing Part 13 (crankshaft) to drive Part 12 (punch) to achieve the punching action.

2) Workpiece resetting process: Part 1 (worm A) drives Part 11 (spline coupling) to move, simultaneously driving Part 2 (worm gear A) to move. Part 2 (worm gear A) drives Part 4 (screw) to move, and then Part 4 (screw) further drives Part 18 (lifting plate) to move upwards, thus resetting the workpiece placed on Part 17 (gear C).

3) Workpiece rotation process: While Part 18 (lifting plate) moves upwards, it simultaneously drives Part 5 (gear A) to move upwards along Part 3 (rack). Part 5 (gear A) drives Part 7 (worm B) to move, and then Part 7 (worm B) drives Part 8 (worm gear B) to move. Part 8 (worm gear B) drives Part 9 (pulley A) to move, and Part 9 (pulley A) drives Part 16 (pulley B) to move. Part 16 (pulley B) drives Part 15 (gear B) to rotate, and Part 15 (gear B) drives Part 17 (gear C) to rotate, thereby achieving the rotation of the workpiece placed on Part 17 (gear C) to avoid local tearing failure of the workpiece.

2.3. Analysis of stress situation during stamping process

Throughout the entire stamping forming process, the strength requirements of the supporting structure, which supports the workpiece, are particularly important. The lifting plate mainly experiences downward pressure load transmitted from the workpiece placed above, while the screw primarily undergoes downward pressure load from the lifting plate and torque transmitted from the worm gear. The stress situation and constraints are illustrated in Figure 2.

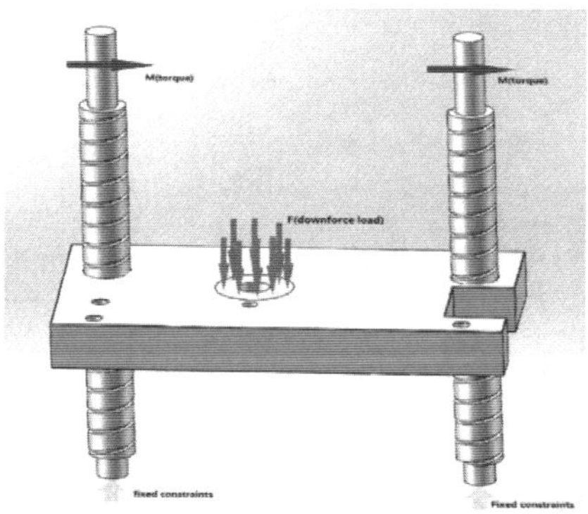

Figure 2. Schematic diagram of stress situation.

3. Strength simulation calculation of supporting structure

Due to the complexity of the loads that the screw experiences during operation, including multiple directions of shear stress and axial pressure, a detailed load calculation will be conducted using ANSYS Workbench for the supporting structure of the thin-walled inverting tube indexing stamping machine. Other mechanisms can be subjected to simple mechanical load calculations, hence they won't be extensively discussed in this paper[8, 12].

3.1. Three-dimensional model processing

In order to facilitate modeling, SolidWorks software was chosen to establish the three-dimensional model of the supporting structure of the thin-walled inverting tube indexing stamping machine. To improve computational efficiency, the model was simplified. In order to ensure the reliability of the analysis results and the efficiency of the analysis model, specializing in the important parts of the model were selected for analysis: the lifting plate + two screws + supporting body. To import the complete helical surfaces from the 3D model, we choose the DesignModeler module in ANSYS Workbench to do that. The import result is shown in Figure 3.

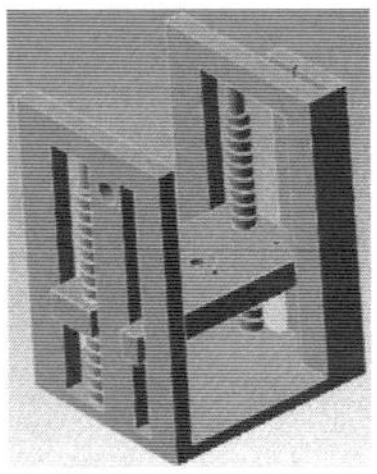

Figure 3. The model of supporting structure.

3.2. Element selection with mesh partition

This article adopts ANSYS Workbench software for stress analysis about the load-bearing components of the thin-walled inverting tube forming machine. The specifications for the lifting plate are as follows: length 440 mm, width 58 mm, height 238 mm. The screw specifications are as follows: trapezoidal thread type, rod length 600 mm, pitch 28 mm, middle diameter 43 mm, small diameter 39 mm. The material for the supporting body is structural steel default by default. The materials and parameters for the lifting plate and screw are listed in the following table:

Table 1. Materials parameters.

Component Names	Material Names	Density	Young's Modulus	Poisson's Ratio
lifting plate	20Mn	7.81g/cm3	190GPa	0.3
screw	40Cr	7.85g/cm3	211GPa	0.277

The model is meshed using tetrahedral elements. The mesh size for the contact area between the lifting plate and the screw is set to 2 mm, while the rest of the model is set to 5 mm. The final mesh contains 373,826 nodes and 209,954 elements.

3.3. Selection of boundary conditions

Displacement boundary conditions were set in the finite element model by applying fixed constraints to the bottom of the two screws and the bottom of the supporting body, as shown in Figure 4.

Based on previous experimental data, it is determined that the maximum downward force required for forming a 2 mm steel tube is around 160 KN, which was obtained from tests. Considering the actual operating conditions and the influence on fatigue work, the load is increased by 25%. Therefore, the load on the lifting plate is set to 200 KN. The 200 KN load is applied to the expected compressed surface of the lifting plate as shown in Figure 5.

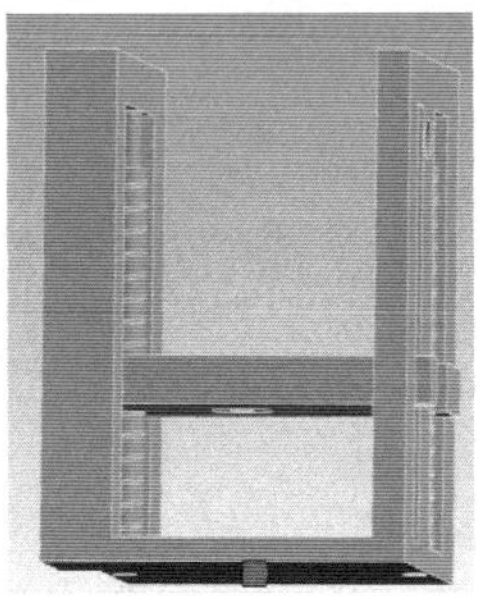

Figure 4. Bottom constraint conditions.

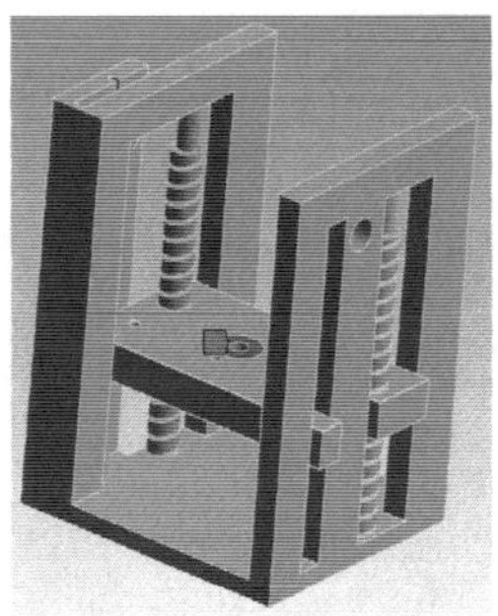

Figure 5. Schematic diagram of applied load.

The contact type between the supporting body and the lifting plate is chosen as frictional contact, with a friction coefficient setting to 0.15. For each corresponding helical contact surface between the screw and the threaded hole in the lifting plate, there will be their own independent contact settings. The contact type chosen is frictional contact, and the friction coefficient is 0.15 either.

4. Results and discussion

4.1. The stress analysis results

The stress distribution and maximum von-Mises stress of the lifting plate, two screws, and supporting body are shown in Figure 6 and Figure 7.

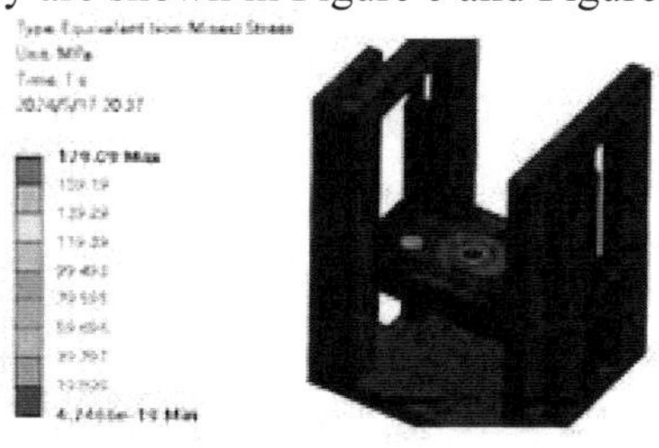

Figure 6. Stress analysis Results (iso view).

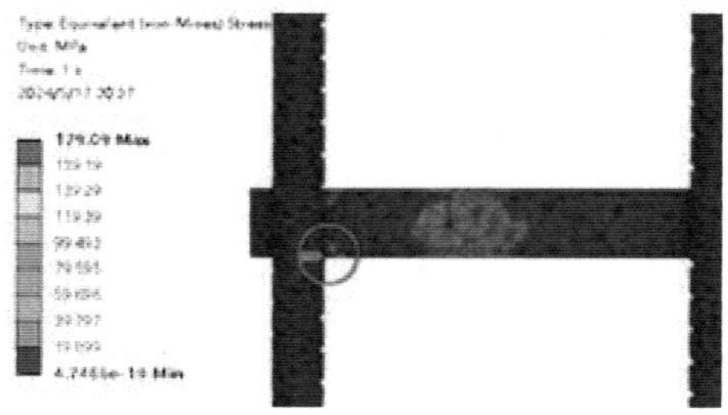

Figure 7. Stress analysis results (left view).

The maximum stress occurs at the connection point between the left screw and the lifting plate. The maximum equivalent stress valueis 179.09 MPa. Due to the downward load remains stable at 200 KN, the allowable stress for the screw can be calculated using the following formula[6, 7]:

$$[\sigma] = \frac{\sigma_s}{3} \tag{1}$$

The allowable stress for the screw is calculated as follows: $[\sigma] = 261.67\text{Mpa}$ with the obtained to compare and deduce:

$$\sigma_{ca} < [\sigma] \tag{2}$$

So, this supporting structure meets the requirements for the thin-walled inverting tube indexing stamping machine.

4.2. Stability safety factor

The calculation formula for the flexibility value of the screw is as follows[6, 7]:

$$\lambda_s = \frac{\mu l}{i} \tag{3}$$

After calculation, the selected screw's flexibility value is obtained $\lambda_s = 61.54$. Due to $\lambda_s < 100$ So, it is concluded that the screw is a medium-length rod, hence the critical load F_{cr} can be obtained using the following formula[6, 7]:

$$F_{cr} = (a - b\lambda_s)\frac{\pi}{4}d_1^2 \tag{4}$$

Furthermore, the calculation safety factor for screw stability can be derived from the following formula, and compared with the safety factor from the requirements[6, 7]:

$$S_{sc} = \frac{F_{cr}}{F} \geq S_s \tag{5}$$

After calculation, the calculated safety factor for screw stability is obtained. $S_{sc} = 3.61$ is the safety factor of screw stability $S_s = 3.5$. So, since the calculated safety factor for screw stability is greater than the safety factor, the screw stability meets the requirements.

4.3. Summary and outlook

Based on the above results, it can be concluded that the screw selected in this article meets the support requirements for the thin-walled inverting tube indexing stamping machine. However, this article focuses on the strength analysis of the support structure of the thin-walled inverting tube indexing stamping machine, and further research is needed for the strength analysis of other mechanisms.

5. Conclusion

This paper presents a self-machining thin-walled coil pipe bending and rolling forming machine, and validates the rationality of its supporting structure through ANSYS Workbench meanwhile. By comparing the analysis results of ANSYS Workbench with the allowable stress and stability safety factor of the screw, the rationality of selecting the designed screw as the main supporting structure of the thin-walled coil pipe bending and rolling forming machine has been validated. Through simulation analysis and theoretical calculations, it was found that the calculated safety factor for screw stability is close to the safety factor that required. Within practical production, it may be appropriate to increase the diameter of the screw to ensure the safety of the equipment. This experiment explored the rationality of the supporting structure of the thin-walled coil pipe bending and rolling forming machine, successfully validating that selecting the screw can support the thin-walled coil pipe bending and rolling forming machine.

Acknowledgements
This Project was supported by the Natural Science Foundation of Zhejiang Province (LY20E050003), & National Undergraduate Innovation and Entrepreneurship Training Program (202310347054).

References
[1] Zheng Yuqing, Yang Minghao, An Yuanyuan, et al. (2021) Study of the External Inverting Process of the Mild Steel Thin-walled Tube. In: Journal of Physics.
[2] Shi Xiaokuan, Xiong Yuxuan, Fei Yiting, et al. (2022) A Comparative Study On the Invertube Processings and Forming Dies. Journal of Huzhou University.Vol.42(4): 69-74.
[3] Hao linpo, Luo Yunhua, Xiong Jialin, et al. (2015) Research on starting load of inversion tube energy absorber. FORGING &STAMPING TECHNOLOGY.40(12): 166-168+175. dio:10. 13330/j.issn.1000-3940.
[4] Gao Qiang, Liao Wei-Hsin. (2021) Energy absorption of thin walled tube filled with gradient auxetic structures-theory and simulation. International Journal of Mechanical Sciences. Volume, Issue prepublish. doi:10.1016/j.ijmecsci.
[5] Wen Hailong, Zheng Yuqing, Fei Yiting. (2021) Optimal Parameter Study of the External Inverting Die. In: Journal of Physics.
[6] Pu Lianggui, Chen Guoding, Wu Liyan. (2019) Mechanical Design. 10th Edition. Beijing: Higher Education Press, Beijing.
[7] Wang Xiaoli, Zhao Jieliang, Zhu Pengzhe. (2022) Principles of Mechanical Engineering. Beijing: Tsinghua University Press, Beijing
[8] Zhou Ju, Su Jinying. (2023) Practical Examples of ANSYS Workbench Finite Element Analysis: Dynamics. Beijing: Posts & Telecom Press, Beijing.
[9] Hasan Sofuoğlu, Salim Çam. (2021) Coupled effect of thickness optimization and plastic forming history on crashworthiness performance of thin-walled square tube. The International Journal of Advanced Manufacturing Technology. Volume, Issue prepublish: 1-14 doi:10.1007/s00170-021-07865-y
[10] Li, Zhaokai, Tian Yuxing, Bai Jiantao, et al. (2021) Bending collapse of treble rectangular thin-walled tubes and its application in conceptual design for automobile crashworthiness. Proceedings of the Institution of Mechanical Engineers. Volume 235, Issue 5: 1269-1284 doi:10.1177/0954407020967869
[11] Yue Gao, Fei Shao, Pengxian Fan, et al. (2020) Effect of Internal Support on the Tensile Properties and Fracture Mode of 304 Stainless Steel Thin-Walled Tubes. Materials. Volume 14, Issue 1: 172-172 doi:10.3390/ma14010172
[12] Ba Yaer, Shen Chengjin, Ji Zhe, et al. (2022) Finite Element Analysis of Bearing Stress-Torque of Steel-Base NiMoCrFeCo-WC Coated Screw [J]. RARE METAL MATERIALS AND ENGINEERING, 51(01): 166-173.

Journal of Physics: Conference Series **2815** (2024) 012012 doi:10.1088/1742-6596/2815/1/012012

Prediction of packing performance of pneumatic control valve based on stem mechanism model and VMD-SSA-LSTM

Limiao Shi[1,2], Bin Yang[2,*], Yujin Wang[2] and Xinze Wang[2]

[1]Tianjin Key Laboratory of Information Sensing & Intelligent Control,Tianjin 300222,China
[2]Tianjin University of Technology and Education, Tianjin 300222, China

[*]Corresponding author's email: 13920940805@139.com

Abstract. Packing performance is one of the most important parts affecting the control operation of pneumatic control valves. In order to avoid the failure caused by packing wear and aging of pneumatic control valves after long-term use, this paper constructs a valve stem mechanism model and combines the advantage of deep learning to predict the packing performance of pneumatic control valves based on Long Short-term Memory (LSTM) and Variable Mode Decomposition (VMD) and Sparrow Search Algorithm (SSA). Firstly, the historical data of the actuator is calculated based on the mechanism model of the valve stem. The time-friction force obtained by the calculation is the parameter set. After the parameter set is subject to the variational mode decomposition, the parameter set is imported into the SSA-LSTM prediction model for prediction, and the prediction data is obtained.

1. Introduction

Pneumatic control valves are widely used due to their simple structure, applicability to various working environments, economy and environmental protection, etc. Proper packing friction is beneficial to the tightness of the control valve. However, with the long-term use of the control valve, frequent control actions will reduce the compression force of the control valve, and the tightness will gradually deteriorate. Therefore, the packing performance change of the control valve can be predicted effectively, which is very important to the performance test and research of the control valve and the later maintenance of the control valve.

In previous studies, most scholars prefer to study the mechanism modeling direction of the control valve. For example, Jin et al. [1] optimized the traditional Karnop friction model of the pneumatic diaphragm actuator, modified the critical speed of the dynamic and static friction judgment condition DA, solved the problem of how to select DV, and improved the parameter identification method of the traditional pneumatic diaphragm actuator model; Gao et al. [2] proposed an online parameter self-tuning compensation algorithm based on the traditional simplex parameter compensation algorithm to reduce the abrasion and basically eliminate the oscillation of loop output, aiming at the characteristics of different viscosity degrees of forward and reverse valve travel and different valve positions in practice; Shang et al. [3] established the mechanism model of the pneumatic stop valve based on the working principle and structural characteristics of the pneumatic stop valve and the thermodynamic and kinematic principles. On the basis of the mechanism model, a kind of air tightness fault diagnosis method was designed and verified, which showed that the model has a high accuracy; Hu et al. [4] standardized and visually preprocessed the small sample data of the acoustic emission signal under different working conditions of the valve through Numpy and Pandas modules and Malfunctions in Python, input the preprocessed data into DCGAN to generate an enhanced data set, input the mixed data set into the 1Dmodel for training, and finally used the confusion matrix to evaluate the model performance; Zhang et al. [5] input the time-frequency domain signal into the constructed DNN model for noise reduction by establishing the valve jet sound field model and DNN noise reduction basic model for the impact of underwater noise on the signal acquisition accuracy, and input the processed acoustic signal into the convolution neural network for fault diagnosis; Shao et al. [6] proposed an intelligent diagnosis model (IHO-LSTM) integrating the improved Harris Hawk Optimization (HHO) algorithm and the Long Short-Term Memory (LSTM) network to solve the problem of rolling bearing fault diagnosis; Zhou et al. [7] proposed the migration diagnosis method combining the Inception module and Long Short-term Memory (LSTM) to monitor and diagnose the running state of roller table bearing and through-shaft bearing in the roller table system of annealing kiln; Shi et al. [8] proposed a life prediction method based on long and short-term memory (LSTM) networks to solve the problem of fatigue failure caused by the cyclic impact of thermal stress and electrical stress on IGBT during operation.

The above pieces of literature have studied the establishment of the mechanism model, the optimization and innovation of the friction model, and the combination of the mechanism model and prediction model. In this paper, the input air pressure of the pneumatic actuator is directly measured by the pressure sensor, and the friction of the control valve is directly calculated by the valve position feedback signal, input current signal, and other characteristic signals in combination with the valve stem mechanism model. The calculated friction is decomposed into variable mode as one-dimensional data. Then, the LSTM parameters are optimized according to SSA, and the decomposed characteristic parameters are input into the LSTM neural network. Finally, each component is added to obtain the predicted friction value.

2. Establishment of actuator structure and mechanism model

2.1. Actuator

Taking the positive-acting diaphragm type pneumatic actuator (i.e. air-to-close type) as an example, this paper explains the structure of the actuator. The pneumatic actuator is divided into the positive-acting pneumatic actuator (i.e., ventilating extension type) and reverse-acting pneumatic actuator (i.e., ventilating retraction type) (as shown in Figure. 1). The working principle of the positive-acting pneumatic actuator is that the air on the top of the diaphragm pushes the diaphragm and the valve stem downward. When the air pressure drops, the spring will return and move upward. Generally, the spring force of the positive-acting actuator is relatively small, and the spring rigidity is relatively low. Only the force that can provide the actuator reset is required.

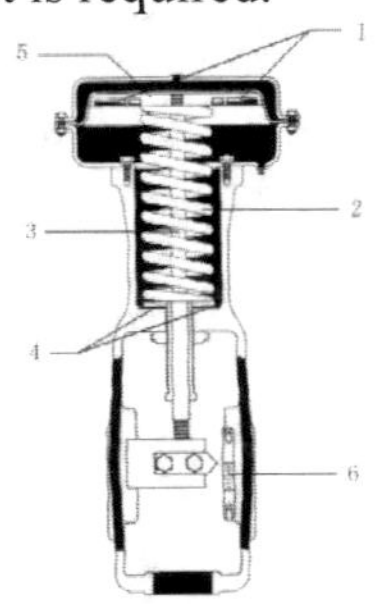

Figure. 1. Actuator of pneumatic control valve.

1-Diaphragm; 2-Spring; 3-Stem; 4-Spring seat; 5-Diaphragm chamber; 6-Valve travel indicator scale

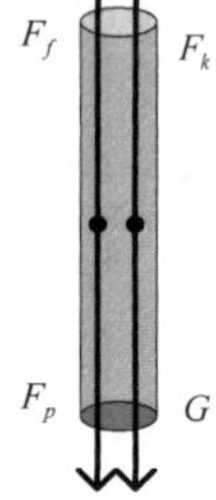

Figure. 2. Stress analysis of valve rod.
F_f -Friction force; F_k -Spring force force; F_k -Diaphragm pressure; G -Stem component combination gravity

2.2. Stem mechanism model

The control valve actuator receives the air pressure signal output by the air relay of the intelligent positioner. As the gas gradually enters the membrane gas chamber, the membrane chamber pressure of the membrane gas chamber gradually increases, and the pressure of the diaphragm will increase until the forces such as spring force and friction force are exceeded. When the force on the valve rod is unbalanced, the valve rod will start to move downward. Figure. 2 shows the stress analysis diagram of the valve rod.

During the process of regulating the valve from full opening to full closing, the static friction force suffered by the valve rod before each movement is opposite to the movement trend, i.e., the valve rod receives downward diaphragm pressure F_p, upward spring force F_k, stem component combination gravity G, and frictional force in the upward direction F_f.

$$F_p + G = F_k + F_f \tag{1}$$

$$(P_0 + \Delta P)S_0 + mg = k\Delta x + F_f \tag{2}$$

$$F_f = (P_0 + \Delta P)S_0 + mg - k\Delta x \tag{3}$$

In the above equation: is the pressure variation of the diaphragm; is the area of the diaphragm; is the combined mass of the valve stem components; is the elastic force coefficient of spring; is the spring compression. The real-time static friction of the control valve can be calculated through the above equation.

3. Prediction model of packing performance based on vmd-ssa-lstm

3.1. Variable Mode Decomposition (VMD)

Variable Mode Decomposition is a fully non-recursive decomposition model [9]. It is very suitable for processing nonlinear time series signals such as historical friction data. This method is mainly to extract historical friction data using the idea of solving variational problems and decomposing the original signal into multiple signals with different center frequencies without losing signal characteristics.

The VMD algorithm assumes that the center frequencies of each mode are different and the bandwidth is limited. By using the alternating direction method of multiplication operators, each mode and its corresponding center frequencies are updated continuously. After the friction curve is decomposed, multiple variational modal components with corresponding center frequencies can be obtained. The constrained variational problem model of this method is:

$$\begin{cases} \min_{\{u_k\},\{\omega_k\}} \{\sum_{k=1}^{K} \left\| \partial_t [(\delta(t) + \frac{j}{\pi t}) * u_k(t)] e^{-j\omega_k t} \right\|_2^2 \} \\ s.t. \sum_{k=1}^{K} u_k(t) = f(t) \end{cases} \quad (4)$$

where $\delta(t)$ is the Dirac distribution; * represents convolution operation; ω_k is the center frequency of $u_k(t)$; $\{u_k\} = \{u_1, ? \cdots, u_k\}; \{\omega_k\} = \{\omega_1, ? \cdots, \omega_k\}$; K is the number of IMF; $(\delta(t) + \frac{j}{\pi t}) * u_k(t)$ is Hilbert transform.

Next, the penalty factor α and Lagrange operator λ are introduced to transform the constrained problem into an unconstrained problem:

$$L(u_k, \omega_k, \lambda) = \alpha \sum_{k=1}^{K} \| \partial_t [(\delta(t) + \frac{\mathrm{j}}{\pi t}) * u_k(t)] \mathrm{e}^{-\mathrm{j}\omega_k t} \|_2^2 + \| f(t) - \sum_{k=1}^{K} u_k(t) \|_2^2 + < \lambda(t), f(t) - \sum_{k=1}^{K} u_k(t) > \quad (5)$$

The alternating multiplier method is used to iteratively solve and obtain K IMF components and their corresponding center frequencies.

3.2. Sparrow Search Algorithm (SSA)

The sparrow search algorithm is a new intelligent optimization algorithm proposed by Xue [10] of Donghua University in 2020, and its central idea originates from the behavior strategy of the sparrow. The sparrow population is mainly divided into two categories: discoverer and joiner. The discoverer is responsible for providing foraging direction for the population, and the joiner uses the discoverer to find food, and the identity of the discoverer and joiner will be exchanged. Once the sparrow finds the predator, the individual will sound an alarm signal, and then the finder will lead the population to other safe areas for feeding. Sparrow search algorithm effectively avoids the situation of getting stuck in the local optimal solution through the identity exchange of individuals.

Algorithm process:

(1) Initial population, iteration times, proportion of discoverer and joiner;

(2) The fitness value is calculated and sorted from large to small;

(3) The discoverer location is updated by using Equation (11);

(4) The location of the joiner is updated by using Equation (12);

(5) Equation (13) is used to update the location of the person issuing the alarm;

(6) Fitness values are calculated and sparrow positions are updated;

(7) If the stop condition is met, it is necessary to exit and output the result if it is met, otherwise, it is necessary to restart from Step (2).

3.3. Long-Short-term Memory Neural Network (LSTM)

Long-short-term memory neural network (LSTM) effectively solves the problem of gradient disappearing and gradient explosion of recurrent neural networks (RNN). LSTM mainly consists of four parts: forgetting gate, input gate, output gate, and unit (cell) state, The LSTM structure diagram is shown in Figure.3. It is assumed that at a certain time t, the input is x_t, the hidden state of the previous moment is h_{t-1}, and the unit state is c_{t-1}.

The forgetting output that controls the status of the unit at the previous moment is:

$$f(t) = \sigma(W_f[h_{t-1} + x_t] + b_f) \quad (6)$$

where f(t) is the sigmoid function; W_f is the weight coefficient of the forgetting gate; b_f is the deviation value of the forgetting door.

The output expression and unit state vector expression after input gate processing are:

$$i(t) = \sigma(W_i h_{t-1} + W_i x(t) + b_i) \quad (7)$$

$$\tilde{c}(t) = \tanh(W_c h_{t-1} + W_c x(t) + b_c) \quad (8)$$

where W_i, W_c, b_i, b_c are the input gate; σ and tanh are the corresponding weight coefficients and deviation values of the activation function.

After the cell update, the equation is:

$$o(t) = \sigma(W_0 h_{t-1} + W_0 x(t) + b_0) \tag{9}$$

$$h_t = o(t) \otimes \tanh(c(t)) \tag{10}$$

where W_0, b_0 are the weight coefficient and deviation value corresponding to the output gate respectively; h_t is the output vector.

The prediction output at the current moment is updated, and the equation is:

$$\hat{y}(t) = \sigma(Vh_t + c) \tag{11}$$

where is the weight coefficient and deviation value of the output layer connection respectively.

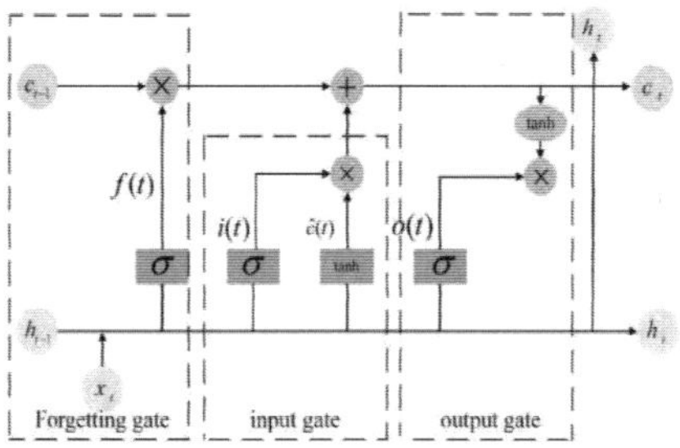

Figure. 3. LSTM structure expansion diagram.

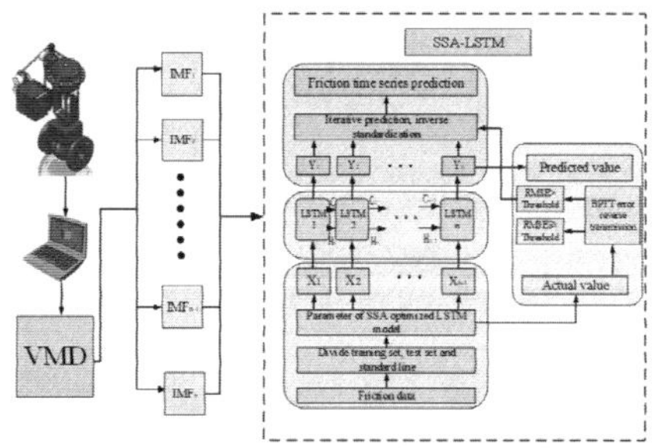

Figure. 4. LSTM flow chart.

4. Experiment and analysis

4.1. VMD-SSA-LSTM prediction model

The construction idea of the VMD-SSA-LSTM model is to first perform variational mode decomposition on the data, and then use the SSA optimization algorithm to optimize the super parameters of the LSTM model [11]. Through error backpropagation of the predicted value, the root mean square error (RMSE) between the calculated predicted value and the actual value is less than the threshold value, the optimization is stopped, and the optimal super parameters of the model are obtained. The SSA-LSTM model calculation process framework is shown in Figure.4, and the specific steps are as follows:

(1) The original friction data of the control valve is divided into a training set and a test set, and the original data is processed by variational mode decomposition and standardization;

(2) The number of sparrow populations is set and the original population is initialized. The number of iterations of the sparrow population is set, and the proportion of predators and participants is determined according to the forecast demand;

(3) The fitness value of the sparrow population is calculated, then any sparrow in the sparrow population is sorted, and the location information of the predator, joiner, and watcher is updated according to the sparrow population sorting results;

(4) The fitness value is recalculated and the sparrow position is updated. If the RMSE < threshold condition is met, it is necessary to exit the optimization process and output the super parameters optimized by the sparrow algorithm. Otherwise, Steps (2) and (3) should be repeated until the stop condition is met.

(5) The super parameters optimized by the SSA algorithm are set as the optimal super parameters of the LSTM model, the standardized training set and test set are input into the SSA-LSTM model for iterative prediction, and the prediction results are destandardized to finally obtain the time series prediction value of the friction force of the control valve.

4.2. Verification of VMD optimal decomposition scale

When the VMD method is used to decompose the original friction data, the value of the number of modes k has a great impact on the decomposition effect. When the number of modes k is large, mixing will occur; When the number of decomposition is small, the information of the original signal is easy to lose. The most appropriate VMD decomposition level can be determined by calculating different k values and observing the center frequency corresponding to different k values.

In this paper, the real-time data of a certain type of pneumatic control valve is taken as the simulation data by calculating the valve stem mechanism model. In order to make the data intuitive and concise, each result is subtracted by 98. The sampling frequency is 1024, and the penalty factor is set to 2000.Simulation analysis was conducted in Matlab, and the center frequency corresponding to the K value is shown in Table 1:

Table 1 . K-value center frequency of VMD.

K (n)	Center Frequency (Hz)
K=9	0.1130 164.68 419.80 795.004 1316.76 2349.88 3476.88 4870.20 5696.84
K=10	0.1120 164.30 416.30 784.48 1227.98 1834.30 2835.52 3934.80 5009.82 5706.41
K=11	0.1090 162.68 396.53 652.23 861.66 1361.39 2324.69 3370.47 4048.59 4985.22 5701.31
K=12	0.1084 162.43 393.88 635.69 840.95 1268.05 1845.55 2713.83 3462.30 4296.24 5069.79 5712.58
K=13	0.1080 162.23 391.75 626.62 830.15 1241.57 1714.91 2282.48 2835.96 3485.25 4292.65 5061.12 5709.75

When k = 11, the frequency change of each center tends to be stable.

Therefore, k = 11 is the optimal decomposition scale, and the time domain waveform of each sub-sequence after the original power is decomposed by VMD at this time is obtained, as shown in Figure. 5.

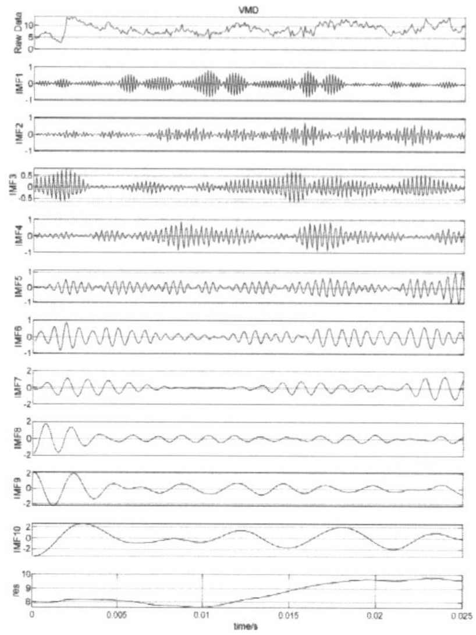

Figure. 5. VMD decomposition and waveform diagram of IMF components.

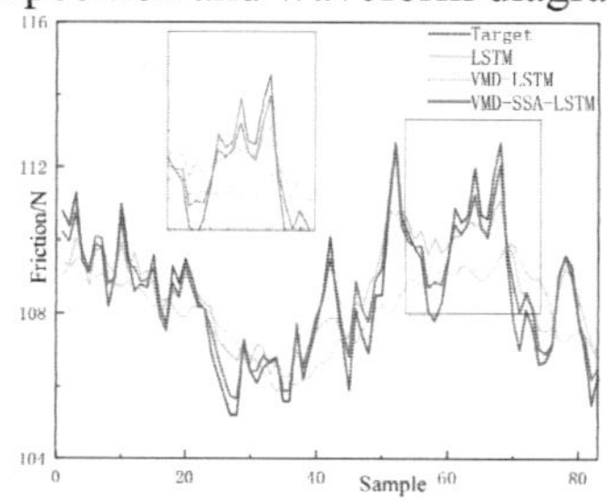

Figure. 6. Comparison of prediction results.

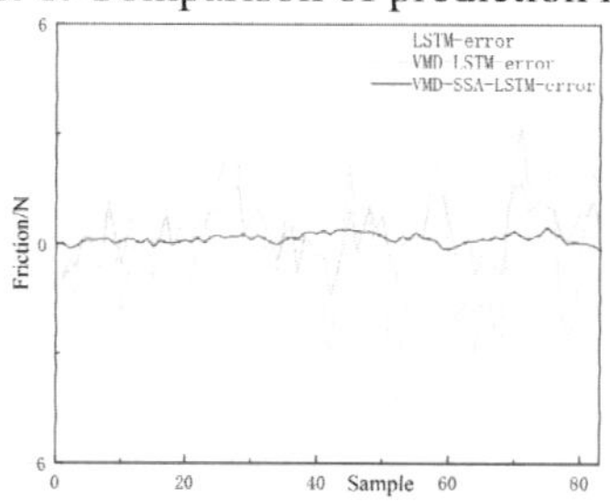

Figure. 7. Prediction error comparison.

4.3. SSA-LSTM Model prediction results

In the SSA-LSTM model, the population number N of sparrow is 6, the maximum iteration number M is 8, the discoverer accounts for 20% of the population, and the rest is the joiner. The early warning value is 0.8. When the early warning value is less than 0.8, there is no predator. Otherwise, there is a predator who is dangerous to the safety of the population, and they need to go to other places to look for food. The specified initial learning rate is 0.005. After 125 rounds of training, the learning rate is reduced by multiplying the factor 0.2. The first 70% of the specified dataset is the training set, and the last 30% is the test set.

It can be seen from Figure. 6 that the predicted results of VMD-SSA-LSTM have the highest degree of fit with the true values of the test set, the predicted results of VMD-LSTM are slightly worse than those of VMD-SSA-LSTM, and the predicted results of LSTM are the worst, with extremely low fit.

According to Figure. 7, the prediction error of VMD-SSA-LSTM in the three algorithms fluctuates gently around the straight line Y=0, while the prediction error of VMD-LSTM and LSTM fluctuates greatly. VMD-SSA-LSTM predicts the best results.

5. Conclusion

This paper proposes for the first time to calculate the friction between the valve stem and the packing through the mechanism model of the regulating valve, and to predict the friction through VMD-SSA-LSTM. Although the predicted results still have some deviation from the actual values, the experimental results still have a positive impact on the follow-up research of the intelligent positioner for valve performance analysis. Secondly, through the variational mode decomposition of the original data, the optimal decomposition scale is obtained from the different degrees of the decomposed center frequency, which prevents the over-decomposition or under-decomposition problems caused by subjective presetting. Finally, although the sparrow search algorithm has better optimization performance than the traditional population optimization algorithm, it still has the problems of too fast convergence speed and easy to fall into local optimization. Therefore, we should try to further expand the research from the perspective of algorithm improvement or model construction in the future.

References

[1] Jin Zongheng, Li Shiwei, Shang Qunli. 2023, 31(09). Model optimization and parameter identification of pneumatic membrane actuator [J]. Computer Measurement and Control.213-219.

[2] Gao Zheng, Shang Qunli, Liu Wei, et al. 2023, 33(09). Research on the self-adjusting compensation algorithm for the viscosity characteristics of regulating valves [J]. High-tech Communication.981-989.

[3] Shang Qunli, Li Mengqiang, Zhang Jingyu. 2020, 48(02). Mechanism modeling of pneumatic stop valve and its application in valve fault diagnosis [J]. Journal of the Zhejiang University of Technology.154-158+216.

[4] Hu Leilei, Wang Xinxin, Lin Zhenyuan, etc. 2023, 49(12). Research on internal leakage fault diagnosis of pipeline valves based on DCGAN-1DCNN [J]. Industrial Safety and Environmental Protection.43-49.

[5] Zhang Xiulong, Liu Guijie, Ning Donghong. 2022, 39(10). Research on valve fault diagnosis based on improved deep neural network [J]. Electromechanical Engineering. 1356-1364+1373.

[6] Shao Liangshan, Zhu Sijia. 2024, 46 (01). Research on fault diagnosis of rolling bearing based on improved HHO-LSTM [J].Mechanical Strength, 17-23.

[7] Zhou Kangqu, Liu Tianchuang, Xin Yu, etc. 2024, 44 (01). Bearing fault diagnosis of annealing kiln roller table system based on Inception LSTM [J]. Noise and Vibration Control.174-180.

[8] Shi Yezhao, Guo Bin, Zheng Yongjun. 2024, 50 (02). Research on IGBT life prediction method based on LSTM network [J]. China Testing.54-58+65.

[9] Dragomiretskiy K, Zosso D. 2013, 62 (3). Variable mode decomposition [J] IEEE transactions on signal processing.531-544.

[10] Xue Jiankai. 2020. Research and application of a new type of swarm intelligence optimization technology [D]. Donghua University.

[11] Dongdong H, Boyuan Z, Ruotong G, et al. 2023, 12. Predicting evolutions of pulse characteristics along cavity position in passively mode-locked fiber laser via SSA-LSTM approach [J].Optics and Laser Technology.

Journal of Physics: Conference Series **2815** (2024) 012013 doi:10.1088/1742-6596/2815/1/012013

The effect of the structure of a water distribution pan in an evaporation drum on a flake ice machine

Qiang Zhou[1,*]

[1] Tianjin University of Commerce, Tianjin, 300000, China

[*] Corresponding author's e-mail: zhouqiangzt@163.com

Abstract: Through experimental research, the performance of the flake ice machine under different structures of the water distribution pan was comparatively analyzed to enhance performance and optimize the design of the water distribution pan in the evaporation drum. An experimental bench for testing the performance of the flake ice machine was constructed, and the optimal scraping cycle under varying water distribution pan structures was determined. The results revealed a 33% improvement in the ice-making area within the water distribution pan structure, along with a 17.5% reduction in power consumption per unit of ice production, achieved through the optimal operational conditions of the flake ice machine using R404A as the refrigerant. Furthermore, the dry area was enhanced by 40%, leading to an 8% reduction in power consumption per unit of ice production. Notably, for this flake ice machine, the optimal ice-making drying ratio was found to be 1.6.

1 Introduction.

The use of ice to extend the storage life of food has a long history [1], dating back thousands of years when people buried their harvest in snow or ice cellars to preserve the freshness of their food. Over time, with the rapid evolution of industrial fisheries, the ice machine industry has undergone various updates and iterations [2]. As the demand for ice has evolved throughout the development of ice machines, it has expanded from its initial use in meat preservation to applications in summer cooling, food preservation, cold beverages for refreshing in hot weather, industrial cooling, and other diverse fields [3]. Consequently, the health needs and utilization scenarios for ice have become more varied, leading to increasingly stringent technical requirements for ice machines. Ice machine types can be classified based on the shape of the ice they produce into various categories, including block ice machines, flake ice machines, tube ice machines, and plate ice machines [4]. Among these, flake ice machines are characterized by their thinner ice thickness [5], large contact area, and rapid cooling capabilities.

The rapid development of flake ice machines has been driven by the increasing market demand for flake ice [6]. Despite this growth, flake ice machines are still plagued by various design challenges, notably the suboptimal design of the evaporation drum and the lack of alignment between the refrigeration system and the machine. In previous research [7], particular attention was given to the design aspects of the evaporation drum. The study focused on optimizing the structure of the water distribution pan within the flake ice machine. This optimization involved establishing distinct areas within the machine, namely the ice-making area, dry area, and ice-scraping area, based on the operational flow. Furthermore, the research introduced a range of dimensionless parameters to enhance the structure of the water distribution pan, ice-scraping cycle, and other critical parameters. Concurrently, Zhao et al. [8] conducted an investigation into the phase-change heat-transfer characteristics of flake ice

Journal of Physics: Conference Series **2815** (2024) 012013 doi:10.1088/1742-6596/2815/1/012013

machines and optimized the ice storage system associated with these machines.

Based on the existing flake ice machine, this paper explores in depth the performance of the flake ice machine under different structures of water distribution panes through experimental research methods. By changing the size of the ice-making area and dry area, the optimal ice-scraping cycle under different structures is found. By introducing the dimensionless parameter ice-making drying ratio, the performance of the flake ice machine can also be evaluated. The experimental reference is provided for the design of the Structure of the water distribution pan in the evaporation drum of the flake ice machine, which facilitates the better optimization of the flake ice machine.

2 Experimental setup and test conditions

2.1 Experimental setups

The experiments covered in this paper were done in a thermostatic laboratory. To monitor and regulate the unit operation, T-type thermocouples, as well as pressure transducers, were installed at the inlet and outlet of the compressor, condenser, throttling device, and evaporator. To control the stable inlet water temperature of the ice machine, T-type thermocouples were installed at the inlet and outlet of the mixing tank. Inverters are installed on the reducer to control different ice scraping cycles and regulate the frequency of the reducer. Figure 1 shows the system diagram of the ice machine.

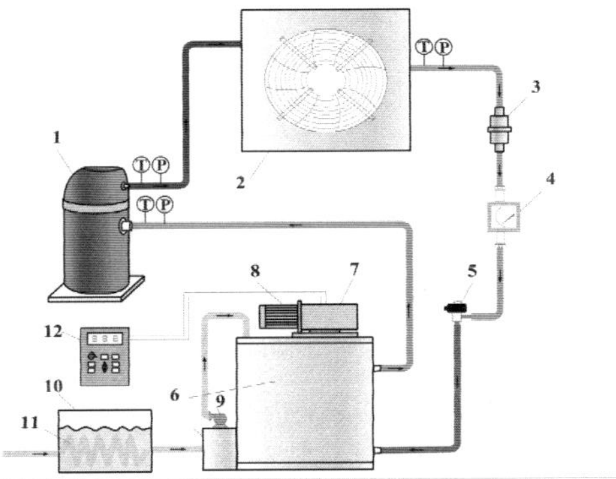

1 Compressor 2 Condenser 3 Filter drier 4 Volumetric flow meter 5 Electronic expansion valve 6 Evaporation drum 7 Reduction gear 8 Reduction gear motor 9 Circulating water pump 10 Constant temperature water tank 11 Heat exchanger coil 12 Frequency converter

Figure 1. System schematic diagram.

The structure of water distribution pan in the evaporation drum is shown in Figure 2. The area occupied by the water distribution pipe is called the ice-making area; according to the order of counterclockwise rotation, the area between the last water distribution pipe and the scraper is called the dry area; the area between the scraper and the first water distribution pipe is called an ice-scraping area. The water distribution pan is connected to the water distribution pipe, which changes the structure of the water distribution pan by preventing the flow of water into the water distribution pipe, thus affecting the percentage of each area occupied.

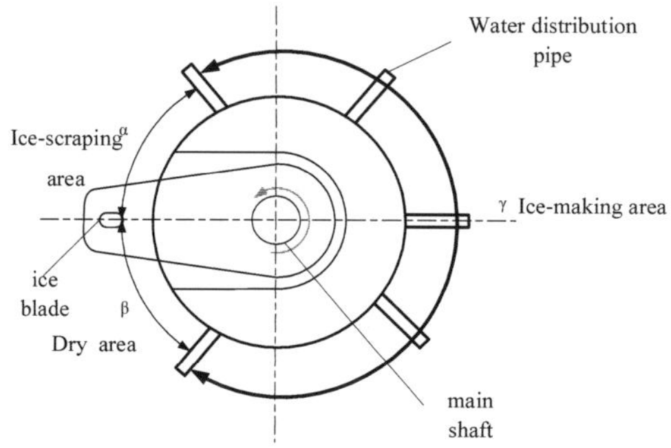

Figure 2. Structure of water distribution pan.

2.2 Test conditions

The ice production test began by setting the experimental conditions according to the domestic trade industry standard, SB/T 10940-2012. The dry bulb temperature was maintained at 32±1°C, and the make-up water temperature was controlled at 21±0.5°C. Once the ice maker reached stabilization under the specified conditions, ice samples were collected at 15-minute intervals during the initial 1.5 hours of operation following machine activation. The operationall stability of the flake ice machine was confirmed based on a criterion that the variation in ice production volume should not exceed ±2.0% in three consecutive measurements. Upon confirming operational stability, sequential sampling was conducted every 15 minutes to measure ice weight and other relevant parameters.

2.3 Data processing

The electrical energy consumed by the flake ice machine for every 100 kg of ice produced is a critical performance indicator, which is calculated as expressed in Equation (1).

$$W = \left(\frac{60\tau}{G} \times 100 \right) * P_{\text{ave}} \Big/ 3600 \,, \tag{1}$$

The ice-scraping cycle is the time taken for one revolution of the ice scraper and can be calculated using the following formula:

$$T = \frac{\text{p} \cdot s}{f} \tag{2}$$

3 Results and discussion

3.1 Optimal operating condition point analysis

As shown in Figure 3.Five sets of experiments were conducted to obtain the optimum refrigerant charge and the optimum evaporation temperature. As shown in the graph, it can be seen that at a refrigerant charge of 1.9 kg, the power consumption per unit of ice produced was close to the minimum at each evaporation temperature. The power consumption per unit of ice production was minimized at an evaporating temperature of -24.5°C for all refrigerant charges tested. Therefore, a refrigerant charge of 1.9 kg and an evaporating temperature of -24.5°C can be taken as the optimal operating point for the tested ice machine.

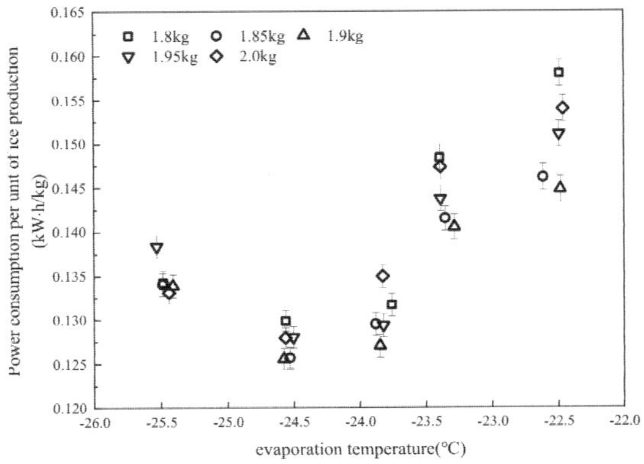

Figure 3. Power consumption per unit of ice production versus refrigerant charge and evaporation temperature.

3.2 The effect of different structure of water distribution pan on flake ice machine

3.2.1 Ice-making area on the flake ice machine impact analysis. In Figure 4, the power consumption per unit of ice production is depicted under the structure of the water distribution pan with a drying area ratio of 0.3125 and ice-making area ratios of 0.375 and 0.5. It is evident from the graph that there is a notable trend in the power consumption per unit of ice production with the increase in the ice scraping cycle. Initially, there is a decrease in power consumption, followed by an increase. The power consumption per unit of the ice-making area varies significantly as it is considered the central region for ice production and heat transfer. Altering the size of the ice-making area impacts the heat transfer process distinctly from that of the overall ice-making area. Consequently, the power consumption per unit of ice production decreases in relation to the expansion of the ice-making area. Specifically, a 33.33% increase in the ice-making area results in a 17.5% rise in the power consumption per unit of ice production.

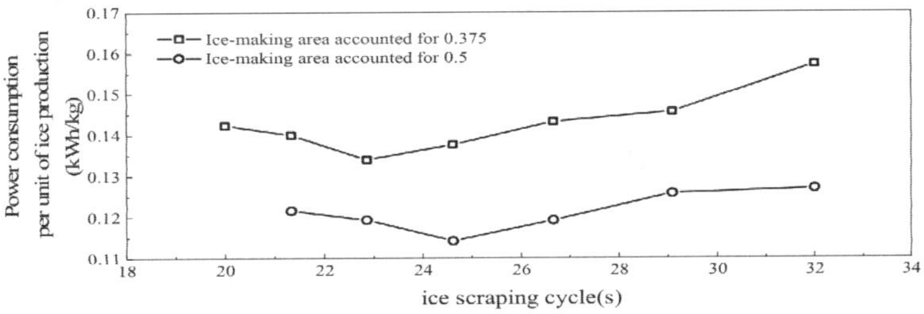

Figure 4. Power consumption per unit of ice production as a function of the scraping cycle.

3.2.2 Analysis of the impact of the dry area on the flake ice machine. In Figure 5, the ice-making area ratio is 0.375, and the dry area ratio is 0.4375 and 0.3125, respectively, under the structure of the water distribution pan. The unit of ice production power consumption per unit of ice production varies with the ice scraping cycle. Observing the data, it is evident that as the scraping cycle increases, the unit of ice production power consumption initially decreases, then increases, ultimately reaching a minimum. For a dry area ratio of 0.4375, the power consumption per unit of ice production is 0.124 kWh/kg at a scraping cycle of 21.33 seconds. In comparison, with a dry area ratio of 0.3125, the power consumption per unit of ice production reaches a minimum of 0.134 kWh/kg at a scraping cycle of 22.86 seconds. Increasing the dry area by 40% can lead to an 8% reduction in power consumption per unit of ice production under optimal conditions. The presence of a dry area facilitates the freezing of water on the ice's surface, enabling further cooling of the ice and allowing for smoother ice scraping by the scraper.

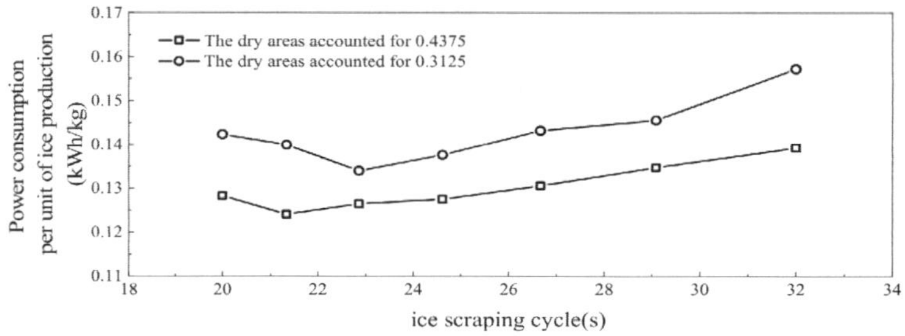

Figure 5. Power consumption per unit of ice production as a function of the scraping cycle.

3.2.3 Analysis of the effect of ice-making drying ratio on flake ice machine. Figure 6 illustrates the variation of power consumption per unit of ice production in relation to the scraping cycle for ice-making drying ratios of 0.4, 0.9, and 1.6. Defined as the ratio of the ice-making area to the dry area, the ice-making drying ratio influences the power consumption. An increase in the ice-making drying ratio is associated with a decrease in power consumption per unit of ice production. This increase is primarily attributed to the expansion of the ice-making area, which leads to higher water supply and consequently results in increased ice production. Optimal ice scraping cycles are observed at different ice-making drying ratios, thereby influencing the design considerations for the water distribution pan structure in the evaporation drum and the reducer frequency. This finding holds significance in enhancing the operational efficiency of ice production systems.

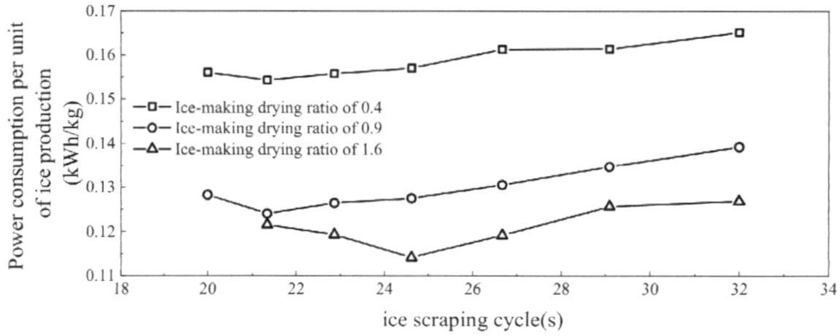

Figure 6. Power consumption per unit of ice production as a function of the scraping cycle.

4 Conclusion

1) In different refrigerant volumes, with different evaporation temperatures, flake ice machine performance will be different, and there will be an optimal operating state.

2) Under the optimal operating condition of the flake ice machine, the structure of the water distribution pan improves the ice-making area by 33%, and the energy consumption can be reduced by 17.5%. A 40% increase in the dry area reduces energy consumption by 8%.

3) With the improvement of the ice-making drying ratio, the energy consumption of the flake ice machine is reduced, according to the flake ice machine; for example, the best ice-making drying ratio is 1.6, at this time, the lowest energy consumption of the flake ice machine.

References

[1] Pan Liangchi. Ancient Chinese rituals and customs of hiding and using ice [J]. Journal of Sichuan College of Arts and Sciences (Social Sciences), 2006, (06): 78-80.

[2] Acha S, Du Y, Shah N. Enhancing energy efficiency in supermarket refrigeration systems through a robust energy performance indicator[J]. International Journal of Refrigeration, 2016, 64: 40-50.

[3] Tian Heqing. Application and practice of flake ice pneumatic conveying in Dagangshan concrete

system [J]. Technology and Enterprise, 2013, (23): 259. DOI: 10.13751/j.cnki.kjyqy.2013.23.245.
[4] Deng Yajing. Ice maker industry: promising growth, promising future[J]. Electrical Appliances, 2023(04): 52-53.
[5] Lee H-K, Choi K-H, Yoon J-I, et al. Performance Characteristics of a Seawater Ice-Making Device Using a Scraped Surface Double Tube Evaporator[J]. Applied Sciences, 2018, 8(11): 2063.
[6] HAN Meishun. China refrigeration equipment market analysis in 2020[J]. Refrigeration Technology, 2021, 41(S1): 67-86+94.
[7] Sun Z, Chen X, Yang Y, et al. Theoretical analysis and experimental study on the performance of ice buckets in a flake ice maker[J]. Case Studies in Thermal Engineering, 2022, 39: 102449.
[8] Zhao J D, Liu N, Kang Y M. Optimization of ice making period for ice storage system with flake ice maker[J]. Energy and Buildings, 2008, 40(9): 1623-1627.

Optimization of sensor mounting position and pose in rotary kiln belt center measurement

Yu Deng[1] and Changliang Luo[1]

[1]School of Mechanical and Electrical Engineering, Wuhan University of Technology, Wuhan, Hubei, 430070, China

[*]Corresponding author's e-mail: 53712165@qq.com

Abstract: To solve the problem that the measurement accuracy of the center of the wheel belt decreases during the adjustment of the center axis of the rotary kiln, a method of sensor installation position optimization based on dynamic simulation analysis and a simple small model is proposed and implemented in this paper. In this method, the coordinates of the laser points on the surface of the belt are obtained by using the sensor laser beam, and then the center position of the belt is obtained by fitting the coordinates of 6 laser points on the surface of the belt. Then V-rep dynamic simulation software is used to analyze the coordinates of the center position of the wheel under different sensor installation positions to find the sensor installation positions closest to the actual center of the wheel. Finally, the simulation results are verified by a simple small model. The experimental results show that this method can reduce the influence of belt deformation on the measurement of the center of the belt and find the most suitable sensor installation position. Taking the wheel with an ellipticity of 0.16% as an example, under the optimized sensor orientation, the error of the Y direction of the center of the wheel is 0.4 mm, and the error of the Z direction of the center of the wheel is 0.73 mm. By this method, the center accuracy of the existing measuring wheel is optimized from about 2 mm to less than 1 mm.

1. Introduction

In the process of measuring the central axis of the rotary kiln, the irregular deformation of the wheel belt is described by ellipticity. In the article on the ovality measurement system of rotary kiln based on laser sensor, Jiang [1] analyzed the causes of the ovality of rotary kiln and studied the relationship between the ovality of rotary kiln and the central axis. In the article on the ovality measurement method of rotary kiln. He [2] assumed that the cylinder was a standard ellipse for convenience of analysis when studying the irregularly deformed cylinder, and on this basis solved the location size of the center of the cylinder. It can be seen that the ellipticity of the belt will affect the size of the center position of the belt, so we should consider the influence of the ellipticity and reduce the influence of the ellipticity of the belt when using the laser point coordinates on the surface of the belt to solve the center of the belt. The three-point cross-diameter direct remote measurement method of rotary kiln developed by Wuhan University of Technology [3] adopts a rectangular coordinate system, establishes a reference plumb plane parallel to the cylinder on both sides of the kiln body, and establishes a spatial coordinate system accordingly. The center point coordinates of each kiln can be directly calculated according to the geometric relationship, and the state of the center line of the rotary kiln can be detected by installing sensors under the wheel belt. At the same time, it is necessary to measure the belt clearance, and the measured Y-axis error of the center of the belt is about 6 mm, and the Z-direction error is about 2 mm.

Phillips established a coordinate system [4-6] by theodolite and level to determine the position of the sensor bracket in the coordinate system and then used three laser displacement sensors to measure the displacement between the measuring point and the surface of the barrel. The coordinates of the three measuring points were calculated by the coordinates of the bracket and the results were measured by the laser displacement sensor, and then the center of the barrel was obtained. If the laser point measured by this method is the deformation area of the cylinder, the fitting accuracy will be affected, and the Z-direction error of the center of the wheel can reach 3 mm maximum. It is difficult to exclude the influence caused by irregular deformation of the wheel belt when solving the existing center position of the wheel belt or compensate by measuring the gap between the wheel belt and the cylinder, and the fitting center of the wheel belt is greatly affected by the deformation of the wheel belt [7]-[9].

To reduce the influence of ellipticity caused by irregular deformation of the wheel belt on solving the center position of the wheel belt, a sensor position installation method based on dynamic simulation analysis and a simple small model [10]-[12] was proposed in this paper. First, the fitting theory of solving the center of the wheel belt by measuring the distance of the laser displacement sensor was described. Then, through simulation analysis, the relationship between sensor posture installation and the center position of the wheel is obtained, the most suitable sensor installation posture is obtained, and the existing center error of the wheel is reduced from about 2 mm to less than 1 mm [13]-[14]. Finally, the error of the center position of the sensor in this position is verified by experiments.

2. Solving the laser point position on the surface of the rotary kiln wheel belt

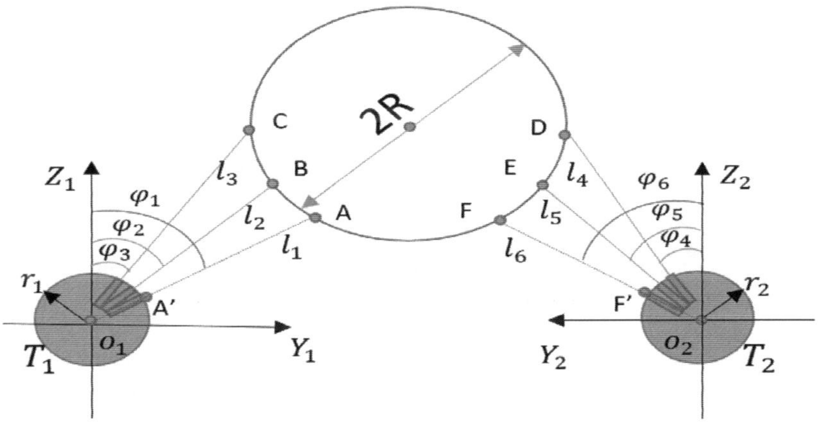

Figure 1. Solution of laser point coordinates on the surface of the wheel belt.

As shown in Figure 1, coordinate systems were established on the left and right sides of the rotary kiln, the laser displacement sensor was fixed on the disk, and the tool coordinate system 1 was established with O_1 as the origin, denoting it as {T1}. and the state coordinate system 2 is established with O_2 as the origin and is recorded as {T2}. Then the pose of the laser displacement sensor is determined by (r_i, φ_i). A', B', C', D', E', and F' are the installation positions of the laser displacement sensor, and A, B, C, D, E, and F are the laser points that the laser displacement sensor hits on the wheel belt. Taking the geometrical relation of the laser beam between $|AO_1|$ and $|FO_2|$ as an example, the coordinates of two points A and F are solved. The actual distance between A and A' is expressed by |AA'|, and the projected distance on the z-axis is expressed by $|AA'|_P$. Other laser points are similar.

Let the unit vectors of the linear equation of the space where the laser beam of A' and F' laser displacement sensor is located by:

$$\vec{v_1}=(0,\ y_1,\ z_1)\quad |v_1|=1 \tag{1}$$

$$\vec{v_6}=(0,\ y_6,\ z_6)\ |v_6|=1 \tag{2}$$

If the Angle between v_1, v_6 and Z axis is φ_1, φ_6, then there is:

$$\cos\varphi_1=\frac{v_1}{|v_1|} \tag{3}$$

$$\cos\varphi_6=\frac{v_6}{|v_6|} \tag{4}$$

Existence relationship:

$$|AO_1|=|AA'|+r_1 \tag{5}$$

$$|FO_2|=|FF'|+r_2 \tag{6}$$

In the equation, r_1 and r_2 are the distances between the laser displacement sensor A' and F' and the coordinate system origin O_1 and O_2, respectively, and their sizes are determined by the disk radius.

$\left|AA'\right|_P$, $\left|FF'\right|_P$ and are projections of $\left|AA'\right|$ and $\left|FF'\right|$ in the direction of the laser beam on the z-axis:

$$|AA'|_P = l_1\cos\varphi_1 \tag{7}$$

$$|FF'|_P = l_6\cos\varphi_6 \tag{8}$$

The solution process of laser Points A and F coordinates on the surface of the wheel belt is as follows:

$$y_A=|AA'+r_1|\sin\varphi_1 = |l_1+r_1|\sin\varphi_1=l_1\sin\varphi_1+r_1\sin\varphi_1 \tag{9}$$

$$z_A=|AA'+r_1|\cos\varphi_1=|l_1+r_1|\cos\varphi_1=l_1\cos\varphi_1+r_1\cos\varphi_1 \tag{10}$$

$$y_F=|FF'+r_2|\sin\varphi_6=|l_6+r_2|\sin\varphi_6=l_6\sin\varphi_6+r_2\sin\varphi_6 \tag{11}$$

$$z_F=|FF'+r_2|\cos\varphi_6=|l_6+r_2|\cos\varphi_6=l_6\cos\varphi_6+r_2\cos\varphi_6 \tag{12}$$

The coordinates of the two laser points are:

$$\text{A}\ (l_1\sin\varphi_1+r_1\sin\varphi_1,\ l_1\cos\varphi_1+r_1\cos\varphi_1)$$

$$\text{F}\ (l_6\sin\varphi_6+r_2\sin\varphi_6,\ l_6\cos\varphi_6+r_2\cos\varphi_6)$$

where l_1 and l_6 are the measuring distances of the laser displacement sensor A' and F' respectively.

3. Simulation experiment based on V-rep

3.1. Solution of laser point position on the surface of rotary kiln wheel belt

According to the design concept of a rotary kiln, a simple model of a rotary kiln wheel belt is built into the dynamic simulation software V-rep. As shown in Figure 2, a circle of D=6000 mm is first constructed, and the pose of the laser displacement sensor is fixed, for example, the attitude of Sensor 1 is (0, 60°, 0), and the position is (-5000, 0, 0), and then the distance of six points on the wheel belt is measured by the laser displacement sensor. The distance measured by the laser displacement Sensors 1-6 is respectively the distance of the measuring Points A-F on the surface of the wheel belt (clockwise rotation).

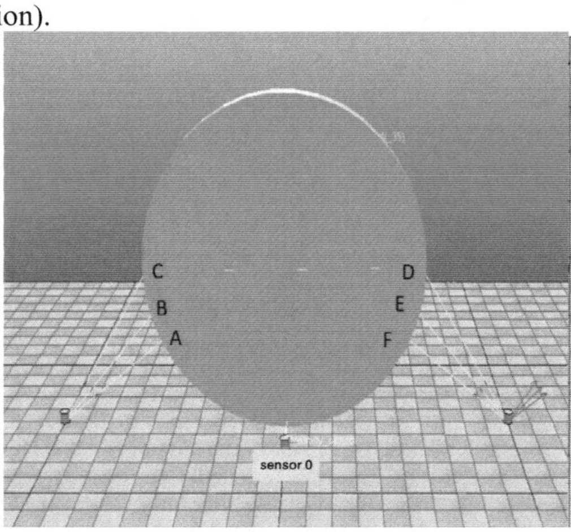

Figure 2. Laser spot simulation system on the wheel surface.

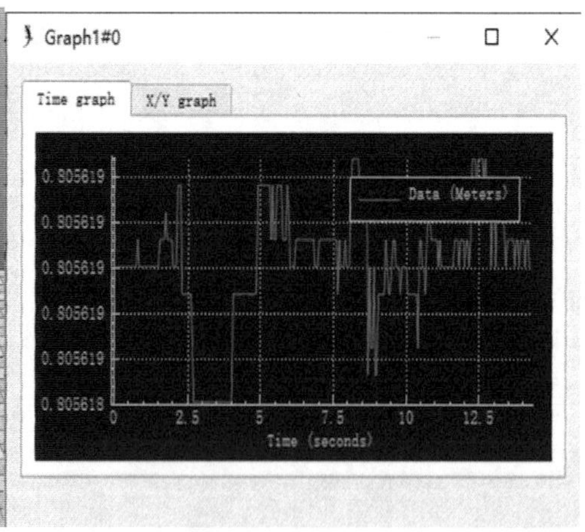

Figure 3. Sensor 0 measurement reading.

As shown in Figure 3, the thickness of the wheel belt model is known. The actual distance between the laser displacement Sensor 0 and the surface of the wheel belt is 800 mm, and the distance measured by the laser displacement Sensor 0 is 805 mm, so the measurement error of the sensor is 5 mm. When the posture of laser displacement Sensors 1-3 is (0, 70°, 0), (0, 60°, 0), and (0, 50°, 0), the measured Distances a-f of laser Sensors 1-6 are 2872 mm, 2846 mm, 3047 mm, 3052 mm, 2852 mm, and 2877 mm, respectively.

Multiple measurements can obtain the actual distance of laser displacement Sensors 1-6:

$$l_i' = l_i - 5 \quad (i=1, 2, 3, 4, 5, 6) \tag{13}$$

3.2. The center position of the rotary kiln wheel belt is solved

According to the distance of the laser displacement sensor obtained in V-REP simulation software and the pose of the laser displacement sensor, the coordinates of 6 laser points on the surface of the wheel are calculated. To simplify the calculation, the position of laser displacement Sensor 1 is taken as the origin of the coordinate system. Since the position of the three laser displacement sensors on the left side can be fixed at the same point in the simulation software, the position of the three laser displacement sensors on the left side is the same, but the posture is different, and the three laser displacement sensors on the right side are symmetrically arranged. If the intermediate distance is L, the coordinates of the 6 laser points on the surface of the wheel belt can be obtained as follows:

$$(y_i,\ z_i) = (l_i' \sin\varphi_i, l_i' \cos\varphi_i) \quad (i=1, 2, 3) \tag{14}$$

$$(y_i,\ z_i) = (L - l_i' \sin\varphi_i, l_i' \cos\varphi_i) \quad (i=4, 5, 6) \tag{15}$$

When the ellipticity of the wheel is equal to 0, the measuring distance of the laser displacement Sensors 1-6 is 2872 mm, 2846 mm, 3047 mm, 3052 mm, 2851 mm, and 2877 mm, respectively. The posture of laser displacement Sensors 1-6 is (0, 70°, 0), (0, 60°, 0), (0, 50°, 0), (0, -50°, 0), (0, -60°, 0), (0, -70°, 0), respectively. When L=10000 mm, the coordinates of the six laser points on the surface of the wheel are A (2763.0, 1005.8), B (2226.7, 1868.0), C (2054.0, 2448.0), D (7946.0, 2448.0), E (7773.3, 1868.0), (7237.0, 1005.8).

3.3. Relationship between wheel ellipticity and the center position of the wheel

In the V-rep dynamic simulation software, the size of the wheel belt was changed, and the model of the rotary kiln wheel belt with an ellipticity of 0.16%, 0.33%, 0.5%, 0.66%, and 0.83% was constructed. We keep the position of the laser displacement sensor unchanged and obtain the center position of the belt according to the above theory. The coordinates of the belt ellipticity and the center position of the belt are shown in Table 1.

Table 1 Location of wheel ellipticity and fitting center.

option	Ovality (%)	Central coordinate/mm	Y-coordinate deviation/mm	Central coordinate/mm	Y-coordinate deviation/mm	Belt radius/mm
1	0	5000.0	0	3000.4	0.4	2996.7
2	0.16	5000.1	0.1	2999.5	-0.9	2996.3
3	0.33	5000.2	0.2	2998.3	-2.1	2996.0
4	0.50	4999.9	-0.1	2996.7	-3.7	2995.5
5	0.66	5000.0	0	2995.3	-5.1	2994.9
6	0.83	4999.7	-0.3	2992.9	-7.5	2991.0

The analysis shows that when the long axis of the wheel is in the Y direction, the influence of the ellipticity of the wheel on the Y coordinate of the wheel center can be ignored, but the influence on the Z coordinate of the wheel center is greater. Taking the wheel ellipticity as 0, the Z-coordinate value of the wheel center decreases with the increase of the wheel ellipticity. When the wheel ellipticity is 0.83%, the Z-coordinate value of the wheel center decreases by 7.5 mm. It can be seen that when we use the coordinates of measuring points on the surface of the wheel to calculate the center position of the wheel, the influence of the ellipticity of the wheel should be taken into account. At the same time, how to reduce the error caused by the ellipticity of the belt to the measurement of the center of the belt is considered.

3.4. Fitting relationship between sensor mounting position and wheel center position

We often use belt ellipticity to describe the irregular deformation of the belt surface. It can be seen from the above that the belt ellipticity has a great influence on the calculation of the Z-coordinate of the center position of the belt, that is, the irregular deformation of the belt surface will affect the measurement of the center position of the belt. In the following, we will solve the influence of the sensor installation position on the fitting of the center position of the wheel after the irregular deformation of the wheel surface. Firstly, the wheel model with 0 ellipticity is measured, the distance parameters are obtained by the laser displacement sensor, and the coordinates of measuring points on the surface of the wheel are obtained by fitting together with the position and pose of the laser displacement sensor. Then the attitude of the laser displacement sensor is changed to obtain different measuring distances, and the center position of the wheel under different sensor installation attitudes is obtained. Finally, the relationship between the sensor installation position and the center of the wheel is calculated when the ellipticity of the wheel is 0.16%, 0.5%, 0.66%, and 0.80%.

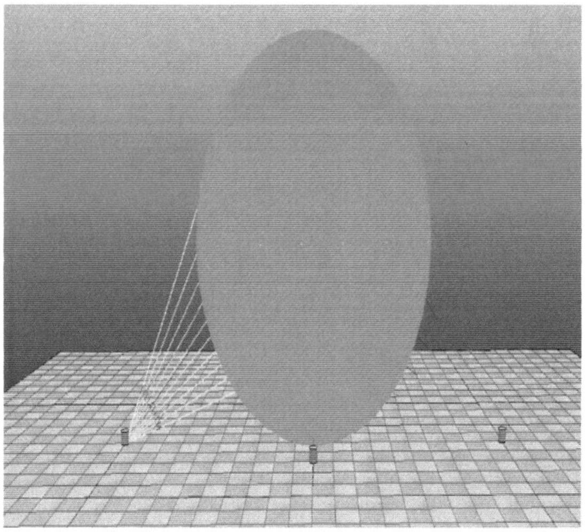

Figure 4. Sensor installation position division.

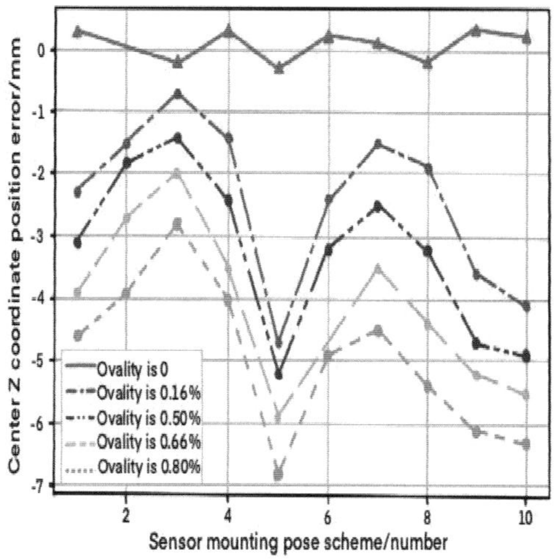

Figure 5. Installation position and center error.

When the ellipticity of the wheel belt is 0, we take the left side of the wheel belt as an example, and divide the wheel belt with 5° as the minimum scale, as shown in Figure 4. Firstly, a set of sensor installation positions are randomly selected for measurement and calculation. For example, the position of laser displacement Sensors 1-3 is (0, 0, 0), the attitude is (0, 55°, 0), (0, 45°, 0), (0, 35°, 0), and the laser displacement Sensors 4-6 is arranged symmetrically. The coordinates of the center position of the wheel under the sensor installation posture are calculated. Then we change the attitude of the laser displacement sensors to (0, 70°, 0), (0, 65°, 0), (0, 50°, 0), calculate the coordinates of the center position of the wheel after changing the attitude, and so on to obtain the center position of the laser displacement sensor between 30°-75°. The relationship between the pose of the laser displacement sensor and the center position of the wheel is shown in Table 2.

When the ellipticity of the wheel is 0, the pose of two laser displacement sensors is fixed, and only the pose of the third laser displacement sensor is changed, the center position and radius of the wheel are basically unchanged, and the obtained center position of the wheel is O (5000, 3000), R=2997. Therefore, when the wheel belt is a circle, the installation position of the sensor has little influence on the fitting result of the center of the circle, that is, when the wheel belt is not deformed, the positioning position of the laser displacement sensors at the measuring point on the surface of the wheel belt can be arbitrarily arranged.

When the ellipticity of the wheel belt is 0.16%, the installation position of the laser displacement sensor is changed to solve the position of the center of the wheel belt. The simulation data are shown in

Table 2. The data of program 1-2-3-4 shows that the pose of laser displacement Sensors 1 and 2 is fixed, and only the attitude of laser displacement Sensor 3 is changed, and the center coordinates and radius of the wheel are calculated at this time. If the attitude of laser displacement Sensor 3 is (0, 40°, 0), the center position and radius of the wheel are closest to the actual parameters. That is, this is the best position of laser displacement Sensor 3. The data of program 3-5-6 shows that the pose of laser displacement Sensors 2 and 3 is fixed, only the pose of laser displacement Sensor 1 is changed, and the center coordinates of the wheel and the radius of the wheel are solved. It can be obtained that when the attitude of laser displacement Sensor 1 is (0, 70°, 0), the center position of the wheel and the radius of the wheel are closest to the actual parameters, that is, the optimal pose of laser displacement Sensor 1 is obtained. The data of program 7-8-9-10 shows that the pose of laser displacement Sensors 1 and 3 is fixed, and only the pose of laser displacement Sensor 2 is changed, and the center coordinates and radius of the wheel are solved at this time. It can be obtained that when the attitude of laser displacement Sensor 2 is (0, 50°, 0), the center position and radius of the wheel are closest to the actual parameters. That is, this is the best position of laser displacement Sensor 2.

Table 2. Sensor attitude and fitting center position when the ellipse is 0.16%.

option	Sensor 1 /°	Sensor 2/°	Sensor 3/°	Central coordinate/mm	deviation /mm	Central coordinate/mm	deviation /mm	Wheel radius/mm
1	70	50	30	5000.2	0.2	2997.7	-2.3	2994.2
2	70	50	35	5000.0	0	2998.5	-1.5	2994.8
3	**70**	**50**	**40**	**5000.0**	**0**	**2999.3**	**-0.7**	**2996.3**
4	70	50	45	5000.0	0	2998.2	-1.8	2999.1
5	75	50	40	4999.9	0.1	2995.3	-4.7	2993.5
6	65	50	40	5000.0	0	2997.6	-2.4	2991.3
7	70	45	40	4999.7	0.3	2998.5	-1.5	2999.7
8	70	55	40	4999.3	0.7	2998.1	-1.9	2999.1
9	70	60	40	5000.0	0.2	2996.3	-3.7	2997.1
10	70	65	40	5000.0	0	2995.9	-4.1	2996.3

3.5. Comparison of sensor installation position and posture under five ellipses of wheel belt

According to the analysis of Table 2, when the long axis of the belt is in the Y direction, the Y coordinate of the central position of the belt is not greatly affected by the ellipticity of the belt, and the ellipticity of the belt mainly affects the Z coordinate of the central position of the belt. Therefore, the relationship between the sensor installation position and the Z coordinate of the central position of the belt will be studied in the following paper. Taking the 10 groups of sensor installation pose schemes in Table 3 as the research object, when the wheel ellipticity is 0, 0.16%, 0.5%, 0.66%, and 0.80%, respectively, the relationship between the fitting center position of the wheel and the actual center position error of the wheel is calculated in Figure 5.

The analysis of Figure 5 shows that when the ellipticity of the wheel belt is 0, 0.16%, 0.50%, 0.66%, and 0.80% respectively, the fitted Z-coordinate of the center of the wheel belt gradually decreases. At the same time, comparing the preset 10 groups of laser displacement sensor installation attitude, when the installation attitude of the laser displacement sensor is scheme 5, the Z coordinate error of the wheel center is the largest, and when the installation attitude of the laser displacement sensor is scheme 3, the Z coordinate of the wheel center is the closest to the actual wheel center.

It can be concluded that when the long axis of the wheel belt is horizontal and the center of the rotary kiln wheel belt is measured and calculated, regardless of the ellipticity of the wheel belt, the installation position of the sensor is arranged according to the state of the third scheme, and the error of the center of the wheel belt is minimal.

Journal of Physics: Conference Series **2815** (2024) 012014 doi:10.1088/1742-6596/2815/1/012014

Taking the wheel ellipticity of 0.16% as an example, when the installation position of laser displacement Sensors 1-6 is (0, 0, 0), and the attitude is (0, 70°, 0), (0, 50°, 0), (0, -40°, 0), (0, -50°, 0), (0, -70°, 0). At this time, the Z coordinate error of the center of the wheel belt is 0.73 mm.

4. Verification experiment based on a simple small model

To verify the effectiveness of the simulation model, the pose-solving experiment based on a laser displacement sensor was carried out. The experimental platform is shown in Figures 6 and 7. The sensor uses a SW-LDS50A laser displacement sensor. The maximum distance is 50 m, the measurement frequency is 2 HZ, the measurement accuracy is ± (2 mm+d1/10000), the cylinder wheel belt used for measurement is a=600 mm, the ellipse is 0.16%, and the diameter of the support wheel is 100 mm.

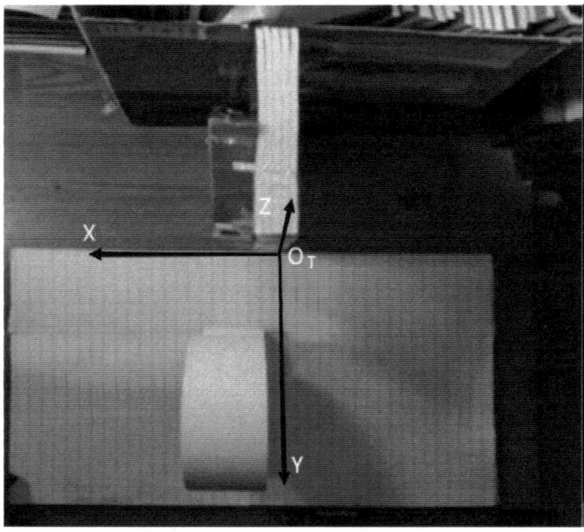

Figure 6. Verification platform.

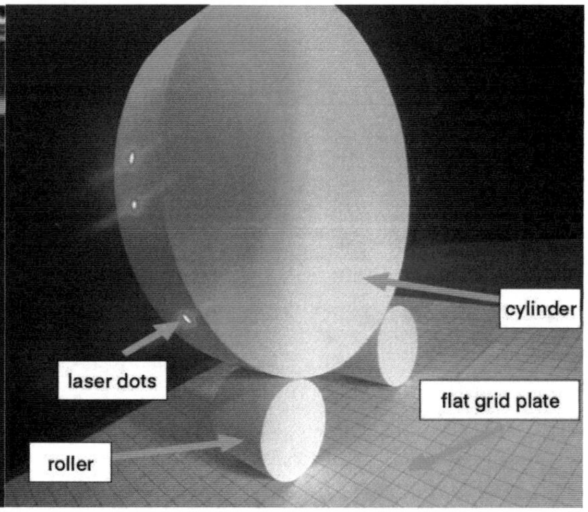

Figure 7. Communication principle scheme.

4.1. Verification procedure

Step 1: Install the sensor and establish the tool coordinate system.

The sensor, sensor mounting frame, plane grid plate, cylinder, and wheel support are installed and fixed. The mounting frame is installed perpendicular to the plane grid plate, and the tool coordinate system {T} is established. OT is the origin of the tool coordinate system.

Step 2: Initial sensor position measurement.

The laser displacement sensor is fixed on the sensor mounting rack, and the position of the laser displacement sensor is fixed according to the position of scheme 1 in the simulation above. The position of the three laser displacement sensors is recorded, and the current position of the laser displacement sensor is recorded as state 1.

Step 3: Obtain the laser point distance on the surface of the cylinder.

The laser displacement sensor, power switch, serial port assistant, etc. are connected according to the regulations. The distances l1, l2, and l3 from the laser displacement sensor to the surface of the cylinder were measured. The displacement data acquisition code of the laser displacement sensor is written in Python, and the distance measured when the posture of the laser displacement Sensors 1-3 is (0, 70°, 0), (0, 60°, 0), and (0, 50°, 0), **respectively.**

Step 4: Calculation of laser point coordinates on the surface of the cylinder

The coordinates A (y1, z1), B (y2, z2), and C (y3, z3) of the laser points on the surface of the three cylinders in the tool coordinate system {T} under state 1 is calculated according to the measured distance and the position information of the sensor itself. Since the coordinates of the 3 laser points on the right side of the cylinder and the 3 laser points on the left side are symmetrically distributed, the distance between the left and right symmetric points can be measured with the help of a pencil and a ruler. We calculate the cylinder on the right side of the three coordinates of the laser spot on the tool coordinate system respectively D (y1 + L1, z1), E (y2 + L1 ', z2), F (y3 + L1 '', z3).

Step 5: Fit the center coordinates of the cylinder.

According to the circle center fitting method prepared by Labview, the coordinates of 6 laser points on the surface of the wheel are substituted into the fitting method, and the coordinates of the center position of the cylinder are obtained. Record the cylinder center coordinates Q1 (Y1, Z1) at this state 1.

Step 6: Adjust the sensor installation attitude.

According to the installation posture of ten groups of sensors obtained by simulation analysis, the installation posture of three laser displacement sensors is changed, respectively recorded as state i. Repeating Step 1 to Step 5, we calculate the coordinates of each laser point in state 1-i, and obtain the fitting center Qi (Yi, Zi) of the cylinder respectively.

Step 7: Data recording and analysis recorded.

The installation position of the laser displacement sensor under the state 1-i, and the center coordinates of the cylinder fitted by the laser point coordinates under the state 1-i, we organize them into a table like Table 3. The experimental results are analyzed, the center size of the cylinder under different positions of the laser displacement sensor is compared, and the most suitable state of the laser displacement sensor is obtained.

4.2. Experimental conclusion and analysis

The center position coordinates of the wheel belt obtained by the simulation are reduced by 10 times and compared with the experimental results. The verification data are shown in Table 3. Taking the wheel simulation data with an ovality of 0.16% as the object, it can be seen that among the 10 groups of experiments, the fitting error of the wheel center obtained under the third group sensor installation attitude scheme is the smallest. The Y direction error of the center position of the wheel is 1.2 mm, and the Z direction error of the center position of the wheel is 1.5 mm. The feasibility of the measurement method and the accuracy of the simulation results are verified.

Table 3. Verification data.

option	Sensor 1 /°	Sensor 2/°	Sensor 3/°	Simulation center/mm	Experimental/mm	error/ mm	Simulation center/mm	Experimental/mm	error/ mm
1	70	50	30	500.0	498.0	-2.0	299.8	297.0	-2.8
2	70	50	35	500.0	498.3	-1.7	299.8	297.4	-2.4
3	**70**	**50**	**40**	**500.0**	**498.8**	**-1.2**	**299.9**	**298.4**	**-1.5**
4	70	50	45	500.0	501.9	1.9	299.8	297.2	-2.6
5	75	50	40	499.9	497.3	-2.7	299.5	295.7	-3.8
6	65	50	40	500.0	501.7	1.7	299.8	297.1	-2.7
7	70	45	40	499.9	498.5	-1.5	299.8	297.9	-1.9
8	70	55	40	499.9	498.4	-1.6	299.8	297.5	-2.3
9	70	60	40	500.0	498.2	-1.8	299.6	296.8	-2.8
10	70	65	40	500.0	502.3	2.3	299.6	296.3	-3.3

5. Conclusion and summary

In the measurement of the rotary kiln belt center, this paper proposes and implements an optimization method of sensor installation position based on dynamic simulation analysis and a simple small model, which improves the fitting accuracy of the rotary kiln belt center position. This paper analyzes the influence of wheel ellipticity on the fitting of wheel center position. It is proved that when the long axis of the belt is Y direction, the Z-coordinate of the central position of the belt will gradually decrease with the increase of the ellipticity of the belt.

At the same time, through simulation analysis, it is obtained that the fitting error of the wheel center is the smallest when the installation attitude of the three laser displacement sensors is (0, 70°, 0), (0, 50°, 0), and (0, 40°, 0). Compared with the actual wheel center, the error of the Y coordinate of the wheel center is 0.4 mm, and the error of the Z coordinate of the wheel center is 0.73 mm.

Finally, a simple small model is used to verify the fitting error of the wheel center when the position and pose of the laser displacement sensor are arranged according to (0, 70°, 0), (0, 50°, 0), (0, 40°, 0), and the solution accuracy of the simulation analysis is less than 1.5 mm. In this paper, the accuracy of solving the belt center is optimized from about 2 mm to less than 1 mm, which improves the fitting accuracy of the belt center of the rotary kiln and the calculation accuracy of the central axis of the rotary kiln and has certain reference value for the research of the operating state of the rotary kiln.

References

[1] Jiang Xuchang. Adjustment of Rotary kiln support wheel [J]. New Century Cement Herald, 2012(09): 49-51+13.

[2] He Xinyong. Related Mechanism of adjusting Rotary kiln support wheel affecting Rotary kiln axis [J]. China Equipment Engineering, 2017(15):114-115.

[3] Xiao-hui Fan, Yi Wang. Mathematical models and expert system for grate-kiln process of iron ore oxide pellet production. Part II: Rotary kiln process control [J]. Journal of South University, 2012(03): 1724-1727.

[4] Jinquan Z, Liang Z, Wenli D. Hybrid model of a cement rotary kiln using an improved attention-based recurrent neural network [J]. ISA transactions, 2022, 129(PB).

[5]. Machine Learning - Intelligent Systems; Findings from University M'Hamed Bougara of Boumerdes Provides New Data about Intelligent Systems (New Reduced Kernel Pca for Fault Detection and Diagnosis in Cement Rotary Kiln) [J]. Information Technology Newsweekly, 2020.

[6] KANG Yihao. Ovality measurement system of rotary kiln based on laser sensor [D]. Wuhan Institute of Technology, 2023.

[7] CAI Changqing. Research on ovality measurement method of Rotary Kiln [D]. Wuhan University of Technology, 2006.

[8] Zhang Yun, Liu Bicheng. Dynamic detection method for the center line of Rotary kiln [J]. Bulk Cement, 2021, (02): 126-128.

[9] LIU Shengchi. Axis measurement and analysis of dynamic rotary kiln based on total station three-point method [D]. Wuhan University of Technology, 2011.

[10] XIE Chan. Feasibility study on online monitoring System of rotary kiln axis [D]. Wuhan University of Technology,2010. [11] O. L I, Akilu Y. Knowledge Criticality Assessment and Codification Framework for Major Maintenance Activities: A Case Study of Cement Rotary Kiln Plant [J]. Sustainability, 2021,13(9).

[11] Weihua W, Wei P, Liquan D, et al. Design and Mechanical Behavior Analysis of Two-Stall Cement Rotary Kiln Cylinder [J]. International Journal of Performability Engineering, 2020, 16(6).

[12] Qcdr, Pas, Fec, et al. Dynamic Analysis of the Temperature and the Concentration Profiles of an Industrial Rotary Kiln Used in Clinker Production [J]. Anais da Academia Brasileira de Ciencias, 2017, 89(4).

[13] Ge Lian-Zheng, Zhao Lijun. Robot Simulation Experiment System and Teaching based on V-REP [J]. Experimental Technology and Management, 2021, 38(03).

[14] Yuan Kangzheng, Zhu Weidong, Chen Lei, et al. Installation Position Calibration Method of Robot End Displacement Sensor [J]. Journal of Zhejiang University (Engineering and Technology), 2015, 49(05): 829-834.

Journal of Physics: Conference Series **2815** (2024) 012015 doi:10.1088/1742-6596/2815/1/012015

Lightweight design and additive manufacturing of one-piece rotating arm axle box for high-speed locomotives

Fuyuan Zhao [1], Xinlin Wang [1,*] and Tianmin Guan [1]

[1] Dalian Jiaotong University, Dalian, Liaoning, 116028, China

[*] Corresponding author's e-mail: fy2452911865@163.com

Abstract: Against the background of the continuous increase in the operating speed of the rolling stock, the working environment of the axlebox is becoming more and more severe. To meet the increasing operational demand, the strength and quality requirements of the axle box are also increased accordingly. To realize lightweight design to ensure structural strength, the traditional casting process has the problems of low densities of molded parts, long processing cycles, etc. In contrast, the laser selective melting technology can mold the workpiece with excellent mechanical properties and realize complex structures, effectively solving the above problems. In this paper, the one-piece rotary arm axle box is selected as the research object, its extraordinary load conditions are calculated, and the feasibility of the SLM axle box is verified by hydrostatic simulation. On this basis, the topology optimization design of the axle box is further optimized, and the optimized structural strength and vibration performance are verified by static and modal simulation. SLM technology is used to print components that are ten times smaller in size, and based on the compression experiments, it is verified that its strength meets the design requirements after printing, and the mass of the optimized components is reduced by about 15.4%.

1. Introduction

As one of the core components of the high-speed bogie, the axle box bears the important responsibility of connecting the bogie wheelset with the frame, transmitting the dynamic and static loads of a series of springs, as well as the longitudinal traction and braking force of the wheel and axle. To ensure the safety and reliability of locomotive operation, the axle box should have sufficient strength, superior vibration performance, and a certain service life. Currently, most axleboxes are manufactured by sand casting, and the casting material is SC450 cast steel [1]. In the train lightweight design goals, the twentieth-century eighties foreign scholars found that aluminum alloy material can greatly reduce the weight of the axle box [2]. At the same time, the modulus of elasticity of aluminum alloy material is low and can attenuate 20-30 Hz high-frequency vibration [3-4]. Therefore, foreign countries are currently using cast aluminum alloy axle boxes to replace cast steel axle boxes. Still, the strength of cast aluminum alloy is weaker than that of cast steel. While casting is a versatile process that is not limited by size or weight restrictions, it can lead to defects such as porosity, inclusions, and shrinkage holes. These defects result in a lack of density in the finished product, leading to stress concentration and cracking. As a result, the service life of the axle box is reduced, which increases the cost of operation and maintenance [5]. Therefore, many scholars have tried to change the axle box from casting to forging, which can increase the density of the axle box and improve its strength. However, forging still produces forging defects such as cracks, and the processing cycle is still long [6]. Among many new molding techniques, selective zone laser melting technology can greatly reduce the design and molding cycle time.

Nowadays, selective laser melting (SLM), or metal 3D printing, is a processing technology that uses a fine-focused laser spot to rapidly melt pre-positioned metal powders to directly obtain arbitrarily shaped parts, which has the advantages of a short processing cycle, small structural constraints on the part, and no need for molds. Due to the lightweight, high performance, and low cost of SLM aluminum alloy, it has a wide range of applications in aerospace, especially for complex lightweight structures and structural parts with shaped curves [7]. Typical structures such as aluminum alloy supports, cabin segments, frame beams, and meshes are used [8]. Compared with the aerospace field, there are fewer cases of SLM application in rail transportation, most of which are applied to non-load-bearing parts of rail transportation [9]. Ltd. through the SLM manufacturing integrated high-voltage ground switch transmission parts so that its transmission rod and transmission disk combined strength and mechanical properties greatly enhanced [10], the German Federation used SLM technology to manufacture a better anti-vibration performance and wear-resistant performance of the bearing cover [11]. However, there are few studies on SLM load-bearing parts in the track, and the topology-optimized anti-snake shock absorber seat of the China Vehicle Industry Research Institute has achieved a weight reduction of 72% compared with the traditional workpiece [12]. Yang et al. [13] investigated the effect of laser power on residual stress and residual deformation of axle boxes processed by SLM through additive simulation of axle boxes. Existing research mainly focuses on the traditional process of axle box manufacturing, and more research is needed on the additive process of manufacturing axle boxes. Thus, it is not possible to ensure the strength of the box to further lightweight design.

In this paper, for the goal of a lightweight axle box, the axle box topology optimization design of the high-speed rolling stock, comparative analysis of the static mechanics, and modal analysis of the axle box structure before and after the optimization to get the structural scheme are in line with the needs. After the test piece SLM manufacturing molding, the compression performance of the molded part is studied and analyzed to verify the feasibility of the molding scheme, which provides a reference for the subsequent axle box SLM manufacturing research.

2. Finite element modeling and load application

2.1. Axlebox finite element modeling

The rotary arm axlebox consists of a rotary arm beam, a spring seat, a shock absorber seat, bearing holes, locating holes, etc. The external dimensions of the model are 822×210×374 mm. To improve the mesh delineation accuracy and the solution rate, the model of the axle box is simplified before mesh delineation, and the chamfering angle, small holes, and other small features that do not affect the accuracy of the solution are removed. A tetrahedral mesh with a size of 10 mm is used. The model is divided into 7769 cells and 13516 nodes, and the structural discrete model is shown in Figure 1.

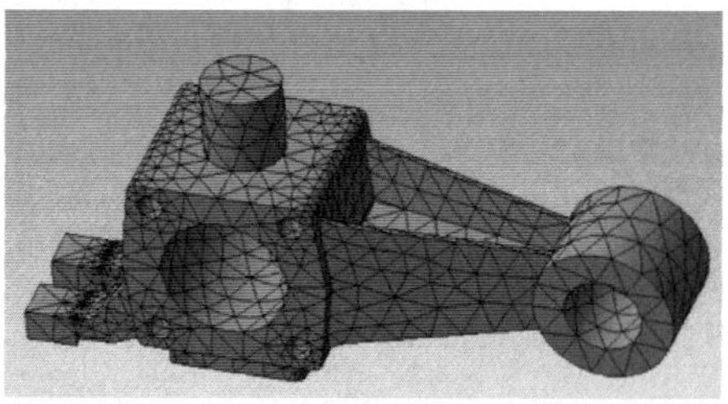

Figure 1. Discrete model of the axle box.

AlSi10Mg has a smaller density, good air tightness, and excellent mechanical, physical, and corrosion resistance properties compared to cast steel materials. Its density is 2.68 g/cm^3, modulus of elasticity is 76.6 Ga, Poisson's ratio is 0.33 [14], and its chemical composition is shown in Table 1:

Table 1. Chemical composition of AlSi10Mg.

Element	Si	Mg	Cu	Mn	Ni	Fe	Ti	C	Al
wt%	≤9-10	≤0.45	≤0.05	≤0.45	≤0.45	≤0.45	≤0.45	≤0.45	Bal.

2.2. Calculation of axle box load conditions

According to EN 13749, the axle box is subjected to vertical loads transmitted by a series of springs and vertical dampers, as well as transverse and longitudinal loads at the positioning of the swivel arm. Extraordinary load is the theoretical maximum load carried by the axle box in train operation, and the following is the calculation of the extraordinary load condition of the axle box:

(1) Vertical overloads

$$F_{\mathrm{Zmax}} = \frac{1.4\mathrm{g}(M - m_{\mathrm{t}})}{2} \quad (1)$$

where F_{Zmax} is the vertical overloads on the axle box; m_{t} is the unsprung mass; g is the acceleration of gravity generally taken as 9.8 m/s^2.

(2) Transverse overloads

$$F_{y\mathrm{max}} = F_{ytr\,\mathrm{min}} + \frac{Mg}{3} \quad (2)$$

where F_{ymax} is the transverse overloads; F_{ytrmin} is a constant value parameter.

(3) Longitudinal overload

$$F_{x\mathrm{max}} = \frac{3gm_{\mathrm{b}}}{2} \quad (3)$$

where F_{xmax} is the longitudinal overload; m_{b} is the weight of the bogie.

(4) Dampers vertical overload

$$F_{\mathrm{jmax}} = 1.5F_j \quad (4)$$

where F_{jmax} is the Dampers vertical overload; F_{j} is the saturation force of the damper.

Set the extraordinary load of the axle box into the ANSYS software and add Z-direction constraints at the axle box bearing mounts and fixed constraints at the positioning points of the swivel arm.

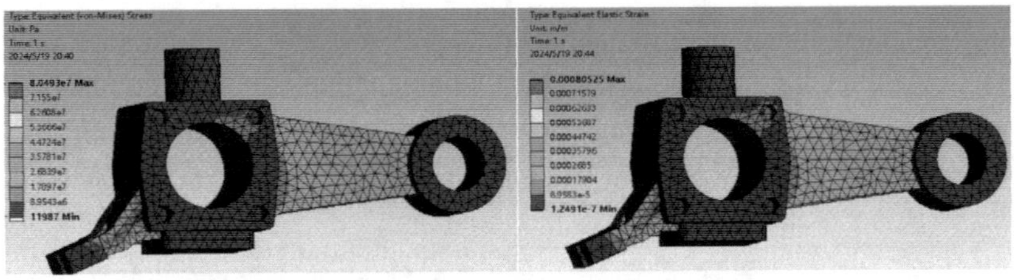

Figure 2. Comparison of stress-strain diagram of SC450 axle box.

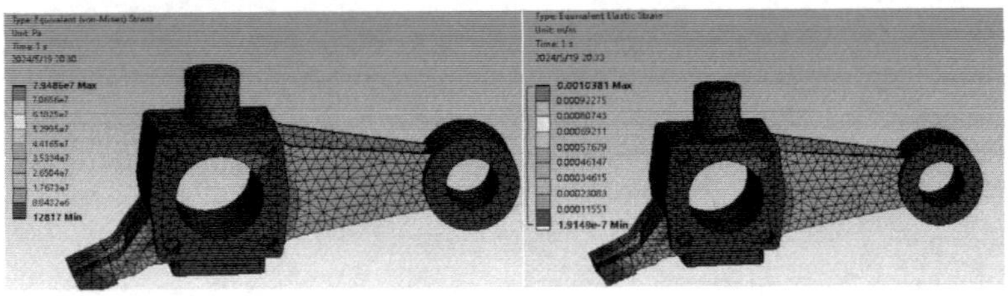

Figure 3. Comparison of stress-strain diagram of AlSi10Mg axle box.

Using ANSYS software, the axle box was given two materials, SC450 cast steel and AlSi10Mg, to solve the problem. As shown in Figure 2 and Figure 3. The equivalent stresses of the cast steel and AlSi10Mg axle boxes are less than the yield strength of the materials, the maximum equivalent stresses and the maximum displacement deformations occur in the lower part of the shock absorber seat connection, and the stresses and deformations in other parts of the AlSi10Mg axle boxes are small. The strength margin of the cast steel axle boxes is smaller than that of the AlSi10Mg axle boxes. Considering the strength, safety, and reliability of the axle box, the AlSi10Mg axle box is better than the cast steel axle box, and its safety coefficient is 3.2, so it has a larger optimization space. The mass of the cast steel axlebox is 106.17 kg, while the mass of the AlSi10Mg axlebox is only 36.344 kg, which is a 65.77% reduction in the mass of the axlebox. The resulting static strength analysis lays the foundation for the subsequent topology optimization of the axle box.

3. Optimized design of axle box topology

Based on the load conditions and design requirements of the axle box model, it is crucial to ensure the basic functionality of the axle box. However, the area where the axle box and bearing contact, as well as the positioning holes, are located is not designed. Also, the area where the spring seat connects to the two series of springs is not designed. These non-designed areas have been taken into consideration during the design process. The design area will be specified as the arm beam and the connecting surface of the arm and the axle box. The mass, volume, and stress responses were established in Ansys, and the equivalent stresses from the static analysis were chosen as the objective function for optimization. At the same time, the relative density of each unit of the axle box was used as the design variable. The whole model was divided into 15480 tetrahedral cells. The mass retention percentage of 76% is defined as the constraint, and the retention threshold is set to 0.43. After 41 iterations, the unit density of 0 appears in the rotating arm of the axle box and the connection of the main body of the axle box rotating arm. According to the optimized density cloud of the axle box, combined with the structural characteristics of the axle box and the actual assembly requirements, the axle box is re-modeled by finite element modeling. As shown in Figure 4. The main changes are to remove part of the material at the crossbeam of the arm and to reduce the wall thickness by 35 mm at the connection between the main body of the axle box and the arm.

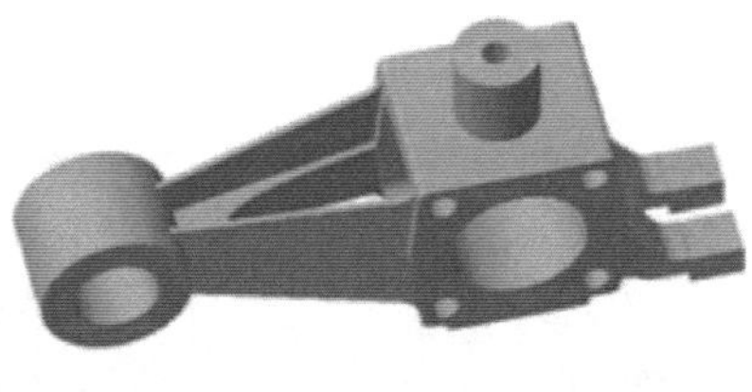

Figure 4. Axlebox Reconstruction Model

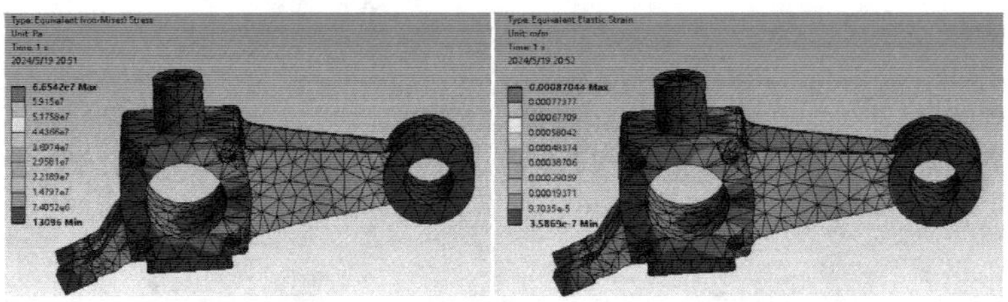

Figure 5. Stress-strain cloud of topology optimized

As shown in Figure 5. The reconstructed model of the axle box is re-conducted static analysis; after optimization, the location of the maximum stress and maximum strain of the axle box is still in the spring seat connection; the maximum stress is 66.5 MPa, the maximum strain is 0.0008704. 13

MPa reduces the maximum stress compared with the pre-optimization axle box, the maximum stress is not more than the maximum stress of the cast steel axle box, the deformation and the cast steel axle box are approximately the same, and the optimized axle box meets the requirements of strength and deformation. The strength and deformation of the optimized axle box meet the requirements. The volume is reduced by 17% compared with the original model, and the optimized axle box is reduced by about 6.24 kg compared with the pre-optimized aluminum alloy axle box and 76.07 kg compared with the cast axle box. After the above analysis, it can be concluded that the mass of the axle box after topology optimization is reduced, and the static characteristics of the axle box can be satisfied at the same time.

4. Experimental verification of compressive properties

After the optimization of the topology of the axle box, the structure of the axle box was analyzed statically. The stress and vibration characteristics of the axle box after the modal analysis were within the safe range of the design, and then the axle box SLM needs to be manufactured to test the optimized structure of the axle box to see if it meets the requirements.

4.1. SLM Axlebox

To reduce the research cost of the SLM axle box, the size of the box of the axle box is reduced by ten times the processing molding to explore the feasibility of the SLM axle box structure. Since the technology of AlSi10Mg material is more mature, the processing parameters with higher mechanical properties are chosen. The processing equipment is AM250, the substrate material is Al30, the laser diameter is 70 um, the laser power is 340 W, and the scanning speed is 1000 mm/s. The mass of the shaft box sample before optimization is 37.76 g, the mass of the optimized shaft box sample is 31.96 g, and the surface of the finished product after printing has a granularity and is gray. The mass reduction of the print after optimization was 15.4%, as shown in Figure 6.

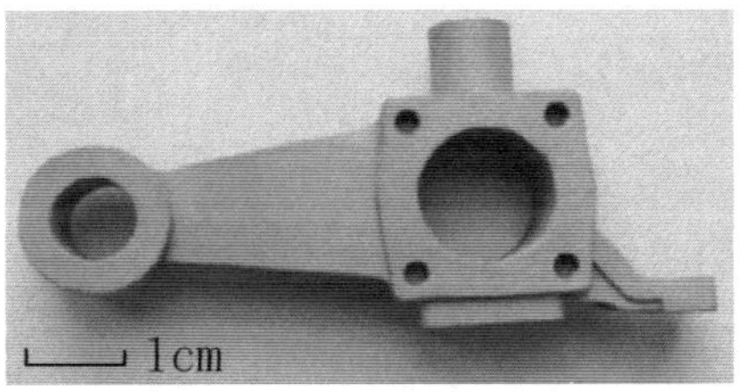

Figure 6. SLM Manufacturing Axlebox.

4.2. Axlebox compression test

Since the main force on the axle box is the vertical pressure at the axle box, one system spring, i.e., at the cylinder above the axle box. To verify whether the optimized axle box can withstand a large enough vertical pressure, the optimized axle box needs to be subjected to a compressive test. Since the spring fixed column above the axle box prevents loading, the upper part of the axle box needs to be cut flat with an EDM cutter and then loaded. The specimen is loaded on the universal mechanical testing machine table, the vertical load is applied, and the displacement speed is set to 5 mm/min until the sample is crushed. As shown in Figure 7

Figure 7. Axlebox compression test.

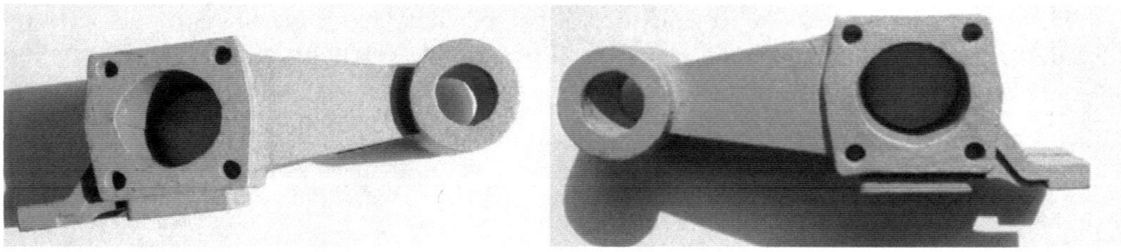

Figure 8. Optimization of front and rear axle box breaks.

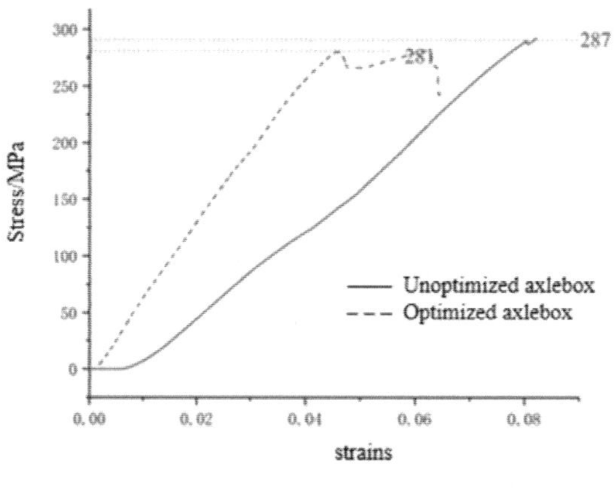

Figure 9. Compressive curve of the specimen.

As can be seen from Figure 8, the fracture of the axle box after the experiment shows that the fracture of the pre-optimization axle box is at the connection between the arm and the box, while the fracture of the post-optimization axle box is at the middle of the box, which is because the post-optimization axle box has been processed with a thin-wall treatment here, to transfer the fracture of the axle box; however, the bearing holes of the main body of the pre-optimization axle box have been deformed before the appearance of the fracture, and the bearings of the axle box in the practical application will be damaged by the compression and lead to the failure. As can be seen from Figure 9, the yield strengths of the SLM axle box specimens before and after optimization are 281 MPa and 287 MPa, respectively, which are higher than that of the cast steel axle box, which is 260 MPa. Through the stress calculations of the deflated axle boxes, the pre-optimized axle boxes can withstand a load of 500 kN. In comparison, the optimized axle boxes can withstand a load of 300 kN, much higher than that of the axle boxes in extraordinary working conditions, which is 107 kN. Both are much larger than the load of 107 kN of the axle box under abnormal working conditions.

5. Conclusion

In this paper, firstly, the forming process of the axle box is determined, its topology is optimized based on hydrostatics, and the hydrostatic and vibration modal analysis of the axle box structure before and after optimization is compared. The relevant experimental analysis of compressive performance is carried out after using SLM forming, and the following conclusions are obtained:

(1) AlSi10Mg material was selected to manufacture the axle box based on the SLM process. Under the supernormal load static mechanical conditions of the axle box, the maximum equivalent stresses of the AlSi10Mg axle box and the cast steel axle box are roughly equal. At the same time, the yield strength of the former is better than that of the latter, and the safety coefficient of the former reaches 3.4, which is a large space for further lightweight optimization.

(2) Based on the extraordinary loading condition of the axlebox, the mass of the axlebox is reduced by 17% through the optimization of the axlebox topology by the variable density method. Compared with the cast steel axle box, the optimized SLM axle box reduces the maximum stress by 14 MPa and the mass by 76.07 kg.

(3) To verify the compressive performance of the SLM axleboxes before and after optimization, the machining parameters were selected to be reduced by a factor of 10 to manufacture the SLM axleboxes. The mass of the formed shaft box sample is 37.76 g before optimization, and the mass of the optimized shaft box sample is 31.96 g, which is a 15.4% reduction in mass. The compression test verifies that the optimized structure meets the requirements of extraordinary loading conditions.

References

[1] Ágh, C. (2019) Comparative analysis of axlebox accelerations in correlation with track geometry irregularities. *Acta Technica Jaurinensis*, 12 2, pp 161-177.

[2] Sun, X., Han, X., Dong, C. and Li, X. (2021) Applications of aluminum alloys in rail transportation. *Advanced Aluminum Composites and Alloys*, 9, pp 251-268.

[3] Thring, J. F. (1954) The Design of Light Alloy Coaches for East African Railways. *Journal of the Institution of Locomotive Engineers*, 44 240, pp 495-540.

[4] Копытько, B.B. and Lu, G. (1993), Operational test of roller bearing housing components with aluminum alloy shaft housings. *Rolling stock technology*, 03, pp 15-18.

[5] Ren, X., Zhang, L., Chen, Y., Sun, F., Zhang, W., Lin, L. and Wang, Y. (2012). Tensile and fatigue properties of 7050 aluminum alloy axle box used for high speed train. *Procedia Engineering*, 27, pp. 914-922.

[6] Rong, L. (2008) Common defects and countermeasures in large forgings. *SME Management and Technology*, 20, pp. 222-223.

[7] Zhihong, J., Yanmei, L. and Xun, C. (2023) Process simulation and performance study of SLM compent with special curved surface of aviation aluminum alloy. *MW Metal Forming*, 06, pp. 23-27

[8] Chun, J., Chaoqi, Q. and Kai, Z. (2021) Additive manufacturing technology for large load-carrying component of aluminun alloy in aeronautics and astronautics. *Electric Welding Machine*, 51 08, pp.39-54

[9] Hongbin, Z. and Yu, L. (2019) Current research status of metal prototyping manufacturing (3D-printing) technology application in rail transit equipment, *Modern Urban Transit*, 10, pp. 77-81

[10] Mingming, M., Maizhi, T. and Dengxing, S. (2018) Fabrication of transmission part in High-voltage earthing switch by selective laser melting. *Electric Locomotives & Mass Transit Vehicles*, 41 01, pp. 76-80

[11] Jing, C., Hongbin, Z. and Fei, B. (2018) Status quo and prospect of 3D printing research for rail transit sector. *Locomotive & Rolling Stock Technology*, 03, pp. 10-11

[12] Bing, Y., Zhen, L. and Shengchuan, W. (2021) Development of additive manufacturing technology and its application prospect in advanced rail transit equipment. *Journal of Traffic and Transportation Engineering*, 21 01, pp. 132-153

[13] Kai, Y., Hai, X. and Wen, D. (2021) Analysis of Influential Parameters Based on Laser Additive Manufacturing of Axle Box for High-speed EMU. *Journal of Lanzhou Jiaotong University*, 40 05, pp. 72-78
[14] Jun, T., Qiancheng, X. and Zhang, L. (2023) Research on the Forming Process and Properties of Laser Selective Melting of AlSi10Mg. *Applied Laser*, 43 09, pp.37-48

Journal of Physics: Conference Series **2815** (2024) 012016 doi:10.1088/1742-6596/2815/1/012016

Research on thermal characteristics of double-nut ball screw

Pingping He*

Sanmenxia Polytechnic, Sanmenxia, Henan, 472000, China

*Corresponding author's e-mail: 24524100@qq.com

Abstract: Preload, external load, and feed speed have significant effects on thermal characteristics of ball screws. In this paper, ground on friction heat generation theory and Hertz contact theory, a thermodynamic modeling of the double-nut ball screw is established, heat generation under different preload forces and feed speeds is analyzed, and temperature field is estimated based on finite difference method. The results show that preload and feed speed have a obvious impact on screw nuts temperature. The finite difference method makes numerical calculation of temperature field more convenient.

1. Introduction

High- efficiency and high-accuracy machine tools are one of the main development trends. The feed system is its hardcore part. At present, double-nut ball screw (DNBS) is diffusely adopted in the feed system because of its large load, strong stability, and high precision. In the research of machine tools, the thermal characteristics are also a non-negligible aspect. In the course of study, the thermal and dynamic properties of the screw nut should be considered comprehensively, and it is necessary to study the numerical calculation method of its temperature field to further predict the service performance of ball screws.

Therefore, researchers have studied the stiffness, life, and thermal properties of the screw nut [1-2]. Brecher et al. [3] introduced a method for calculating the load distribution of a ball screw drive, which laid the foundation for thermal modeling. Ulacia et al. [4] presented a numerical simulation method of the temperature of DNBS by the finite element method and discussed the variation reason of thermal preload. Wang et al. [5] builded the thermo-mechanical model of ball screws and proposed the solution scheme. Liu et al. [6] established a parametric thermal model of ball screw feed system by using finite difference method. Song et al. [7] found a frictional thermal effect model of ball screws according to the principle of relative motion of planetary transmission. Tanaka et al. [8] developed a wireless multipoint series temperature sensor system to directly measure temperature distribution of ball screw feed system.

The above research analyzes the prediction method of heat generation and temperature field of leader-nut. However, the commonly used finite element method to estimate temperature field needs to extract data from simulation software, which makes it difficult to calculate temperature rise numerically. Therefore, based on Hertz contact theory and friction heat generation theory, this study establishes heat generation model of DNBS and estimates the temperature field by finite difference method. The proposed method makes thermal characteristic analysis of screw nuts more efficient and convenient, providing a basis for the reasonable setting of screw nut preload assembly.

Journal of Physics: Conference Series **2815** (2024) 012016 doi:10.1088/1742-6596/2815/1/012016

2. Force analysis of DNBS

Figure 1 shows the structure of DNBS studied in this paper. Figure 2 sketches contact force between the raceway and the ball. Table 1 gives some parameters of ball screw.

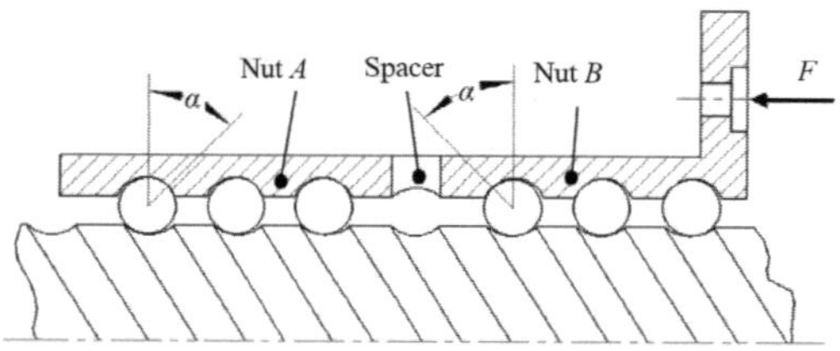

Figure 1. Cross section of DNBS.

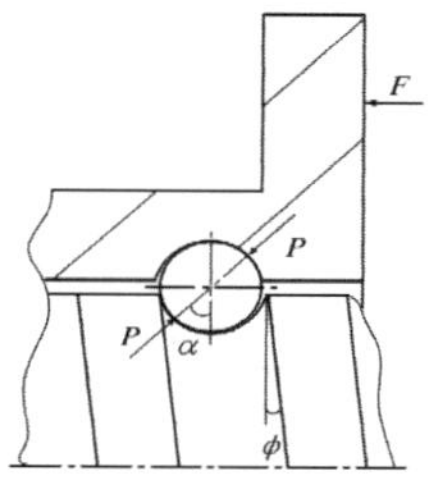

Figure 2. Contact load diagram.

where F is the axial force; P is the normal contact force.

Table 1. Parameters of DNBS.

Parameter	Figure	Parameter	Figure
Screw diameter d	25 mm	Lead angle φ	3.7°
Ball diameter d_b	5 mm	Table quality m	16 kg
Contact angle α	45°	Pitch P_h	8 mm
Total number of balls N	30	Length of screw L	340 mm
Thermal conductivity λ	36.5 W/(m·°C)	Feed speed v	m/min
Specific heat capacity C	460 J/(kg·°C)	Elastic modulus E_S	210 GPa
Heat conduction coefficient h	100 W·°C	Density ρ_S	7820 kg/m^3
Thermal expansion coefficient	1.12e^{-5} K^{-1}	Poisson's ratio ξ_S	0.3

The normal contact load is decomposed into axial component P_a and radial component P_r along the axial and radial directions, and its mathematical expression is as follows:

$$\begin{cases} P_a = P\sin\alpha\cos\varphi \\ P_r = P\cos\alpha \\ F = NP\sin\alpha\cos\varphi \end{cases} \tag{1}$$

Assuming that only elastic deformation occurs between the ball and screw (nut) under preloading, contact stiffness between the two can be obtained by Hertz contact theory. Considering the effects of external load, preload, acceleration and friction on DNBS, the normal contact load P between the nut and ball A is:

$$P = \frac{F + F_a + ma}{\left(i \times \dfrac{\pi d}{d_b \cos\phi}\right)\sin\alpha\cos\phi} \tag{2}$$

where F_a is the nut preload, and a is the acceleration.

3. Thermal characteristics analysis of DNBS

3.1 Heat calculation of DNBS

Heat is the source of thermal deformation in the feed system, which results in thermal error. Heat of DNBS mainly derives from the friction between raceway and balls. The nut-generated heat Q_S is:

$$Q_S = 0.12\pi f_{S0} \upsilon_0 n M_S \tag{3}$$

where f_{S0} is the coefficient associated with nut type and lubrication way; n is rotational speed.

Total friction moment M_S is composed of the driving moment M_D of ball screw and spiral drag moment M_P of balls, which can be defined as:

$$\begin{cases} M_S = M_D + 0.94M_P \\ M_D = \dfrac{P_a P_h}{2\pi\eta} \\ M_p = \dfrac{F_a P_h}{2\pi\eta}\left(1-\eta^2\right) \end{cases} \quad (4)$$

where η stands for the transmission efficiency of DNBS , taken as 0.92.

3.2 Heat transfer of DNBS

During machine operation, forced convection heat dissipation will occur on the outer surface of the screw nut. Heat transfer by convection Q_C is:

$$Q_C = hA\left(T - T_0\right) \quad (5)$$

According to the Nusselt criterion, the heat conductivity factor h of heating surface is:

$$h = \frac{N_u k}{L_C} \quad (6)$$

where N_u denotes Nusselt number; k stands for heat conduction coefficient of air, W/(m•K); L_C is Characteristic size.

For forced convection heat dissipation, Nusselt number N_u is:

$$\begin{cases} N_u = 0.133R_e^{\frac{2}{3}}P_r^{\frac{1}{3}} \\ R_e = \dfrac{\omega d^2}{\upsilon'} \end{cases} \quad (7)$$

where R_e is Reynods number; P_r is Plandtl number; ω is screw angular speed, rad/s; υ' is motion viscosity of air, mm^2/s.

3.3 Temperature field analysis of DNBS

3.3.1 Finite element method. Solidworks software was used for 3D modeling of ball screws. Compared with the right nut, the left nut generates more heat under axial external load, inertial load, and preload, so the right nut is ignored in modeling. A 3D model was imported into Ansys-Workbench software, and the grid was divided, as shown in Figure 3. The material and thermal conductivity of the model are defined. Table 1 gives some material parameters of DNBS.

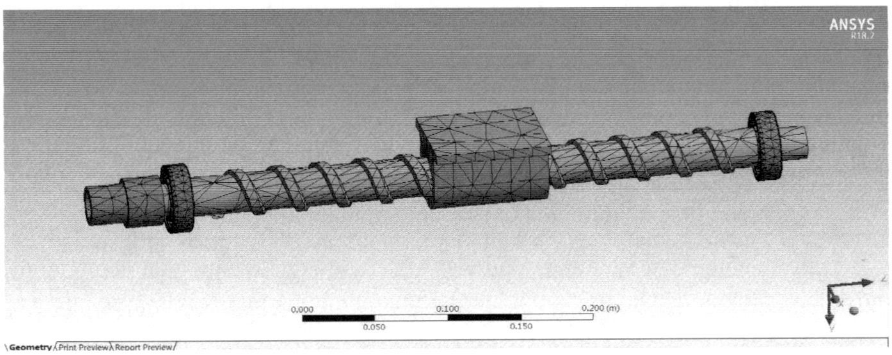

Figure 3. Finite element model of ball screw.

Thermal flux at screw-nut was calculated according to the heat generation by Equation (3) and was loaded as thermic load. Convection transfer rate of lead screw nut surface was calculated respectively according to Equation (5) and was applied on all outer surfaces as boundary conditions. The bearing heat generation is shown in [9]. Then steady state thermal field of DNBS is gained by solving.

3.3.2 Finite difference method. A ball screw system is a discontinuous structure composed of several components. Arithmetic solution of steady-state heat transfer can also be employed by finite difference method to solve the system temperature. The numerical solution of the ball screw thermal conductivity problem is divided into three steps: node division, node difference equation establishment, and node equation solution. Figure 4 gives a simplified discretization model of DNBS. The temperature of the node and its adjacent nodes is shown in Figure 5.

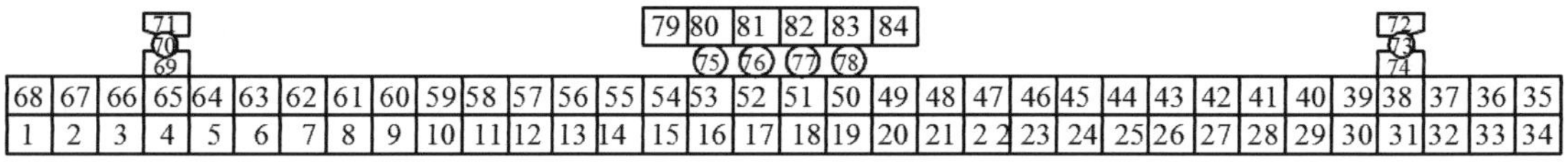

Figure 4. Grid discretization of ball screw.

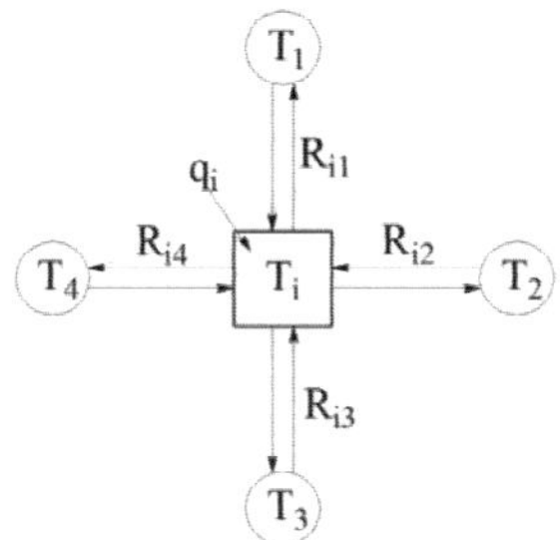

Figure 5. Temperature of the node and its adjacent nodes.

The difference equation for each node is gained by using energy conservation:

$$\sum_{n=1}^{4}\frac{T_n - T_i}{R_{in}} + q_i = m_i C_i \frac{\partial T_i}{\partial t} \tag{8}$$

where T_n is the nth node temperature around node i; T_i is node i temperature; R_{in} is heat resistivity between node i and node n; m_i is quality; q_i is the heat rate of node i; C_i is heat capacity.

Partial differential equations of all nodes are established, and the steady-state temperature solutions of all nodes are gained by the Runge-Kutta method.

4. Results and discussion

Table 1 gives some parameters of DNBS. When the preload F_a=1000 N and the acceleration is zero, the variation of heat generated by DNBS with feed speed under different external loads is presented in Figure 6. It can be seen that when external load is constant, heat generation of screw nut increases linearly with feed speed augment. When feed speed is constant, heat generation increases with external load augment.

When the acceleration is zero, heat generation changes with the preload of the screw nut, as shown in Figure 7. It is observed that a linear increase in heat generation can be caused by the increase in feed speed, preload, and external load. When the preload is constant, heat generated by the nut increases gradually with external load augment. Feed speed and axial force have a great influence on heat generation.

1000 N preload is applied to the front and rear bearings respectively, 2000 N preload and 800 N axial force are applied to the nut, the initial temperature is 24°C, feed speed is 0.25 m/s, and thermal field of DNBS is presented in Figure 8.

Figure 9 displays thermal field of DNBS by finite difference method. It is observed that the temperature of the screw connected with the nut is the highest, followed by the bearing connection, and the axial thermal distribution of DNBS is uneven. A comparison between the numerical solution and software simulation results shows that there are some differences in node temperature, but the temperature distribution trend is very similar. Temperature rise of DNBS can be predicted by the finite difference method.

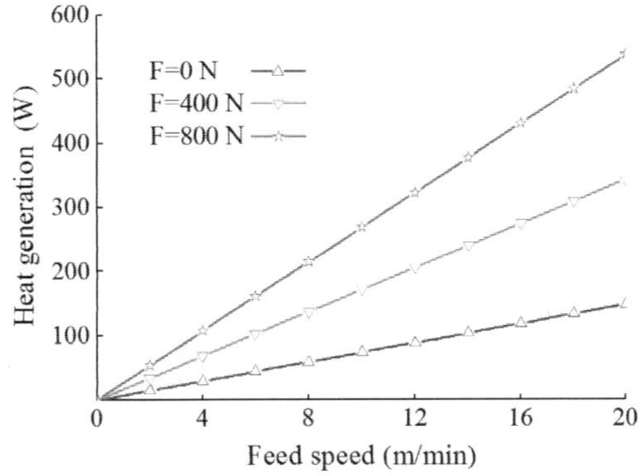

Figure 6. Heat generation with feed speeds.

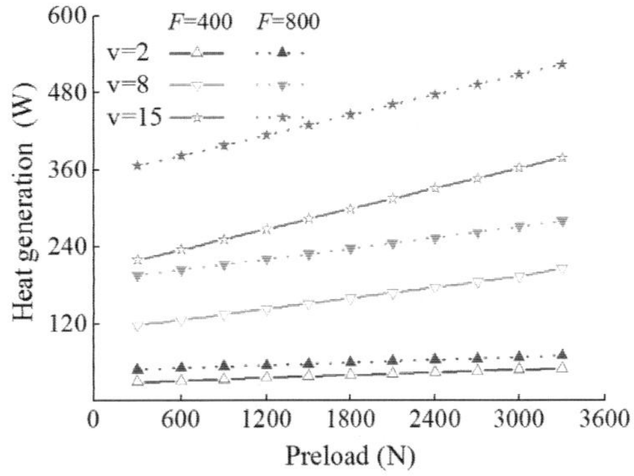

Figure 7. Heat generation with preloads.

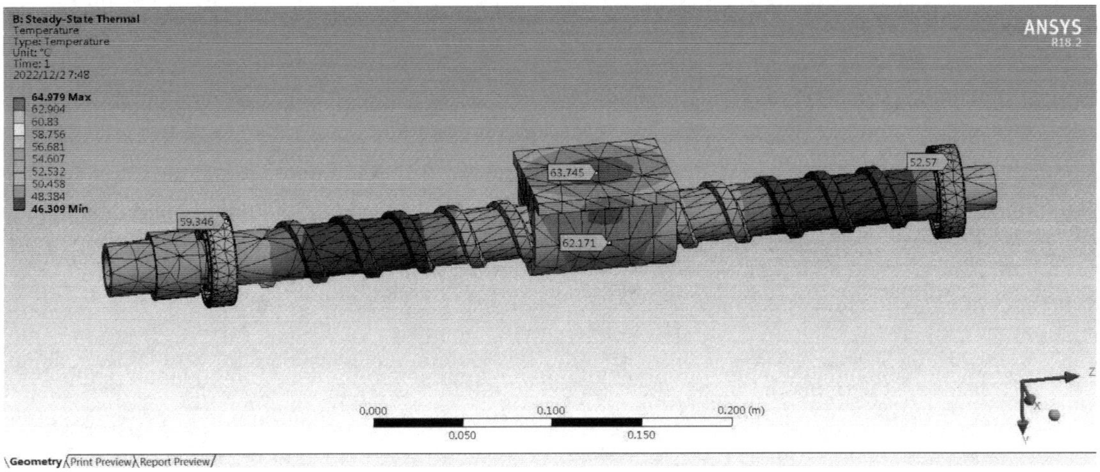

Figure 8. Temperature field distribution of ball screw at different feed speeds.

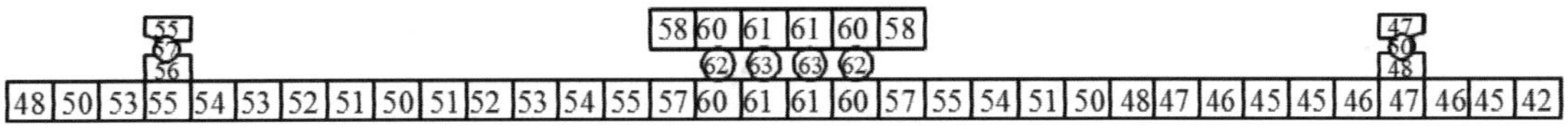

Figure 9. Temperature distribution of feed system nodes.

Figure 10 sketches the influence of feed speed and preload on nut temperature rise. It is observed that a preload increase leads to a gradual increase in temperature rise. The increase of feed speed may cause inertia force of the nut to grow, which may cause a high-temperature rise.

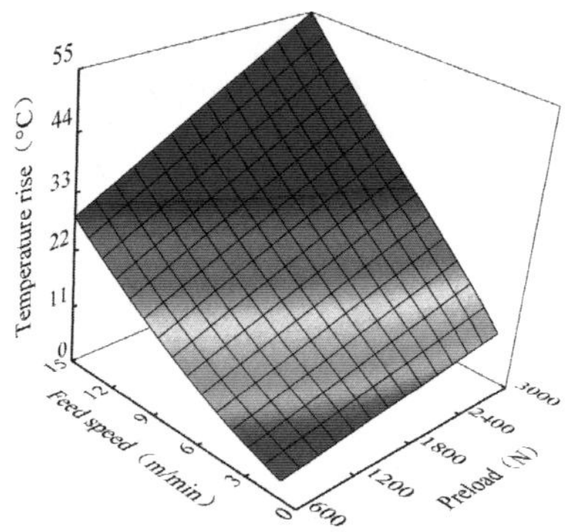

Figure 10. Influence of preload and feed speed on temperature rise of the screw nut.

5. Conclusion

In order to analyze the service performance of DNBS, based on Hertz contact theory and friction heat generation theory, stiffness, life, and temperature field models of DNBS are established in this paper. The temperature rise of DNBS is analyzed using different feed speeds and preload. The conclusions are as follows:

(1) Among the three factors of preload, feed speed, and axial external load, the axial external load has the greatest effect on the heat generation of DNBS, and the increase of three factors result in high temperature rise of DNBS.

(2) There is little difference between the estimation results of the temperature field by finite difference method and finite element method, and the former makes numerical calculation of the temperature field more convenient.

Acknowledgments

The research is financially supported by the Henan Province Science and Technology research project (202102210279) and the Sanmenxia Polytechnic doctor special fund (SZYGCCRC-2022-002).

Reference

[1] Kundu P, Miller M, Gore P, et al. (2023) Detection of inception of preload loss and remaining life prediction for ball screw considering the change in dynamics due to worktable position. Mechanical Systems and Signal Processing, 189: 110075.
[2] Denkena B, Bergmann B, Schmidt A. (2021) Preload monitoring of single nut ball screws based on sensor fusion. CIRP Journal of Manufacturing Science and Technology, 33(3), 63-70.
[3] Brecher C, Eer B, Falker J, et al. (2018) Modelling of ball screw drives rolling element contact characteristics. CIRP Annals, 67(1): 409-412.
[4] Ulacia I, Oyanguren A, Larrañaga J. (2018) Thermo-mechanical modeling of ball screw preload force variation in different working conditions. International Journal of Advanced Manufacturing Technology, 97(1): 723-739.
[5] Wang H, Li F, Cai Y, et al. (2020) Experimental and theoretical analysis of ball screw under thermal effect. Tribology International, 152(2): 106503.
[6] Liu H L, Rao Z F, Pang R D, et al. (2021) Research on Thermal Characteristics of Ball Screw Feed System Considering Nut Movement. Machines, 9: 249.
[7] Song L, Liu K, Liu H Q, et al. (2024) Frictional heating effect of ball screw pairs for machine tools: A new calculation method. Tribology International, 194: 109459.
[8] Tanaka S, Kizaki T, Tomita K, et al. (2022) Direct observation of temperature distribution in ball screw feed drive system using wireless multi-point series temperature sensor. Procedia CIRP, 107: 876-881.
[9] He P P, Gao F, Li Y, et al. (2019) Study on thermo-mechanical coupling characteristics of angle contact ball bearing with fix-position preload. Industrial Lubrication and Tribology, 71(6): 795-802.

Journal of Physics: Conference Series **2815** (2024) 012017 doi:10.1088/1742-6596/2815/1/012017

Integrated torque sensors for humanoid robot joints

Hongwang Zhao, Lin Bo, Changyan Peng, Weikai Gu, Peng Wang and Bingkui Chen*

Chongqing University, Chongqing, 400000, China

*Corresponding author's e-mail: bkchen@cqu.edu.cn

Abstract. Torque sensors are becoming increasingly important in the joints of humanoid robots, serving as an indispensable part of the joint system. This paper designs a strain-based torque sensor integrated into the output flange of a cycloidal reducer. Through mechanical analysis, the relationship between strain and torque is derived. Then, through model simulation analysis, the optimal placement for the strain gauge is determined. A combination of experimental design and response surface analysis is used to optimize the structure's parameters. Finally, finite element simulation is conducted using ANSYS to obtain the optimized performance parameters. Subsequent steps include the static calibration of the integrated sensor. Based on the measured data points, a linear relationship between the sensor's output signal and torque is established.

1.Introduction

In humanoid robot joints, the research on perception systems and information processing systems is particularly important to avoid causing unnecessary harm to humans. Currently, there are two main methods to measure torque sensors [1] : 1) The motor current method estimates the joint torque through the motor drive current, but this method is not very accurate and is prone to deviation. 2) The torque sensor can achieve accurate torque measurement by adding an extra torque sensor. Although the second method is more accurate, the addition of extra parts causes problems such as larger overall volume and insufficient rigidity. Zhu et al. [2] integrated the torque sensor on the output shaft of the harmonic reducer, providing a new direction for sensor integration. Tian et al. [3] designed an ultra-compact torque sensor, proposing solutions to eliminate the cross-talk interference caused by harmonic drives, among other issues. It achieves modularization, lightweight, and low-cost characteristics for joint torque sensors. Han et al. [4] designed a compact robotic joint torque sensor, achieving high sensitivity of the joint through the use of floating and supporting beam configurations. Chiaradia et al. [5] used silicone as the spring material, enhancing the measurement resolution at low torque values and expanding its torque measurement range. It integrates its application with exoskeleton joints. Wang et al. [6] developed a six-dimensional torque sensor and derived the relationship between strain and torque to obtain its measurement results. Park et al. [7] investigated a new method for dynamic shaft torque measurement using polyvinylidene fluoride (PVDF) films. The feasibility of measuring torque through PVDF output signals was experimentally validated. Shi et al. [8] introduced a fiber optic torque sensor that leverages a unique structural design, combining a hyperboloid hollow cruciform and cross-beam structure. This innovative design facilitates a linear measurement range and solid resistance to torque loading, providing valuable insights for the development of subsequent sensor structures.

Most robot joints currently use harmonic reducers, which have low stiffness and efficiency, leading to significant efficiency losses and inaccurate signals when used in humanoid robot joints. Addressing

Journal of Physics: Conference Series **2815** (2024) 012017 doi:10.1088/1742-6596/2815/1/012017

this issue, based on the research of previous scholars, this paper designs a strain-based torque sensor integrated into the output flange of a cycloidal reducer. Compared to harmonic reducers, cycloidal reducers have a high transmission ratio, compact structure, and strong load-bearing capacity. Using mechanical theory, a mechanical model of the sensor is established, and the response function expression of the torque sensor is obtained through finite element simulation and response surface experimental design. The structure of the torque sensor is optimized, and the optimized elastomer is combined with the subsequent signal-processing circuit to complete the signal acquisition. Finally, the static calibration of the torque sensor is used to verify the performance of the sensor in various aspects.

2.Integrated joint structure design and establishment of mechanical model

Figure 1 presents the newly designed electromechanical control integrated joint of this paper, where the sensor is integrated into the output flange of the cycloidal reducer. The integrated joint torque sensor has a measurement range of $0 \sim 100\mathrm{N \cdot m}$, and the integrated joint reducer is designed to consider a 3.3-fold overload, $330\mathrm{N \cdot m}$, an instantaneous load, under which the integrated torque sensor can still function normally. The joint has a cylindrical shape with a diameter of 98 mm and a length of 120 mm. Strain gauges are affixed to the elastic beams on the output flange. Rotary encoders are installed at the end of the motor shaft and the input flange of the cycloidal reducer, facilitating the control of joint speed and position.

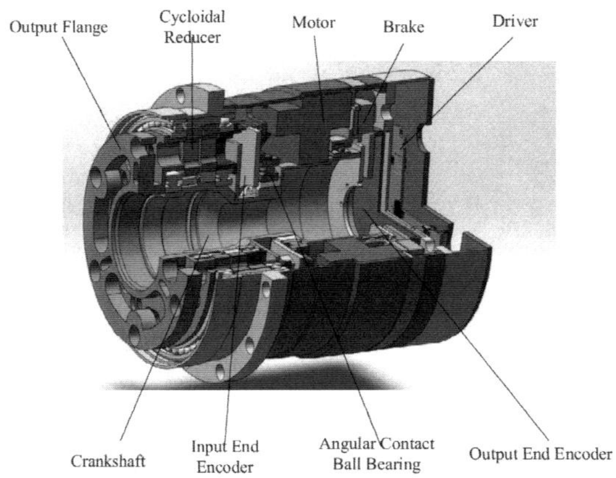

Figure 1.Integrated Joint Structure Diagram.

The structure of the integrated torque sensor is shown in Figure 2. This paper adopts a force mode with three strain beams. Due to the symmetry of the torque on the sensor itself and because the patch method adopted is a spoke type, a single beam can be equivalently modeled as a cantilever beam. Based on the force situation in Figure 3, we can establish a mechanical equation:

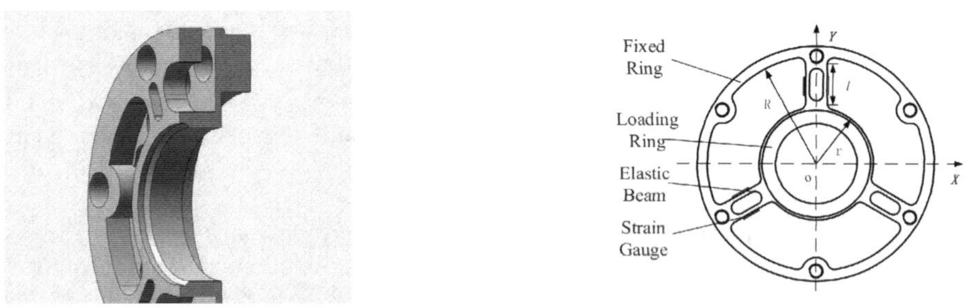

Figure 2. Integrated Torque Sensor Structure.

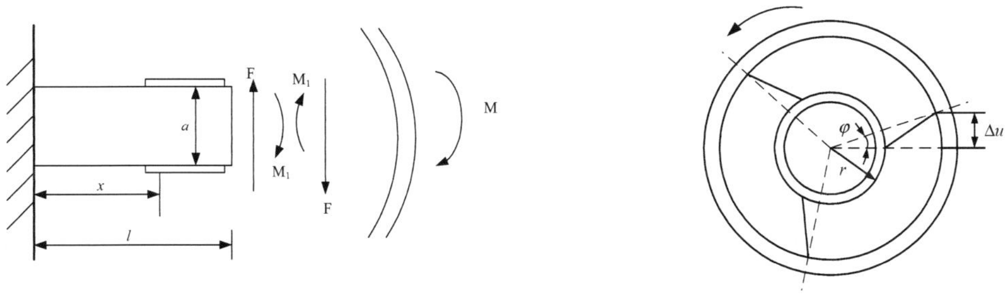

Figure 3. Elastomer Force Diagram.

Based on the force balance equation, it can be satisfied as follows [9]:

$$M_1 - F(r+l) = \frac{M}{3} \tag{1}$$

When subjected to a torque M, the fixed ring is regarded as a rigid body, remaining stationary, while the loading ring moves in a circular motion with the elastomer, as depicted in the diagram above. The angle between the tangent direction of the elastomer and the horizontal direction is φ. The magnitude of this angle is the same as the deflection angle of the elastomer, and Δu represents the angular displacement occurring under the effect of torque M. Because the angular deformation of the elastomer is very small, its magnitude can be considered equal to the deflection at the elastomer's end.

$$\begin{cases} \varphi = \theta_F - \theta_{M_1} \\ \Delta u = \varphi(r+l) \end{cases} \tag{2}$$

When only subjected to a force F, the rotation angle and the deflection curve equation are:

$$\begin{cases} \theta = \dfrac{Fl^2}{2EI} \\ y = \dfrac{Fx^2}{6EI}(3l - x) \end{cases} \tag{3}$$

When only subjected to a torque M, the rotation angle and the deflection curve equation are:

$$\begin{cases} \theta = \dfrac{Ml}{EI} \\ y = \dfrac{Mx^2}{2EI} \end{cases} \tag{4}$$

It can be concluded that under the combined action of force F and torque M1, the displacement at the end of the elastomer and the total rotation angle are, respectively:

$$\begin{cases} \Delta u = \dfrac{M_1 l^2}{2EI_z} - \dfrac{Fl^3}{3EI_z} \\ \varphi = \dfrac{M_1 l}{EI_z} - \dfrac{Fl^2}{2EI_z} \end{cases} \tag{5}$$

By combining Equations (2) and (5), it can be concluded that:

$$\begin{cases} F = -\dfrac{M(2r+l)}{2l^2+6rl+6r^2} \\ M_1 = -\dfrac{Ml^2+3Mrl}{3(2l^2+6rl+6r^2)} \end{cases} \tag{6}$$

Substituting Equation (6) can derive the strain and rotation angle equation:

$$\begin{cases} y(x) = \dfrac{2Mx^2(3rl-6xr+5l^2-6lx)}{Ea^3b(2l^2+6rl+6r^2)}(0 \le x \le l) \\ \varepsilon(\mathrm{x}) = \dfrac{3M(-l^2-3lr_1+3xl+6xr_1)}{4E\mathrm{ba}^2(l^2+3lr_1+3r_1^2)}(0 \le x \le l) \end{cases} \tag{7}$$

Through force analysis, the mapping relationship between strain and torque is derived, which provides a theoretical basis for the relationship between output voltage and torque in the following text.

3.Integrated torque sensor structural optimization analysis

The sensitivity of a torque sensor is an important indicator of its quality, and how to ensure that sensitivity is maximized under the condition of satisfactory stiffness becomes the key to sensor design. To study the effect of the keyway size on the torsional angle and the magnitude of strain in the elastic beam, the following six sets of parameter variables were established for this model, namely the inner radius x_1, beam length x_2, keyway length x_3, keyway width x_4, beam width x_5, and the thickness of the sensor beam x_6.

$$\begin{cases} 80mm \le x_1 \le 88mm \\ 36mm \le x_2 \le 40mm \\ 13mm \le x_3 \le 15mm \\ 9mm \le x_4 \le 11mm \\ 15mm \le x_5 \le 17mm \\ 6mm \le x_6 \le 8mm \end{cases} \tag{8}$$

The objective function with strain is set as y_1 and stress is set as y_2:

$$\begin{cases} \max y_1 \\ 0 < y_2 < 785 \end{cases} \tag{9}$$

This paper establishes mathematical expressions for stress and strain in relation to dimensions using the response surface methodology (RSM) [10]. By employing the Box-Behnken Design (BBD) experimental method, 54 sets of parameter models are designed. Subsequently, the corresponding stress and strain magnitudes are derived through ANSYS finite element simulation. Ultimately, two response surface models are obtained as follows:

$$\begin{aligned} y_1 = {} & 2979.3 - 12.97x_1 - 13.98x_2 + 84.41x_3 + 813.96x_4 \\ & -882.62x_5 - 431.02x_6 + 0.1875x_1x_2 + 0.35x_1x_3 \\ & -10.3x_1x_4 + 6.39x_1x_5 + 5.97x_1x_6 + 73.49x_2x_3 \\ & +0.1625x_2x_4 - 15.67x_2x_5 - 34.03x_2x_6 - 343.76x_4x_5 \\ & -107x_4x_6 + 152.47x_5x_6 - 6.1x_1^2 - 7.48x_2^2 + 42.67x_3^2 \\ & +136.2x_4^2 + 204.02x_5^2 + 4.48x_6^2 \end{aligned} \tag{10}$$

$$\begin{aligned} y_2 = {} & 595.86 - 0.275x_1 - 2.54x_2 + 21.89x_3 + 160.48x_4 - 179.1x_5 \\ & - 88.63x_6 - 6.64x_1x_2 + 0.0663x_1x_3 - 5.4x_1x_4 + 1.27x_1x_5 + 1.2x_1x_6 \\ & - 0.2975x_2x_3 + 6.72x_2x_4 + 0.7325x_2x_5 + 0.4675x_2x_6 + 1.51x_3x_4 \\ & - 5.16x_3x_5 - 3.21x_3x_6 - 68.75x_4x_5 - 21.4x_4x_6 + 23.22x_5x_6 \\ & + 1.11x_1^2 + 0.4949x_2^2 - 3.77x_3^2 + 29.58x_4^2 + 36.11x_5^2 + 10.85x_6^2 \end{aligned} \tag{11}$$

Table 1. Fitted Curve Evaluation Table.

Response	Correlation Coefficient	Signal-to-Noise Ratio	Adjusted Correlation Coefficient	Coefficient of Variation
Stress	0.9968	66.8246	0.9934	2.29
Strain	0.9915	40.8484	0.9827	3.71

Based on **Table 1**, it is evident that the fitted response surface models are quite accurate. Subsequent structural optimization is conducted through the response surface model, and **Table 2** shows the final results after optimization. Considering a 3.3-fold overload, according to the simulation results in Figure 4, it can be concluded thatits strain is $\max y_1 = 0.00384mm$, and stress is $y_2 = 772.1MPa < 785MPa$. Further overload experiments demonstrate its excellent overload resistance capability.

Table 2. Optimized Structural Dimensions.

	x_1	x_2	x_3	x_4	x_5	x_6
Optimal Solution	86.575	40	15	9.736	15.067	6.691
Final Dimensions	86.6	40	15	9.7	15.1	6.7

Figure 4. Optimized Strain Contour Map.

4. Torque sensor static calibration and performance analysis

A static calibration device is designed according to the working environment of the torque sensor. Since the torque sensor transmits a voltage signal, it is necessary to calibrate it with the torque. The experimental setup is shown in the figure, and the fitting curve between the output signal and the torque is obtained as Figure 5:

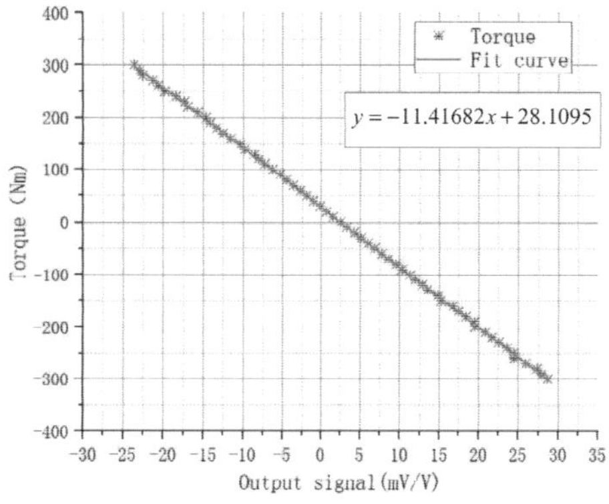

a)

b)

Figure 5. a) Fitting Curve b) Calibration Platform.

5. Conclusion

This paper presents the design of a torque sensor integrated into the output flange of a cycloidal reducer. A systematic analysis of its working principle, force analysis, structural optimization, and subsequent circuit design has been conducted. The feasibility of the sensor design was validated through experimental verification, providing a fundamental approach for future related research.

References

[1] Jin, Y. and Cui, J. W.. (2022). Design and Research of Joint Torque Sensors for Collaborative Robots. Sensors and Micro Systems, 41(02): 21-23+28.

[2] Zhu, D. F., Zhang, E. Y., & Han, K., et al. (2021). Research on Integrated Torque Sensor for Collaborative Robot Joints. Instrument Technology and Sensor, 2021(10): 6-9+22.

[3] Tian, X., Cheng, Z., Bi, M., & Zhu, J. (2024). Ultra-Compact Joint Torque Sensor Units with Enhanced Resolution for Collaborative Robots.IEEE Sensors Journal.

[4] Han, K., Chen, L., Xia, M., Wu, Q., Xu, Z., & Wang, G. (2020). Design and optimization of a high sensitivity joint torque sensor for robot fingers. Measurement,152, 107328.

[5] Chiaradia, D., Tiseni, L., & Frisoli, A. (2020). Compact series visco-elastic joint (svej) for smooth torque control. IEEE transactions on haptics,13(1), 226-232.

[6] Wang, Y. J., Hsu, C. W., & Sue, C. Y. (2020). Design and calibration of a dual-frame force and torque sensor. IEEE Sensors Journal, 20(20), 12134-12145.

[7] Park, S. H., Lee, D. H., Lee, S. E., & Kim, G. W. (2021). Kalman filter-based loading rate-dependent hysteresis compensation of flexoelectric torsional responses in polyvinylidene fluoride films for shaft torque sensors. Mechanical Systems and Signal Processing, 147, 107112.

[8] Shi, C., Lai, D., Tang, Z., & Yang, Z. (2021). Development of an optic fiber-based torque sensor with a torsion-translation conversion flexure. IEEE Sensors Journal, 22(1), 344-351.

[9] Wang, X. D. & Deng, A. (2014). Mechanics of Materials. China Water Power Press, Beijing: 371.

[10] Liu, H., Yin, Z., & Zhang, Q., et al. (2019). Mathematical Modeling of Bovine Hides Swelling Behavior by Response Surface Methodology for Minimization of Sulfide Pollution in Leather Manufacture. Journal of Cleaner Production, 2019, 237: 117800.

Journal of Physics: Conference Series **2815** (2024) 012018 doi:10.1088/1742-6596/2815/1/012018

Study on ballast test of underwater pile gripper

Dawei Zhang[1,*], Jinyu Chen[1], Meng Zhang[1], Yongqiang Yin[1] and Xiaolei Zhao[1]

[1] Offshore Oil Engineering Co., Ltd., Tianjin 300461, China

[*] Corresponding author's e-mail: zhangdw15@cooec.com.cn

Abstract. The underwater pile gripper plays an important role in the installation of the offshore jacket platform. With the further development of offshore oil, it is gradually moving into deep water. Therefore, people pay more attention to the reliability test of pile grippers. The Gripper of underwater pile gripper is the key component of the supporting pipe pile, which determines the quality of the underwater pile gripper to a great extent. Two sets of underwater pile grippers are made of imported and domestic materials, respectively. The maximum loading capacity is 600 tons. The pile gripper is clamped on the test steel pile, and the load is applied by the hydraulic ballast machine. After the experiment is completed, whether the scratch of the tooth mark of the gripper exceeds the expected amount is detected, so as to prove the clamping performance of the pile gripper. The results show that all the components of the pile gripper are not damaged, the hydraulic components are not leaking, the teeth of the imported material gripper and the domestic material gripper are all intact after testing, and the maximum width of tooth marks on steel piles of the two kinds of grippers is about 4 mm and 5 mm, respectively, less than the requirement of 10 mm. The pile gripper made in China can meet the requirements of the actual project for the holding performance, and the holding effect is good.

1. Introduction

With the exploitation of offshore oil, various exploitation platforms have been created. At present, offshore jacket platforms are widely used in the offshore oil industry. The pile grippers are an essential part of the installation of platforms, which are used to maintain a steady level between the skirt pile and jacket during grouting. The environment of offshore oil exploitation is becoming more and more severe, so the reliability and precision of the pile gripper are particularly important.

Domestic research on the pile gripper is backward and shallow. At present, there is no mature product ready for the mass market, and the products we are currently using are all imported from abroad [1]. However, with the development of China's ocean industry and foreign technical restrictions, we have raised attention to the development of pile grippers and have also achieved some research results. For example, Sun et al. analyzed the stiffness and strength of the structure of the pile gripper in the design of its body based on finite element analysis and provided a feasible theoretical basis for the design of the pile gripper [2]. Zhang et al. introduced the hydraulic circuit of the pile gripper [3]. Li et al. developed a test prototype of an underwater jacket leveling tool and completed relevant mechanical experiments on the gripper [4]. Jiang et al. developed an engineering prototype of the underwater pile gripper and conducted a simulation experiment [5]. Zhang completed the mechanical analysis of the foundation pile clamping load under different conditions [6]. Gong et al. based on the finite element analysis, completed the force analysis of the pile gripper with different numbers of holes and optimized the selection of the number of holes according to the tension and stress of the body [7]. Wang et al. analyzed the embedded clamping mechanism and its damage to the pile and proposed a kind of device for the upending of the pile [8].

In this paper, the stress analysis of the clamping tooth is carried out by using the semi-inverse stress function method, and the experimental test of the clamping tooth simulating the actual working condition is completed under the certification of the China Classification Society (CCS), which provides a theoretical basis for the design of the pile clamp and promotes the localization of the pile clamp.

2. Working principle of the pile gripper

2.1. Composition

The Pile Gripper is mainly composed of a water hydraulic system, underwater control system[9], and mechanical system. The water hydraulic system is the hydraulic station of the Pile Gripper. Its function is to provide the driving force for the Pile Gripper. The function of the underwater control system is to control the opening and closing of the hydraulic circuit, which mainly consists of a ball valve, hand-plug valve, ROV control panel, pressure gauge, etc. The mechanical system is used to achieve the required task of clamping the pipe pile, which is also the most virtual part of the whole pile gripper[10]. It is mainly composed of a frame ring and several hydraulic cylinders, a piston rod, and a clamping claw. The frame ring mainly provides support for hydraulic cylinders and other devices. The operation of holding the steel pile is carried out by the piston in several hydraulic cylinders, which drives the clamping claw to embed into the pipe pile to complete the clamping action, as shown in Figure 1.

2.2. Working process

Taking the Wenchang oilfield deep-water as an example, the offshore jacket platform has to go through the process of vessel positioning, jacket lifting and launching, jacket support, jacket positioning, steel pile lifting, pile insertion/piling, leveling, and grouting during the installation. The underwater pile clamp is mainly used after leveling and before grouting[11].

Before the operation, the pile gripper is welded to the jacket sleeve. During jacket installation, after the completion of each piling operation, the jacket needs to be leveled and the hydraulic system of the pile gripper on the water is activated[12]. Driven by the hydraulic system, the claw in the body of the pile gripper will extend and hold the steel pile vertically to complete the clamping task, as shown in Figure 2. Since the gripper material is harder than the steel post material, the gripper will be embedded in the steel post. At this point, according to the anti-bending and shear ability of the claw embedded in the steel pile, it provides sufficient lifting force in the axial direction. After the gripper clamped on the steel pile, we could withdraw the leveling machine to carry out the grouting operation. At this point, the steel pile would not shake during the grouting. At the end of the grouting, the pile clamp is integrated with the offshore jacket platform, and its function is completed.

The underwater pile gripper plays an important role in the installation of the jacket. In the pile gripper, the claw is the main structure to carry the gravity of the steel pile.

To ensure that the underwater pile gripper can work without failure, simulated ballast tests of clamping teeth are carried out to verify the clamping force performance and to test the stability and reliability of the underwater pile gripper, thus reducing the possibility of grouting failure and ensuring normal and safe construction.

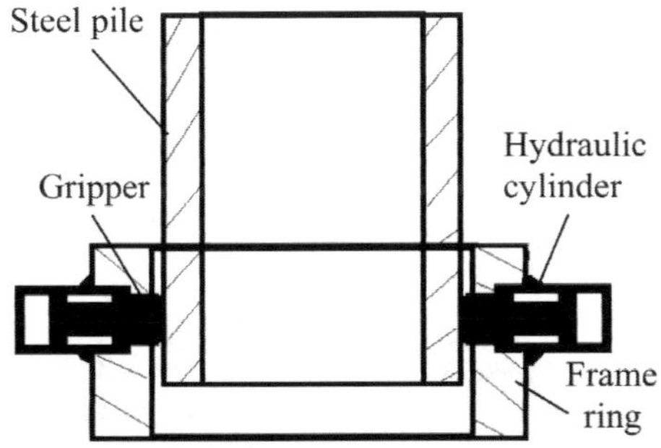

Figure 1. The clamping principle.

Figure 2. Working diagram of underwater pile gripper.

2.3. Claw material
The claws are the executive part of the pile clamp and their mechanical properties are inextricably linked to the ability of the pile gripper to provide sufficient clamping force[13]. The material of the claws is an important factor in the stable and reliable operation of the pile gripper. The claw materials of this experiment are the imported UNIMAX and the domestic FS438. Through the composition analysis of foreign claw product materials, and combined with the current domestic high-strength stainless steel composition, the composition of claw material with excellent performance is shown in Table 1. UNIMAX and FS438 material composition is shown in Table 2. In terms of hardness, after 1030°C heat preservation, oil quenching, and 550°C heat preservation and tempering, the hardness of FS438 die steel can be 52.5 HRC; The hardness of UNIMAX die steel can reach 56~58 HRC after two cycles of gas quenching at 1025°C and tempering at 525°C. In terms of strength, at room temperature, FS438 die steel has a tensile strength of 1629 MPa and a yield strength of 1505 MPa, while UNIMAX die steel has a tensile strength of 1780 MPa and a yield strength of 2150 MPa. In terms of the chemical composition and physical properties of the claw alloy steel, UNIMAX and FS438 have excellent properties and can be used in the jaw section of the pile clamp[14].

Table 1. Table of Alloy Compositions for Excellent Performance Claws.

Ingredients	C	Si	Mn	Cr	Ni	Mo
Content	0.49~0.55	0.17~0.23	0.38~0.44	1.57~1.63	4.00~4.06	0.28~0.34

Table 2. Composition of UNIMAX and FS438 Alloys.

Materials	Chemical Composition (wt%)					
	C	Si	Mn	Cr	Mo	V
UNIMAX	0.5	0.2	0.5	5	2.3	0.5
FS438	0.4	0.28	0.57	5.03	1.9	0.54

2.4. Purpose of the experiment
The clamping ability of the pile clamp is verified by testing the stability and non-failure of the pile clamp under the condition that the load of the steel pile reaches 600 T, so as to determine whether it meets the requirements of the localization of the pile clamp. The main evaluation criteria are: in the axial direction, the maximum displacement does not exceed 10 mm, and the structure of the hydraulic cylinder, the claw, the body, and other parts of the pile clamp are not damaged. The experimental verification will provide a basis for the rational design of the pile clamp.

2.5. Experimental scheme
The experimental equipment mainly includes a THP11-3000B special hydraulic press for liquid die forging with a maximum loading capacity of 3000 T, a load testing device for jacket pile gripper, steel

pile, pile gripper, and its hydraulic system. The pile gripper is inverted on the test device, and the steel pile is inserted into the pile gripper. After assembly, the steel pile is put under the hydraulic press head. The test site is shown in the Figure 3. When the liquid die forging hydraulic press is started, it is necessary to first load to 100, 200, 300, 400, and 500 T and keep for 5 minutes each time, and then load to 600 T and keep for 2 hours. When maintaining the pressure, we need to check the pressure gauge of the pile gripper to make sure there is no obvious change in pressure and observe whether the steel pile slips significantly. If there is an obvious slip of the steel pile, we should stop loading immediately and terminate the test instead of continuing the subsequent pressure test.

Figure 3. Ballast test site.

2.6. Experimental results

After a series of pressure-holding tests, it can be observed that there is no loosening of the pile gripper and no obvious slip of the steel pile. The claw test records of the two materials are shown in Tables 3 and Table 4 below. The claw marks of the two different materials are shown in Figure 4 and Figure 5. The maximum width of the teeth indentation of two different materials on steel piles is about 4 mm and 5 mm respectively. After the ballast test, the slip momentum of the tooth mark does not exceed 10 mm. Through observation, it is found that there is no deformation of the hydraulic cylinder and claw of the pile gripper, no deformation or crack of the pile gripper, and no visible damage of other parts, which proves that the pile clamp can meet the actual requirements. After the test, the appearance of the claw and hydraulic cylinder is shown as follows.

Table 3. Test Records for Claws Made of Domestic Material Type FS43.

No	Value (T)	Actual Loading Force (T)	Keeping Time (Min)	Hydraulic Pressure (MPa)	Slippage in Range (YES or NO)
1	100	108	5	25.5	YES
2	200	205	5	24.8	YES
3	300	305	5	24.8	YES
4	400	405	5	24.5	YES
5	500	497	5	24.3	YES
6	600	600	120	23.8	YES

Journal of Physics: Conference Series **2815** (2024) 012018 doi:10.1088/1742-6596/2815/1/012018

Table 4. Test Records for Jaws Made of Imported Material Type UNIMAX.

No	Value (T)	Actual Loading Force (T)	Keeping Time (Min)	Hydraulic Pressure (MPa)	Slippage in Range (YES or NO)
1	100	113	5	24.82	YES
2	200	193	5	24.82	YES
3	300	300	5	24.13	YES
4	400	395	5	24.13	YES
5	500	500	5	24.13	YES
6	600	614	120	24.13	YES

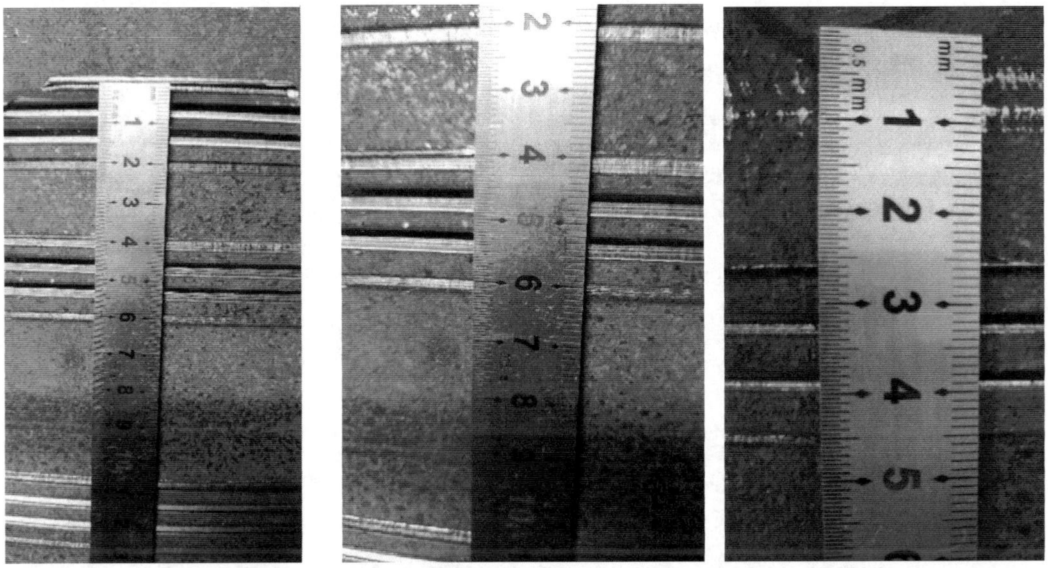

Figure 4. Record of claw marks of domestic material FS43.

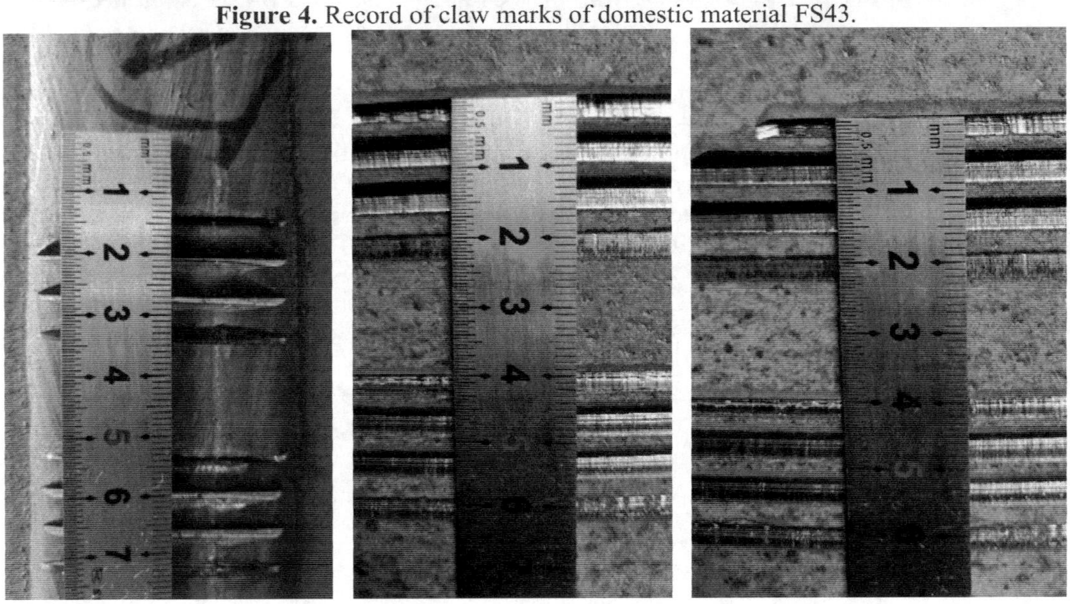

Figure 5. Record of claw marks of imported material UNIMAX.

3. Conclusion

Through the ballast test, the clamping force performance of the underwater jacket pile gripper is analyzed and verified. The test results show that the maximum width of claw marks on steel piles of the two kinds of grippers is about 4 mm and 5 mm, respectively. It indicates that the slippage of the gripper after ballast is less than 10 mm. At the same time, through the test of 600 tons of ballast, the member of the pile clamp has not been damaged. The pile clamp tested can meet the standard requirements of the ballast test. It shows that the gripper can meet the requirement of steel pile clamping force during the installation of offshore oil jackets and can be used in practical construction. The domestic material gripper can meet the requirements of actual engineering for pile-holding performance, and the pile-holding effect is good.

References

[1] Sun Z J, Meng Q X and Wang L Q 2007 Optimised design of hydraulic cylinder structure of underwater pile clamp for offshore oil platform *Petroleum Eng. Constr.* (01) 27-30+84

[2] Sun Z J, Meng Q X, and Zhou Z 2007 Structural analysis of underwater pile clamp body for offshore oil platform *Oil Mining Machinery* (01) 36-38

[3] Zhang L 2009 Working principle of the hydraulic system of underwater pile gripper for offshore oil platform *Sci. Techno. Wind* (09): 102

[4] Li H L, Yu W T, Wang L Q and et al. 2009 Experimental study on leveler jaws of underwater conduit rack *China Offshore Oil & Gas* **21** (01) 57-60

[5] Jiang P R, Fang X M 2011 Analysis and experimental study on the clamping performance of underwater pile clamp *J. Harbin Eng. Univ.* **32** (09) 1228-1232

[6] Zhang B 2012 Analysis of clamped bearing capacity of base piles for offshore platforms and research on a series of devices Harbin Engineering University

[7] Gong H, Li H, Yu W et al. 2018 Structural design and optimization of an underwater skirt pile gripper *Int. J. Perform. Eng.* **14** (12) 3033-3042

[8] Wang Z, Zhang B, and Wang T 2019 Analysis and experiments of the embedded gripping mechanism used in large-scale tools holding up pile foundation in ocean *Pol. Mar. Res.* **26** (3) 15-21

[9] Zhu S H 2003 Installation Technology of deepwater conduit rack in Wenchang oilfield *China Offshore Oil & Gas Eng.* (01) 12-14+38-60

[10] Dong C F 2012 Research on optimization and synchronous control technology of toothed jaws of pipe rack leveling gripper Harbin Engineering University

[11] Jiang P R 2012 Mechanical analysis and experimental research on submerged leveling clamping system of conduit rack Harbin Engineering University

[12] Mou F, Kang A J, Liu M, et al. 2019 Research on high-temperature tensile properties of new die casting mold steel FS438 *Mould Manufacturing* **19** (06) 85-88

[13] Anonymous. FastBallast Technology Undergoes Canadian Ballast Testing[J].Sea Technology,2018,59(12):31-31.

[14] Pile grippers secure two-phase jacket installation[J].Offshore: Incorporating the oilman,2003,63(8):150.

Journal of Physics: Conference Series **2815** (2024) 012019 doi:10.1088/1742-6596/2815/1/012019

Development and application of pressure transmitter based on IMC chip

Kean Zhao [1, *], Mengjie Hu [1] and Yaoqing Chen [1]

[1]Zhejiang SUPCON Instrument Co., Ltd. Hangzhou, China

[*] Corresponding author: zhaoka@supcon.com

Abstract—This paper proposes a design scheme for an FF bus pressure transmitter with a domestic IMC chip, using an ECS-700 instrumentation control system to apply the pressure transmitter in refining and chemical projects. Through planning and designing, configuration, and debugging, it improves the convenience and accuracy of the instrumentation system operation.

1. Introduction

Foundation Fieldbus (FF) is an all-digital, bi-directional transmission, multi-point communication powered by the bus, used to connect intelligent equipment and automation system communication link; its communication rate is usually 31.25 Kbit/s, used to connect a variety of instruments, valve positioners, and other intelligent equipment [1-2]. Foundation fieldbus technology has entered a new stage of application in China. With the domestic users' knowledge and familiarity with FF technology and the gradual maturity of the FF market, this technology has been widely used in the oil and gas, petrochemical, metallurgical, electric power, papermaking, pharmaceutical, and other industries, which saves a lot of debugging, driving time and manpower costs for the customers, and improves the operation efficiency of the instrument [3-5].

Wu and Wang [6] used a microcontroller to design and implement a pressure measurement module that can be embedded and installed inside a pressure transmitter, with small volume, high accuracy, and digitalization. It is used to transform traditional pressure transmitters, which can effectively improve the efficiency and reliability of information exchange between on-site equipment and control systems. Jin [7] adopted the MSP430 series microcontroller as the core. It designed a miniaturized intelligent pressure transmitter, which had the advantages of high cost-effectiveness, small size, strong anti-interference ability, and wide application range. Bi [8] used a low-power M3030 series microprocessor, a 24-bit ADS1241 to collect pressure signals, and digital filtering to suppress interference effectively. The transmitter operates flexibly and quickly and is more cost-effective than similar transmitters [8]. It can be understood that the core microcontroller chips of domestic instruments mainly rely on imports, and there are "bottleneck" problems such as uncertain supply cycles and poor substitutability.

2. Development of pressure transmitter

2.1 Hardware design of pressure transmitter

Common sensors used in industrial pressure transmitters are piezoelectric pressure sensors, capacitive pressure sensors, and piezoresistive pressure sensors [9]. Piezoelectric pressure sensors are mainly based on the piezoelectric effect to measure pressure; piezoelectric sensors cannot measure the static pressure signal. They can only measure the dynamic pressure signal. The capacitive pressure sensor is a kind of pressure sensor that utilizes capacitance as a sensitive element to convert the measured pressure into a change in capacitance value. Capacitive pressure sensors can measure static and dynamic pressure; stability and reliability are better, but the capacitive signal processing circuit is more complex and generally larger. Piezoresistive pressure sensors detect pressure based on the piezoresistive effect. Monocrystalline silicon pressure sensors, such as piezoresistive pressure sensors, have high sensitivity, fast response, wide range of controllability, can make accurate and timely responses to small pressure changes, and at the same time, have very high stability and reliability. In long-term use, the sensor's performance almost does not change. However, the sensor is more sensitive to temperature, but through the temperature, compensation can be very accurate. So, the sensor, in this case, uses a monocrystalline silicon piezoresistive pressure sensor.

An instrument Microcontroller Chip (IMC chip) is a microcontroller chip integrated with an FF communication controller, 32-bit ARM Cortex-M3 processor, 16-bit analog-to-digital converter, DMA module, FLASH memory, SRAM memory, timer, serial port, and other peripherals. The microcontroller chip is specialized for instrumentation. The National Key Research and Development Program supports the chip, led by Shenyang Microcyber Technology Co., Ltd., and developed a supporting development kit.

The main characteristics of IMC chips:

Embedded ARM Cortex-M3 32-bit processor.

Support 384 Kbyte Flash and 256 Kbyte SRAM.

Support 16-bit analog-to-digital conversion.

Support FF communication controller.

Support DMA function.

Support I2C/SPI/UART standard peripherals.

Support universal timers and protocol-specific timers.

Support Watchdog timers.

Support SysTick timers and timestamp timers.

Working frequency 32 MHz, supporting 1/2/4/8 frequency division.

Support SW debugging mode.

The block diagram of the design of the pressure transmitter based on the IMC instrumentation-specific micro-controller chip is shown in Figure 1. Among them, the pressure sensor signal passes through signal conditioning and enters into A/D conversion. IMC, a dedicated microcontroller for instrumentation, acquires the sampling data of the pressure sensor and finally gets the pressure measurement value through calculation. The instrument-specific microcontroller IMC simultaneously acquires the key input and communicates with the display unit through the SPI interface to realize the display function of pressure data. The instrumentation microcontroller IMC communicates with the media-associated unit MAU through the serial interface. The media-associated unit MAU meets all the requirements of the IEC1158-2 international standard for the signal transmission quality of the physical layer of the field. At the same time, it can efficiently obtain energy from the bus to supply power for the whole system to maintain the instrumentation operation. The instrument-specific microcontroller IMC has an RS232 communication interface for setting the common and data parameters via the configuration utility software.

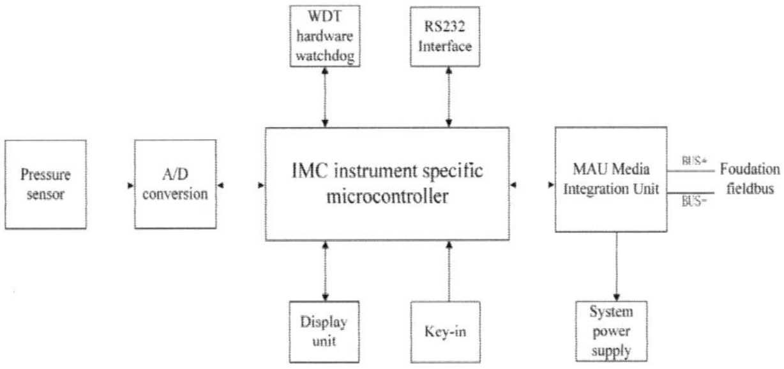

Figure 1. Block diagram of pressure transmitter structure.

2.2 Software design of pressure transmitter

The pressure transmitter software design adopts the main task cycle and interrupt service architecture, as shown in Figure 2. After system initialization, it enters the main task cycle scheduling, including configuration update service, PV data operation service, LCD, key press, and system detection service. The communication service function includes AD data acquisition and FF real-time communication. At the same time, the system enables the watchdog to ensure that the system can be reset in time when the operation is abnormal.

2.3 Pressure transmitter testing

Single-crystal silicon pressure sensors are greatly affected by temperature, so it is necessary to calibrate multiple sets of temperature points within the pressure range. The temperature range of industrial pressure sensors is usually -40.0~80.0°C. Based on these data, we establish the relationship between sensor output and actual physical quantities, usually using linear fitting methods to obtain calibration curves. After calibration and pressure experiment testing, the pressure transmitter obtained ideal experimental data, as shown in Figure 3, with a maximum measurement error of 0.02%, fully meeting the design requirements of the transmitter.

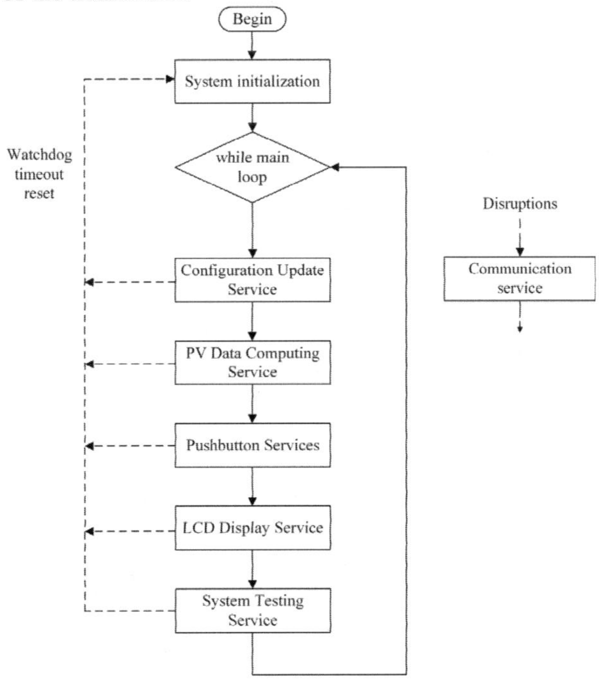

Figure 2. Pressure transmitter software flow chart.

Journal of Physics: Conference Series **2815** (2024) 012019 doi:10.1088/1742-6596/2815/1/012019

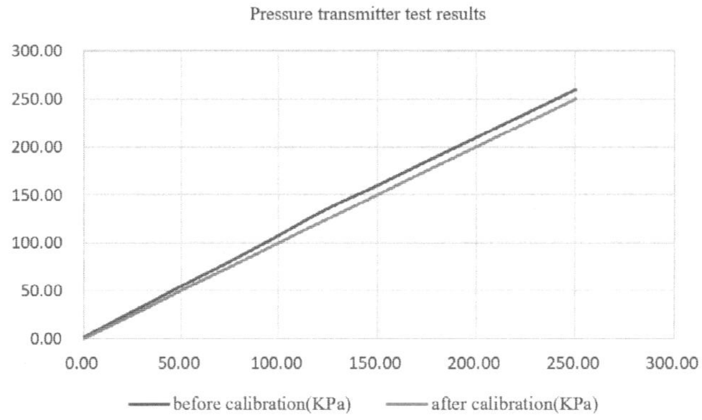

Figure 3. Pressure transmitter pressure test curve.

3. Pressure transmitter applications

3.1 FF system components

Field instrumentation control systems based on FF bus have been widely used in oil and gas, refining, gold, electric power, paper, pharmaceutical, and other industries. ECS-700 system is a large-scale WebField series instrumentation control system of Zhejiang SUPCON Technology, which is committed to helping users realize enterprise automation. The system is designed by the principle of reliability to ensure the safety and reliability of the system fully; all components of the system support redundancy; in the event of any single component failure, the system can still work normally. The system integrates the latest fieldbus technology and network technology. It supports access to international standard fieldbuses such as PROFIBUS, PROFINET, MODBUS, FF, HART, EtherNet/IP, etc., and the comprehensive integration of various heterogeneous systems.

ECS-700 system structure is shown in Figure 4. The control system is mainly composed of three parts. The upper level is the engineer station, which can access real-time and historical information and issue operation commands through the data server of each operation domain. The middle layer is the control station, which is the core unit of the system where samples I/O data come directly from the field and carry out control operations to complete the real-time control function of the whole industrial process. The controller is the core of the hardware and software of the control station, coordinates the relationship between hardware and software in the control station, and completes various control tasks. The controller can automatically complete various functions such as data acquisition, information processing, control operation, etc. It is connected to the engineer station through the process control network, receives the management information from the upper level, and transmits the characteristic data of the processing device and the collected real-time data upward; it is connected to the I/O connection module through the expansion I/O bus (E-BUS) and the local I/O bus (L-BUS) to realize the information exchange with the I/O module and complete the on-site signal and the control operation. The lower level is the FF H1 bus. The lower level is the field measurement and control equipment connected to the FF H1 bus.

Journal of Physics: Conference Series **2815** (2024) 012019 doi:10.1088/1742-6596/2815/1/012019

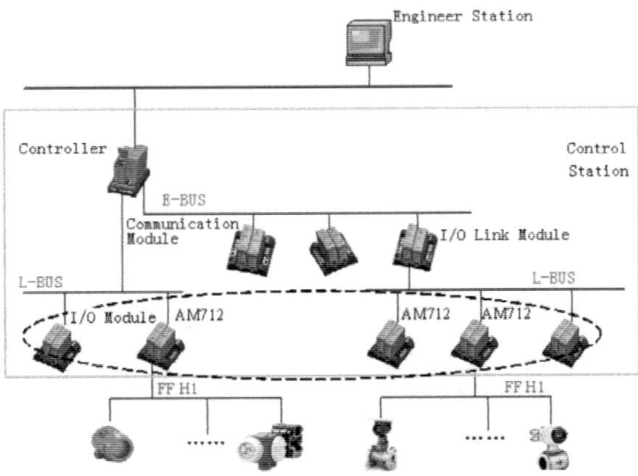

Figure 4. ECS-700 system structure.

3.2 FF bus connection
The equations are an exception to the prescribed specifications of this template. You must determine whether your equation should be typed using either Times New Roman or the Symbol font (please, no other font). Treating the equation as a graphic may be necessary to create multilevel equations and insert them into the text after styling your paper.

The FF H1 interface module AM712 provides a communication interface between FF-compliant field instruments and the ECS-700 system. AM712 supports collecting real-time input (output) data from FF instruments and uploading it to the controller to realize centralized control of FF instruments. It supports control down to field instruments to realize decentralized control. It supports standard FF fieldbus function blocks and manages all block parameters of field devices through FF configuration software and SAMS device management software.

The main components of the FF H1 segment are shown in Figure 5, including the H1 interface module, trunk distributor, branch module (wiring module), surge protector, FF cable, terminator, instrumentation, etc. The pressure transmitter developed in this paper is connected to the FF H1 bus.

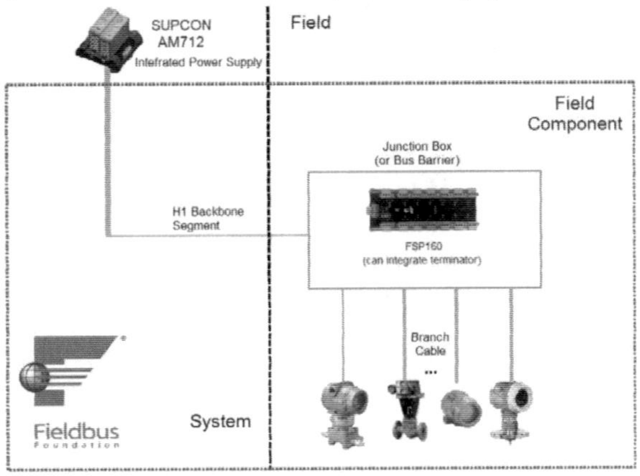

Figure 5. FF H1 components.

H1 card used in control AM712-S card, in which each pair of AM712-S (redundant) can be connected to two FF segments H1 card redundant configuration. A pair of H1 cards connected to the FF segments are in the same control cabinet. H1 card can communicate with the Fieldbus device diagnostic information and can be non-control data transfer to the Fieldbus device resource management application. The H1 card has a hot-swappable function, as shown in Figure 6.

The fieldbus segments are powered by a redundant FF power supply regulator (FFPS), and a power regulator chassis that supplies power to 4 segments is selected. The output voltage of the FFPS should be 24 V to 32 V DC, and the power supply voltage at the Fieldbus device end of each FF segment should not be lower than 13 V DC; the output current is 500 mA/segment, and the FFPS should be equipped with an internal short-circuit protector and a current limiting feature.

3.3 FF instrument configuration

FF instrument configuration can complete the management of FF bus device node address allocation, the configuration of the link between Fieldbus devices, and the link between device bit number and master controller and obtain the device alarm diagnosis information and block alarm diagnosis information of FF instrument. FF instrument configuration can be divided into two parts: offline configuration and online configuration. In the project implementation process, the offline configuration can be performed in the factory first. Then the online configuration can be performed after completing the installation of the system in the field and the physical connection of FF devices. The specific FF instrumentation configuration process is shown in Figure 7.

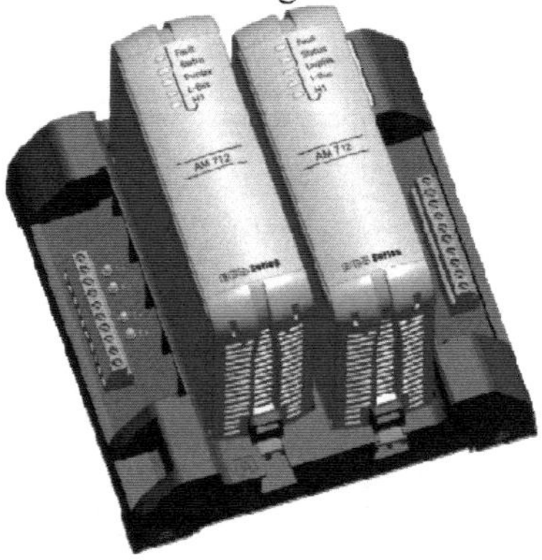

Figure 6. AM712-S card.

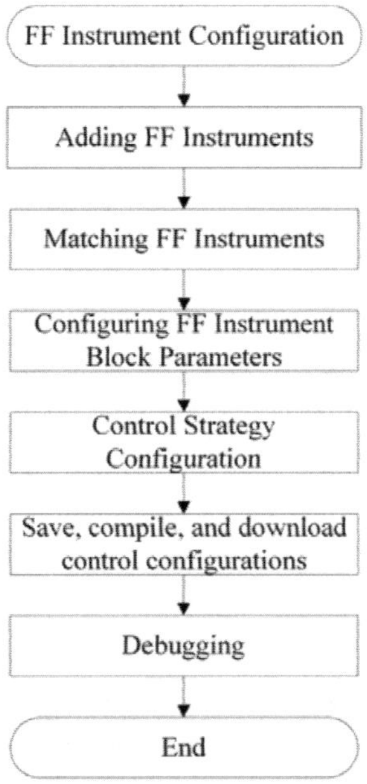

Figure 7. FF instrument configuration flowchart.

During the online configuration process, you can match the configuration segment FF devices with the FF devices online in the field through FF device matching, as shown in Figure 8. Only the matched FF devices can be used for device management and monitoring.

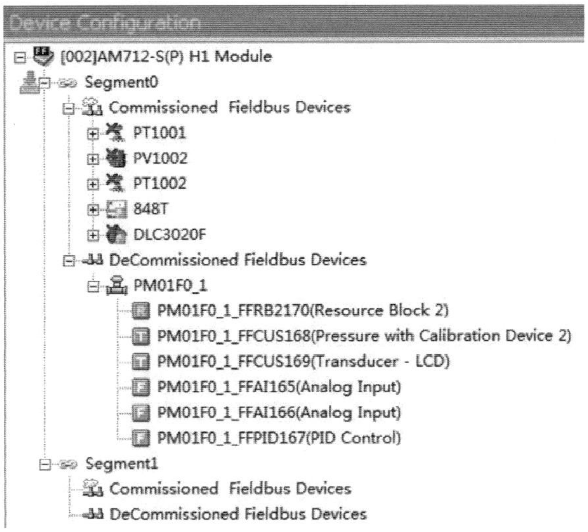

Figure 8. FF device configuration window.

3.4 FF system commissioning

FF system debugging can realize the debugging of the FF-H1 network segment. In the system debugging interface, we double-click the FF H1 interface module AM712-S location in the cabinet, then double-click the AM712 card to enter the debugging interface. As shown in Figure 9, the interface can be seen in "segment 0" and "segment 1" on each instrument. You can see the meters connected to "Network Segment 0" and "Network Segment 1" in the interface.

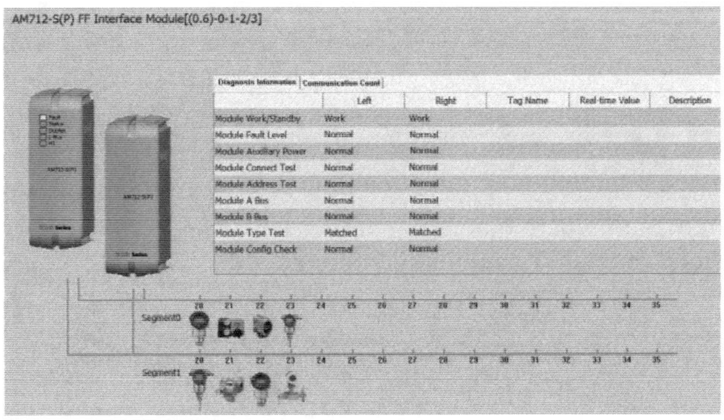

Figure 9. FF system debugging interface.

The scheduling period of the FF network segment is related to the number and type of devices mounted on that network segment. To ensure the real-time performance of the system, the FF system has a limited requirement for the number of FF devices. Each network segment can be designed with a maximum of 6-8 devices, including 1 regulating valve. In this case, the FF devices include Temperature, Pressure, Flow, and Control Loops, with a scheduling cycle set at 1000 ms, which can meet the control requirements. The control response cycle of the equipment is detailed in Table 1.

Table 1. FF device control response cycle.

FF Device Category	Maximum control response period (ms)
Flow, Pressure, Differential Pressure	500
Temperature	1000
Level	1000
Designated Fast Process Control Loops	150
ALL Others (Analysis, Density, PH, etc.)	1000

The diagnostic basis for the FF network segment is the packet count statistics on the network segment, as shown in Table 2. Through these statistical data, it is possible to diagnose the connection quality of FF networks in real time. Among them, NumRequestTimeOut represents the number of request timeouts. If 0, it indicates normal communication, while non-zero indicates abnormal communication. StackError indicates an error in parsing the received backpack. If it is non-zero, it indicates a consistency error in the communication stack of the FF device. NumAllPDUsSent and NumAllPDUsRxd indicate sending and receiving packets, and if there is no continuous growth, it indicates abnormal network communication. When NumErrorPDURxd/NumAllPDUsRxd is less than 0.001%, it meets the FF communication requirements. In Table 2, 45/22563908=0.0002%, which meets the indicator requirements.

Table 2. FF network segment message count statistics.

Diagnose Item	Total number	Description
NumServRequestSent	662244	The acknowledgment service request count sent by the FF H1 interface module
NumServPosConRxd	586073	The total number of positive response packets sent by the FF device to AM712-S
NumServNegConRxd	76146	The total number of reply negative response packets sent by the FF device to the AM712-S service request
NumRequestTimeOut	0	The FF device data link layer is online, but there is no corresponding number of service requests
StackError	0	The total number of error packets received by AM712-S for the FF device
NumAbortRxd	0	The total number of packets sent by the FF device to the AM712-S for the disconnect request
NumAckEventNotifySent	76221	The total number of alarm packets that the FF device broadcasts to the bus
NumAllPDUsSent	48154224	The count continues to increase under normal circumstances
NumPDUsSentFail	0	The number of failures that the FF H1 interface module sends packets to the FF H1 network
NumAllPDUsRxd	22563908	The count of packets on the FF H1 network received by the FF H1 interface module.
NumErrorPDURxd	45	The count of error packets received by the FF H1 interface module

4. Conclusions

This paper proposes an FF bus pressure transmitter design scheme based on an IMC chip. The scheme is based on designing an IMC instrumentation microcontroller, which integrates multiple functional units such as data processing, data storage, serial communication, and FF bus communication into one chip, thus realizing a high degree of integration, miniaturization, and intelligence. It has the advantages of a simple circuit structure, high measurement accuracy, good stability, etc. The developed FF bus pressure transmitter is applied in the refining and chemical integration project to improve the intelligent production, maintenance, and pressure transmitter of the plant. The developed FF bus pressure transmitter is applied in the refining and chemical integration project, which improves the intelligent production, maintenance, and management level of the plant. It solves the problems of the traditional 4-20 mA signal using a lot of wires, the workload of on-site installation and debugging, the high investment, the transmission accuracy, and the low anti-interference ability. It can better help the operator to understand the actual situation of the on-site instrumentation, the adjustment, and troubleshooting, and greatly improves the quality of the instrumentation, and also improves the quality of the instrumentation. It can better help operators understand the actual situation of the field instrument, adjust its parameters, and troubleshoot the problems, which greatly improves the convenience and accuracy of the instrument system operation.

Acknowledgment

This paper is supported by the National Key Research and Development Program of China (2019YFB2005900). Project name: Instrument high-density integration design and module development based on IMC chip. Sub-project: design and application of key technologies of the micro-controller chip for the instrument.

References

[1] David A. Glanzer. (2008) Foundation Fieldbus Technical Overview (R3). J. 37(14): 83-84.
[2] Thomesse J P. (2005) Fieldbus Technology in Industrial Automation. Proceedings of the IEEE. J. 93(6): 1073-1101.
[3] LIU P F and DU X T. (2022) Application and Optimization of New Pressure Transmitters in the Coal Chemical Industry. Process. J. 04: 34-36, 39.
[4] LIU X F. (2018) Application of FF Fieldbus Technology in Refining Equipment. Petrochemical Industry Technology. J. 09: 325-326, 322.
[5] ZHUANG Y. (2015) Application of Foundation Fieldbus Technology in Polyethylene Project. Process Automation Instrumentation. J. 36(07): 12-14, 17.
[6] WU G P and WANG H. (2023) Design and Realization of Pressure Measurement Module in Pressure Transmitter. Ship Electronic Engineering. J. 43(01): 209-211.
[7] JIN L H. (2013) Design and Implementation of Intelligent Pressure Transmitters. Dalian University of Technology. D.
[8] BI J B. (2014) Design and Implementation of High Performance Intelligent Pressure Transmitters. Control and Instruments in Chemical Industry. J. 41(05), 583-586.
[9] GAO R. (2023) Design and Process Research of Piezoresistive Pressure Sensor Based on Ultra Shallow Junction Doping. North University of China. D.

Simulation analysis of CFD of composite pressure-resistant vessel pressure resistance and uplift process

Qi Feng[1, *], Zhen Su[1], Li Sheng[2] and Di Jia[2]

[1]Jiangsu University of Science And Technology, Jiangsu, Zhenjiang, 212003, China
[2]The PLA Unit 92578, Beijing, 100161, China

[*]Corresponding author's e-mail: 211210301408@stu.just.edu.cn

Abstract. A pressure-resistant vessel is the core equipment of deep-diving equipment, which needs to have good pressure-resistant performance. In this paper, due to the problem that it is difficult to monitor the pressure-resistant vessel in the real diving process, we design a lightweight carbon fibre pressure-resistant vessel, construct a finite element model of the pressure-resistant vessel in deep diving, and carry out simulation and analysis, so as to provide a simulation test method for the real diving work. Specifically, the pressure-resistant vessel is subjected to finite element analysis at typical depths, and the floating process of the pressure-resistant vessel under different working conditions is simulated by using Computational Fluid Dynamics (CFD) simulation and orthogonal tests. Comparison of the CFD simulation results with the orthogonal test results proves that the selected pressure-resistant vessel can meet the working requirements under different working conditions.

1. Introduction

The deep-diving pressure vessel is the core component of a manned submersible, and with the continuous development of manned submersible and underwater vehicle technology, ensuring its structural safety is the key point for manned submersibles to successfully complete underwater tasks, so it is imperative to ensure that it has sufficient hydrostatic pressure-resistant capacity and buoyancy capacity. Currently, it is time-consuming and costly to build pressure-resistant vessels for real underwater deep-diving tests. Therefore, there is an urgent need to carry out research on safety simulation test methods.

At present, many scholars have carried out a large number of studies on the safety and load-carrying capacity of deep-diving pressure-resistant vessel structures. Among them, DNV, BV and RS all have requirements on the load-carrying capacity of manned vessels, requiring that the ultimate load-carrying capacity should be at least 1.5 times the maximum working pressure[1][2][3]. Ke C et al. derived an analytical solution for the self-tensioning pressure of a spherical external pressure shell based on elasto-plastic mechanics for the problem of optimizing the strength of deep-sea submersible vessels [4]. Kelvin et al. proposed a new approach to optimize the design of composite pressure vessels based on the classical stacking theory analysis method for carbon fibre composite pressure vessels[5]. Wenbo et al. carried out a reliability study for hydrogen storage composite pressure-resistant vessels (CPV) and used the uncertainty of two types of random intervals to carry out uncertainty analysis and reliability analysis of CPV to find out the critical factors of CPV bursting[6]. Jianbo T et al. proposed a three-dimensional winding diameter modelling method for cladding pressure-resistant vessels based on a novel mesh-oriented projection and normal adaptive convex helix algorithm[7]. However, few studies have addressed the difficulty of monitoring deep-diving pressure-resistant vessels during real dives.

This paper proposes a CFD-based simulation test method for deep-diving pressure-resistant vessels. Firstly, a lightweight carbon fibre pressure-resistant vessel model is constructed, and secondly, based on the model test data, a finite element model of the deep-diving pressure-resistant vessel is constructed, and simulation analysis is carried out. Specifically, the finite element analysis of the pressure-resistant vessel is carried out at typical depths, and the floating process of the pressure-resistant vessel under different working conditions is simulated by using Computational Fluid Dynamics (CFD) simulation and orthogonal tests. Finally, according to the results of CFD simulation and orthogonal test analysis, it is proved that the selected pressure-resistant vessel can meet the working requirements under different working conditions.

2. Simulation modelling and test parameters of deep submerged pressure-resistant vessel

2.1. Structural design of three-dimensional model of pressure-resistant vessel

According to the pressure resistance, buoyancy, volume and other requirements of the container, in order to meet the requirements of the use of functional requirements, taking into account the material, weight, manufacturing process and other aspects of the factors, pressure-resistant containers are designed into the middle of the column with two rounded ends of the disc-shaped structure, the use of the upper and lower parts of the welded formation. As shown in Figure 1 (a), the container's centre is the origin of the spherical coordinate system *O*, and the Z-axis in the vertical direction, as well as the X-axis and Y-axis in the horizontal direction, are perpendicular to each other to set up the spatial right-angle coordinate system. We set point *B* as the floating centre of the container, *G* as the centre of gravity of the container, *h* as the distance between the centre of gravity and the floating centre, and *M* as the overturning moment of the container. If the pressurized container has n points of mass, the mass of the points of mass are m_i, the position of each point of mass is (x_i, y_i, z_i), and the position of the centre of gravity of the pressurized container B is (x_s, y_s, z_s). The details are as follows:

$$x_s = \frac{\sum_{i=1}^{n} m_i x_i}{M_s}, y_s = \frac{\sum_{i=1}^{n} m_i y_i}{M_s}, z_s = \frac{\sum_{i=1}^{n} m_i z_i}{M_s} \tag{1}$$

The outer diameter of the shell is 1250 mm, and the total height is 960 mm. The structural dimensions of the shell are 1250 mm outer diameter, 960 mm total height, and 30 mm wall thickness. In order to ensure that the weight of the vessel is reduced with sufficient strength and stiffness, the shell is reinforced with carbon fibre, the epoxy resin system is used as the base resin, and the liquid moulding structural solution is adopted, as shown in Figure 1(b).

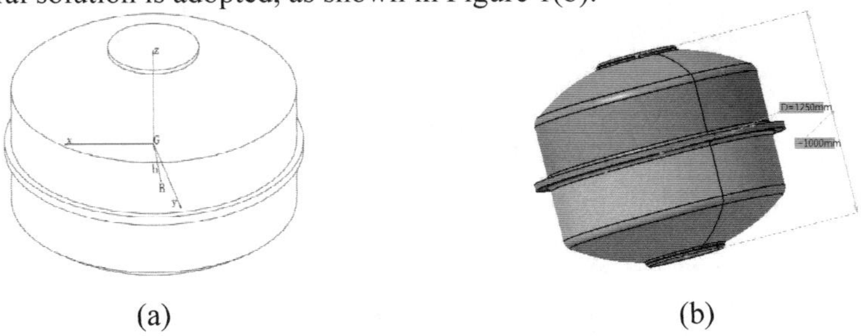

(a) (b)

Figure 1. Schematic diagram of pressure-resistant vessel shell structure (unit: mm).

Pressure-resistant containers with good compressive properties shall have appropriate tensile strength, modulus, and elongation. In this paper, the pressure-resistant container test plate material properties are shown in Table 1.

Table 1. Material Properties Table.

	Tensile strength (MPa)		Tensile modulus (GPa)		Elongation at break (%)		Interlayer shear strength (MPa)
Direction	radial	axial	radial	axial	radial	axial	N/A
Value	678	641	51.1	62.2	1.22	1.11	62.4

2.2. Multiphase flow modelling and control model

When the pressure-resistant vessel is in the underwater state, without considering the pulling force of the cable, it is only acted by gravity and its own buoyancy in the y-direction, and its equation of motion is:

$$\int_{t_0}^{t} dv = \int_{t_0}^{t} (F/m)\,dt \qquad v_t = v_{t-\Delt} + (F/m)\Delta t \tag{2}$$

The transverse centre of stability height and longitudinal centre of stability of the pressure-resistant vessel in the underwater state are expressed as the distance between the centre of buoyancy and the centre of gravity. Since the pressure-resistant vessel is a uniform butterfly structure, the transverse and longitudinal tilt test is only done for the transverse tilt 0°~45° related test, and its underwater transverse stability movement process is:

$$\begin{cases} \overline{GM} = \overline{GM_L} = z_B - z_G \\ M_R = \Delta(z_B - z_G)\sin\phi = \Delta\alpha\sin\phi \end{cases} \tag{3}$$

where $\overline{GM}$ denotes high transverse stability, also called high initial stability, M_R denotes restoring moment, and ϕ denotes transverse inclination angle.

The underwater transverse stability formula is not limited by the tilt angle and can be directly applied to small tilt angle (0°~15°) problems and large tilt angle (15°~45°) problems.

The relationship between the moment, rotational inertia and angular velocity of a vessel rotating underwater has:

$$M = I\alpha = I\frac{d\omega}{dt} = I\frac{d^2\theta}{dt^2} = I\frac{d^2(r\omega)}{dt^2} \tag{4}$$

where M denotes the tilting moment acting on the object, I denotes the rotational inertia of the object, α denotes the acceleration of motion, θ denotes the angular displacement of the object, and r denotes the radius of rotation of the object around the centre.

When assessing the mass of the pressure vessel, the inertia tensor L can be obtained, whose diagonal elements are defined as the moment of inertia around the X-axis, Y-axis and Z-axis respectively, and the Z-axis is taken as the centre of rotation in the CFD simulation.

In summary, the equations of motion for the automatic underwater overturning of the pressure vessel can be obtained:

$$\frac{d\omega}{dt} = \frac{M}{I} = \frac{M_R = \Delta(z_B - z_G)\sin\phi}{I_{zz}} = \frac{\Delta a\sin\phi}{I_{zz}} \tag{5}$$

Previous studies have shown that the steady-state CFD calculations of the Realizable type k-ε turbulence model used in this paper can meet the accuracy requirements in the preliminary engineering analyses, which will not be repeated here.

2.3. CFD simulation model

Numerical simulation analysis was used to reshape the deformed deep-diving pressure-resistant vessel on the basis of photographic data from experimental studies. Under hydrostatic conditions, experiments and CFD simulations were carried out using a real-scale prototype in order to eliminate the scale effect, proving that the results of numerical calculations can reach a high agreement with the experimental results.

The computational model for this simulation is triangularly meshed with local encryption near the boundary layer and inside the pressure-resistant vessel. The total number of cells was verified to be 2.17 million, and the skewness of the measured grid metric was 0.26112. The grid model was divided into two fluid domains, i.e., the external flow field and the internal flow field of the vessel. The dimensions of the computational domain of the external flow field are 900m*550m*400m. The computational domain of the external flow field is divided into two fluid domains by the multiphase flow model, which are the seawater domain (900 m*500 m*400 m) and the air domain (900 m*50 m*400 m). The distance between the pressure-resistant spherical shell and the bottom of the fluid domain in CFD simulation is at least 50 m, and the maximum distance is 400m. Figures 2(a), (b), and (c) represent the CFD model's fluid domain model, multiphase flow model, and mesh details in the vicinity of the vessel, respectively. The pressure-resistant vessel hardly deforms under hydrodynamic and aerodynamic loads, and the similarity with the behaviour of the rigid body mainly comes from the new structure and new materials.

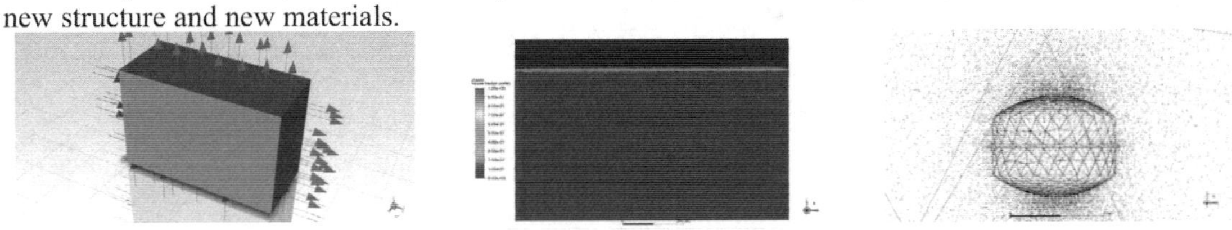

(a) (b) (c)

Figure 2. CFD simulation model.

The boundary conditions used in the model are: velocity inlet for the inlet boundary of the external flow field; outflow boundary for the outlet boundary; pressure outlet for the top boundary; fixed wall conditions for the bottom boundary of the flow field and the pressure-resistant vessel; and symmetric boundaries for the rest of the external boundaries.

The initial conditions used for calculation are as follows:

- The velocity inlet is set up with four groups of initial velocities, which are 0 m/s, 0.77 m/s, 1.54 m/s and 5.14 m/s in the X-axis direction; the velocity in the Y-axis direction of the pressure-resistant vessel is determined by the fluid force calculated from the surface circulation; the angular velocity of the pressure-resistant vessel rotating around the Z-axis is determined by the change of the tilt angle.
- The initial transverse tilt angle of the pressure-resistant vessel is set at 0°, 15°, 30° and 45°. The initial conditions for the four working conditions are shown in Figure 3.
- The initial value of ambient temperature is set to 288.16k, the initial value of turbulent kinetic energy is set to 3.75e-07m^2/s^2, and the initial value of turbulent dissipation rate is 8.664304e-11 m^2/s^3.

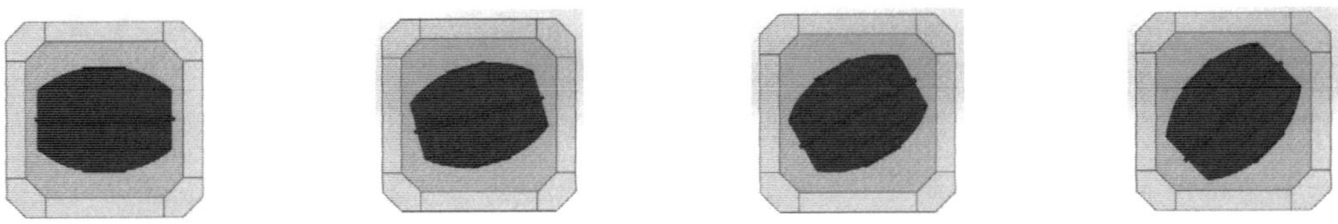

Figure 3. Range of transverse inclination of pressure-resistant vessel (0°, 15°, 30°, 45° from left to right).

3. Analysis of test results

3.1. Analysis of compressive test results

The test selects the underwater speed of the underwater vehicle (1.5 kn and 20 kn) to represent the normal and maximum values of water flow velocity on the surface of the pressure-resistant vessel, respectively. In underwater vehicle normal underwater navigation, the mechanical release device will pressure the vessel fastened to the outside of the vehicle. Therefore, a typical depth of 500m is chosen to analyze the pressure resistance of the pressure-resistant vessel, where the static pressure of seawater is nearly 5MPa.

The dynamic test results are shown in Table 2.

Table 2. Table Dynamic test results.

Underwater vehicle speed (kn)	Maximum surface water flow velocity (m/s)	Maximum surface pressure (MPa)
1.5	1.248	0.506
20	4.651	9.654

The compressive test results are shown in Table 3.

Table 3. Table Compressive test results

Constraints and loads	Static analysis methods	Results (maximum)
5MPa seawater static pressure 3 displacement constraints	Total deformation (mm)	9.2120
	Deformation in the x-axis direction (mm)	9.1798
	Deformation in the y-axis direction (mm)	1.0819
	Equivalent force (MPa)	404.15
	Maximum principal stress (MPa)	320.04

Figures 4(a) and (b) show the water flow streamlines and pressure clouds on the surface of the pressure-resistant vessel at a speed of 20 Kn, and Figures 4(c) and (d) show the water flow streamlines and pressure clouds on the surface of the pressure-resistant vessel at a speed of 1.5 Kn. As can be seen from Figure 4, when the underwater vehicle sails at an average speed of 1.5kn, the maximum velocity of the water impact on the outer surface of the pressure-resistant container fixed outside the underwater vehicle is 1.248 m/s, and the maximum pressure of the water flow borne by the surface is 0.506MPa; when the underwater vehicle sails at a maximum speed of 20kn, the maximum velocity of the water impact on the outer surface of the pressure-resistant container fixed outside the underwater vehicle is 4.65 m/s. The maximum velocity of the water impact on the outer surface of the pressure-resistant container is 4.65 m/s, and the maximum velocity of the water flow impact is 4.65 m/s. When the underwater vehicle is sailing at the maximum speed of 20kn, the maximum speed of water impact on the outer surface of the pressure-resistant container fixed outside the body of the underwater vehicle is 4.651m/s, and the maximum water pressure on the surface is 9.654MPa.

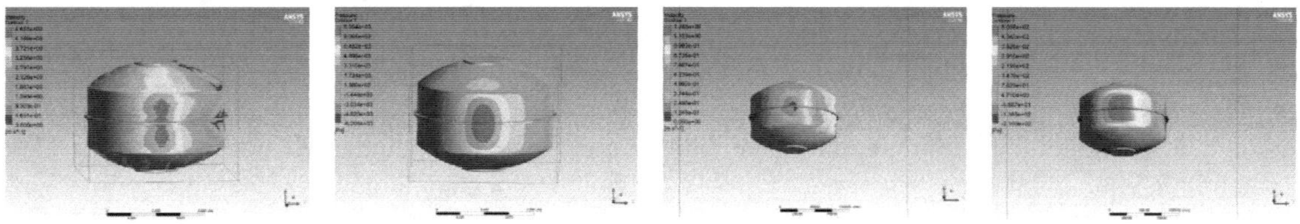

(a) (b) (c) (d)

Figure 4. Dynamic test results of pressure-resistant vessel.

The results of the compressive test of the pressure-resistant vessel when the simulated underwater vehicle is at a typical depth are shown in Figure 5. Figures 5(a), (b), (c), (d), (e) and (f) represent the total deformation, maximum principal stress, equivalent force, equivalent elastic strain, deformation in x-axis direction and deformation in y-axis direction of the pressure-resistant vessel at seawater static pressure of 5 MPa, respectively. As can be seen from Figure 5, when the underwater vehicle is subjected to hydrostatic external pressure of 5MPa, the maximum deformation of the outer surface of the pressure-resistant vessel is 9.212mm, and the maximum axial deformation and circumferential deformation are only 1.0819 mm and 9.1798 mm, respectively, which is a very small deformation and stable structure.

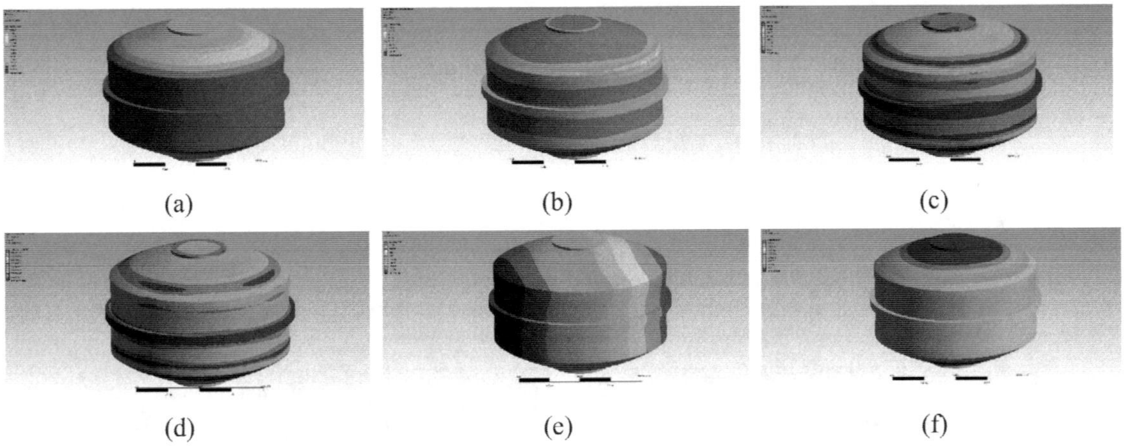

Figure 5. Compressive test results of pressure-resistant vessel.

Pressure-resistant vessel in 5.0MPa hydrostatic external pressure, static analysis of the maximum equivalent force of 404.15MPa, the maximum principal stress of 320.04MPa, are less than the material's compression strength of 464MPa, the material strength, stability is sufficiently safe to meet the use of pressure-resistant vessel.

3.2. Analysis of uplift test results

In this study, orthogonal tests will be designed to screen the parameters that have an influence on the stability of underwater release of pressure-resistant vessels. Three working conditions, namely, underwater vehicle speed, pressure-resistant vessel release depth, and release attitude of the release device, are determined as the influencing factors of the orthogonal test. The speed of the underwater vehicle is set to be 0kn, 1.5kn, 3kn and 10kn, which represent the stopping, low speed, medium speed and high speed of the underwater vehicle in case of an accident, respectively. The typical depths of the underwater vehicle are set to be 100m, 200m, 300m and 450m, which represent the shallow water and deep water state of the underwater vehicle in the event of an accident, respectively. The release angles of the pressure-resistant vessel release device were set to 0°, 15°, 30° and 45°, representing different release postures. A total of 16 sets of orthogonal tests were formed.

Table 4 shows the orthogonal test results obtained from the CFD simulation results, in which the uplift results represent whether the pressure vessel can be uplifted to the surface, and the self-righting results represent whether the rotation of the pressure vessel meets the final transverse inclination angle of less than 5°. As can be seen from Table 4, 16 groups of results for typical working conditions show that the pressure vessel can meet the requirement of releasing to the surface at 450m water depth, while 15 groups of results meet the requirement of self-correcting the rotation of the pressure vessel, and 1 group of rotation angle does not reach the correct range.

Table 4. Table Dynamic test results.

Speed of the underwater vehicle (kn)	Release depth of the pressure-resistant vessel (m)	Release attitude of the release device (°)	Uplift simulation results	Automatically flipping results
1.5	200	45	√	√
1.5	100	30	√	√
10	100	45	√	×
3	200	0	√	√
3	100	15	√	√
0	300	45	√	√
0	100	0	√	√

0	450	15	√	√
3	450	45	√	√
0	200	30	√	√
10	200	15	√	√
10	450	30	√	√
1.5	450	0	√	√
3	300	15	√	√
10	300	30	√	√
1.5	300	30	√	√

Figure 6 shows the results of the CFD simulation of the vessel versus the turbulent viscosity ratio (from left to right) for a set of tests (underwater vehicle speed of 0 kn, pressure resistant vessel release depth of 100 m underwater, release attitude of the release device of 0°).

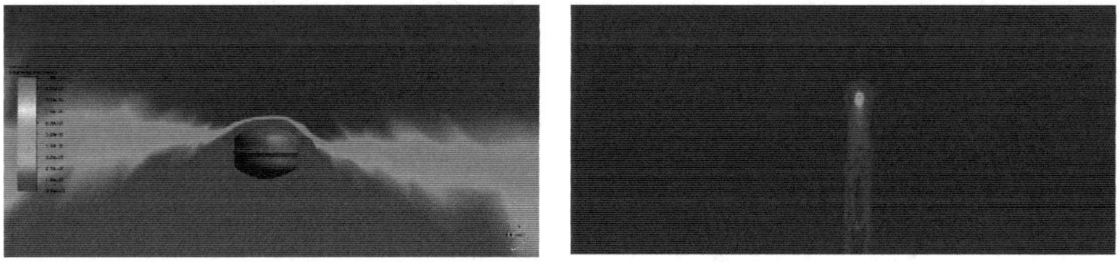

Figure 6. Simulation results for one set of tests.

Figure 6 shows that the pressure-resistant vessel squeezes the grid of multiphase flow interfaces apart when it floats to the water surface, and the turbulent viscosity ratio gradually increases during the uplift process, with the turbulent viscosity in the centre of the vessel being much higher than that on the outer surface of the vessel.

The instantaneous rate change curves of uplift caused by the influence of the speed of the underwater vehicle and the release angle of the release device when the pressure-resistant vessel is released at the underwater release depths of 100 m and 200 m, respectively, are selected, and there are two sets of them, which are formed into Figure 7 and Figure 8, respectively.

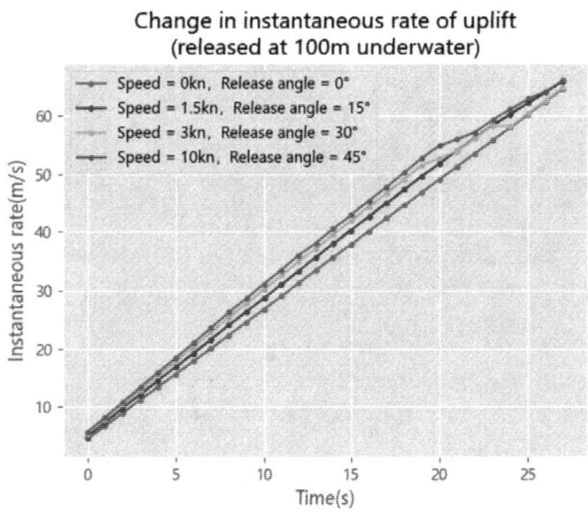

Figure 7. Change in instantaneous rate of uplift of pressure-resistant vessel at a release depth of 100 m under water.

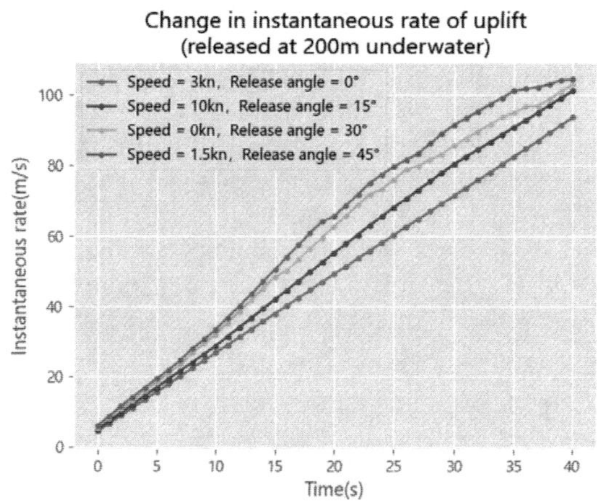

Figure 8. Change in instantaneous rate of uplift of pressure-resistant vessel at a release depth of 200 m under water.

From Figures 7 and 8, it can be seen that in the flotation test, the turbulence caused by the horizontal velocity of the underwater vehicle and the overturning caused by the automatic overturning process of the pressure-resistant vessel will affect the flotation rate of the vessel.

The reasons for the problems in the CFD simulation results are analyzed as follows:

- When calculating the automatic overturning process of the pressure-resistant vessel, the distance between the centre of buoyancy and the centre of gravity of the vessel is set on the small side, resulting in a slightly smaller height of the transverse stabilizing centre so that the automatic overturning requirements are not met in a set of short-distance uplift.
- When the pressure-resistant vessel uplift process, fluent calculates the buoyancy force using the surface cycle to calculate the force area, resulting in a smaller force area when the vessel is turned over, and the instantaneous rate is increased without affecting the normal work of the vessel.

4. Conclusions

In this paper, the underwater release stability of the pressure-resistant vessel of a life raft is simulated and verified through 16 working conditions, and the corresponding orthogonal test results are compared and analyzed. The main conclusions are as follows:

- For this type of pressure-resistant vessel, the axial deformation and circumferential deformation under hydrostatic pressure of 5MPa are very small, the structure is stable, and the material strength and stability are safe enough to meet the requirements of a pressure-resistant vessel.
- The test results indicate that the pressure-resistant vessel is capable of autonomously surfacing and essentially satisfies the automatic righting requirements. This performance is consistent regardless of various release attitudes (0°, 15°, 30°, 45°), release depths (100 m, 200 m, 300 m, 450m), or speeds of the underwater vehicle (0 kn, 1.5 kn, 3 kn, 10 kn).
- For this type of pressure-resistant container, the distance between the floating centre and the centre of gravity is slightly short. Consequently, the underwater transverse and longitudinal stability is diminished, resulting in a limited underwater automatic turning moment. Therefore, during short-distance floating movements, the automatic turning effect may not be significant.

References

[1] Zhiqi X. (2014) Analysis of the Pressure-Bearing Capacity of Spherical Hulls Using the Elastic Foundation Beam Theory[J]. Journal of Harbin Engineering University, 35: 690-695.

[2] Kohnen W. (2008) Manned Underwater Vehicles: A Relevant and Cornerstone Technology for the Future Assessment of Ocean Eco-Systems[C]. OCEANS. IEEE.

[3] Wei C. (2011) A Comparison of Different Rules for the Spherical Pressure Hull of Deep Manned Submersibles[J]. Journal of Ship Mechanics, 15: 276-285.

[4] Ke C and Jianling D. (2023) Strength optimization design of spherical hulls for deep-sea submersibles: A hydraulic autofrettage approach in external pressure vessels, Ocean Engineering, Volume 287, Part 2, 115853, ISSN 0029-8018.

[5] Kelvin M K I, Pedro R and Willian K R. (2023) Methodology to analyze carbon fibre/epoxy composite pressure vessel based on finite element modelling and classical laminated theory, Journal of the Brazilian Society of Mechanical Sciences and Engineering 45(6).

[6] Wenbo L, Leiqi Z, Hong L, Lijun Z, Min L, Cunman Z and Pengfei H. (2023) Reliability analysis of hydrogen storage composite pressure vessel with two types of random-interval uncertainties, International Journal of Hydrogen Energy, Volume 48, Issue 81, Pages 31685-31699, ISSN 0360-3199.

[7] Jianbo T, Jun W, Ximing X, Xin Z, Wei Y, Shuxin L, Haitao Y, Shiqing X and Jian X. (2023) 3D winding path modeling of overwrapped pressure vessels with novel mesh-based directional projection and normal adaptive convex helix algorithms, Composite Structures, Volume 323, 117447, ISSN 0263-8223.

Journal of Physics: Conference Series **2815** (2024) 012021 doi:10.1088/1742-6596/2815/1/012021

Hydrodynamic performance analysis of dual vertical axis turbine

Aiping Wu[1], Tianli Ma[2], Shiming Wang[3,*], Chengling Ding[4] and Boting Liu[5]

[1]College of Engineering Science and Technology, Shanghai Ocean University, Shanghai, China
[2]School of Artificial Intelligence, Shanghai Normal University Tianhua College, Shanghai, China
[3]College of Engineering Science and Technology, Shanghai Ocean University, Shanghai, China
[4]School of Artificial Intelligence, Shanghai Normal University Tianhua College, Shanghai, China
[5]College of Engineering Science and Technology, Shanghai Ocean University, Shanghai, China

[*] Corresponding author's e-mail: smwang@shou.edu.cn

Abstract—This article adopts a biomimetic approach and uses the profile of the fastest sailfish in the ocean as the airfoil profile of a water turbine. Integrating centrifugal and axial flow turbines on one drive shaft can effectively integrate wave and current energy (wave current integration). The wake of a centrifugal turbine flows into the flow field of an axial flow turbine, and based on the Kantar effect, the energy harvesting coefficient of the axial flow turbine is improved. The five blades of the axial flow turbine and centrifugal turbine are subjected to uneven forces, resulting in torsional and radial forces on the drive shaft, causing periodic series motion of the turbine and reducing the overall lifespan of the machine.

1. Introduction

Marine renewable energy, as inexhaustible and clean energy with ultra-low carbon emissions, is known as the "ultimate renewable energy." The latest assessment from the Open Energy Information website of the US Department of Energy shows that the carbon emissions from the entire lifecycle of marine renewable energy use are 0.008 kg/Kwh, which is less than 1% of the unit carbon emissions generated from coal use [1]. By 2030, the global installed capacity of renewable energy will reach 10 GW and reduce 1.7 billion tons of CO_2 emissions [2].

According to the Opinion on Fully Implementing the New Development Concept and Doing a Good Job in Carbon Peaking and Carbon Neutrality issued by the Central Committee of the Communist Party of China in September 2021, we will vigorously develop marine energy and continuously increase the proportion of nonfossil energy consumption. In October 2021, the State Council issued the Action Plan for Carbon Peak before 2030, proposing to explore the development and utilization of new marine energy sources such as wave energy, tidal energy, and temperature difference energy. Marine energy, as an important intersection of China's three strategic areas of renewable energy, high-end equipment, and advanced manufacturing, has brought unprecedented challenges and opportunities to adjust the energy structure and increase the proportion of marine renewable energy consumption through the "dual carbon" goal [3].

Wave energy is the cleanest renewable resource, with advantages such as large energy storage, high energy density, and minimal environmental impact. The trend has strong regularity and stable energy and has high development value. China's marine energy resources are quite abundant. According to surveys, the theoretical average power of wave energy in China's nearshore areas is about 12.85 million kW, and the theoretical average power of tidal energy is 13.94 million kW. More than 90% of these resources are distributed along the coast of Shanghai, Zhejiang, Fujian, and Hainan, where conventional energy is severely lacking. The wave and tidal energy in the ocean is even more abundant [4].

2. Dual vertical axis water turbine model

The wing profile of the dual vertical axis water turbine adopts a sailfish profile, as shown in Figure 1. The swordfish is considered to be the fastest fish in short distances. In a series of speed tests conducted in Langchi Fishing Village, Florida, a swordfish swam 91 meters in 3 seconds, equivalent to a speed of 109 kilometers per hour. The maximum speed of a sailfish can reach 190 kilometers per hour, and its speed can also reach 90-110 kilometers per hour on weekdays. When swordfish swim rapidly, they retract their dorsal fins to reduce water resistance, so the contour curve of their dorsal fins is also removed when collecting swordfish profiles.

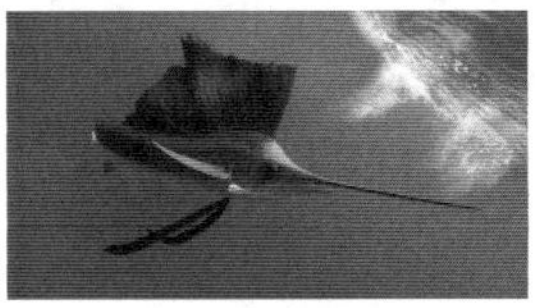

(a)

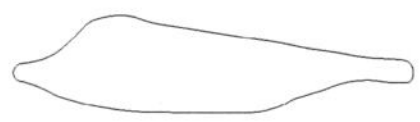

(b)

Figure 1. Profile of Water Turbine Blade:(a) Picture of Sailfish; (b) Profile Curve with Dorsal Fins Removed.

During the process of tidal flow, the centrifugal water turbine is subjected to the action of seawater, causing a pressure difference between the upper surface (suction surface) and lower surface (pressure surface) of the blades, driving the centrifugal water turbine to rotate and converting the kinetic energy of seawater into rotational mechanical energy. During the upward and downward movement of waves, seawater flows through an axial flow turbine, creating a pressure difference between the upper and lower surfaces of the turbine. This drives the turbine to rotate, converting the gravitational potential energy of seawater into rotational mechanical energy. Centrifugal and axial flow turbines share the same drive shaft to drive the generator, as shown in Figure 2.

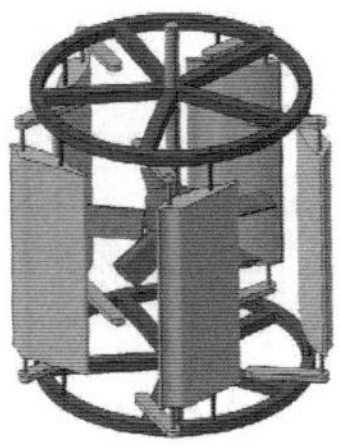

Figure 2. Dual Vertical Axis Water Turbine Model.

The lift coefficient of the blade is defined as F_L, the drag coefficient is F_D, and the pitch moment is M [5]:

$$F_L = \frac{1}{2}\rho C_L A v^2 \tag{1}$$

$$F_D = \frac{1}{2}\rho C_D A v^2 \tag{2}$$

$$M = \frac{1}{2}\rho C_M A v^2 L_a \tag{3}$$

The energy harvesting coefficient of the water turbine is C_p and the torque coefficient is C_T.

$$C_p = C_{p1} + C_{p2} = \frac{P_{cent}}{E_1} + \frac{P_{axia}}{E_2} = \frac{M_{cent} n_{cent} \pi}{30 \rho R_1 L_{b1} v_1^3} + \frac{M_{axia} n_{axia}}{30 \rho R_2^2 v_2 g h} \tag{4}$$

$$C_T = \frac{T}{0.5 \rho v^2 \pi R^3} \tag{5}$$

where ρ is the fluid density, C_L is the lift coefficient, C_D the drag coefficient, A is the blade swept area, v is the fluid inflow velocity, C_M is the pitch moment coefficient, L_a is the blade chord length, C_{p1} is the centrifugal turbine energy harvesting coefficient, C_{p2} is the axial turbine energy harvesting coefficient, R_1 is the centrifugal turbine radius, L_{b1} is the centrifugal turbine blade extension, v_1 is the inflow velocity relative to the centrifugal turbine, R_2 is the axial turbine radius, v_2 is the inflow velocity relative to the axial turbine, g is the gravitational acceleration, and h is the wave height relative to the axial turbine.

According to the momentum theorem, the torque T and energy harvesting coefficient C_p of the water turbine can be obtained [6]:

$$T = \int_0^R 4\pi \rho V \Omega b(1-a) R^3 dR \tag{6}$$

$$C_p = \frac{8\lambda^2}{R^4} \int_0^R b(1-a) R^3 dR \tag{7}$$

3. Numerical simulation calculation

3.1 Grid division

The dual vertical axis wave current power generation system, in this case, adopts fluent mesh for grid division and the polyhedral mesh method. The polyhedral mesh has lower requirements for geometric simplification and does not require too much time for geometric simplification work. Polyhedral mesh division can conveniently achieve multi-core parallel acceleration, fully utilize hardware performance, and save waiting time for grid division. At the same mesh resolution, the number of elements in a polyhedral mesh is smaller. According to the geometric complexity, polyhedra can reduce the number of elements by more than 30% compared to tetrahedra, achieving a significant reduction in computational time. The grid image of the dual vertical axis water turbine is shown in Figure 3.

(a) (b)

Figure 3. Grid Division of Dual Vertical Axis Water Turbine: (a) XY profile grid; (b) XZ profile grid.

3.2 Boundary condition settings

The turbulence model in this article adopts SST k- ω Model. The SST model combines the advantages of κ-ε Model and κ-ω model. Near the wall surface is adopted in the κ-ω Model. Free shear flow in the distance is adopted in the κ-ε Model. The model solved two variables: turbulent kinetic energy k and turbulent flow energy dissipation rate ε (epsilon). This model uses a wall function therefore it does not simulate the flow in the buffer zone. The k- ε model has a good convergence rate and relatively low memory requirements, making it popular in many industrial applications. However, it does not accurately calculate the flow field that displays the inverse pressure gradient and strong curvature in the flow or jet. The κ-ω model is similar to the κ-ε model; however, it solves for the specific rate of kinetic energy dissipation ω (omega). It is a low Reynolds number model but can be combined with wall functions. It is better than k- ε. The nonlinearity of the model is greater, making it more difficult to converge and quite sensitive to the initial guess value of the solution. At k- ε, in many cases where the model is not precise enough, k- ω Models can be very helpful, such as internal flow, flow exhibiting strong curvature, separated flow, and jet flow. Finally, the SST model combines k- ε in free flow and the k- ω near the wall Model. It is a low Reynolds number model and a "universal" model in industrial applications. In terms of resolution requirements, this model is consistent with k- ω, and the model is similar to the low Reynolds number model, but its formula eliminates k- ω Model and k- ε weaknesses exhibited by the model.

A velocity inlet is selected for the model to simulate the inflow of tidal currents. A pressure outlet with a reference pressure of 101325 Pa and a relative pressure of 0 Pa is chosen without considering the factor of multiphase flow. When simulating the ups and downs of waves on the upper and lower walls, the inflow and outflow of seawater flow are simulated. The model selects the upper surface as the inlet for wave inflow and the lower surface as the outlet for wave outflow, with a reference pressure of 101325 Pa and a relative pressure of 0 Pa. Left and right are set as wall surfaces. Axial and centrifugal water turbines are set with no sliding wall surface and a friction coefficient of 0.5. This model adopts a SIMPLE algorithm and second-order upwind surface, with turbulent kinetic energy of 0.8, turbulent viscosity of 1, and residual convergence standard less than 10-4.

In order to save computing resources, the calculation model discards the fixed bracket and peripheral fixed structure of the water turbine and only retains the centrifugal and axial flow water turbine models.

4. Simulation and hydrodynamic analysis

4.1 Impact of attack angle and external characteristics of water turbine

Figure 4 shows the relationship curve between the angle of attack and the energy harvesting coefficient of the water turbine. As the angle of attack increases, the energy harvesting coefficient gradually increases. When the attack angle of the centrifugal water turbine is 25°, the energy harvesting coefficient reaches its maximum, with a Cp1 of 56.47%. When the attack angle of the axial flow turbine is 35°, the energy harvesting coefficient reaches its maximum, with a Cp2 of 45.75%. The angle of attack simultaneously affects the self-starting and energy-harvesting efficiency of the water turbine. A too-small angle of attack makes it difficult for the water turbine to self-start, while a too-large angle of attack can cause a stall of the airfoil, greatly reducing the energy harvesting coefficient of the water turbine. Therefore, it is necessary to choose an appropriate angle of attack for the water turbine. In order to verify the matching effect of different attack angles on dual vertical axis water turbines (centrifugal and axial flow turbines), Figure 5 shows the total energy harvesting coefficient of the system under different attack angle matching. Table 1 shows the simulation data of the energy harvesting coefficient for dual vertical axis water turbines with centrifugal and axial flow impellers using different attack angle matching. Finally, the angle of attack for centrifugal turbines is 25°, and for axial turbines it is 35°.

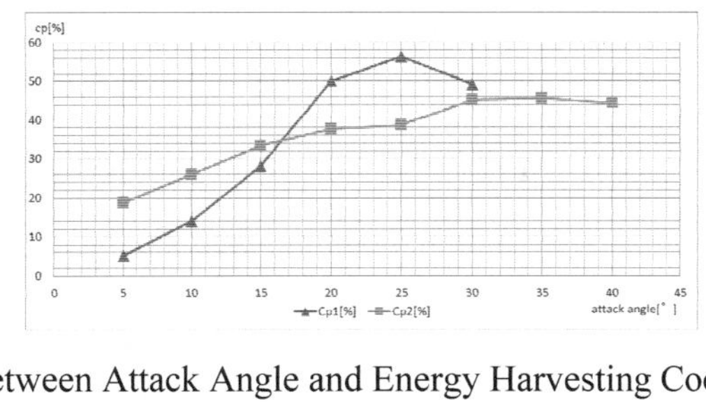

Figure 4. Relationship between Attack Angle and Energy Harvesting Coefficient of Water Turbine.

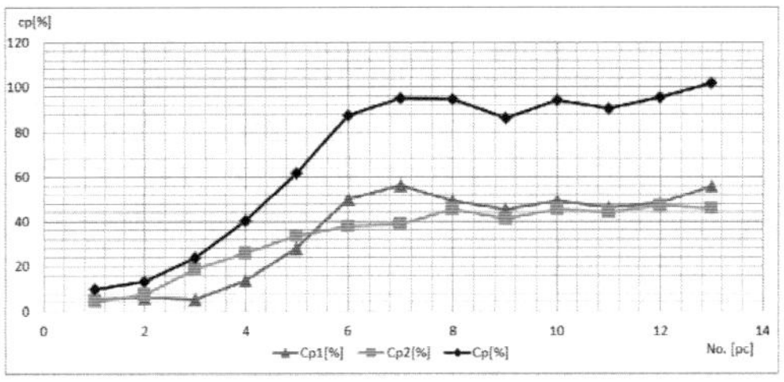

Figure 5. Relationship between Attack Angle and Energy Harvesting Coefficient of Dual Vertical Axis Water Turbine.

Table 1. Water Turbine Attack Angle and Performance Data.

Item	1	2	3	4	5	6	7	8	9	10	11	12	13
attack angle_axial[°]	5	10	5	10	15	20	25	30	25	30	35	40	35
attack angle_centri[°]	5	10	5	10	15	20	25	30	20	20	20	20	25
blade No._axial[pc]	3	3	5	5	5	5	5	5	5	5	5	5	5
blade No._centri[pc]	5	5	5	5	5	5	5	5	5	5	5	5	5
torque_axial[Ncm]	1.68	2.67	6.82	9.47	12.13	13.64	14.05	16.42	14.91	16.42	16.07	17.03	16.58
torque_centri[Ncm]	-32.66	-38.41	-32.70	-86.46	-174.06	-308.53	-348.05	-303.55	-278.94	-303.02	-284.00	-298.89	-345.75
shaft power_axial[W]	0.42	0.67	1.70	2.37	3.03	3.41	3.51	4.10	3.73	4.11	4.02	4.26	4.14
shaft power_centri[W]	8.16	9.60	8.18	21.62	43.51	77.13	87.01	75.89	69.74	75.75	71.00	74.72	86.44
E1[W]	154.08	154.08	154.08	154.08	154.08	154.08	154.08	154.08	154.08	154.08	154.08	154.08	154.08
E2[W]	9.06	9.06	9.06	9.06	9.06	9.06	9.06	9.06	9.06	9.06	9.06	9.06	9.06
Cp1[%]	5.30	6.23	5.31	14.03	28.24	50.06	56.47	49.25	45.26	49.17	46.08	48.50	56.10
Cp2[%]	4.64	7.37	18.81	26.13	33.47	37.64	38.78	45.32	41.14	45.33	44.36	47.01	45.75
Cp[%]	9.94	13.60	24.12	40.16	61.72	87.70	95.25	94.57	86.40	94.50	90.44	95.51	101.85

4.2 Effects of tip speed ratio and reynolds number on the external characteristics of hydraulic turbines

The expression for Reynolds number is [7]:

$$R_e = \frac{\rho C \omega R}{\mu} \tag{8}$$

The expression for the tip speed ratio is:

$$\lambda = \frac{\omega R}{U} \tag{9}$$

The relationship between the coefficient of harvesting coefficient C_p and the tip speed ratio and torque coefficient is as follows [8]:

$$C_p = \lambda C_T \tag{10}$$

where R_e is the Reynolds number, ρ is the fluid density, C is the chord length of the blade, ω is the angular velocity of the water turbine, R is the radius of the water turbine, μ is the fluid dynamic viscosity, and U is the fluid velocity at infinity. Figures 6 and 7 below show the relationship between the energy harvesting coefficient C_p, tip speed ratio, and Reynolds number for centrifugal and axial flow turbines. As the tip speed ratio and Reynolds number increase, the energy harvesting coefficient of the water turbine also gradually increases. When the tip speed ratio reaches 1.25, and the Reynolds number is 1.3e^5, the energy harvesting coefficient of the centrifugal water turbine reaches 69.1%. When the tip speed ratio reaches 1.5 and the Reynolds number reaches 5.5e^4, the energy harvesting coefficient of the axial flow turbine reaches 57.6% [9].

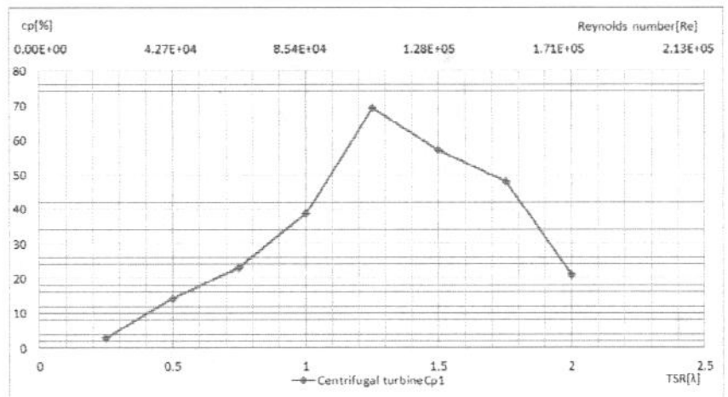

Figure 6. Relationship between the Energy Harvesting Coefficient and The Tip Speed Ratio and Reynolds Number of a centrifugal Water Turbine.

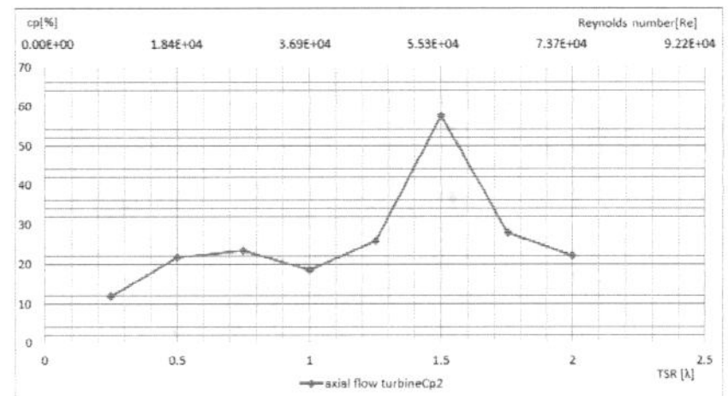

Figure 7. Relationship between the Energy Harvesting Coefficient and The Tip Speed Ratio and Reynolds Number of An Axial Flow Turbine.

4.3 Vortex intensity analysis

The following Figure 8 shows the distribution of eddy current intensity in the XY section of a dual vertical axis water turbine, with tip speed ratios for 8a to 8h, respectively λ= Distribution at 0.25, 0.5, 1, 1.25, 1.5, 2, 2.5, 3. The 252-degree azimuth angle blades of a centrifugal water turbine can maintain good flow at a tip speed ratio of 0.25~1.25 due to being located upstream of the water flow. There is no boundary layer separation phenomenon, and there is no obvious vortex shedding phenomenon at the trailing edge. When the tip speed ratio is greater than 1.5, the trailing edge falls off and flows downstream along the axial flow rotation domain. The stagnation point in the leading edge region moves forward, causing the vortex to move forward, which affects the hydrodynamic performance of the blades at that azimuth angle. Under 180 degrees, when the tip speed ratio is less than 1.5, vortices are generated on the suction surface. As the mainstream spreads downstream, boundary layer separation occurs when the tip speed ratio is greater than 1.5. As the tip speed ratio increases, boundary layer separation becomes more severe. The 108-degree blade is located in the downstream area and relies on the Kantar effect for work. There is a significant vortex shedding phenomenon at both the leading and trailing edges, and the shedding vortices at the trailing edge and the 180-degree blade converge to flow towards the blade at an azimuth angle of 36 degrees. The blade with a 36-degree azimuth angle is also located in the downstream area, and the fluid flows from the trailing edge to the leading edge. Both the leading and trailing edges have obvious boundary layer separation and vortex shedding phenomena and are also affected by 180-degree and 108-degree trailing edge vortices, resulting in low efficiency. The 324-degree azimuth angle blade is located in the upstream region, but the fluid also flows from the trailing edge to the leading edge. Both the leading and trailing edges have obvious boundary layer separation and vortex shedding phenomena. In this state, the blade cannot effectively use the lift drag principle to do work, which is equivalent to the principle of an impulse turbine to do work. There is a direct positive correlation with speed. The higher the speed is, the greater the work done by the blade at this azimuth angle will be, which confirms the results in Figure 7. When the tip speed ratio is 0.25, the energy harvesting efficiency is the lowest under these four tip speed ratios.

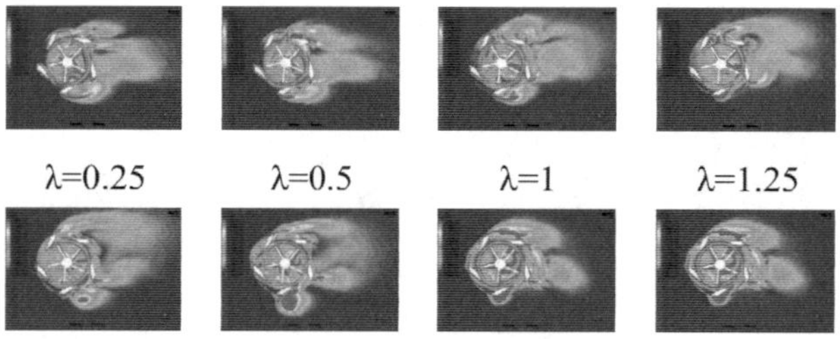

λ=0.25 λ=0.5 λ=1 λ=1.25

λ=1.5 λ=2 λ=2.5 λ=3

Figure 8. Eddy Current Intensity Distribution in XY Section of Water Turbine.

4.4 Pressure distribution analysis of water turbine

As shown in Figure 9 is the pressure distribution on the blade surface of an axial flow turbine, where a and b are the pressure cloud maps of the inflow and outflow of the axial flow turbine when the tip speed ratio is 1.25; c and d are the pressure cloud maps of the inflow and outflow of the axial flow turbine when the tip speed ratio is 1.5. According to the analysis in section 3.2, the pressure cloud map was selected at the tip speed ratio with the highest energy harvesting coefficient at λ= 1.25 and λ=1.5. The working principle of an axial flow turbine is to collect the potential energy of waves, convert it into the rotating mechanical energy of the turbine, and then convert it into the electrical energy of the generator. The pressure difference between the upper and lower surfaces of the blades is maximum at 30° and 60°. In contrast, the pressure difference at other angles is relatively small compared to these two angles. This is because a portion of the wake generated by the peripheral centrifugal turbine is carried backward by the main flow. A portion enters into the flow field of the axial flow turbine, interacting with the blades at these two angles, based on the principle of the Kantar effect, as well as Equations (5) and (10), the wake of a centrifugal water turbine can improve the energy harvesting efficiency of an axial flow water turbine; At the same time, the force on the blades in the first and fourth quadrants is greater than that in the second and third quadrants, causing the drive shaft to produce torsional oscillation and periodic series movement during operation, reducing the overall lifespan of the machine.

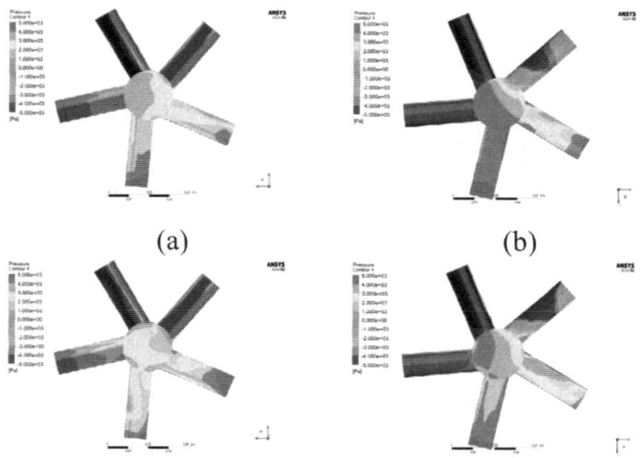

(a) (b)

(c) (d)

Figure 9. Cloud Map of Surface Pressure on Axial Flow Turbine Blades:(a) λ= 1.25 inflow side;(b) λ=1.25 outflow side;(c) λ=1.5 inflow side;(d) λ= 1.5 outflow side.

As shown in Figure 10, the pressure distribution on the surface of the centrifugal water turbine blades is shown, where a and b are the suction side pressure cloud maps of the centrifugal water turbine when the tip speed ratio is 1.25; c and d are the pressure side pressure cloud maps of the centrifugal water turbine when the tip speed ratio is 1.5. At azimuth angles of 180° and 108°, the pressure difference between the upper and lower surfaces of the blades is greater than that of the other three blades. Similar to coaxial flow turbines, the driving shaft of centrifugal turbines also experiences an uneven radial force due to the pressure imbalance of the five blades, resulting in periodic series motion during operation and reducing the overall lifespan of the machine.

5. Conclusions

How to use a single device to collect wave and current energy in the ocean has always been a challenge for the academic community, and when and how to integrate them is an urgent problem that needs to be solved. The paper integrates centrifugal and axial flow turbines, driving a rotating shaft together. The centrifugal turbine collects the kinetic energy of tidal energy, while the axial flow turbine absorbs the potential energy of wave energy. The airfoils of the two water turbines are designed using biomimetic principles, adopting the contour of a sailfish during rapid swimming. By using the Fluent batch processing function, the following conclusions were obtained:

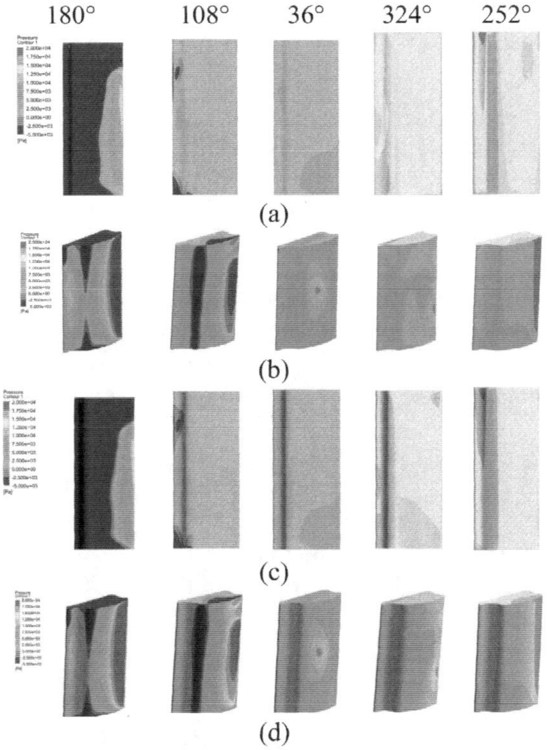

Figure 10. Cloud Map of Surface Pressure on Centrifugal Turbine Blades:(a) λ=1.25 suction side;(b) λ=1.25 pressure side;(c) λ=1.5 suction side;(d) λ=1.5 pressure side.

- With the increase of the tip speed ratio and Reynolds number, the energy harvesting coefficient of the water turbine shows a process of first increasing and then decreasing. When the tip speed ratio is 1.25, the active energy coefficient of the centrifugal water turbine reaches its maximum of 69.1%. When the tip speed ratio is 1.5, the maximum energy harvesting coefficient of the axial flow turbine is 57.6%. Both torque coefficient and power coefficient are related to angular velocity ω, flow velocity U, and blade tip speed ratio λ. The product of the torque coefficient and tip speed ratio is the energy gain coefficient.

- The trend of Reynolds number and tip speed ratio is similar. In the initial stage, the energy harvesting coefficient increases with the increase of Reynolds number. When the Reynolds number reaches the optimal state, the energy harvesting coefficient reaches the highest, and in the later stage, it decreases with the increase of the Reynolds number. When the Reynolds number is 1.3e^5, the maximum energy harvesting coefficient of the centrifugal water turbine reaches 69.1%. When the Reynolds number is 5.5e^4, the maximum energy harvesting coefficient of the axial flow turbine reaches 57.6%.
- Based on the principle of the Kantar effect, the wake of a centrifugal water turbine can improve the energy harvesting efficiency of an axial flow water turbine. The uneven force on the five blades of an axial flow turbine can cause the drive shaft to produce torsional swing force, while the pressure imbalance on the five blades of a centrifugal turbine can lead to uneven radial force on the drive shaft. During the operation of torsional swing force and radial force, periodic series motion occurs, affecting the overall service life of the machine.

References

[1] ParkYH.Analysis of characteristics of dynamic tidal power on the westcoast of Korea[J]. Renewable and Sustainable Energy Reviews, 2017,(68):461-474

[2] Tian Qiyun. The Definition of Marine Energy in the Energy Law[J]. Academic Journal of Zhongzhou. 2022,(8):65-68.

[3] Wang Xiangnan,Jia Ning,Xue Caixia,Wang Ji,Ma Changlei. The Industrialization Development of Marine Renewable Energy[J]. Marine development and management,2019,36(12):14-18

[4] MBarzegar,MJKetabdari,Kayhan,D.Palaniappan.Boussinesq Modelling of Waves and Currents in the Presence of Submerged Detached/ Discontinuous Breakwaters[J].2020,4(18):1-18

[5] Chen Wenpeng, Liu Yingchao, Chen Liwei. Study on hydrodynamic performance of horizontal tidal turbine rotating passively based on UDF [J]. The Ocean Engineering, 2018, Vol. 36 No. 2, 119-126

[6] Yan Cheng. Design of Floating Horizontal Axis Power Flow Turbine Blades and Studt of Hydrodynamic Characteristics[D]. Harbin: Harbin Institute of Technology, 2019.

[7] FERNÁNDEZ JIMÉNEZ, Aitor; BLANCO-MARIGORTA, Eduardo; GHARIB-YOSRY, Ahmed; FERNÁNDEZ-PACHECO, Víctor-Manuel; ÁLVAREZ-ÁLVAREZ, Eduardo. EVALUATION OF A VERTICAL AXIS HYDROKINETIC TURBINE FOR WATER CHANNELS[J]. DYNA. March 2021, vol. 96, no. 2, 6 pp

[8] Bian Peixiang, Yang Zhihong, Wang Yong, Ma Penglei, Wang Shiya. Hydrodynamic performance of Savonius water turbine[J]. Journal of Zhejiang University(Engineering Science). 2018,52(2):268-272

[9] Omar D. Lopez Mejia , Jhon J. Quiñones and Santiago Laín. RANS and Hybrid RANS-LES Simulations of anH-Type Darrieus Vertical AxisWater Turbine[J]. Energies 2018, 11, 2348

Journal of Physics: Conference Series 2815 (2024) 012022
doi:10.1088/1742-6596/2815/1/012022

Centerline measurement of rotary kilns based on the free station method

Yuhan Niu[1,*], Jingbing Wu[1] and Yun Zhang[1]

[1]School of Mechanical and Electronic Engineering, Wuhan University of Technology, Wuhan, Hubei 430070, China

[*]Corresponding author's e-mail: nyh@whut.edu.cn

Abstract: The rotary kiln is an important production equipment in the building materials and metallurgy industry, and its operating conditions, safety, and efficiency largely determine the economic benefits of its users and the safety of workers on site. At present, the testing methods for each rotary kiln part are relatively mature and can be used to realize accurate measurements in a dynamic environment. However, the main means of measurement is limited by the site conditions, and the actual detection speed is slow; the preparation work is cumbersome and will be subject to the measurement of the site obstacles and other environmental factors constraints. In this paper, we propose and design an optimized rotary kiln centerline measurement method using the free station method and verify its feasibility with respect to the optimization of the centerline offset, which is an important measurement item in rotary kiln measurement.

1. Introduction

Rotary kiln is a kind of large-scale, low-speed, and heavy-duty production equipment widely used in the construction industry, which is mainly applied to calcine cement clinker as well as calcine clay, limestone, and slag drying, production of alumina, etc., among which the cement industry is particularly extensive use of rotary kiln. The technical performance and operation of the rotary kiln, to a large extent, determines the quality, output, and cost of the enterprise products, which is the enterprise's valuable equipment, usually a rotary kiln annual mechanical maintenance cost of about 10 million yuan [1]. Ensuring the safe operation of the rotary kiln and minimizing manufacturing costs are important issues nowadays. Therefore, rotary kiln abnormalities need to be detected and diagnosed quickly to avoid kiln performance degradation or even damage [2]. The most tedious and important measurement item in the field measurement is kiln centerline offset measurement. The measurement of this item is based on the establishment of the baseline. Usually, this project and the establishment of the baseline takes several days or even dozens of days, and due to the inconsistency of different site conditions, it is difficult to generalize the difficulty, accuracy, and complexity of the baseline establishment. Most rotary kilns will have a series of devices that are difficult to move or remove, such as blowers on both sides of the base for the tires and supporting wheels. These devices become obstructions in the process of establishing the datum line, greatly affecting the speed and accuracy of the datum line.

Therefore, the main purpose of this paper is to develop a set of new standard procedures for the above measurement items: centerline offset measurement. The measurement process is free from dependence on the datum line and is not subject to the constraints of the site's special environment. It can effectively solve all the problems existing in the current kiln site measurement, improve the measurement efficiency and accuracy, and reduce implementation difficulties.

2. Centerline measurement method

2.1. Instruments used for measurement and theoretical methods

In the current rotary kiln measurement method, some of the more critical measurement projects use the total station to determine, such as centerline offset measurement, the space position of the supporting wheels detection, and other projects. In this paper, we also choose to use a total station for on-site measurement.

The total station free setup method, also known as the total station rear rendezvous method (hereinafter referred to as free station method), is based on the following principle: the total station at the location to be determined is set up, the distance and angle between the point to be determined and the known control point are measured, the directional error equation and side length error equation are established based on the directional observation value and the side length observation value, and then the coordinate of the point to be determined is calculated in accordance with the principle of least squares [3]. Its outstanding advantage is that this method combines the flexible, efficient, and accurate features of a total station. In a site with certain obstacles, it can be convenient and flexible to set up the station according to the needs of the actual project so as to achieve the purpose of a more flexible location of the selected measurement site [4]. This method has been widely used in other measurement areas [5]. This method can well make up for the problems of the current measurement scheme described above, so it is chosen to design the optimized centerline measurement method based on this method.

2.2. Measurement of centerline horizontal position offset

When measuring the horizontal position deviation of the rotary kiln centerline, the points that need to be measured are the outermost points of the tires for each stall. In the traditional method, these points are obtained by means of a datum line. However, this method does not have a datum line, so this paper proposes a method to find the outermost point of the roller belt and verify its accuracy by calculation. The method is divided into the following two steps:

1) The theoretical height difference between the center of the tire and the center of the supporting wheel can be obtained by consulting the installation standard of the rotary kiln being measured to obtain the height of the center of the tire from the base, the height of the center of the supporting wheel from the base, and other installation data.

2) The horizontal knob of the total station is locked, and the measuring angle of the total station is adjusted up and down to measure the height difference obtained in the previous step. The flat distance in the data returned by the total station is observed; when the light emitted by the total station is perpendicular to the surface of the rotary kiln tire, the position with the smallest flat distance is the outermost point of the tire.

However, in the actual measurement process in the field, it is not possible to ensure that the total station optical path is perpendicular to the side of the tire, so this method needs to be verified and adjusted. It is known that the total station can be leveled according to the horizontal bubble before measurement, so the total station in locking the horizontal knob and adjusting the longitudinal measurement angle, the measurement of the optical path of the measurement of the possible spatial location of the measurement of a plane perpendicular to the horizontal plane. Assuming that the rotary kiln tire is a horizontally placed cylinder, the tangent line between the two is shown in Figure 1.

Journal of Physics: Conference Series **2815** (2024) 012022 doi:10.1088/1742-6596/2815/1/012022

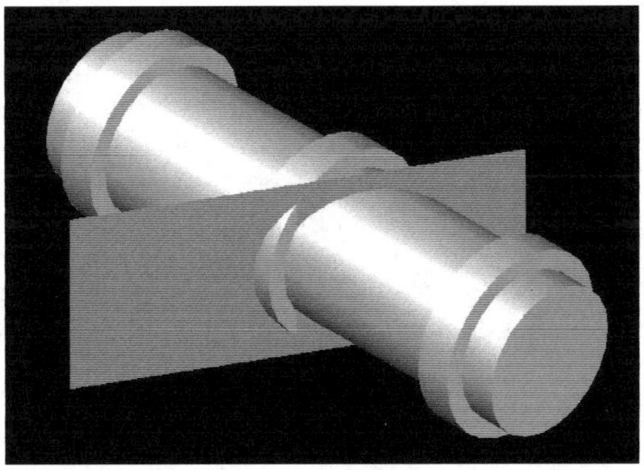

Figure 1. Tangent of the optical path of the total station to the tire.

The rotary kiln is actually placed at an angle. The total station position is set as O point, the z-axis is perpendicular to the horizontal plane, the x-axis is parallel to the axis of the tire, and the y-axis is parallel to the diameter of the tire. The optical path plane can be expressed as:

$$y = kx \tag{1}$$

The rotary kiln tilt angle tan value is θ, and the axis passes through the point P1 (A, B, C) and parallel to the x-axis. The axis is set, then the rotary kiln axis can be expressed as:

$$\begin{cases} z - C = \theta(x - A) \\ y = B \end{cases} \tag{2}$$

The radius of the roller belt is R (the absolute value of R is less than the absolute value of B), and the surface of the tire can be expressed as:

$$R = \left[(y - B)^2 + \frac{(z - \theta x - C + \theta A)^2}{1 + \theta^2}\right]^{\frac{1}{2}} \tag{3}$$

Combining Equation (1) with Equation (3), the equation of the curve at the point of intersection of the optical path and the surface of the tire can be solved:

$$\begin{cases} z = \sqrt{[R^2 - (kx - B)^2](1 + \theta^2)} \\ \quad +\theta x + c - \theta A \\ y = kx \end{cases} \tag{4}$$

The projected distance (HD) of this curve line is calculated from the origin on the horizontal plane. We can find the point where the minimum value of $(x^2 + y^2)$ is located, and obtain the result:

$$\begin{cases} \left(\frac{B - R}{k}, B - R, \theta\frac{B - R}{k} + c - \theta A\right) & B > 0 \\ \left(\frac{B + R}{k}, B + R, \theta\frac{B + R}{k} + c - \theta A\right) & B < 0 \end{cases} \tag{5}$$

It can be seen that the location with the smallest flat distance is the outermost point of the tire.

After getting the required outermost point of the tire, the next step is to determine the ideal centerline, here to the most common three-stage rotary kiln as an example, the ideal centerline benchmark for the first and third gear tire horizontal center. In determining the position of the horizontal center of the stalls on the way, the baseline three-point cross-direction method is used to calculate the width of the baseline and the outermost point of the tire to the baseline to determine the distance. The method used by the Swiss company TomTom is to obtain data for the entire surface of the tire by scanning multiple points on the entire side of the belt and by fitting a circle [6]. This approach requires making sure that the scanned points are in the same position in the axial direction, which is difficult to achieve in practice. In view of the current shortcomings of these two methods, this paper proposes a method of determining the ideal horizontal centerline because regardless of the position of the points measured at both ends of the tire in the axial direction, the midpoint of the line between the two points will always be in the centerline of the radial position of the two sides of the tire. Thus, it is possible to use the midpoint of the coordinate line of the outermost point of the tire on both sides of the tire of the first gear as the origin, and the point points to the direction of the midpoint of the line of the outermost point of the two sides of the tire of the third gear as the x-axis direction to establish a spatial coordinate system as shown in Figure 2. Since the original coordinate system of the free station and this coordinate system have a common z-axis direction, the measured points can be positioned in the new coordinate system by simple translation and horizontal rotation.

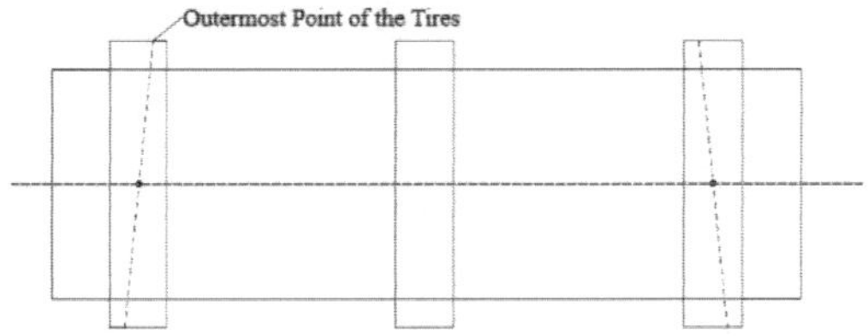

Figure 2. Ideal horizontal centerline determination method.

2.3. Centerline longitudinal offset measurement

After obtaining the diameter of the tire in each gear, you can start the measurement of the longitudinal position of the tire; this method is selected as the measurement of the lowest point of the tire, recorded as point C [7]. We measure the spatial coordinates of the point C of each gear and transform them into the new coordinate system mentioned above, as shown in Figure 3.

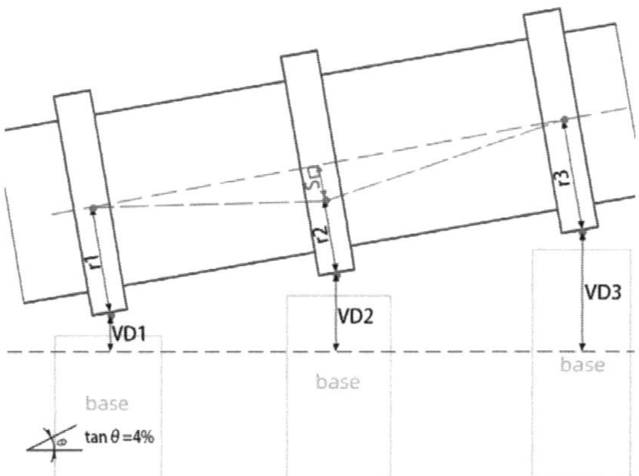

Figure 3. Centerline longitudinal offset measurement.

3. Adjustment program for centerline offset

The adjustment of centerline offset is mainly through the adjustment of the supporting wheels on both sides of the kiln so that the position of the center point of the tires changes, thus achieving the purpose of adjusting the center axis of the rotary kiln as a straight line [8]. In order to quantitatively analyze this adjustment process, the kiln center axis adjustment process needs to be modeled [9].

The main way of adjusting the supporting wheels is generally to move the pallet in a plane perpendicular to the kiln axis in the horizontal plane by rotating the pallet adjusting screw [10]. Modeling is carried out on the cross-section perpendicular to the rotary kiln axis, and based on the relevant data already obtained, the model is shown in Figure 4.

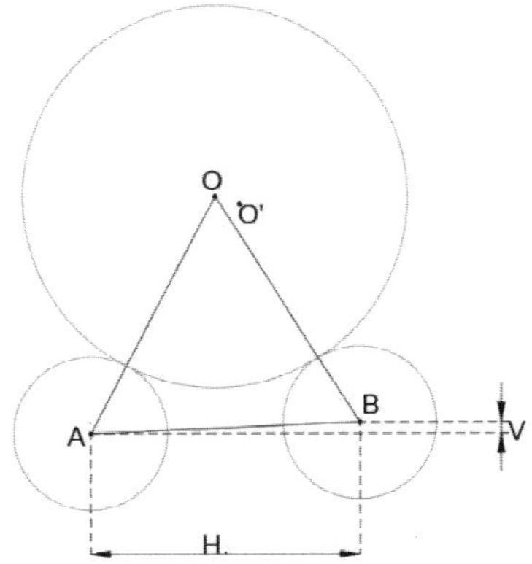

Figure 4. Centerline adjustment program.

Point O is the current center of the roller belt, with this as the coordinate origin, the x-axis vertically upward, and the y-axis horizontally to the right to establish a right-angle coordinate system. v is the difference in elevation of the center of the right pallet with respect to the left, H is the distance between the centers of the two pallets in the direction of the y-axis, R is the radius of the roller belt, which has been obtained above, and r is the radius of the pallets on both sides, which can be obtained from other measurements made in the field. Point O' is the target of adjustment for this gear.

$$\begin{cases} x_2 - x_1 = V \\ y_2 - y_1 = H \\ L_1 = R + r_1 \\ L_2 = R + r_2 \\ {x_1}^2 + {y_1}^2 = {L_1}^2 \\ {x_2}^2 + {y_2}^2 = {L_2}^2 \end{cases} \tag{6}$$

To facilitate the programmatic nature of the calculation process, the equation representation of y_1 and each constant is calculated:

$$\begin{cases} T = {L_2}^2 - {L_1}^2 - H^2 - V^2 \\ \left(\frac{H^2}{V^2} + 1\right){y_1}^2 - \frac{TH}{V^2} y_1 + \frac{T^2}{4V^2} - {L_1}^2 = 0 \end{cases} \tag{7}$$

According to Equation (6), the value of y_1 can be found more simply, and then several other coordinate values can be found based on the relationship of y_1 to x_1, x_2, and y_2.

According to the rotary kiln, the centerline adjustment process is generally through the radial position of the supporting wheels in the rotary kiln. The adjustment process is mainly to adjust the y value of the center of the two supporting wheels; before and after the adjustment, the model in the L_1, L_2, x_1, and x_2 four parameters will not change, so we can use the last two equations in Equation (6):

$$\begin{cases} (x_1 - x_0)^2 + (y_1' - y_0)^2 = {L_1}^2 \\ (x_2 - x_0)^2 + (y_2' - y_0)^2 = {L_2}^2 \end{cases} \tag{8}$$

y_1' and y_2' are the values of the two supporting wheels after adjustment, which can be calculated according to Equation (8). The distance to be adjusted can be obtained by comparing it with the position before adjustment obtained by Equation (6).

4. Feasibility experiments on measurement methods

In order to ensure the practicality of the measurement method, two sets of experiments were carried out on the method, one on the measurement points of an ideal three-stage rotary kiln model placed in a laboratory environment and the other on the rotary kiln in a field measurement environment and compared with the results obtained by the three-point crosshatching method commonly used today.

4.1. Experimental results in a laboratory setting

Since the laboratory environment does not have vibration, high temperature, and interference from various other environmental factors during field measurements, this experiment only tests the feasibility of the calculation process of the method itself. The placement of the measurement point is set as a two-block left deviation of 2 mm and a downward deviation of 1 mm. The placement of the point position is tested, and the measurement results are shown in Table 1.

Table 1. Centerline offset measurement methods tested in a laboratory environment.

Number	Horizontal (mm)	Vertical (mm)
1	2.037	-0.922
2	2.044	-0.931

In Table 1, the measurement results are the offset value to the left and the offset value to the down, and it can be seen that the measurement accuracy can be controlled within 0.2 mm.

4.2. Experimental results in the field measurement environment

The results of the field measurements are shown in Table 2.

Table 2. Centerline offset measurement methods in the field measurement environment.

Measurement items	Measurement results of traditional methods (mm)	Measurement results of this method (mm)
1st gear pulley diameter	5147.4	5145.4
2nd gear pulley diameter	5165.9	5164.1
3rd gear pulley diameter	5160.7	5158.9
Centerline left offset value	1.9	2.1
Centerline up offset value	0.6	0.6

The optimized free-standing method is within 0.2 mm of the traditional three-point crosshatch in terms of actual results, where the same target is used for the C-point measurement. Hence, the results are as expected, i.e., exactly the same as the traditional method. The measured values of all three roller belts are smaller than the traditional three-point crosshair method by about 2 mm, which may be due to the traditional free-standing method depending on the baseline width to project the diameter of the tires. When the baseline width of the error occurs to a certain extent, the diameter of the tires of all the gears will be all the same size as the error occurs. However, because the same size of diameter error occurs in the three-point cross alignment method, it will not affect the final measurement results, so the diameter of the tires is only as a process quantity, but not concerned about its final conclusion. This method can get the accurate diameter of the tires.

5. Conclusion

For the current rotary kiln centerline field measurement problems, this paper proposes the use of an optimized free station method centerline measurement method, and compared with the traditional method through experiments and field tests, the conclusions are:

(1) The method of using the free station-setting method for spatial substitution of measurement points to combine the measurement points into the same coordinate system for analysis can eliminate the establishment of a datum line and quickly and accurately derive the results of centerline offset.

Journal of Physics: Conference Series **2815** (2024) 012022 doi:10.1088/1742-6596/2815/1/012022

(2) The optimized centerline measurement method can accurately derive some process quantities that are not required in the traditional three-point cross-alignment method for technicians' reference.

This method is a simple measurement process, which can avoid some cumbersome preparations in the centerline offset measurement of rotary kiln and reduce the impact of the field environment on the measurement process to a certain extent. It has the value of popularization and application in rotary kiln measurement projects.

References

[1] Liu Z and Bao X Y. (2019) Study on Thrust On-line Monitoring System of Rotary Kiln Roller. Mechanical Engineer, 8:11-12+15.
[2] Rusinski, E, Stamboliska, Z and Moczko, P. (2013) Proactive control system of condition of low-speed cement machinery. Automation in Construction, 31: 313-324.
[3] Cheng X J and Miu D. (2008) Discussion on Accuracy of Liberal Station Method with Total Station Machine. Railway Investigation and Surveying, 6: 1-4.
[4] Yan B D, Wang Y P, Tian H Z and Ma L Z. (2020) Discussion and analysis of the free station method with total station. Geotechnical Investigation and Surveying, 1: 47-52.
[5] Yan P Y, Zhao X J and Lin W M. (2021) Application of free station side-angle measuring intersection method in mine survey and C# programming. Mine Surveying, 49: 56-60.
[6] TomTom. (2020). Kiln axis alignment system. https://tomtomtools.com.
[7] Peng Q and Zhang Y. (2020) Research on Fault Detection Method for Cylinder Fault of the Rotary Kiln. Machinery Design & Manufacture,6: 87-90.
[8] Kuang M S, Luo H B and Zhang Y. Measurement and adjustment of the rotary kiln and roller position on line. Cement Engineering, 5: 42-45.
[9] Kussel U, Reiter M, Abel D, Schumacher M and Wen M. (2010) Observer-Based Model Predictive Control of Rotary Kilns - Application to Limestone Calcination. IFAC Proceedings Volumes, 43: 309-314.
[10] Zhang X J and Zhang Y. (2021) Research on Fault Extraction of Rotary Kiln Shell Based on Wavelet Packet Decomposition. Mechanical Engineer, 9: 64-67.

Journal of Physics: Conference Series **2815** (2024) 012023 doi:10.1088/1742-6596/2815/1/012023

An envelope statistic features and MFO-SVM fusion method for tidal stream turbine blades impact fault diagnosis

Zhen Wu [1,*] and Tian-zhen Wang [1]

[1] Logistics Engineering College, Shanghai Maritime University, Shanghai, 201306, China.

[*] Corresponding author's e-mail: 202130210108@stu.shmtu.edu.cn

Abstract: The fault diagnosis of a Tidal Stream Turbine (TST) blade impact fault benefits its stable operation. However, when the stream velocity changes, it is hard to distinguish between different fault severities directly by observing the changes in the signal feature. To address this problem, this paper proposes a blade impact fault diagnosis method based on envelope statistical features and MFO-SVM. The method is divided into two parts. In the first part, the Teager-Kaiser energy operator (TKEO) and sliding window technique are introduced to extract the envelope statistic features of the current signal, and then the local outlier factor (LOF) values of fault sample points are calculated to form a new set of feature samples; in the second part, the fault feature samples are input into the support vector machine (SVM) optimized by moth fame optimization (MFO) for fault diagnosis. The experimental results show that the proposed method is more accurate than traditional fault diagnosis methods.

1. Introduction

In the ocean, the TST blades are impacted by organisms of different masses, and the fault severities of blades vary under different impact intensities. Under the effect of small mass impact objects, the vibration and noise of the TST body will be intensified. When the mass of the impact objects increases, the blades under a certain fatigue accumulation may be cracked or even fractured [1]. Therefore, it is necessary to perform fault diagnosis of the blade impact faults of different severities and then judge the TST downtime and repair according to the result of diagnosis [2].

The studies on the diagnosis of TST blade impact faults are still limited to fault detection, and the diagnosis of impact faults of different severities has yet to be conducted. Staines et al. [3] used sound sensors and cameras installed on the TST to observe the impact behavior of fish, but the sensors are susceptible to corrosion. Therefore, utilizing the current signal output from the TST to diagnose blade impact fault is an effective way. Xie et al. [4] extracted the stator current signal envelope spectrum by Hilbert Transform (HT) and Fast Fourier Transform (FFT). Finally, the impact fault is detected by Principal Component Analysis (PCA), but the boundary dispersion in HT leads to low accuracy. Xie et al. [5] proposed a MEGK-means and PCA method to detect TST blade impact fault at variable stream velocity. Still, the method needs a reasonable basis for classifying TST conditions into three cases.

To diagnose the TST blade impact faults of different severities, this paper proposes a method based on envelope statistical features and MFO-SVM, which includes two parts. In the first part, the TKEO is first introduced to demodulate the stator current signal to obtain the envelope. The sliding window is used to split the envelope samples matrix with equal length to obtain multiple submatrices. The range, mean deviation, standard deviation, and coefficient of variation in each submatrix are calculated as the envelope statistical features. Finally, the four-dimensional envelope statistical features matrix is put into a LOF model. The LOF values of the fault sample points are selected to form a new set of feature samples. The second part proposes the MFO-SVM algorithm to diagnose and classify the new feature samples. The main innovations of the paper are as follows:

(1) Since the current signals are interfered with by variable stream velocity, it is hard to distinguish between different severities of blade impact faults by directly observing the current changes. This paper proposes to use the envelope statistic features to highlight the differences under different severities of faults.

(2) The traditional fault diagnosis methods applied to TST blades impact fault show low accuracy, so this paper proposes an MFO-SVM method to improve diagnosis accuracy.

2. Problem description

The variable stream velocity directly affects the rotation of the TST blades. When marine organisms impact the blades, the rotating speed will change rapidly, and the current signals will fluctuate [6].

The blade mechanical torque is affected by the stream velocity and can be expressed as:

$$T_m(t) = \frac{2\rho C_p S U_0^2(t) R}{\lambda} \tag{1}$$

where ρ is the density of the stream, C_p is the power coefficient, S is the swept area of blades, $U_0(t)$ is the stream velocity, R is the radius of blades, and λ is the blade tip speed ratio.

Due to $T_m \propto \omega_m$, ω_m is the mechanical angular velocity, and $\omega_m \propto \omega_e$ is in the direct-drive TST, ω_e is the current angular velocity, so the current angular velocity is affected by the stream velocity. Suppose an object of mass m impacts the TST blades, when the blades impact fault occurs, according to the angular momentum conservation equation, the following relationship exists:

$$\Delta\omega_m = \omega_m(t_2) - \omega_m(t_1) = -\frac{mr^2}{J + mr^2}\omega_m(t_1) \tag{2}$$

where J is the blades rotation inertia, r is the distance from the impact position to the impeller. $\Delta\omega_m$ is the change of blades angular velocity due to impact.

The angular velocity change will directly cause the stator current amplitude and phase variation. Therefore, the stator current is denoted as:

$$i(t) = [A(t) - A(t_1)]\cos[\omega_e t + p\Delta\omega_m t + \zeta] \tag{3}$$

where p is the number of pole pairs, and ζ is the initial angle. Since $J \gg mr^2$, the $\Delta\omega_m$ is small, when m changes, the current signal will produce different changes, but these changes are easily hidden under variable stream velocity. Therefore, it is necessary to propose a method for impact fault diagnosis.

3. The envelope statistic features and MFO-SVM method

The fault diagnosis method proposed in this paper is divided into envelope statistic features and the MFO-SVM algorithm.

3.1 Envelope statistic features

The TKEO performs well in extracting the signal envelope [7]. Assuming that the discrete single-phase stator current signal is $x(i) = A(i)\cos(\Phi(i)), i = 1,2,3,\cdots,N$, N is the length of the signal. Its energy operator is defined as:

$$\Psi[x(i)] = [x(i)]^2 - x(i-1)x(i+1) = A^2(i)\sin^2\Phi(i) \tag{4}$$

We define $y(i) = x(i) - x(i-1)$, and the instantaneous amplitude of the signal is calculated as follows:

$$A(i) = \frac{2\Psi[x(i)]\cdot\sqrt{\Psi[x(i)]}}{\sqrt{4\Psi[x(i)]\Psi[y(i)] - \Psi^2[y(i)]}} \tag{5}$$

Then, the sliding window divides the envelope samples matrix $A=[A(1),A(2),\cdots,A(N)]$ into n submatrices of equal length. The length of the window is ω, the subsamples matrix obtained is denoted as:

$$A_{n\times w}=[A_1,A_2,\cdots,A_n]^T \tag{6}$$

where $n\times w=N$, then four statistic features of the range R_k, the mean difference M_k, the standard deviation S_k, and the variation coefficient V_k are computed for each sub-matrix $A_k(k=1,2,\cdots,n)$ to form the envelope features samples matrix P.

$$P_{n\times 4}=[P_1,P_2,\cdots P_n]=\begin{bmatrix} R_1 & M_1 & S_1 & V_1 \\ R_2 & M_2 & S_2 & V_2 \\ \vdots & \vdots & \ddots & \vdots \\ R_n & M_n & S_n & V_n \end{bmatrix} \tag{7}$$

In the end, as an outlier detection model, the LOF model can highlight the degree of abnormality of the fault sample points. In this paper, a new set of feature samples is formed by calculating the LOF values of multiple fault sample points in the matrix P. The specific steps can be found in [8].

3.2 MFO-SVM algorithm

The SVM performs well in solving multi-classification problems. Since the new set of feature samples based on LOF values is non-linear, the penalty factors and slack variables are added to the SVM to enhance the generalization ability of the model. Meanwhile, the samples in the low-dimensional space are mapped into the high-dimensional space by using nonlinear mapping, which requires a suitable kernel function. The specific steps of SVM can be found in [9]. However, the penalty factor C and the kernel parameter γ affect the fault diagnosis of SVM, so this paper proposes the MFO algorithm to select the optimal parameters (C,γ).

In the MFO algorithm, the positions of moths and flames are used as candidates and optimal solutions, respectively, and then these positions are updated by calculating fitness. In this paper, the positions of moths are $D_j=(C_j,\gamma_j)$, the positions of flames are optimal parameters $F_l=(C_{b,l},\gamma_{b,l})$, and the fault diagnosis accuracy is set as a fitness function F_n, which is expressed as follows:

$$F_n=\frac{n_1}{n_1+n_2}\times 100\% \tag{8}$$

where n_1 and n_2 are the number of correctly and incorrectly diagnosed samples, respectively. The D_j are updated in the following manner:

$$\begin{cases} D_{j+1}=G(D_j,F_l) \\ G(D_j,F_l)=de^{gv}\cos(2\pi t)+F_l \\ d=\left|F_l-D_j\right| \end{cases} \tag{9}$$

where G is spiral flight function of the moths, and d is the distance between the moths and the flames. g is a spiral constant, and v is a constant in [-1, 1]. To avoid the algorithm falling into a local optimum, the number of flames is iteratively reduced. The specific process in MFO is described in [10].

Based on the above analysis, the specific steps of the MFO-SVM algorithm are shown below:

(1) Variable initialization: They are the number and dimension of moths, the maximum number of iterations, the maximum number of flames, spiral constants, etc.

(2) Calculation of fitness values: The SVM diagnosis accuracies are calculated as the fitness values based on the initialized moth positions (C,γ), which are ranked from smallest to largest and saved.

(3) We update the number of flames and then update the positional information of the moths and flames.

(4) We judge whether the algorithm has finished iterations; if so, the optimal parameters are output; if not, we return to Step (2) to continue iterative updating until the end of the algorithm.

Combining the above analysis, the overall flow of the proposed method is shown in Figure 1.

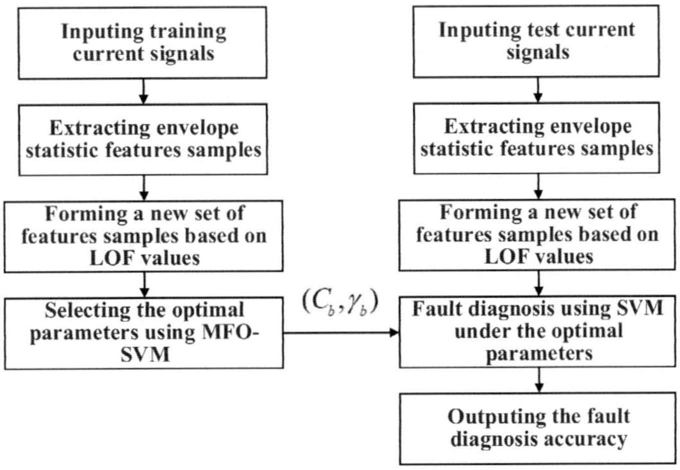

Figure 1. The overall flow of the proposed method.

4. Experimental results analysis

This work uses a direct-drive 230 W permanent magnet synchronous TST to collect the stator current signals. A specific description of the experimental platform can be found in [4-6]. The mean stream velocity is set to be 1.0 m/s, 1.1 m/s, 1.2 m/s, 1.3 m/s, 1.4 m/s, 1.5 m/s, and 1.6 m/s, respectively. At each stream velocity, the TST blade is impacted by 50 g, 90 g, and 130 g balls to simulate the three fault types under different fault severities. Three fault types are named as Fault Type 1, Fault Type 2 and Fault Type 3, respectively. Then, the stator current signals of the health state and three fault types are collected at different mean stream velocities. Figure 2 is the envelope under Fault Type 1 when the mean stream velocity is 1.1 m/s, and Figure 3 is under Fault Type 2 at 1.6 m/s. At the moment of the 90 s impact fault, the envelope will have a transient drop and then rise rapidly. Still, the amplitudes of the envelope are different at different mean stream velocities and for the different fault types, so the fault diagnosis cannot be achieved by directly observing transient changes in the envelope.

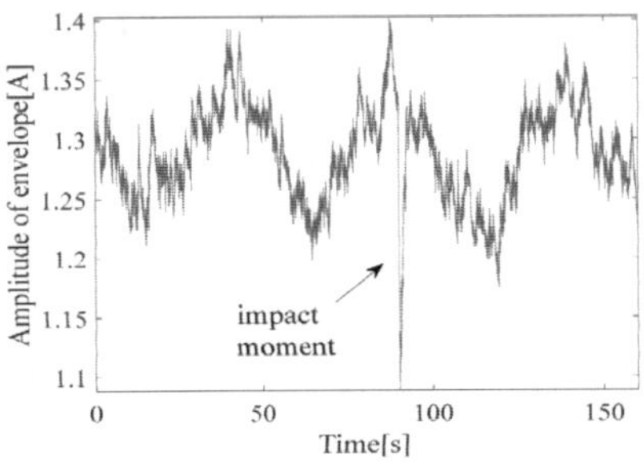

Figure 2. The envelope under Fault Type 1 at 1.1 m/s.

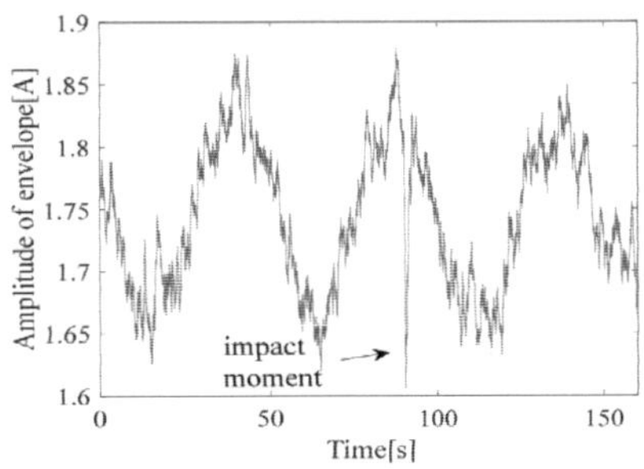

Figure 3. The envelope under Fault Type 2 at 1.6 m/s.

The window length is set to 200, 400, and 500, respectively, and then the set of feature samples based on LOF values under different window lengths is calculated. In Figure 4, for the healthy state and three fault types, the overall LOF values of the fault sample points are different. For example, when the window length is 200, the LOF values in the healthy state are all around 1, the LOF values of Fault Type 1 are all slightly larger than 4, the LOF values of Fault Type 2 are all around 6, and the LOF values of Fault Type 3 are all significantly larger than 7.

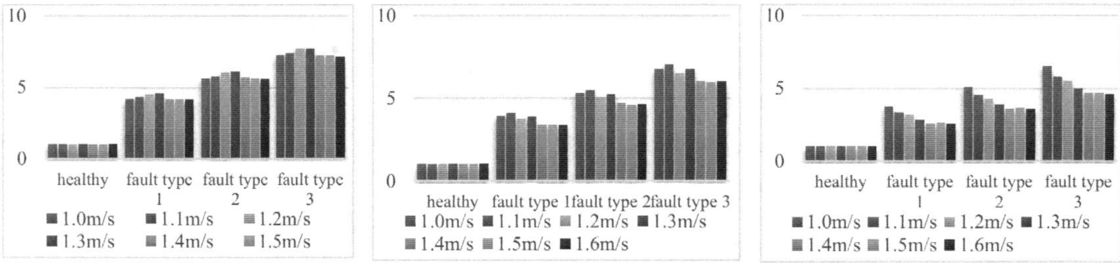

(a) (b) (c)

Figure 4. The histogram of LOF values: (a) $w=200$, (b) $w=400$, (c) $w=500$.

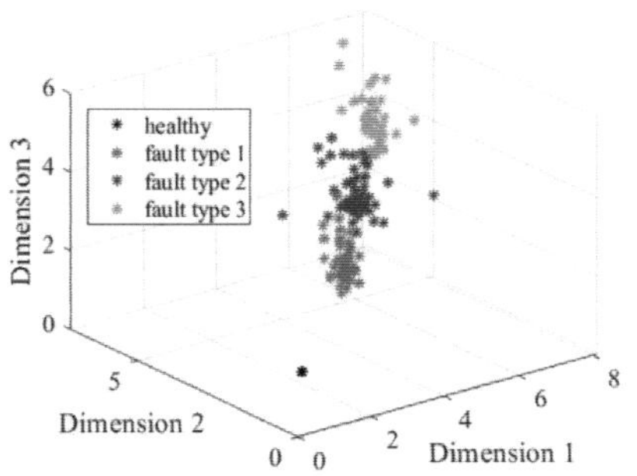

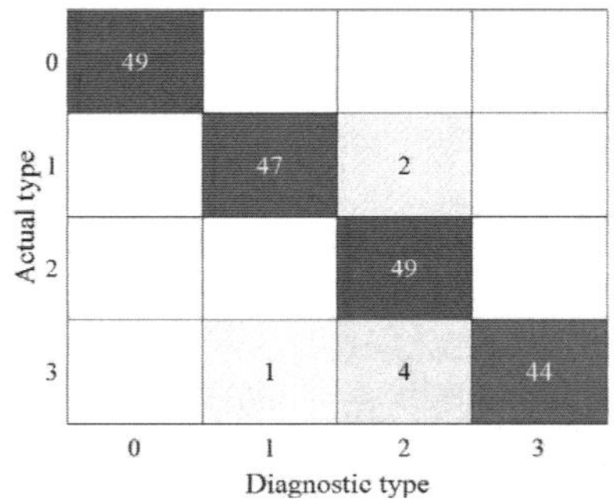

Figure 5. Diagnosis result of proposed method. **Figure 6**. Confusion matrix for the proposed method.

As seen in Figure 5, there is a high degree of separation of sample points from different types and a high degree of closeness of sample points from the same type. In Figure 6, there are only two sample points incorrectly diagnosed as Type 2 in Type 1. One and four sample points in Type 3 are incorrectly diagnosed as Type 1 and Type 2, respectively, so the fault diagnosis accuracy of the proposed method is 96.43%. Table 1 shows the impact fault diagnosis accuracy of different conventional methods such as the Extreme Learning Machine (ELM), the Long Short Term Memory Network (LSTM), the Random Forest (RF), the original SVM, and the Particle Swarm Optimised SVM (PSO-SVM). The results show that the proposed method gives the best diagnosis results.

Table 1. Impact fault diagnosis accuracy of different traditional methods.

	ELM	LSTM	RF	SVM	PSO-SVM	MFO-SVM
Accuracy	85.20%	64.8%	90.82%	88.27%	95.41%	**96.43%**

5. Conclusion

Aiming at different severities of TST blade impact faults, this paper proposes a fault diagnosis method based on envelope statistic features and MFO-SVM. In the method, the envelope statistical features are used to form a set of feature samples based on LOF values; then the MFO-SVM classifies the samples. The proposed method has a higher diagnosis accuracy than the traditional methods.

Reference

[1] Zhang M, Tang T and Zhou F. (2021). A Mechanical Impact Fault Detection Method Based on PCA for Marine Current Turbine [C]. In: The CAA Symposium on Fault Detection, Supervision, and Safety for Technical Processes. Chengdu, China. pp. 1–5.

[2] Zhang M, Wang T and Tang T. (2017). A multi-mode process monitoring method based on mode-correlation PCA for Marine Current Turbine [C]. In: IEEE International Symposium on Diagnostics for Electrical Machines. Tinos, Greece. pp. 286-291.

[3] Staines G J, et al. (2022). Capabilities of an Acoustic Camera to Inform Fish Collision Risk with Current Energy Converter Turbines [J]. Journal of Marine Science and Engineering, 10: 483.
[4] Xie T, Wang T, et al. (2021). Impact Fault Detection for Marine Current Turbine Blade via Hilbert Envelope Spectrum and PCA [C]. In: The CAA Symposium on Fault Detection, Supervision, and Safety for Technical Processes. Chengdu, China. pp. 1-6.
[5] Xie T, Zhang W D, et al. (2023). Impact fault detection for MCTs blade via MEGK-means and PCA under variable marine conditions [J]. Measurement Science and Technology, 13: 095906.
[6] Li Z, Wang T and Wang Y. (2020). A Wavelet Threshold Denoising-Based Imbalance Fault Detection Method for Marine Current Turbines [J]. IEEE Access, 8: 29815–29825.
[7] Gałęzia A and Gryllias K. (2021). Application of the Combined Teager-Kaiser Envelope for bearing fault diagnosis [J]. Measurement, 182: 109710.
[8] Chen Z, Xu K and Wei J. (2019). Voltage fault detection for lithium-ion battery pack using local outlier factor [J]. Measurement, 146: 544–556.
[9] Shi Q and Zhang H. (2021). Fault Diagnosis of an Autonomous Vehicle with an Improved SVM Subject to Unbalanced Datasets [J]. IEEE Transactions on Industrial Electronics, 68(7): 6248–6256.
[10] Mirjalili S. (2015). Moth-flame optimization algorithm: A novel nature-inspired heuristic paradigm [J]. Knowledge-Based System, 89.

Journal of Physics: Conference Series **2815** (2024) 012024 doi:10.1088/1742-6596/2815/1/012024

Current status and application of welding automation technology for bridge steel structure

Jiaying Ji, Xingxing Wang [*], Junyi Li, Shuaishuai Lv and Hongjun Ni

School of Mechanical Engineering, Nantong University, Nantong 226019, China

[*] Corresponding author: wangxx@ntu.edu.cn

Abstract: In bridge construction, welding automation technology has become an indispensable and important technical means for developing related industries. The paper mainly introduces the current status of welding automation technology in bridge steel structures, elaborates on its existing technology from the control of welding deformation, U-frame welding process, robot welding technology, etc., and gives full play to the positive role of advanced steel structure welding automation technology, and forms a good working mode and technical system. The difficulties in welding are analyzed, which plays a role in promoting the development of welding automation technology for bridge steel structures.

1. Introduction

With the improvement of China's economic development and infrastructure capacity, high-speed railways and expressways have been greatly developed, which has greatly promoted the development and improvement of China's transportation system. Among them, steel structure bridges have also made great progress. More and more steel structures have been built, and bridges with complex technology, large scale, and diverse functions have been built [1]. The steel structure is composed of steel materials; the entire structure contains a large number of steel columns and steel beam components, and all components are connected by screws, welding, and so on. With the development of the times, China's welding industry is bound to develop in the direction of intelligence, automation, and high efficiency.

2. Status quo and application of automatic welding technology for bridge steel structure

2.1 Research and development of special anti-deformation tire frame to control welding deformation

Xia [2] proposed an anti-deformation welding method to reduce or eliminate the welding deformation of components by optimizing the groove form and designing the anti-deformation tooling. When the anti-deformation jig device is used, it can realize the automatic disassembly of welded components, and the submerged arc automatic welding car or carbon dioxide protection welding is used.

By combining the advantages of traditional anti-deformation tire frames and semi-automatic welding anti-deformation tire frames, Shi [3] put forward the improvement. A good hydraulic anti-deformation tire frame should be used, and the rocker arm and hydraulic device should be increased as shown in Figure 1.

(a) (b)

Figure 1. Improved hydraulic anti-deformation tire frame and its upper rocker arm: (a) Tire frame, (b) Rocker arm.

Fu and Zhang [4] adopted the rigidly fixed anti-deformation method and determined the anti-deformation curve through empirical formula calculation and finite element analysis. Through the correction of the curve and the follow-up measurement of the finished product of the plate unit, it was found that the application of the anti-deformation jig can control the welding deformation of the plate unit well.

2.2 Automatic assembly and welding process of U-shaped floor

Qin [5] designed a U-shaped stiffener lining bracket. The lining bracket extends into the inner side of each U-shaped stiffener before welding the U-shaped stiffener stiffened bridge deck, and then opens the bracket and closes to the two inner sides of the U-shaped stiffener and the bridge deck.

Bei et al. [6] adopted the production mode of double-sided submerged arc welding through the U-shaped rib fillet weld of the plate unit. The inner U-rib weld uses a special welding system for internal U-rib submerged arc welding, and the outer U-rib weld uses external U-rib submerged arc welding equipment, as shown in Figure 2.

(a) (b)

Figure 2. U-shaped rib submerged arc welding special welding system: (a) Internal welding equipment, (b) External welding equipment.

Liu et al. [7] proposed a new process method for U-rib fillet welding of plate elements. By adopting advanced welding equipment such as a U-rib welding machine, YD-500GS deep penetration welding machine, and welding robot, double-sided penetration welding of U-rib fillet welding was realized, and appearance inspection, non-destructive testing, and macro section test were carried out.

Yang et al. [8] used I-rib to simulate U-rib to carry out the welding process test of the U-rib plate unit, formed and optimized the welding process of full penetration angle welding of the U-rib plate unit, which was applied to the welding of U-rib plate unit of a super large bridge.

2.3 Robot welding technology of plate unit

Janez Tušek [9] studied the influence of welding current and welding speed on the welding rate of low alloy steel and obtained the empirical formula of weld cross-sectional area. Tang and Drews [10] adopted an interactive three-dimensional virtual teaching mode to realize welding path planning and programming of robot single-pass welding. They adopted macro technology to realize automatic planning and programming of robot multi-layer and multi-pass welding motion paths.

Li et al. [11] proposed applying a rotating arc sensor to a robot welding curve fillet weld. The curve fillet weld tracking and torch attitude adjustment model were established by studying the collected weld deviation information and the kinematics model of the six-degree-of-freedom welding robot.

Chen and Tang [12] introduced a U-rib submerged arc external welding robot welding system and attached an OSL-50 laser weld tracking system to each robot. The system could give feedback on the weld position data in real-time and guide the robot to correct the position so that the welding gun could accurately weld, as shown in Figure 3.

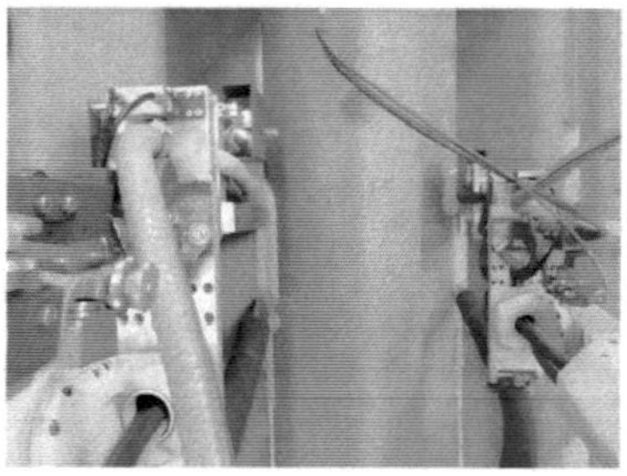

Figure 3. U-rib submerged arc welding robot welding system.

The main weld identification methods based on visual tracking technology are horizontal projection and slope analysis [13]. Lee et al. [14] used the median filter to remove the noise, used the Otsu method for binary segmentation, and proposed an improved Hough transform for feature detection and information extraction of weld images. Zhang [15] developed an intelligent welding robot system that can realize preliminary pre-welding trajectory planning and initial welding position guidance by visual sensing and realize the height tracking of the weld by arc sensing.

2.4 Automatic welding process of the board unit

Tang and Wu [16] completed the preliminary research on the intelligent welding system of steel bridge plate units by studying an enterprise based on the industrial Internet, transforming the automatic welding system, upgrading the welding robot system, and applying the corresponding software system to realize the reading, transmission, storage, calculation, and display of the real-time data of the equipment.

Fan [17] and China Railway Shanhaiguan Bridge Group Co., Ltd. (CRSBG) organized the research on the integrated technology of assembly and welding of U-rib unit, developed the integrated machine tool of automatic assembly and welding of U-rib unit, and realized the functions of automatic panel grinding, assembly of U-rib, inner and outer welding of U-rib unit.

Zhu et al. [18] studied the mechanized production of welding seam grinding and longitudinal stiffener assembly of bridge plate elements. They proposed integrated equipment that realized the grinding of 4 stiffener weld positions in a single time and assembled two stiffeners simultaneously, satisfying the grinding and assembly of longitudinal stiffeners of straight and curved plate elements simultaneously.

2.5 Reasonable selection of automatic welding raw materials

High-quality welding materials must be selected to ensure their good process characteristics in the welding of steel structure bridges. When selecting the welding wire, it is necessary to ensure that the cladding effect of the electrode meets the specified requirements, prevents sandblasting, and avoids the danger of slag [19].

Zhou et al. [20] calculated the corrosion resistance and carbon equivalent of Q500 qENH high-performance weathering steel and corresponding welding materials. The results show that the corrosion resistance of steel and welding materials is good. The strength of the steel increases while its welding performance is affected to a certain extent.

According to Jiang and Yang [21], when welding bridge steel, a high-temperature resistant welding foundation must be used with an embedded arc welding wire and submerged arc welding flux. This requires analysis and development of a welding rod, as well as proper protection of the welding wire gas during the process.

3. The problems existing in the welding automation technology of bridge steel structure

3.1 Influence of component structure

Due to the limitation of component structure, long-span truss steel bridge is difficult to automate because of its unique shape and complex and changeable form. We take the MS26 node rod of Nanjing Dashengguan Bridge as an example. It is connected not only to the stiffening chord and the lower chord but also to the beam, vertical rod, bridge panel, connecting tie rod, inclined rod, etc. The connection hole group composed of inclined rod and connecting tie rod body is 45, and the connection method is more complicated. Due to its large stress, its maximum thickness reached 52 mm, and most of it is the use of deep slope angle and penetration angle of the solder joint, as shown in Figure 4. So, it needs a lot of solder. Because of its spatial distribution, the gap between the weld inside the box bar and the whole joint is narrow, and automatic welding cannot be realized.

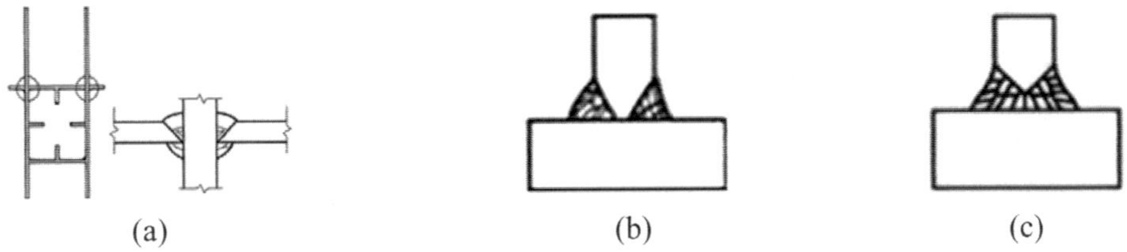

Figure 4. Complex welding operation: (a) Cross-seam penetration welding, (b) Deep groove welding, (c) Penetration welding.

3.2 Poor welding position

In the actual use of automatic welding technology, a more horizontal position method, affected by the welding position in production practice, cannot guarantee the quality of each working link, so it is difficult to achieve an automatic welding process. We take the steel box girder of the Hong Kong-Zhuhai-Macao Bridge as an example, as shown in Figure 5. In assembling the whole beam section, automatic welding technology cannot directly use the welding seam of the roof elevation angle, transverse diaphragm, and web plate. There are still many problems and defects in the automatic welding of bridge steel structures in China, which seriously affects the quality and level of the overall welding.

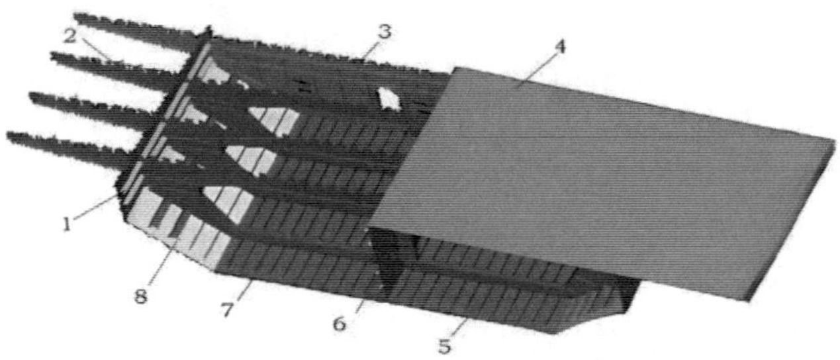

Figure 5. Steel box girder segment of Hong Kong-Zhuhai-Macao Bridge: 1. Side web, 2. Cantilever partition, 3. Transverse partition, 4. Top plate, 5. Transverse rib, 6. Middle web, 7. Bottom plate, 8. Inclined bottom plate.

4. Summary and prospect

China's automatic welding technology has made great progress and development and is widely used to construct many large bridges. This paper mainly introduces the application of welding automation technology in bridge steel structures. Although there are many advantages compared with traditional welding technology, there are still some difficulties in providing reference for the next research direction.

In the automatic welding work of bridge steel structures, the robot automatic welding technology has been used to produce plate units. However, due to the excessive number of butt welds, the proposed automatic welding quality standards are very high, so it is not reasonable to use advanced automation technology to carry out welding work, and the work is very difficult. Therefore, relevant enterprises should, according to the current development situation, reform the welding production mode, use advanced welding robots to carry out work, effectively enhance the overall quality of welding work, train composite welding professionals, and constantly develop and explore new welding processes to ensure that the automatic welding of bridge steel structure is achieved.

Acknowledgment

The research was funded by the Priority Academic Program Development of Jiangsu Higher Education Institutions (PAPD), the Jiangsu Innovation Support Program (International Science and Technology Cooperation) Project (Grant No. BZ2023002), and the Natural Science Horizontal Research Project of Nantong University (Grant No. 22ZH643).

References

[1] Xu, X J. (2015) Application and development of Welding Automation Technology for Bridge Steel Structure [J]. Metal Processing, (22): 14-15. (In Chinese)

[2] Xia, Y L. (2022) Research on Anti-deformation Technology of Steel box Structure Bridge Construction Welding [J]. Science and Technology Innovation and Application, 12(32): 58-61+65.DOI: 10.19981/j.CN23-1581/G3.2022.32.015. (In Chinese)

[3] Shi, X. (2022) Technical Improvement of Hydraulic Anti-deformation tire frame for roof and floor unit of Steel Structure Bridge [J]. Fujian metallurgy, 1(02): 55-58. DOI: 10.19574/j.cnki.issn1672-7665.2022.02.005. (In Chinese)

[4] Fu, L P and Zhang, Z Y. (2018) Control Measures for Welding Deformation of Steel box girder plate Element [J]. Hong Kong Engineering Technology, 55(06): 64-66. DOI: 10.16403/j.cnki.ggjs20180614. (In Chinese)

[5] Qin, S L. (2023) Research on Single-side Full Penetration Welding Technology of U-shaped stiffener bridge Panel [D]. Harbin University of Science and Technology. DOI: 10.27063/d.cnki.ghlgu.2023.000977. (In Chinese)

[6] Bei, Y C, et al. (2022) Research on Welding Technology of U-shaped Rib Fillet Weld for Bridge Panel Unit [J]. Metal Working (Hot Working), (05): 64-67. (In Chinese)

[7] Liu, Z X, et al. (2020) Research on full Penetration Welding Technology of U-rib fillet weld of plate Element [J]. Metal Working (Hot Working), (02): 45-47. (In Chinese)

[8] Yang, W F, et al. (2023) U floor unit welding process simulation test and application [J]. Journal of welding technology, 52 (02): 47-50+108. DOI: 10.13846/j.cnki.cn12-1070/tg.2023.02.026. (In Chinese)

[9] Janez Tušek D S. (2000) Mathematical modeling of melting rate in twinwire welding [J]. Journal of Materials Processing Technology, 100: 250-256.

[10] Tang, X H and Drews P. (2005) 3D Visualization Offline programming and simulation system for robots [J]. Journal of Welding Technology, 26(2): 64-68.

[11] Li, Y, et al. (2018) Modeling and simulation of robot fillet weld tracking based on rotating arc [J]. China Mechanical Engineering, 29(3): 348-352. (In Chinese)

[12] Chen, J and Tang, C Y. (2021) U-rib Submerged Arc Welding Robot Welding System and Its Application [J]. Mechanical Engineering & Automation, (02): 164-166. (In Chinese)

[13] He, Y, Xu, Y, Chen, Y, et al. (2016) Weld seam profile detection and feature point extraction for

multipass route planning based on visual attention model [J]. Robotics and Computer-Integrated Manufacturing, 37: 251-261. (In Chinese)
[14] Lee J, Wu Q, Park M, et al. (2015) A study on modified Hough algorithm for image processing in weld seam tracking system [J]. Advanced Materials Research, 1088(11): 824-828.
[15] Zhang, Q L. (2013) Welding robot system based on vision guidance and arc tracking [D]. Shanghai Jiao Tong University. (In Chinese)
[16] Tang, C Y and Wu, D L. (2023) Intelligent Welding System of Steel bridge plate Unit based on Industrial Internet [J]. Metal World, (04): 41-44. (In Chinese)
[17] Fan, J Q. (2023) Create a new model for U-rib plate unit assembly welding [J]. Journal of China Highway, (11): 124. DOI: 10.13468/j.cnki.chw.2023.11.023. (In Chinese)
[18] Zhu, W W, Wang, J F and Chen, Z J. (2023) Research on integrated Equipment for grinding and assembling plate elements of Large Steel components [J]. Industrial Building, 53(S1): 317-319. (In Chinese)
[19] Huang, G. and Li, X Y. (2022) Technical Analysis of Bridge Steel Structure Fabrication and Installation [J]. Chinese Science and Technology Journal Database (Citation Edition) Engineering Technology, 7: 84-87. (In Chinese)
[20] Zhou, L et al. (2022) Welding Technology and application of Q500qENH High Performance Weathering Steel in Viaduct Engineering [J]. Welding, (10): 54-58. (In Chinese)
[21] Jiang, Y W and Yang, T L. (2021) Development of Welding Material of High Toughness Weather Resistant Bridge Steel [J]. Shihezi Science and Technology, (03): 14-15. (In Chinese)

Journal of Physics: Conference Series **2815** (2024) 012025 doi:10.1088/1742-6596/2815/1/012025

Rolling bearing fault detection method based on teager energy operator

Jie Zhou[1,*]

[1]School of Electrical Engineering, Shanghai DianJi University, Shanghai, China

[*]Corresponding author's e-mail:1439754092@qq.com

Abstract. This study presents a method for detecting subtle faults in rolling bearings by integrating singular value decomposition (SVD) and local modal decomposition (LMD). Initially, the original vibration signals undergo reconstruction in phase space and subsequent denoising using SVD. Following this, LMD is employed to generate multiple local modal functions (IMFs), with the most sensitive IMF components selected based on kurtosis-mean square deviation criteria. The Teager energy operator (TEO) is then computed for these selected IMFs. Ultimately, Fourier transform analysis of TEO is utilized to identify faults in the inner and outer races of rolling bearings. Test results conducted on publicly available data from Case Western Reserve University validate the accurate identification of inner and outer race faults using this approach.

1. Introduction

Damage to rolling bearings often results in periodic impact vibration signals, which reflect repeated contact between the damaged point and other components. Therefore, these periodic impacts and their time intervals are important features for diagnosing local damage to bearings. However, in practical applications, detecting early bearing faults becomes more challenging due to the complexity of vibration propagation paths and environmental noise interference. Therefore, effective analysis and extraction of impact features in the vibration signals of rolling bearings, as well as identifying their periodic frequencies, are crucial for fault diagnosis [1].

Empirical mode decomposition (EMD) is an effective adaptive denoising method for handling nonlinear and non-stationary signals, but it faces issues such as end effects and mode mixing [2]. To address these problems, local modal decomposition (LMD) has been developed, which iteratively extracts local signal characteristics to reduce mode mixing, particularly suitable for extracting high-frequency modulation information in rolling bearing vibration signals. The Teager energy operator (TEO), as a tool for tracking the instantaneous energy of signals, outperforms traditional Hilbert transforms, especially in monitoring impact components. However, given the limitations of TEO in handling signals with multiple components and low signal-to-noise ratio, researchers suggest combining LMD and maximum correlated kurtosis deconvolution methods with TEO to improve fault detection accuracy. Additionally, the method of intrinsic time-scale decomposition (ITD) morphology filtering followed by TEO application has also been proven effective. Singular value decomposition (SVD) is used for signal denoising and extracting periodic components, demonstrating excellent performance in eliminating noise interference while retaining fault information. These studies indicate that denoising the signal before applying TEO is crucial.

Considering the challenges of existing technologies in extracting weak fault features, this study proposes a method that combines singular value decomposition (SVD) and local modal decomposition (LMD) to enhance the accuracy of the TEO energy spectrum in detecting weak faults in rolling bearings. Through SVD and LMD techniques, this method effectively improves the signal-to-noise ratio of weak fault features in noisy environments and enhances the ability of the TEO energy spectrum to identify fault features. The application of this method in actual rolling bearing experiments confirms its efficient practicality, paving the way for accurate extraction and identification of rolling bearing fault features in complex noise environments, indicating its promising prospects for widespread application in industrial fault diagnosis.

2. Basic principles introduction

2.1. Principles of LMD algorithm

Local Modal Decomposition (LMD) stands as an adaptive signal decomposition method that revolves around the extraction of local signal features. At its core, LMD considers a signal as a composite of various local oscillatory modes. It follows an iterative approach to progressively extract these local oscillatory features, ultimately decomposing the signal into multiple local modal functions (IMFs). The procedural steps entail: initially, extracting local features from the signal, such as identifying extremum points; subsequently, through iterative iterations, continuously extracting the local oscillatory modes of the signal and deriving the corresponding local modal functions; within the iterative process, determining the iteration count based on preset stopping criteria to prevent excessive decomposition; finally, amalgamating all local modal functions to yield the decomposed signal outcome. In comparison to traditional Empirical Mode Decomposition (EMD), LMD boasts enhanced locality and convergence, facilitating a more precise depiction of the signal's local characteristics, thereby rendering it suitable for the analysis and processing of nonlinear and non-stationary signals.

2.2. Definition of TEO

The Teager energy operator (TEO) is a nonlinear operator specifically designed for extracting the instantaneous energy of a signal [3]. For continuous-time signals, the definition of TEO [4] is given by Equation (1),

$$\varphi(y(t)) = (\dot{x}(t))^2 - x(t)\ddot{x}(t) \tag{1}$$

where $\dot{x}(t)$ and $\ddot{x}(t)$ represent the first and second derivatives of the signal with respect to t, respectively.

In the case of a discrete-time signal $y(t)$, the TEO definition equation is as shown in Equation (2),

$$\varphi(y(t)) = (\mathrm{y}(t))^2 - \mathrm{y}(t+1)y(t-1) \tag{2}$$

With only three sample data points, TEO can compute the signal energy at any given moment, endowing it with excellent time resolution for monitoring transient changes in the signal.

When dealing with noisy vibration signals from rolling bearings, singular value decomposition (SVD) and its extended method, the singular value differential spectrum, play a crucial role. By leveraging phase space reconstruction theory [5], a $m \times n$ dimensional Hankel matrix can be created as shown in Equation (3),

$$H_m = \begin{bmatrix} x_1 & x_2 & \cdots & x_n \\ x_2 & x_3 & \cdots & x_{n+1} \\ \vdots & \vdots & & \vdots \\ x_m & x_{m+1} & \cdots & x_{m+n-1} \end{bmatrix} \tag{3}$$

where $m+n$-1 is the length of the signal, and $m > n$. Performing SVD on H_m yields Equation (4) [6],

$$H_m = A\Lambda B^T \tag{4}$$

where A and B are orthogonal matrices of order $m\times m$ and $n\times n$ respectively. Λ is an $m\times n$ diagonal matrix with diagonal elements $a_1,a_2,\cdots,a_z$ and $a_1 \ge a_2 \ge \cdots \ge a_z$, $z=\min(m,n)$, and $a_1,a_2,\cdots,a_z$ are the singular values of H_m.

By retaining the first k singular values of H_m and setting the remaining singular values to zero, a more optimal approximation of the matrix H_m can be obtained, effectively improving the signal-to-noise ratio [7]. In this process, selecting the appropriate denoising order is crucial. To this end, the concept of the singular value differential spectrum is introduced [8], shown in Equation (5),

$$\varepsilon_i = \sigma_i + \sigma_{i+1}, (i=1,2,\cdots,m-1) \tag{5}$$

This method captures significant changes between singular values, aiding in determining the appropriate denoising order. According to Equation (5), the larger differences between adjacent singular values in the differential spectrum indicate differences in correlation between the signal and noise, serving as a basis for distinguishing between them. In SVD processing, the components corresponding to singular values before a major peak are considered valid signals, while the portion after the peak is regarded as noise [9]. By performing the inverse transformation of SVD, a denoised optimized signal matrix can be reconstructed, thereby enhancing the signal-to-noise ratio. Additionally, by utilizing evaluation criteria for IMF components obtained through LMD decomposition, key features containing important fault information are selected to achieve effective noise reduction. In high signal-to-noise ratio environments, TEO energy spectra can effectively identify transient changes in the signal. Enhancing the detectability of subtle fault features using SVD and LMD and refining and confirming features through the high-resolution capability of TEO energy spectra provide an effective tool for accurate fault diagnosis of rolling bearings.

3. Feature extraction method design

3.1. Evaluation metric setting

In this study, kurtosis and mean square deviation are selected as evaluation metrics. Kurtosis is utilized to identify impact pulses and early fault signals, which is particularly effective for diagnosing surface damage-related faults. Mean square deviation serves as a crucial indicator for measuring data dispersion and is essential for accuracy representation. Rolling bearing faults typically involve enhanced impact pulses and signal amplitude changes, making these two metrics particularly sensitive in fault identification. By utilizing these metrics to filter data, the complexity of feature extraction computation is significantly reduced while also achieving effective noise reduction.

3.2. Feature extraction method

To enhance the accuracy of fault feature extraction, this study integrates singular value decomposition (SVD) and local modal decomposition (LMD) techniques. The purpose of this integration is to highlight impact pulses in the original vibration signals, ensuring that the selected intrinsic mode functions (IMFs) are more clearly defined physiologically. This is beneficial for accurately identifying fault frequencies under bearing operation using the Teager energy operator (TEO) spectrum method. Based on this, a fault feature extraction method combining the SVD-LMD-TEO energy spectrum is proposed. The method steps are as follows:

- Phase space reconstruction of the original vibration signals is performed, and the appropriate denoising order is determined [10]. SVD is used for denoising and the optimal approximation signal is obtained through its inverse transformation process.
- LMD decomposition is implemented on the signals processed by SVD denoising to obtain a series of IMF components. IMF components containing rich fault information are selected based on kurtosis and mean square deviation criteria.

- TEO is used to extract the periodic impacts generated by bearing faults, calculate the TEO of the selected IMF components, and analyze and identify bearing fault types through the Fourier spectrum of TEO.3.

In this study, fault data provided by the Bearing Data Center at Case Western Reserve University in the United States was utilized, focusing on testing the end-face vibration signals of SKF6205 deep groove ball bearings. The experimental conditions involved the rotation of the inner race while the outer race remained fixed, with a set speed of 1797 revolutions per minute (rpm). The sampling frequency was set at 12 kilohertz (kHz), and a total of 4096 samples were collected.

The experiment primarily targeted the collection of weak fault signals occurring at the points of pitting on both the inner and outer races. According to the bearing fault frequency calculation equation cited in [1], the calculated fault frequencies were 162.186 Hz for the inner race and 107.364 Hz for the outer race. The experimental results revealed significant interference of background noise in the original waveform of both inner and outer race faults, as depicted in Figure 1. To address this, the signals underwent preprocessing using Singular Value Decomposition (SVD) for denoising. SVD effectively filtered out irrelevant noise while preserving useful signals, as illustrated in Figure 2.

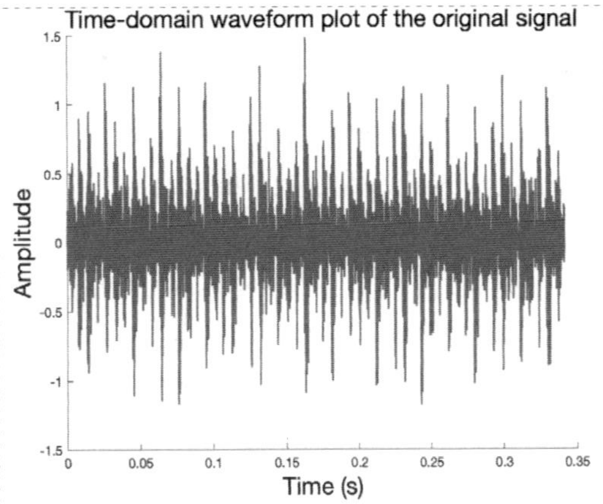

Figure 1. Time-domain waveform of the original signal.

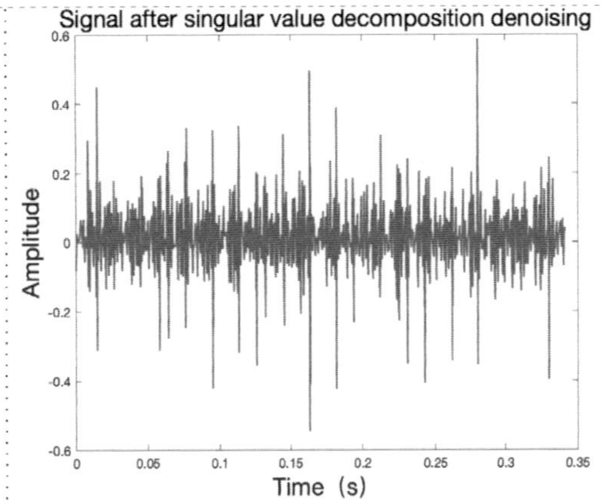

Figure 2. Denoised signal using Singular Value Decomposition.

Subsequently, the denoised signal undergoes adaptive decomposition using LMD. LMD progressively disassembles various vibration modes from high to low frequencies, as depicted in Figure 3. Following this, the most sensitive Intrinsic Mode Functions (IMFs) are meticulously chosen based on kurtosis and mean-square deviation criteria. Subsequently, the Teager Energy Operator (TEO) is employed to extract the energy spectrum of inner race faults (Figure 4) and outer race faults (Figure 5). In Figure 4, distinct signals of the outer race fault frequencies' first harmonic (108.398 Hz), second harmonic (213.867 Hz), third harmonic (322.266 Hz), fourth harmonic (430.664 Hz), and fifth harmonic (539.062 Hz) are vividly displayed. Similarly, Figure 5 illustrates unmistakable signals of the inner race fault frequencies' first harmonic (161.133 Hz), second harmonic (322.266 Hz), third harmonic (486.328 Hz), fourth harmonic (662.109 Hz), and fifth harmonic (823.242 Hz). Clearly, the periodic impact components of both inner and outer race faults can be effectively extracted. In summary, this research method has been demonstrated to be accurate and efficient in the analysis of inner and outer race faults.

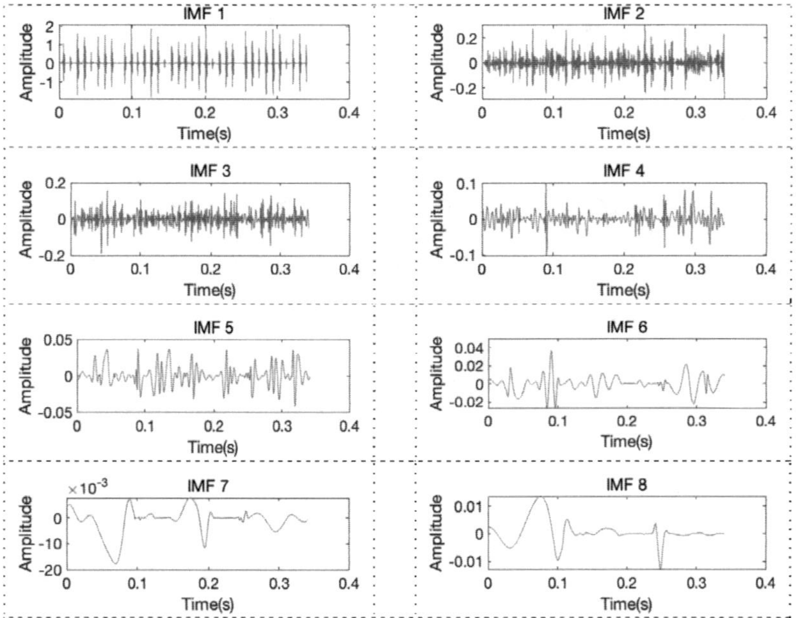

Figure 3. LMD decomposition plot of the signal after denoising using Singular Value Decomposition (SVD).

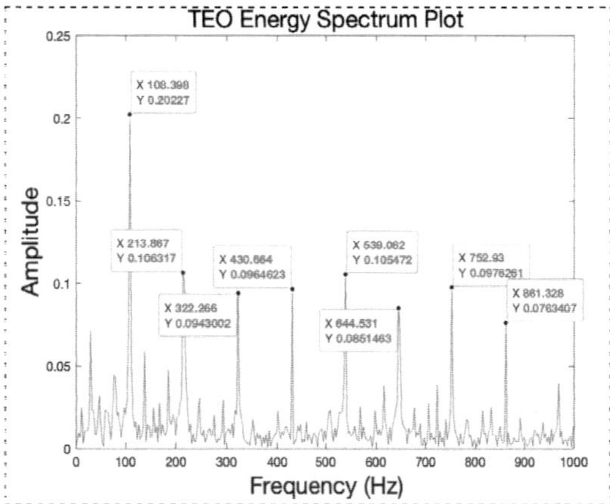

Figure 4. Teager Energy Operator (TEO) energy spectrum plot for the outer race fault.

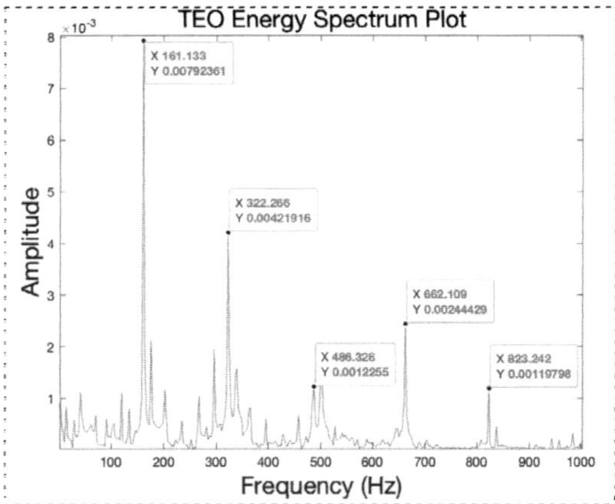

Figure 5. Teager Energy Operator (TEO) energy spectrum plot for the inner race fault.

4. Conclusion

This study introduces a novel technique that integrates Singular Value Decomposition (SVD) and Local Modal Decomposition (LMD) to precisely detect subtle faults in rolling bearings. Initially, the original signal is preprocessed using SVD to effectively eliminate random noise. Subsequently, the application of kurtosis-mean-square deviation criteria facilitates the selection of essential features containing fault information from the Intrinsic Mode Function (IMF) components, thereby enhancing the accuracy of fault detection. Moreover, employing the Teager Energy Operator (TEO) energy spectrum method allows for the precise identification of inner and outer race faults in rolling bearings.

References

[1] Zhu Y, Jiang S. Method for Extracting Impact Characteristics of Rolling Bearing Faults Based on Generalized S-Transform Time-Frequency Spectrum and SVD Denoising [J]. Bearing, 2016, (11): 53-57. DOI:10.19533/j. issn1000-3762.2016.11.014.

[2] Zhang Y. Research on Signal Processing and Diagnostic Methods for Mechanical Faults [D]. Lanzhou Jiaotong University, 2019. DOI: 10.27205/d.cnki.gltec.2019.000519.

[3] Dong N, Huang W.Q, Zhou C, et al. Rapid Detection Method of Voltage Sag Based on Improved Energy Operator [J]. Computer Application Research, 2013, 30 (10): 3011-3014+3021.

[4] Kwak D. H, Lee D .H, Ahn J.H, et al. Fault Detection of Roller-Bearings Using Signal Processing and Optimization Algorithms [J]. Sensors, 2013, 14(1): 283-298.

[5] Wang J.G, Li J, Wan X.D. Method for Extracting Fault Characteristics of Rolling Bearings Based on Singular Value Decomposition and Local Mean Decomposition [J]. Journal of Mechanical Engineering, 2015, 51 (03): 104-110.

[6] Zhang J.D, Guo X.N, Luo X.Q, et al. Improved modal parameter identification method based on particle swarm algorithm [J]. Journal of Vibration and Shock, 2022, 41(02): 255-264. DOI: 10.13465/j.cnki.jvs.2022.02.031.

[7] Sun X.H, Zhang L.M, Wang T. Denoising Method for Frequency Response Functions Based on Singular Value Decomposition [J]. Vibration, Testing and Diagnosis, 2009, 29(3): 325-328.

[8] Zhao X.Z, Ye B.Y, Chen T.J. Theory of Singular Value Differential Spectrum and Its Application in Fault Diagnosis of Lathe Spindle Boxes [J]. Journal of Mechanical Engineering, 2010, 46(1): 100-108.

[9] Niu H.Q, Song T.H, Luo X, et al. Denoising Method for Partial Discharge Signals Based on S-Transform and Singular Value Decomposition [J]. Journal of South China University of Technology (Natural Science Edition), 2020, 48(02): 9-15.

[10] Zhang C. Research on Feature Extraction Method of Rotating Machinery Fault Based on Ensemble Empirical Mode Decomposition and Entropy Characteristics [D]. Lanzhou University of Technology, 2018.

Valve positioner control based on active disturbance rejection control technology

Yike Ding [1], Na Wang [2], Zhibo Sun [2, *], Xiaolei Shang [1], Shuwang Rui [1] and Zhongdi Wang [1]

[1] School of Automation Science and Electrical Engineering, Beihang University, Beijing 100191, China
[2] Engineering Training Center, Beihang University, Beijing 102206, China

[*] Corresponding author's e-mail: sunzb@buaa.edu.cn

Abstract: The pneumatic regulating valve is a commonly used actuator in industrial settings. The smart valve positioner, acting as the "brain" of the pneumatic regulating valve, dictates the control performance of the regulating valve, primarily relying on control algorithms. However, the valve positioner is a typical nonlinear system, making it difficult to establish an accurate mathematical model that can reflect the system's nonlinear characteristics. Furthermore, due to variations in valves, installations, and processes, the model parameters differ significantly, making it challenging to self-adjust the parameters. Additionally, various disturbances in process control will affect the positioner's control performance. Currently, commonly used control algorithms rely on precise mathematical models, resulting in poor control performance and robustness. Therefore, this article proposes a smart valve positioner control algorithm based on self-disturbance control. Initially, a steady-state model and system identification of the valve positioner system are obtained through open-loop testing, resulting in a simplified second-order system model. Based on the simplified second-order model, a second-order self-disturbance control algorithm is designed. Finally, a test platform based on the rapid control prototype (RCP) controller is established to verify that the introduction of the self-disturbance control algorithm can effectively enhance the control performance and robustness of the smart valve positioner.

1. Introduction

With the advancement of industrial automation in China, process control, which refers to the automatic control of industrial production, has gained increasing prominence in the national economy. Among these technologies, pneumatic control valves hold a dominant market share of over 60% [1]. They are widely utilized as actuators in various industrial applications and consist of pneumatic actuators and valves. Pneumatic actuators receive input pressure signals, generate thrust, and drive the valve stem to move up and down or rotate. This movement enables precise control over valve openings, facilitating adjustments in flow rates for achieving automated production processes.

To achieve precise control of the flow rate of the regulating valve, a positioner is commonly installed on the actuator to improve the performance of the regulating valve. This paper adopts a nozzle-flapper-type smart valve positioner, and its control performance mainly depends on the control algorithm used [2]. Due to the compressibility of air and the negative effects of friction, the entire system is a typical nonlinear system [3], making it difficult to establish an accurate mathematical model that can reflect the system's nonlinear characteristics. Currently used PID control algorithms rely on mathematical models, and due to the differences in valves, installations, and processes, the model parameters vary significantly, making it difficult to self-tune the control algorithm parameters, and challenging to achieve precise control for various valves. In [4], a fuzzy logic-based PID control algorithm was introduced for valve regulation, which accommodates system nonlinearity and parameter variations, resulting in enhanced stability and precision. In [5], predictive control into pneumatic actuators was incorporated, effectively mitigating the impact of nonlinearities and elevating control accuracy. In [6], a pneumatic actuator control method based on fuzzy adaptive control was proposed, utilizing a fuzzy controller combined with an adaptive control strategy to achieve precise control of the nonlinear dynamic characteristics of pneumatic actuators. In [7], an adaptive neural network control model was presented with a disturbance observer for pneumatic actuators, which approximates unknown nonlinearities and compensates for external disturbances, as confirmed by simulation. In [8], a model-based and sliding mode control strategy was proposed for servo-pneumatic systems, addressing their nonlinearity and surpassing the constraints of conventional controllers to achieve accurate position tracking with minimal steady-state errors. It can be seen that most of the mentioned control strategies are still unable to escape the requirement for precise mathematical models, with complex algorithm structures, large computational demands, and difficulties in practical implementation. Additionally, disturbances acting on the valves during actual load operation are unavoidable. Therefore, this paper proposes a smart valve positioner control algorithm based on active disturbance rejection control (ADRC), incorporating ADRC into the smart valve positioner. This control method has a simple structure, low reliance on precise mathematical models, strong disturbance rejection capability, and robustness to model parameters [9][10]. It fundamentally solves the difficulties of self-tuning smart positioner control parameters and poor robustness, improving control performance. The proposed method holds significant theoretical and practical significance and represents a challenge. It is also the driving force behind this research.

This paper first conducts open-loop tests to establish the steady-state modeling and system identification of the valve positioner system, obtaining a simplified second-order system model. Then, based on the simplified second-order model, a second-order Active Disturbance Rejection Control (ADRC) algorithm is designed, including a tracking differentiator, extended state observer, and nonlinear state error feedback controller. Finally, an experimental platform is built by using a Rapid Control Prototype (RCP) controller, and it is verified that the introduction of the ADRC algorithm can effectively improve the control performance and robustness of the smart valve positioner.

2. Valve positioner system model

The smart valve positioner system mainly consists of the nozzle-flapper type electric-to-pneumatic transducer (I/P transducer), valve position feedback mechanism, pneumatic amplifier, pneumatic actuator, valve, and controller. Due to the compressibility of air and the negative effects of friction, the system exhibits significant nonlinearity. To simplify the model, we assume that the working medium in the system is an ideal gas and do not consider the gas flow model.

The principle block diagram of the smart valve positioner system is shown in Figure 1. In this system, the controller receives the control signal and feedback signal, performs control calculations within the controller, and outputs the control signal, which is then converted into a current control signal to drive the I/P transducer. The I/P transducer receives the current control signal, and the torque motor within it generates a corresponding electromagnetic torque to drive the movement of the flapper, controlling the distance between the flapper and the nozzle, thereby adjusting the size of the orifice to control the pressure in the pneumatic amplifier's feedback chamber. The pneumatic amplifier compares the pressures in its various chambers. Due to the small membrane area in the output pressure chamber and the large membrane area in the feedback pressure chamber, it achieves control of the small pressure information from the pneumatic amplifier to drive the large pressure signal for the pneumatic actuator. Ultimately, the pneumatic actuator converts the air pressure into the displacement of the push rod.

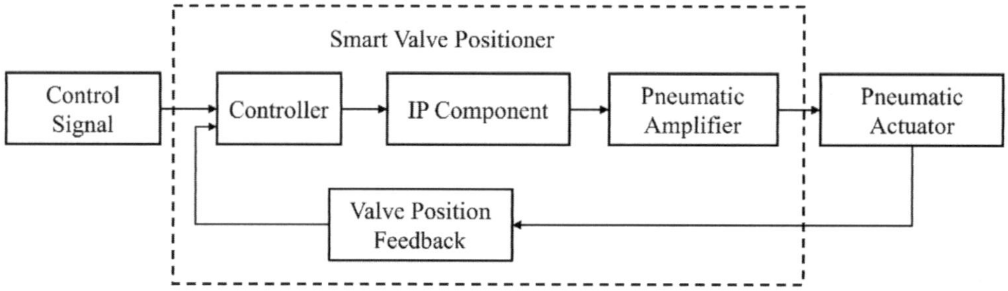

Figure 1. Structural diagram of the smart valve positioner system.

2.1 Steady-state models for I/P transducer and pneumatic amplifier

The action process of the I/P transducer relative to the pneumatic actuator can be considered instantaneous, as the torque motor within the I/P transducer converts electrical energy into magnetic field energy. The flapper in the magnetic field generates electromagnetic torque, causing small oscillations in the flapper due to electromagnetic torque, spring torque, jet force torque, and damping torque, as well as changes in air pressure. Similarly, the motion of the valve seat within the pneumatic amplifier is minimal and has a very rapid response speed. Therefore, the relationship between the excitation current of the I/P transducer and the back pressure chamber pressure, as well as the relationship between the back pressure chamber pressure and the output pressure within the pneumatic amplifier, can be simplified into steady-state models.

To establish the steady-state relationships between the excitation current, back pressure chamber pressure, and output pressure, a steady-state test platform, as shown in Figure 2, can be constructed. This platform includes a pressure-reducing valve, signal generator, pressure gauges, and valve positioner components (I/P transducer, valve position feedback mechanism, and pneumatic amplifier). It enables simultaneous measurement of the back pressure chamber pressure (P_b) and the output pressure (P_{out}). The pressure-reducing valve is used to stabilize the supply air pressure (P_s), and the signal generator outputs 0-1 mA of driving current. The pressure gauges are used for dual measurement, with one measuring the output pressure and the other measuring the back pressure chamber pressure.

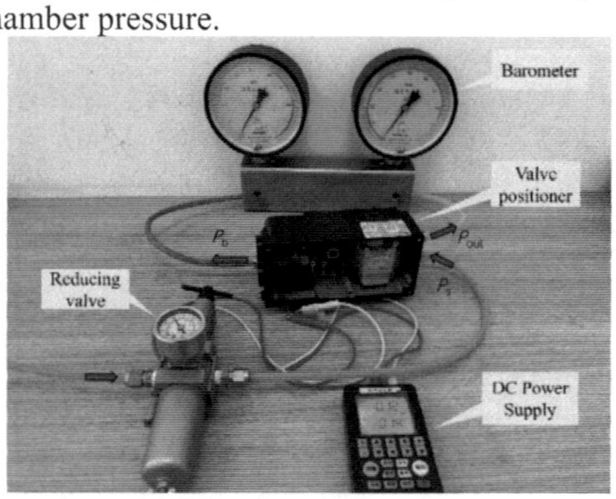

Figure 2. Input and output steady-state testing platform.

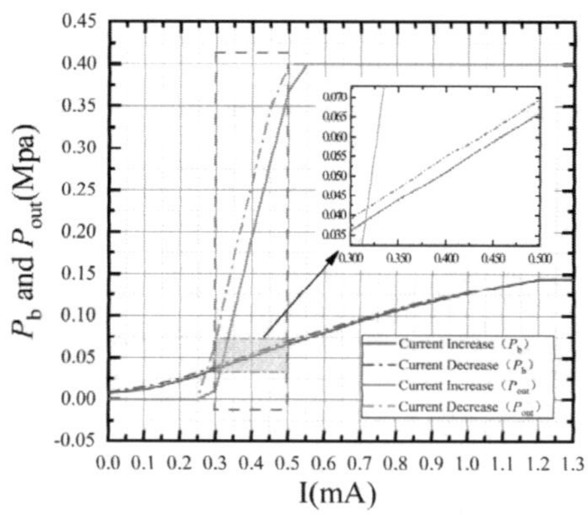

Figure 3. The fitting results of the steady-state experiments of I with P_b and P_{out}.

To meet the subsequent gas supply requirements of the valve, the supply pressure was set to 0.4 MPa, and steady-state testing experiments were conducted. The driving current was linearly increased and linearly decreased within the range of 0-1 mA. Simultaneously, the magnitude of the driving current I , P_b , and P_{out} was recorded during a steady state. The experimental results, obtained by fitting the test data, are shown in Figure 3.

Figure 3 demonstrates that the driving current (I) exhibits a linear relationship with both P_b and P_{out} . Therefore, it can be concluded that P_b and P_{out} are also linearly related. This implies that the steady-state models of the I/P transducer and the pneumatic amplifier can be simplified to zero-order models. By connecting and simplifying two zero-order models in series, the relationship between I and P_{out} can be expressed as a linear relationship, as shown in Equation (1).

$$P_{out} = 1.78\mathrm{I} - 0.525 \tag{1}$$

2.2 Motion model of pneumatic actuator

The valve stem is mainly affected by gravity, frictional force, and input pressure. Due to the initial condition where the forces of gravity, friction, and pressure are balanced, we take x as the displacement of the valve stem relative to the equilibrium point. Therefore, the motion model of the pneumatic actuator can be represented by a second-order motion equation, as shown in Equation (2).

$$m\ddot{x} + c\dot{x} + nkx = P_{out} A_e \tag{2}$$

where c is the damping coefficient of the actuating push rod (N*s/mm); n is the number of springs in the actuator; k is the elasticity coefficient of the springs (N/mm); x is the relative displacement (mm); A_e is the effective area of the diaphragm.

2.3 Valve positioner system model

The I/P transducer, pneumatic amplifier, and pneumatic actuator are connected in series, so the entire system can be simplified as a second-order model. To determine the model parameters, an open-loop step test is conducted on the valve positioner system, and the input excitation current signal and valve position feedback signal are collected as identification data. The MATLAB system identification module is used to identify the open-loop second-order model of the entire system, and the system transfer function is obtained, as shown in Equation (3).

$$G(\mathrm{s}) = \frac{x}{I} = \frac{4.164}{\mathrm{s}^2 + 4.195\mathrm{s} + 3.113} \tag{3}$$

where the input is the excitation current I , and the output is the relative displacement of the valve stem x , with an identification fit of 93.11%.

We introduce the control input variable u and define $4.164I = \mathrm{b}_0 u + \Delta u$, where Δu is the nonlinear part of the control input and $\Delta u = f(x_1, x_2) + 4.195x_2 + 3.113x_1$, and $f(x_1, x_2)$ is the unknown nonlinear model and all disturbances in the system model. Therefore, Equation (3) can be rewritten as:

$$\begin{cases} \dot{x}_1 = x_2 \\ \dot{x}_2 = \mathrm{b}_0 u + f(x_1, x_2) \\ x = x_1 \end{cases} \tag{4}$$

where b_0 serves as a tuning parameter, participating in subsequent control debugging to improve the estimation capability and generality of the algorithm.

3. Controller design

To address the nonlinearity and the difficulty in establishing an accurate mathematical model for the smart valve positioner, and to improve its control performance, disturbance rejection capability, and robustness, this paper introduces the Active Disturbance Rejection Control (ADRC) technique into the smart valve positioner system and proposes an ADRC-based control algorithm for the smart valve

Journal of Physics: Conference Series **2815** (2024) 012026 doi:10.1088/1742-6596/2815/1/012026

positioner. The ADRC algorithm consists of three parts: Tracking Differentiator (TD), Extended State Observer (ESO), and Nonlinear State Error Feedback (NLSEF) controller. Based on the simplified second-order model of the smart positioner, a second-order ADRC controller is designed.

3.1 Design of tracking differentiator (TD)

Based on the system model, this paper first designs a second-order nonlinear tracking differentiator to provide a transitional stage and extract the differential signal. The discrete form of the TD is as follows:

$$\begin{cases} v_1(k+1) = v_1(k) + hv_2(k) \\ v_2(k+1) = v_2(k) + h\mathrm{fh} \\ \mathrm{fh} = \mathrm{fhan}(v_1 - v_0, v_2, r, h_0) \end{cases} \tag{5}$$

where v_0 represents the given signal, v_1 represents the transitional stage that tracks v_0, v_2 represents the differential signal of v_1, and h represents the computation step. The parameter r determines the speed of the transitional stage. The larger the value of r is, the shorter the transition time is. The parameter h_0 is the filtering factor. The larger the value of h_0 is, the stronger the low-pass filtering is. $\mathrm{fhan}(v_1 - v_0, v_2, r, h_0)$ is the specific expression of the second-order discrete tracking differentiator's fastest control synthesis function [11].

3.2 Design of extended state observer (ESO)

The smart valve positioner system is a typical nonlinear system. By simplifying and system identification, we obtain a simplified second-order system model, as shown in Equation (5). In this model, due to the continuity and boundedness of the $f(x_1, x_2)$ term, it can be treated as an extended state x_3. Therefore, the model can be rewritten as:

$$\begin{cases} \dot{x}_1 = x_2 \\ \dot{x}_2 = x_3 + b_0 u \\ \dot{x}_3 = w(t) \\ x = x_1 \end{cases} \tag{6}$$

According to the model, the discrete form of the second-order nonlinear ESO is designed to estimate the states and disturbances of each order, as shown in Equation (7).

$$\begin{cases} e(k) = z_1(k) - y(k), \mathrm{fe}_1 = \mathrm{fal}(e(k), \alpha_1, \delta_1), \mathrm{fe}_2 = \mathrm{fal}(e(k), \alpha_2, \delta_2) \\ z_1(k+1) = z_1(k) + h(z_2(k) - \beta_{01} e(k)) \\ z_2(k+1) = z_2(k) + h(z_3(k) + b_0 u(k) - \beta_{02} \mathrm{fe}_1) \\ z_3(k+1) = z_3(k) + h(-\beta_{03} \mathrm{fe}_2) \end{cases} \tag{7}$$

where β_{01}, β_{02}, and β_{03} are parameters that need to be tuned. By selecting appropriate parameters, it is possible to achieve a precise estimation of the states of each order of the system, and z_1 is utilized to estimate the feedback signal x_1 of the valve position. The variable e represents the estimation error of z_1 relative to x_1. By introducing this error as feedback, it ensures precise estimation of the states of each order. To prevent the occurrence of high-frequency oscillations near the origin during discrete calculations, a continuous power function $\mathrm{fal}(e, \alpha, \delta)$ with a linear segment is introduced [12].

3.3 Design of nonlinear state error feedback (NLSEF)

We combine e_1 and e_2 by using the nonlinear function fal to obtain the commonly used feedback control signal.

$$\begin{cases} e_1(k+1) = v_1(k+1) - z_1(k+1) \\ e_2(k+1) = v_2(k+1) - z_2(k+1) \\ u_0(k+1) = \beta_1 \text{fal}(e_1(k+1), \alpha_3, \delta_3) + \beta_2 \text{fal}(e_2(k+1), \alpha_4, \delta_4) \end{cases} \tag{8}$$

where α_3 and α_4 is the parameter of the nonlinear function $\text{fal}(e, \alpha, \delta)$ to ensure that the feedback function has the characteristic of "small error, large gain", and its typical value range is (0, 1); Parameters β_1 and β_2 are adjustable parameters for the nonlinear combination. The size of Parameter β_1 mainly affects the steady-state error of the system, while the size of Parameter β_2 mainly affects the dynamic response performance of the system.

To suppress the uncertainty caused by the nonlinearity of the system and the inaccuracy of the model, as well as the unknown disturbances in the output signal caused by external influences, the total sum of the internal and external disturbances estimated by the extended observer z_3 is compensated into the feedback control signal u_0, jointly determining the final actual control quantity u.

$$u(k) = u_0(k) - \frac{z_3(k)}{b_0} \tag{9}$$

Through the compensation of z_3, the complex system can be transformed into a pure integral series system consisting of two integrators with the feedback control signal u_0 as the control signal. It is equivalent to integrating the error signals e_1 and e_2 in the feedback control signal twice, transforming proportional-derivative control into integral control. Therefore, it can be deduced that in the steady state, both error signals e_1 and e_2 will converge to zero, achieving precise control without steady-state error.

4. Experimental results

To verify the practical effect of the ADRC algorithm in the smart valve positioner, we set up an experimental test platform for the smart valve positioner based on the Rapid Control Prototype (RCP) controller as shown in Figure 4. The platform mainly includes a DC power supply, controller, voltage-to-current circuit, actuator, AEP I/P transducer, pneumatic amplifier, and valve position feedback potentiometer. The actuator chosen is the MF2R-76B valve manufactured by Wuzhong Instrument Co., Ltd., with a stroke of 16 mm, PTFE packing, and a supply air pressure of 0.4 Mpa. The controller used is the Model MT-1050 RCP controller developed by ModelingTech.

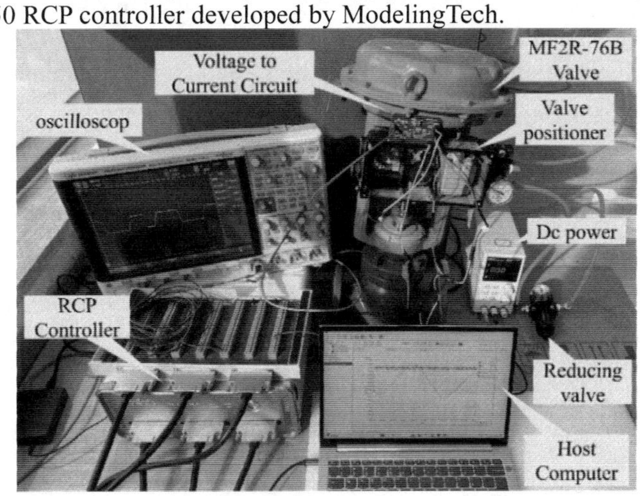

Figure 4. Experimental platform for smart valve positioner based on RCP.

The RCP controller collects the feedback voltage from the valve position feedback potentiometer and compares it with the set signal. The control algorithm processes this input and outputs a control signal. During the test process, the supply air pressure is stabilized at 0.4 Mpa by using a pressure-reducing valve according to the requirements of the actuator's supply air pressure. The range of valve movement feedback stroke on this test platform is 0.78 V-1.6 V. Design testing was carried out, using a step signal with an amplitude of 0.4 V and a sine signal with an amplitude of 0.2 V and a frequency of 0.4π rad/sec as the set signals. The size of the final output signal u is restricted to 0-2.4 V.

Under the step input, we debug and design the appropriate transition period to determine the parameter TD as $r = 0.45$ and $h_0 = 4 * h = 0.0004$. Combined with the bandwidth method [13], we design the parameters of ESO as $\beta_{01} = 3\omega_0, \beta_{02} = 3 * {\omega_0}^2$, and $\beta_{03} = 0.05 * {\omega_0}^3$, where $\omega_0 = 3.2$. To ensure the convergence of steady-state error and consider the speed of response and the magnitude of fluctuations, the parameters β_1 and β_2 in NLSEF need to satisfy a suitable proportion, and its parameters are designed as $\beta_1 = 13 * k, \beta_2 = k$, and $k = 0.6$.

We implement the above active disturbance rejection control algorithm in the RCP controller and build a Simulink-based model algorithm as shown in Figure 5. The "input1" is the valve position feedback signal input interface, and "output1" is the control signal output interface.

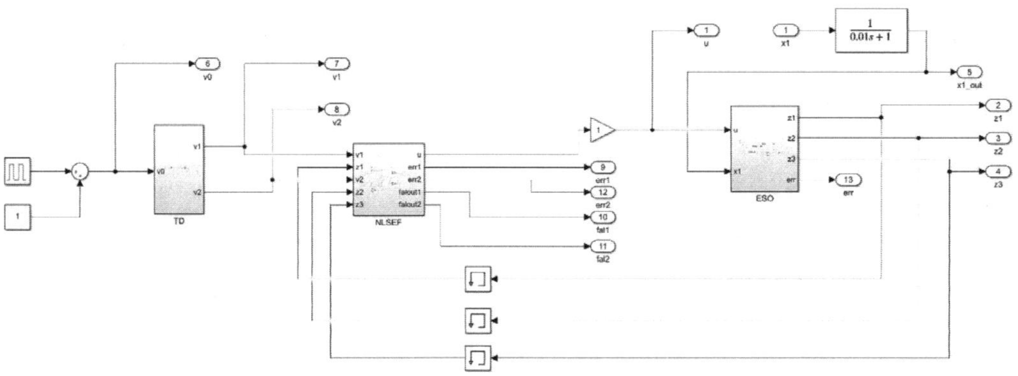

Figure 5. Implementation of ADRC algorithm based on Simulink.

Under the final ADRC control, the step response is shown in Figure 6, while the sine input signal only requires expanding the parameter $r = 1.5$ in TD, and the tracking effect is shown in Figure 7. Under PID control, the optimal parameters for step input are $k_p = 1.41$ and $k_i = 0.36$. For the sine input signal, the optimal parameters are $k_p = 7.42$ and $k_i = 8.73$, and the control effect is shown in Figure 8.

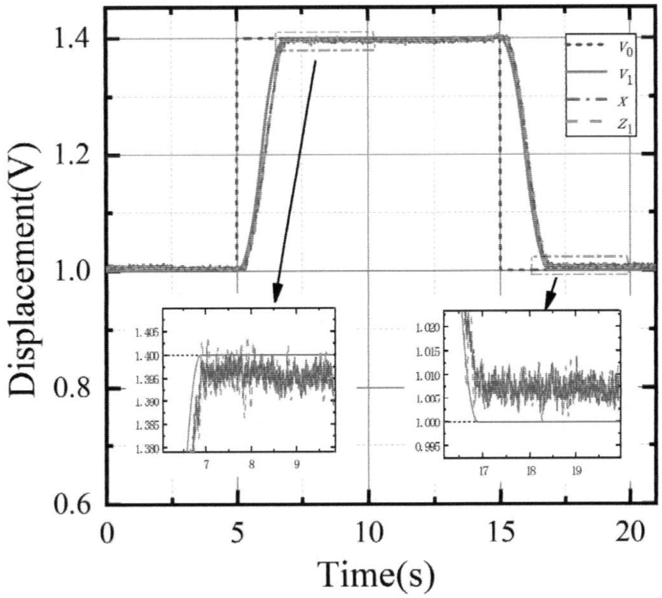

(a)

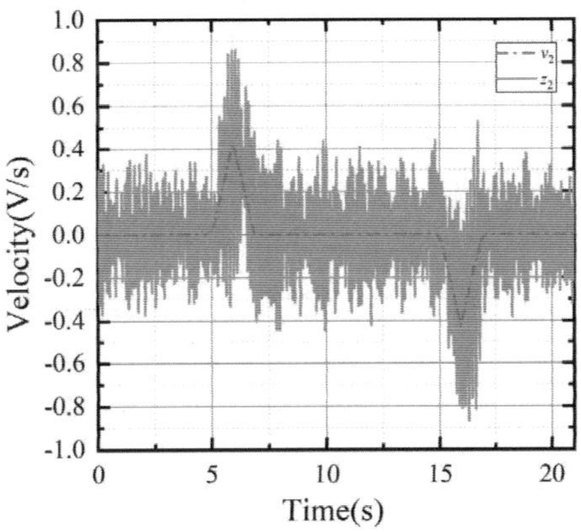

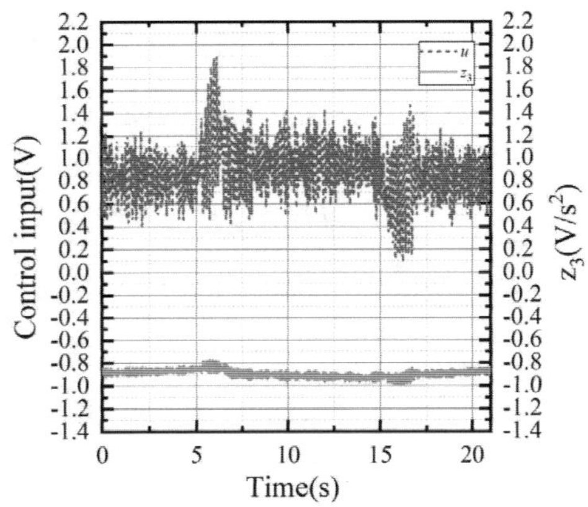

(b) (c)

(a) Displacement; (b) Velocity; (c) Control input and z_3

Figure 6. Step response experiment results based on ADRC control.

Figure 6 shows the experimental results of tracking a step signal with an amplitude of 0.4 V by using the ADRC control strategy. As shown in Figure 6 (a), the displacement feedback signal x tracks the transition signal v_1 very well, with a rise time of 1.9 s, a steady-state error of 0.6%, and no overshoot. The response time for the downward stroke is also 1.9 s, with a steady-state error of 0.9% and no overshoot. The estimated displacement state variable z_1 overlaps almost completely with the displacement feedback x, indicating a good estimation and tracking of x. Figure 6 (b) shows the waveform of the velocity state variable z_2 and the differential signal v_2 of the transition section. The average value of z_2 almost coincides with v_2 and converges to zero in a steady state. Although z_2 has fluctuations, it is reasonable due to the hysteresis characteristics of the system and does not affect the ultimate stability of the valve position. Figure 6 (c) records the waveforms of the disturbance

Journal of Physics: Conference Series **2815** (2024) 012026 doi:10.1088/1742-6596/2815/1/012026

estimation variable z_3 and the final control variable u, where z_3 provides a good estimation of the disturbance, and through feedback, compensates the entire experimental system to an integral concatenation system, enabling the errors $e_1 = v_1 - z_1$ and $e_2 = v_2 - z_2$ to converge to zero. In steady state, z_3 tends to stabilize, and Errors e_1 and e_2 converge to zero. The steady-state output of the control variable u mainly depends on z_3. In Figure 6 (c), the control variable u shows high-frequency fluctuations within the range of 0-2.4 V, mainly resulting from $v_1 - z_1$ and $v_2 - z_2$, with the steady-state convergence value coming from z_3. It can be seen that z_3 provides an effective estimate of the total disturbance and feedback compensation for the entire system, achieving fast, non-overshooting, and precise positional control of the valve.

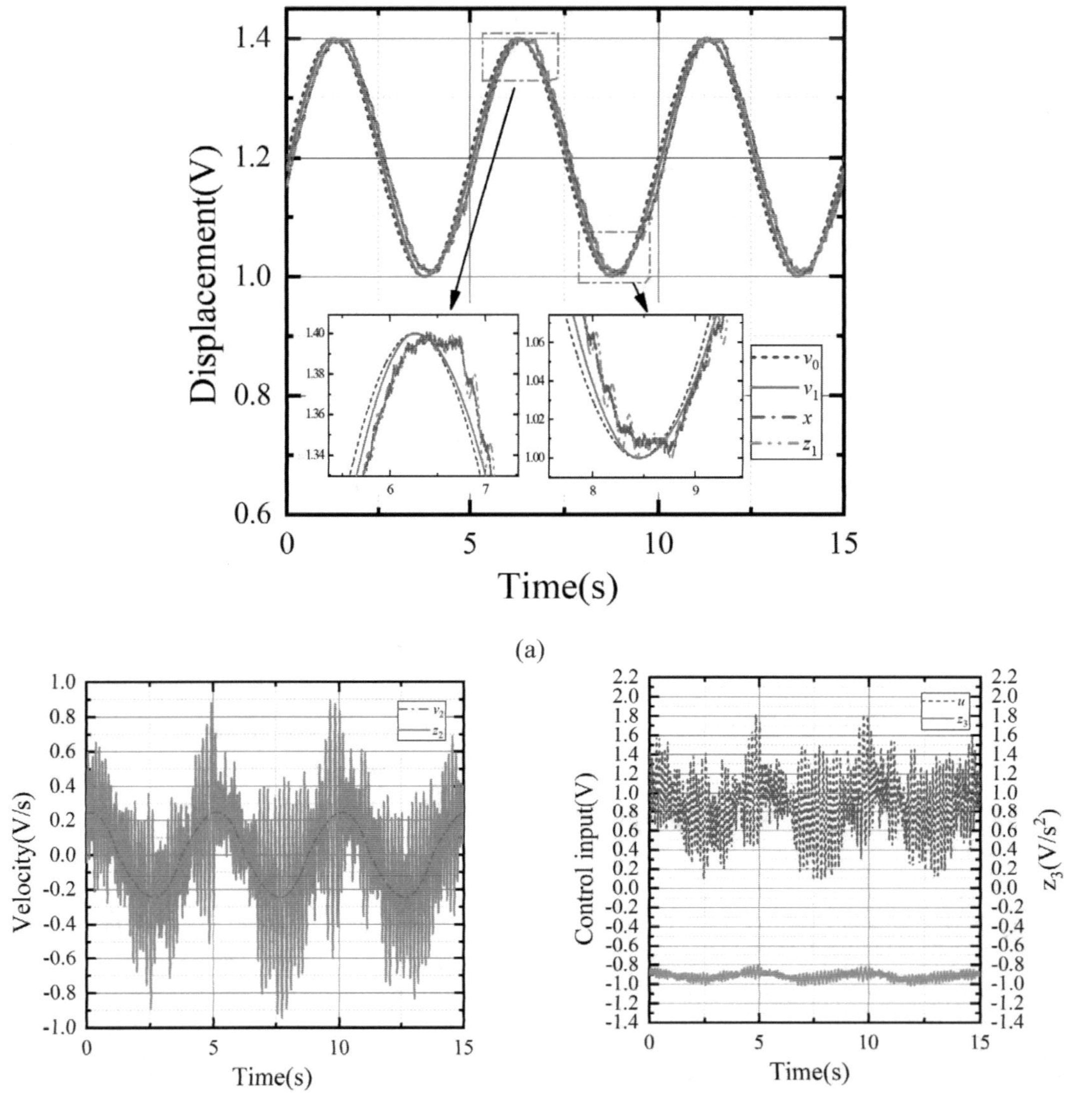

(a)

(b) (c)

(a) Displacement; (b) Velocity; (c) Control input and z_3.

Figure 7. Experimental results of sine wave tracking based on ADRC control.

Figure 7 shows the experimental results of tracking a sine signal with an amplitude of 0.2 V and a frequency of 0.4π rad/sec by using ADRC control. Figure 7 (a) shows that the displacement feedback signal x smoothly tracks the transition signal v_1, with only 0.16 s of lag and minor fluctuations. At the same time, the displacement state estimator z_1 provides a good estimation and tracking of x. Figure 7 (b) illustrates that the average value of the velocity state estimator z_2 almost coincides with the differential signal v_2 of the transition section, with a phase difference of 90 degrees from the transition signal v_1. Figure 7 (c) demonstrates that both z_3 and u are in a dynamically stable state as the given signal changes. The magnitude of the disturbance estimator z_3 directly determines the magnitude of the final control variable u, once again confirming the effectiveness of z_3 in estimating the total disturbance and the feedback compensation of the entire system, and achieving high-performance tracking of the valve to the sine reference signal.

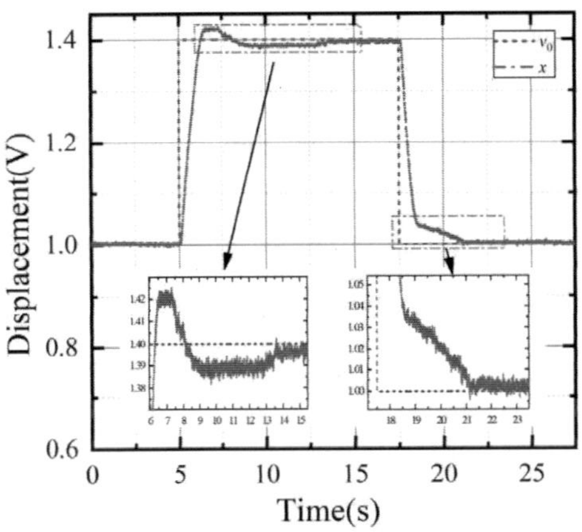

(a)

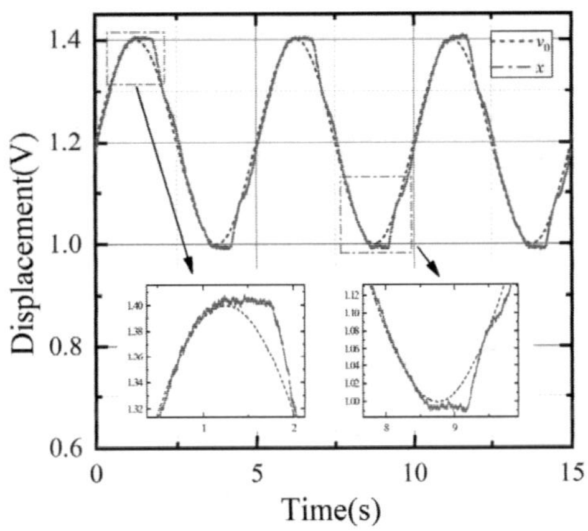

(b)

(a) Step response results; (b) Sine wave tracking results

Figure 8. Experimental results based on PID control.

Figure 8 shows the experimental results of tracking a step signal with an amplitude of 0.4 V and a sine signal with an amplitude of 0.2 V and a frequency of 0.4π rad/sec by using PID control. Figure 8 (a) shows the step response results, with a rise time of 8.6 s, a steady-state error of 0.6%, and an overshoot of 5%. The upward stroke has a rise time of 3.6 s, a steady-state error of 0.3%, and no overshoot for the downward stroke. Figure 8 (b) shows the sine wave tracking results, with a high tracking speed but significant overshoot and severe waveform distortion.

A comparison of Figures 6, 7, and 8 shows that the dynamic responsiveness and tracking performance of the smart valve positioner under ADRC control are superior to those under traditional PID control. ADRC effectively resolves the contradiction between rapid response and overshoot. Furthermore, under ADRC control, the dynamic performance of the upward and downward strokes is almost identical, and the parameters under both step and sine inputs are highly compatible, demonstrating the ADRC controller's compatibility and robustness with different inputs. Therefore, introducing ADRC control into the smart valve positioner can significantly improve its control performance and robustness.

5. Conclusion

This paper has established a system model for the nozzle-flapper valve positioner through system identification and, based on this, proposed a second-order Active Disturbance Rejection Control (ADRC) method for the nozzle-flapper valve positioner. This method effectively overcomes the nonlinearity and inaccuracy issues of the model. Experimental results demonstrate that the extended observer provides a good estimation of the total uncertainty, including nonlinearity, disturbances, etc., and effectively compensates for the system. Under step response, the upward stroke response time is 1.9 s, with a steady-state error of 0.6% and no overshoot; the downward stroke response time is 1.9 s, with a steady-state error of 0.9% and no overshoot. Meanwhile, with almost no need for parameter changes, the system also provides good tracking of a sine signal. Compared to traditional PID control, the system's control performance and robustness are significantly improved.

Acknowledgments
This paper was supported by the National Key Research and Development Program under Grant (2022YFF0708903), Ningbo Key Technology Research and Development Program under Grant (2023Z018), and Youth Fund of National Natural Science Foundation of China under Grant (52105046 and 52205043).

References

[1] Kayihan A. and Iii F. J. D. Friction compensation for a process control valve [J]. Control Engineering Practice, 2000, 8 (7): 799-812.
[2] Zhang B., Jiang A., Jiang J., et al. A New Positioning Strategy Based on Parameter Tuning and Optimal Control Technique for Pneumatic Control Valve [C]// Actuators. MDPI, 2022, 11 (10): 279.
[3] Zhu M., Chen S. Xu Z., et al. Design of Two-Degree-of-Freedom Fractional-Order Internal Model Control Algorithm for Pneumatic Control Valves [C]// Actuators. MDPI, 2023, 12 (5): 214.
[4] González J. A., et al. A fuzzy logic PID control algorithm for pneumatic actuators [J]. Control Engineering Practice, 2019, 83: 113-122.
[5] Haslinda, et al. Research on predictive control applied to pneumatic actuators [J]. Automation Technology, 38, 5 (2019): 891-899.
[6] Zhang C., et al. Fuzzy adaptive control of pneumatic actuators [J]. IEEE Transactions on Industrial Electronics, 2016, 63 (1): 468-479.
[7] Yang Y., Li Y., Liu X., et al. Adaptive neural network control for a hydraulic knee exoskeleton with valve deadband and output constraint based on nonlinear disturbance observer [J]. Neurocomputing, 2022, 473: 14-23.
[8] Ramezani S. and Baghestan K. Observer-based nonlinear precise control of pneumatic servo systems [J]. Proceedings of the Institution of Mechanical Engineers, Part E: Journal of Process Mechanical Engineering, 2019, 233 (2): 165-176.
[9] Wei Q., Wu Z., Zhou Y., et al. Active Disturbance-Rejection Controller (ADRC)-Based Torque Control for a Pneumatic Rotary Actuator with Positional Interference [C]// Actuators. Multidisciplinary Digital Publishing Institute, 2024, 13 (2): 66.
[10] Song J., Guo J., Qin C., et al. The Design of a Reaction Flywheel Speed Control System Based on ADRC [J]. Automation, 2023, 4 (3): 246-262.
[11] Han Jingqing. Active Disturbance Rejection. Control Technique: The Technique for Estimating and Compensating the Uncertainties [M]. Beijing: National Defense Industry Press, 2008 (in Chinese).
[12] Li S., Wu Q., Liu Y., et al. Research on Three-Closed-Loop ADRC Position Compensation Strategy Based on Winch-Type Heave Compensation System with a Secondary Component [J]. Journal of Marine Science and Engineering, 2024, 12 (2): 346.
[13] Gao Z. Scaling and bandwidth-parameterization based controller tuning [C]// ACC. 2003: 4, 989-4, 996.

Journal of Physics: Conference Series **2815** (2024) 012027 doi:10.1088/1742-6596/2815/1/012027

Fuzzy-pid control design of biped robot based on motion capture technology

Rongchang Fu [1, *]**, Jinan Yu** [2] **and Xiaoyu Yang** [3]

[1-3] School of Mechanical Engineering, Xinjiang University, Urumqi 830047, China

[*] Corresponding author's e-mail: 2781642414@qq.com

Abstract. To improve the precision of the biped robot's walking control system, a fuzzy PID controller with a variable theory domain was added to the design of the biped robot's lower limb control system to optimize the tracking process of joint angular displacement and improve the motion effect. Firstly, the motion capture experiment is carried out to obtain the angular displacement data of the joint. The theoretical angle was calculated according to the biped robot's lower limb diagram model and kinematics equation, and the expected trajectory was obtained by simple gait planning and using MATLAB to verify the validity of the angle parameters. Then the fuzzy PID controller of the lower limb joint was established, and the variable theory domain fuzzy controller was established by introducing the expansion factor. Finally, the Simulink test bench is built to simulate the two controllers, and the expected data and simulation data are compared. The results show that the error range of the variable theory domain fuzzy controller is only ±2°, and the variable theory domain adaptive fuzzy controller can accurately track the desired trajectory, reduce the joint angular displacement error of the biped robot, ensure the precision and stability of the biped robot motion, and basically meet the requirements of practical production.

1. Introduction

At present, the control system of a biped robot is the most widely used method is the traditional PID control [1]. However, the traditional PID under the influence of uncertain factors leads to poor parameter calibration easily, the control effect is not stable, and there are many other problems, leading to the incapability of achieving the ideal state of accurate control effect [2]. Especially when traveling in a complex environment, due to the influence of external factors, the joint angular motion of the biped robot tracking the desired trajectory will lead to a large angular displacement tracking error and can not achieve the desired operation and control effect. Li [3] proposed the variable domain fuzzy control in 1995 in which fuzzy control rules are adjusted online with the change of the domain scaling factor, which improves the accuracy of the control. Wiput and Manukid [4] and Wang et al. [5] researched the design of the robot structure and the control method, and the fuzzy control algorithm was verified, improving the robot's working efficiency. However, the angle data of the above studies are obtained by calculation, and verifying its reliability is more difficult to realize.

In this paper, based on the above studies, we obtain the point data as a basis through motion capture experiments, use the Lagrange equation to establish the dynamics equation of the biped robot and design simple gait planning to get the desired trajectory to verify the angle parameters. Through the three-step method of fuzzy control: fuzzification - fuzzy inference - clarity to design the fuzzy PID controller, the expansion factor is introduced to design the variable domain adaptive fuzzy controller of the biped robot.

Simulation experiments are carried out on both fuzzy as well as variable-theory domain controllers in MATLAB's Simulink simulation lab bench. By comparing the simulation data, the experimental results show that both of them can realize the control of the joints of the biped robot, but the error of using the variational domain fuzzy PID control algorithm is smaller than that of the fuzzy PID controller, and the error range is ±2°, so it can accurately meet the design requirements. The accuracy and effectiveness of the designed variable-domain fuzzy controller are further verified, which has strong practical significance and reference value in engineering.

2. Three-dimensional motion capture experiment

2.1 Purpose of the experiment
We obtain data on the joints of the lower limbs during normal gait walking, as a basis for comparison.

2.2 Experimental equipment and subjects
The experimental equipment was an optical motion capture Vicon system manufactured by OML (UK) [6]; the experimental subjects were healthy adult men between 173-176 cm in height.

2.3 Experimental data
We cleaned up the experimental bench before the start of the experiment, to ensure that the experimental camera acquisition area did not appear any interference marking points of reflective equipment. To get more accurate experimental data, we set up a total of 10 MX cameras around the test bench at different angles and different heights, as shown in Figure 1, with a T-type correction bracket placed on the force plate to determine the origin, and finally open the Vicon Nexus program to check that the hardware device is connected to the correct. According to the distribution of each joint of the human body, 41 points were designed to paste reflective balls as shown in Figure 1. Experimenters first carry out a simple warm-up and several groups of gait exercises to ensure the flexibility and accuracy of the movement process. The experimenter in the gait process controls the rhythm of walking through the pedometer, to ensure the validity of the experimental data. Each experimenter for each group of experiments carried out 15 times, and at the end of the experiment, they collect the parameters on each subject's height, weight, leg length, shoulder length, elbow width, wrist width, palm thickness, knee width, and ankle width parameters, which are recorded for later data processing. The system collects data with an accuracy of 0.1 mm, shoots 100 frames per second, and finally utilizes the Vicon system to process the experimental data to obtain the velocity, acceleration, and moment values of the reflective points.

Experimenters utilize the three-dimensional optical motion capture system acquisition to obtain the people's normal walking limb point data and utilize the space vector corresponding to the coordinate relationship of the use of MATLAB processing of each joint angle to obtain the left and right hip, knee, and ankle joint angle change curve, as shown in Figure 1, for the dynamic walking process of the lower limb of each joint angle change curve.

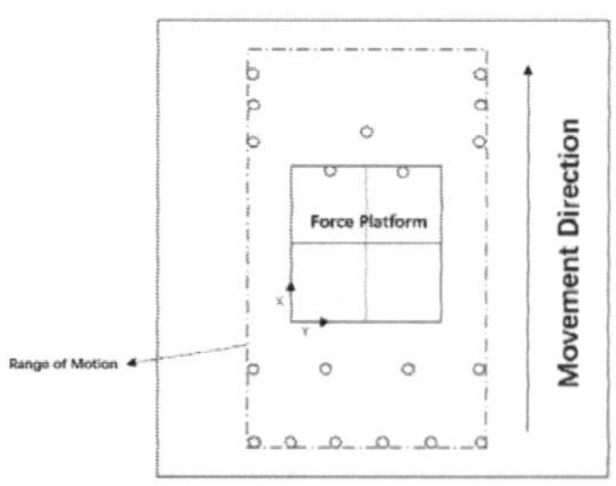

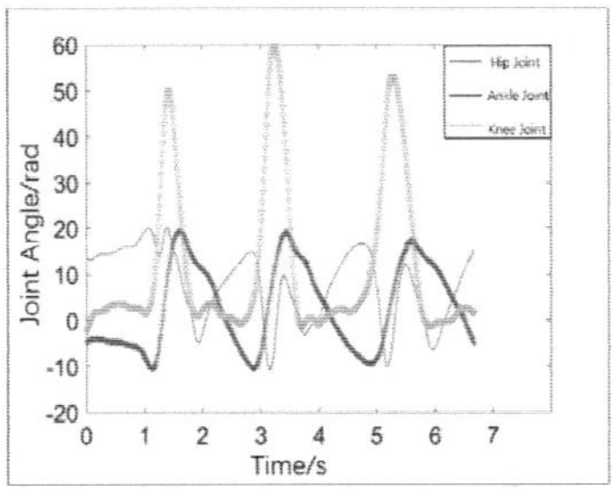

Figure 1. Calibration of the test range.

As shown in Figure 1, the ranges of angular changes for each joint in the experimental results were obtained as follows: knee (-3°, 59°), ankle (-11°, 23°), hip (-11°, 19°).

3. Gait planning for bipedal robots

3.1 Kinematic modeling of bipedal robots

The leg of the bipedal robot is modeled as shown in Figure 2(a) and split into single legs as shown in Figure 2(b):

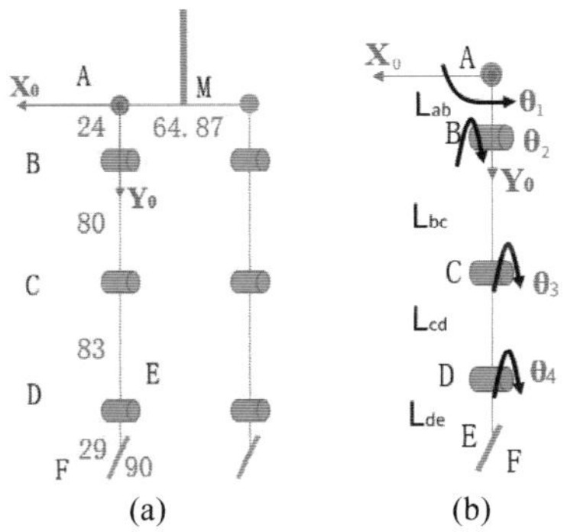

Figure 2. Kinematics model of the lower limb of biped robot.

The rotation angles of the three joints of the robot's single leg are θ1, θ2, and θ3. Assuming that the coordinates of the robot's end D are O_end, and assuming that $\theta_1=0$, we know that the robot's end D is located in the X_0Y_0 plane and the coordinates of D are D_proj, and the x-coordinate of D_proj is 0. During the coordinate transformations, we know that there exists the following relationship between O_end and D_proj, as shown in Equation (1):

$$O_{end}=ROT_z(\theta_1)*D_{proj} \tag{1}$$

$$\begin{bmatrix} Ox_{end} \\ Oy_{end} \\ Oz_{end} \end{bmatrix} = Rot_z(\theta_1)\times\begin{bmatrix} 0 \\ Dy_{proj} \\ Dz_{proj} \end{bmatrix} \tag{2}$$

$$\begin{bmatrix} Ox_{end} \\ Oy_{end} \\ Oz_{end} \end{bmatrix} = \begin{bmatrix} -Dy_{proj}*\sin(\theta_1) \\ Dy_{proj}*\cos(\theta_1) \\ Dz_{proj} \end{bmatrix} \tag{3}$$

By comparing the first two terms in the above equations, the following equations can be derived:

$$\frac{O_{Xend}}{O_{Yend}} = \frac{-Dy_{proj}*\sin(\theta_1)}{Dy_{proj}*\cos(\theta_1)} \tag{4}$$

$$\theta_1 = atan2(\frac{-O_{xend}}{O_{yend}}) \quad (5)$$

We convert a robot's single-leg angle problem to a planar solution of the angle problem:

$$D_{proj} = \begin{bmatrix} Dx_{proj} \\ Dy_{proj} \\ Dz_{proj} \end{bmatrix} = Rot_z(\theta_1)^T \begin{bmatrix} Ox_{end} \\ Oy_{end} \\ Oz_{end} \end{bmatrix}$$

$$Dx_{proj} = 0. \quad (6)$$

Now, the solution for the angles of the rear two joints of the robot becomes a planar two-degree-of-freedom robotic arm solution process as follows in Figure 3:

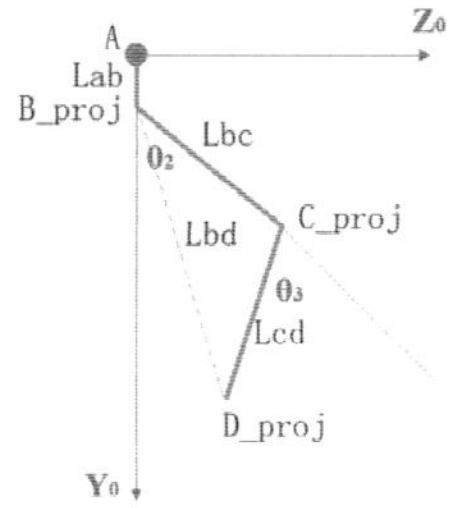

Figure 3. Plane solution diagram.

We move the coordinate system Y0Z0 down, at which point the coordinates of D_proj are relative to B_proj. Therefore, the expression for the two joint θ_1 and θ_2 angles are as follows:

$$\theta_2 = atan2(\frac{D_{proj}(3)}{D_{proj}(2)-Lab}) + acos(\frac{L_{bc}^2+L_{bd}^2-L_{cd}^2}{2L_{bc}*L_{bd}}) \quad (7)$$

$$\theta_3 = \pi - arcos(\frac{L_{bc}^2+L_{cd}^2-L_{bd}^2}{2L_{bc}*L_{cd}})$$

In Figure 3, the lengths of the connecting rods between the joints of the robot are denoted by Lab, Lbc, and Lcd, respectively; and the angular displacements of the robot joints are denoted by θ1, θ2, and θ3, respectively. From the Lagrange's theorem, the robot dynamics equation is:

$$M(q)q'' + C(q,q')q' + H(q,q') = u \quad (8)$$

where q denotes the joint angle displacement vector of the lower limb of the biped robot; q' is the velocity vector of this joint angle; q'' is the acceleration vector of this joint angle; $M(q)$ is the inertia matrix; H is the gravity and friction constituent vectors; $C(q,q')$ is the Coriolis matrix; u is the control moment matrix.

3.2 Biped robot walking program

Based on the imitation of people's daily walking habits, the motion process and walking steps of the bipedal robot can be derived.

The whole step is divided into two parts: walking and stopping. Walking is carried out by alternately supporting the left and right feet and circulating the center of gravity, and the center of gravity is shifted to the middle position and then shifted upward when stopping, as shown in Figure 4.

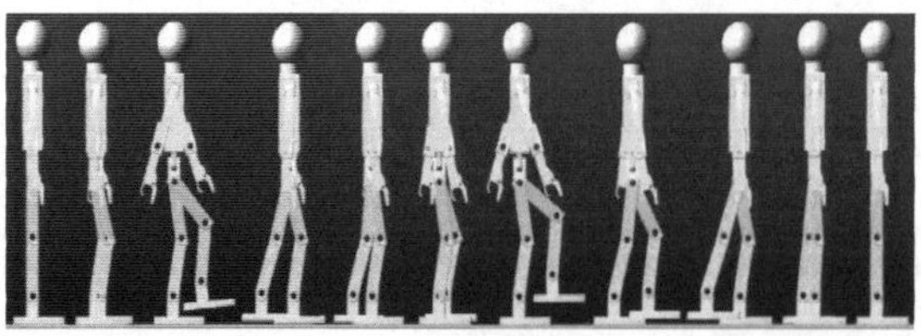

Figure 4. Side view of forward motion of biped robot.

3.3 Gait planning for bipedal robots

Bipedal robot foot end motion process gait planning is in the case of the starting point coordinates and the slope, the end point coordinates and the slope are known, ensuring that the end trajectory from the initial point to the end point between the fitting of a smooth curve, and the requirement of the curve of the least constraints [7]. The position form of the end joint actuator at the start and end time is obtained by using the following inverse motion theory, and the rotation angle of each joint can be calculated. We do the interpolative trajectory of the planning of the smooth operation on the angle of rotation of each joint. Each joint needs to have a speed and acceleration being 0 joints from the beginning. In addition to the known conditions, the following cubic polynomials can be derived, as shown in Equations (9)-(10):

$$x(t) = a+bt+ct^2+dt^3 \tag{9}$$

$$\begin{cases} x_1=a+bt_1+ct_1^2+dt^3 \\ k_1=b+2ct_1+3dt_1^2 \\ x_2=a+bt_2+ct_2^2+dt_2^3 \\ k_2=b+2ct_2+3dt_2^2 \end{cases} \tag{10}$$

The matrix is shown in Equation (11):

$$\begin{bmatrix} 1 & t_1 & t_1^2 & t_1^3 \\ 0 & 1 & 2t_1 & 3t_1^2 \\ 1 & t_2 & t_2^2 & t_2^3 \\ 0 & 1 & 2t_2 & 3t_2^2 \end{bmatrix} \begin{bmatrix} a \\ b \\ c \\ d \end{bmatrix} = \begin{bmatrix} x_1 \\ k_1 \\ x_2 \\ k_2 \end{bmatrix} \tag{11}$$

Therefore, it is sufficient to solve the subfunctions of the polynomial coefficients directly in MATLAB. The trajectory of the motion can be obtained by calculating the position of the intermediate nodes, and the velocity and acceleration of the motion must be careful to ensure that the trajectory is continuous and smooth. Afterward, the third-degree polynomial interpolation points of each joint of the robot are derived through MATLAB programming, the linear heights of the interpolation points, the control parameters of the movement time, and the longitudinal direction are defined, the Cartesian coordinates of the endpoints of the biped robot corresponding to each interpolation point are derived through the fkine function in the Robotics Toolbox [8], and the desired trajectory as in Figure 5 can be made in MATLAB. The image shows the spatial trajectory map of the robot end-effector. The desired trajectory matches with the walking scheme, and the gait planning simulation is realized without error, then the angle parameter verification is valid.

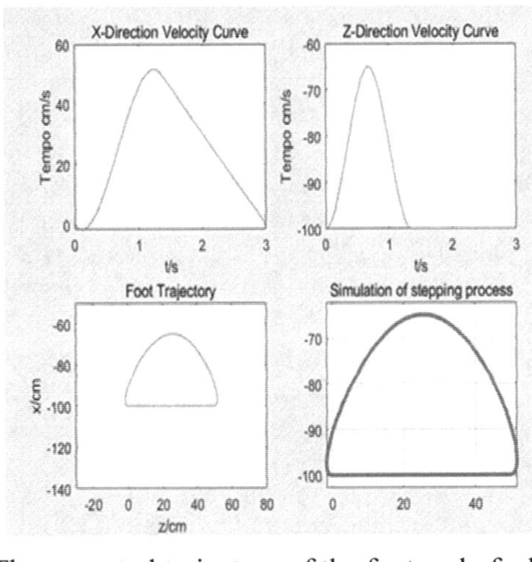

Figure 5. The expected trajectory of the foot end of a biped robot.

4. Fuzzy rule controller

4.1 Fuzzy controller

The fuzzy controller can feedback on joint position, speed signal, and other related parameters through fuzzy control, to overcome the shortcomings of traditional PID without adaptive online adjustment. Therefore, its parameters can be adjusted in real-time optimization, to meet the requirements of system performance diversification [9, 10]. The fuzzy PID control structure and the specific steps are shown in Figure 6.

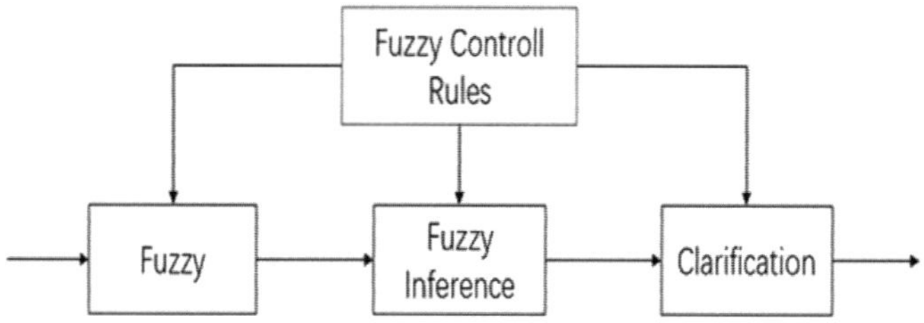

Figure 6. Fuzzy control structure.

4.1.1 Fuzzification. We assume that the fundamental domain of error is [$-e_{max}$, e_{max}], the domain of error variation is [$-ec_{max}$, ec_{max}], and the range of control domain variation is [$-u_{max}$, u_{max}]. The error fuzzy subset domain can then be defined as X = [-n, -n+1, ..., 0, ..., n-1, n], the error change fuzzy subset domain can also be defined as Y = [-m, -m+1, ..., 0, ..., m-1, m], and the definition of the control quantity of the fuzzy subset domain Z = [-l, -l+1, ..., 0, ..., l-1, l]. Typically, the range of error theories n ≥ 6, the range of error variation theories m ≥ 6, and the range of the theories of control quantities l ≥ 7.

The expression for the quantization factor of the error, error variable, and control variable is given by Equations (12)-(14):

$$K_e = \frac{n}{e_{max}} \tag{12}$$

$$K_{ec} = \frac{m}{ec_{max}} \tag{13}$$

$$K_u = \frac{u_{max}}{1} \tag{14}$$

Therefore, in this design, each of the above conditions of the lower limb joints of the bipedal robot should use the method of transforming intervals, so all the above are attributed to the standard dimensionless domain, and when determining the intervals, the accurate data are input to ensure the accuracy of the whole calculation. The intervals are generally chosen as numerically symmetric intervals around 0. Each range is calibrated as [-7, 7]. The fuzzy sets of angular errors e_1, e_2, e_3 and error change rates e_{c1}, e_{c2}, e_{c3} are taken as {NB, NM, ZE, PM, PB}; the fuzzy sets of output control quantities are taken as {NB, NM, NS, ZE, PS, PM, PB}.

We open the command window in MATLAB and enter the character command 'fuzzy' to open the fuzzy designer, define E and E_C as input variables and Δk_d, Δk_p, and Δk_i as output quantities, and select the triangular affiliation function, as shown in Figure 7:

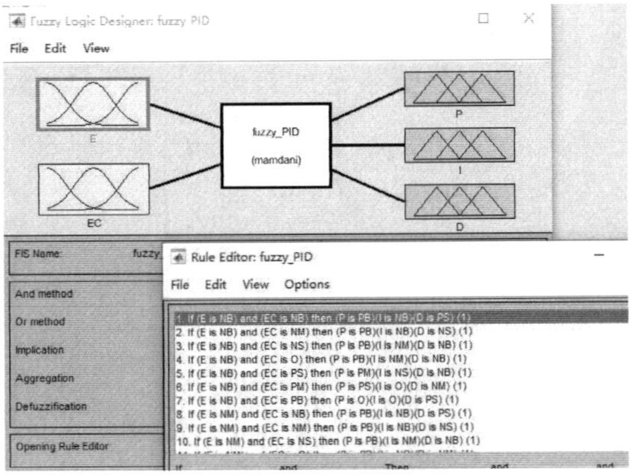

Figure 7. Subordinating degree function.

4.1.2 Fuzzy inference. The fuzzy control rules [11, 12] are as follows: when the system error is large, the control quantity should be mainly to reduce the error; when the error is small, to ensure the stability of the system and avoid the sudden change of the control state, the prevention of overshooting should be taken as the goal of selecting the control quantity, i.e., if e is too small, the Δkp and Δki should be increased and Δkd should be reduced at the same time. The equation of the regulation law is in Equation (15):

$$\begin{aligned} k_p' &= k_p + \Delta k_p \\ k_i' &= k_i + \Delta k_i \end{aligned} \tag{15}$$

where kp and ki are set values; Δkp and Δki are variable values.

The above regulation law can define the joint fuzzy rule table shown in Table 1 (e and ec are inputs). Similarly, the fuzzy rule control table of ki and kd can be obtained:

Table 1. Rule table of the fuzzy control.

U		*e*						
		NB	NM	NS	ZE	PS	PM	PB
ec	NB	PB	PB	PM	PM	PS	ZO	ZO
	NM	PB	PB	PM	PS	PS	ZO	NS
	NS	PM	PM	PM	PS	ZO	NS	NS
	ZE	PM	PM	PS	ZO	NS	NM	NM
	PS	PS	PS	ZO	NS	NS	NM	NM
	PM	PS	ZO	NS	NM	NM	NM	NB
	PB	ZO	ZO	NM	NM	NM	NB	NB

The fuzzy inference method is chosen to reason by the Mamdani method. The specific algorithm is to perform the direct product operation for the contained fuzzy relation Qij ($E, E_C, \Delta kp$) according to the set generated by the minimum constraint relation on the sets E'i, EC'i, K'pij, where i and j are the fuzzy set linguistic values, respectively, and the operation relation [13] is in Equation (16):

$$Q_{ij}(E, EC, \Delta K_p) \in E'i \times EC'i \times \Delta K'pij \tag{16}$$

The fuzzy sets E'i, EC'i, and K'pij can be characterized in the form of matrices if they all have discrete domains and are denoted by (17):

$$D_{ij}(E, EC) = E'_i \times EC'_j = E_i'^T \wedge EC'_j \tag{17}$$

where $D_{ij}(E, E_C)$ denotes that the elements of this row are represented as columns, followed by two to seven rows as described above. If the controller consists of n fuzzy rules, then the fuzzy relation Q of this controller is concatenated by the above Q_{ij} as follows:

Journal of Physics: Conference Series **2815** (2024) 012027 doi:10.1088/1742-6596/2815/1/012027

$$Q = U_{i=1,j=1}^{j=7,j=7} Q_{ij}(E, EC, \Delta k_p) \tag{18}$$

Based on the fuzzy rules determined above, the range of the basic thesis domain can be obtained:

$$D_1 = E'^T(PB) \times E'C(NS) = \begin{bmatrix} 0 \\ 0 \\ 0 \\ 0 \\ 0 \\ 0.5 \\ 1 \end{bmatrix}_{7\times1} \times$$

$$[0\ 0.5\ 1\ 0.5\ 0\ 0\ 0]_{1\times7} = \begin{bmatrix} 0 & 0 & 0 & 0 & 0 & 0 & 0 \\ 0 & 0 & 0 & 0 & 0 & 0 & 0 \\ 0 & 0 & 0 & 0 & 0 & 0 & 0 \\ 0 & 0 & 0 & 0 & 0 & 0 & 0 \\ 0 & 0 & 0 & 0 & 0 & 0 & 0 \\ 0 & 0.5 & 0.5 & 0.5 & 0 & 0 & 0 \\ 0 & 0.5 & 1 & 0.5 & 0 & 0 & 0 \end{bmatrix}_{7\times7}$$

$$Q_1 = D_1 \wedge k'_{P,KP} = \begin{bmatrix} 0 \\ \vdots \\ 0.5 \\ 0.5 \\ 0.5 \\ \vdots \\ 0.5 \\ 1 \\ 0.5 \\ \vdots \\ 0 \end{bmatrix}_{49\times1} \times \begin{bmatrix} 0 \\ 0 \\ 0 \\ 0 \\ 0.2 \\ 0.6 \\ 1 \end{bmatrix}_{1\times7} = \begin{bmatrix} 0 & \cdots & 0 & 0 & 0 \\ \vdots & \vdots & \vdots & \vdots & \vdots \\ 0 & \cdots & 0.2 & 0.6 & 0.5 \\ 0 & \cdots & 0.2 & 0.6 & 0.5 \\ 0 & \cdots & 0.2 & 0.5 & 0.5 \\ \vdots & \vdots & \vdots & \vdots & \vdots \\ 0 & 0 & 0.2 & 0.2 & 0.5 \\ 0 & 0 & 0.2 & 0.6 & 1 \\ 0 & 0 & 0.2 & 0.5 & 0.5 \\ \vdots & \vdots & \vdots & \vdots & \vdots \\ 0 & \cdots & 0 & 0 & 0 \end{bmatrix}_{49\times7} \tag{19}$$

The fuzzy inference for Δk_p is realized as above, and similarly, Q2 ~ Q49 and Δk_i and Δk_d are thus obtained.

4.1.3 Clarification. After fuzzification and fuzzy inference, the controlled object, i.e., the leg joints of the biped robot, is transformed into the actual control quantity according to the affiliation function of the output quantity, and then the appropriate output value is clarified to obtain the accurate output quantity [14]. In this paper, we adopt the method of multiplying the affiliation degrees of two input variables belonging to the same fuzzy subset and then carry out the center-averaged inverse fuzzy, and the fuzzy expressions are designed as follows:

$$y^{k+1} = F(x^k, y^k) = \frac{\sum_{j=1}^{5} \bar{y}_j^k \prod_{i=1}^{2} \mu_{A_{ij}^k}(x_i^k)}{\sum_{j=1}^{5} \prod_{i=1}^{2} \mu_{A_{ij}^k}(x_i^k)} \tag{20}$$

where k is the time series; A_{ij}^k is the fuzzy set; $u_{A_{ij}^k}(x_i^k)$ is the corresponding affiliation function on A_{ij}^k; y_j^k is the center of the fuzzy subset of the output thesis.

We substitute all the above known conditions into the simulation system, and then the fuzzy control system of the lower limb of the bipedal robot can be established, which is realized by MATLAB programming.

4.2 Variable theory domain fuzzy controller

The variable thesis domain fuzzy control system is a control system obtained by feeding back error information to influence the change of the thesis domain under the premise of unchanged fuzzy rules. The variable argument domain is essentially an interpolator for dynamic point-by-point convergence.

The selection of a scaling factor is the key to designing the variable domain fuzzy controller [15], which directly affects the accuracy of the controller. Through the theoretical study, after adding the exponential expansion factor, the domain changes linearly with the error and has reliable stability, so this paper chooses the exponential domain expansion factor: $a(x) = 1\text{-}\lambda e^{-kx^2}$, $\lambda \in (0, 1)$, $k > 0$, takes $\lambda = 0.001$ and $k = 0.65$. Based on the designed fuzzy controller, we add the suitable exponential domain expansion factor as above, which can be evolved to variable-theory-domain fuzzy control.

As mentioned above, a block diagram of the variable-theory-domain fuzzy controller structure for a single leg of a three-joint biped robot is designed as in Figure 8:

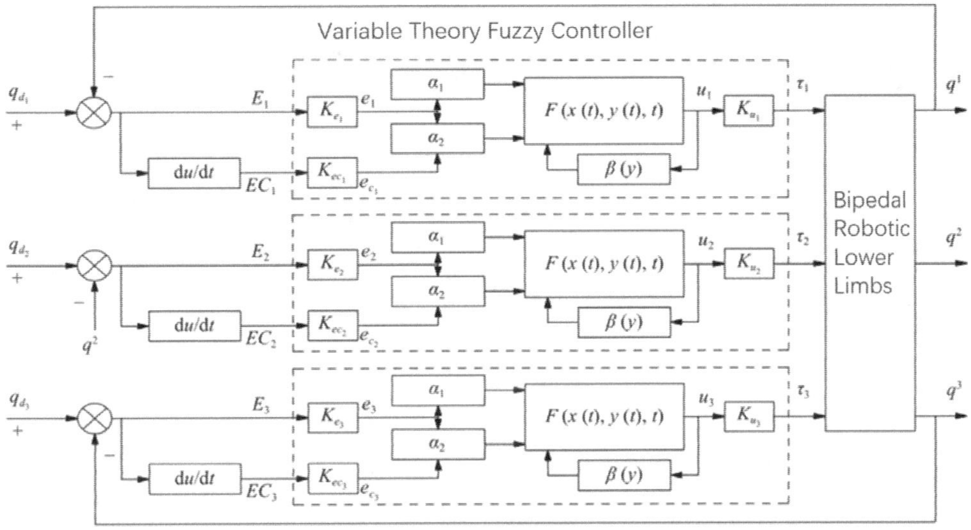

Figure 8. Structural block diagram of variable theory domain fuzzy controller.

In Figure 8, the input q_{dn} (n=1, 2, 3, all the following n are the same as above) is the desired joint angle of the single leg joint of the lower limb of the bipedal robot; the output q_n is the actual number of joint angles for the three joint angles of the ankle, knee, and hip joints; e_n is the positive and negative absolute value of the difference value of the angles of the three joints; e_{cn} is the rate of change of the angular error of each joint; K_{en} and Ke_{cn} are the quantization factors of the error e and the rate of change of the error ec, respectively; e_n and e_{cn} are the dimensionless rates of change of e and ec after the quantization factors in the previous step, respectively; K_{un} is called the scaling factor; τ_n is the torque that is acted by the controller on the lower limb of the biped robot.

5 MATLAB simulink

5.1 Platform

The 3D model file of the biped robot in .xml format is exported from solidworks using Simscape Multibody Link plug-in to MATLAB and opened by calling the import command, which shows the block diagram of the model of the biped robot, as shown in Figure 9(a). The parameter adjustment of each joint is shown in Figure 9(b).

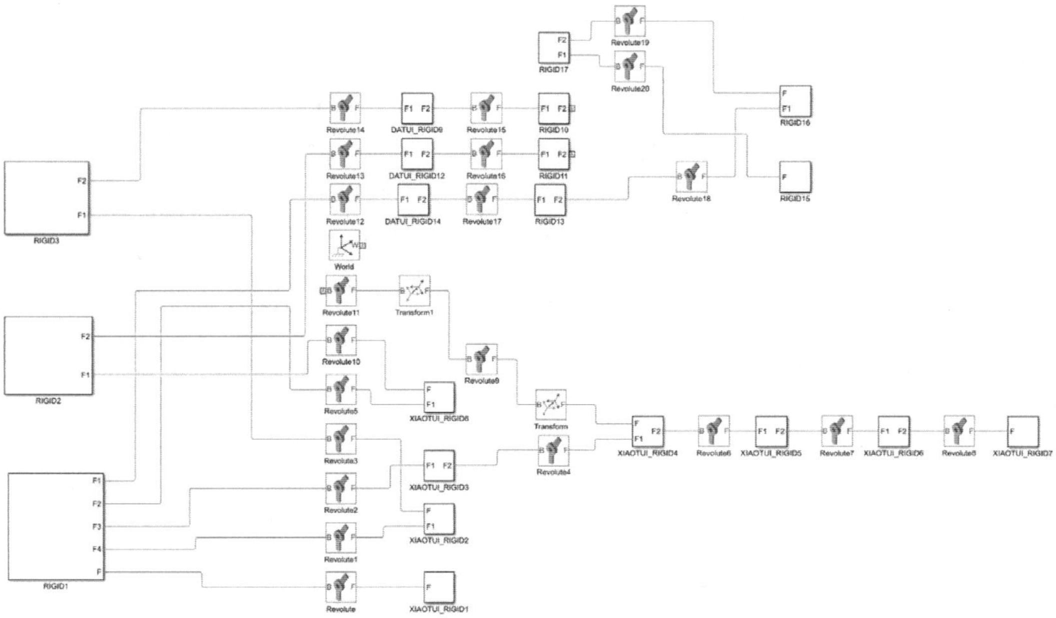

a. Block diagram of a 3D model of a bipedal robot

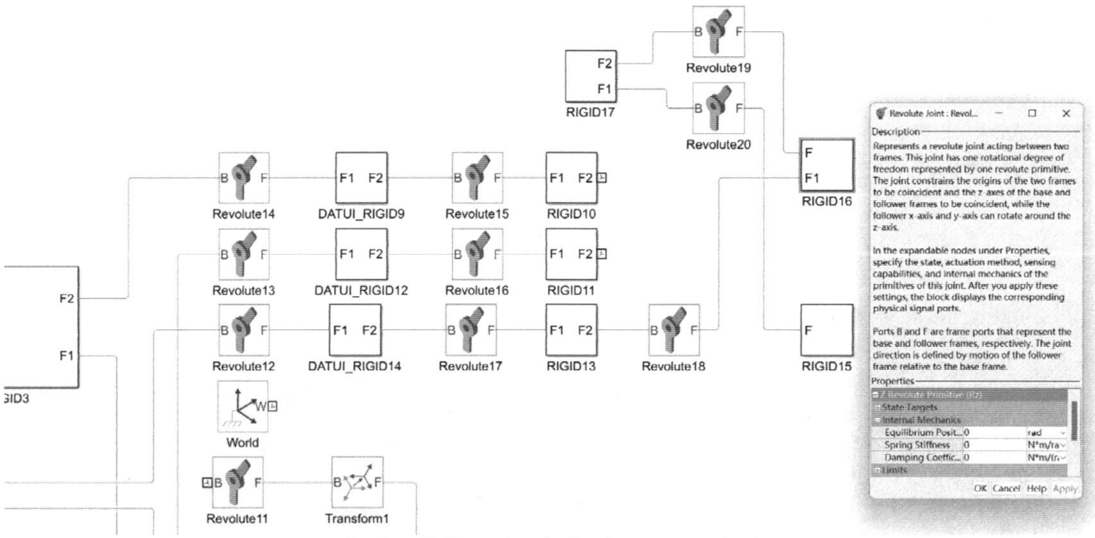

b. Simulink mechanical subsystem testbed

Figure 9. Simulink mechanical subsystem test bench.

The simulation block diagram of the variable-theory domain fuzzy control system for a single leg of the lower limb of a bipedal robot built afterward is shown in Figure 10, where the controllers of the three joints are encapsulated into a single module, so the angular outputs of the joints of the lower limb quickly track the inputs of the desired joints, which are given beforehand, with as high an accuracy as possible.

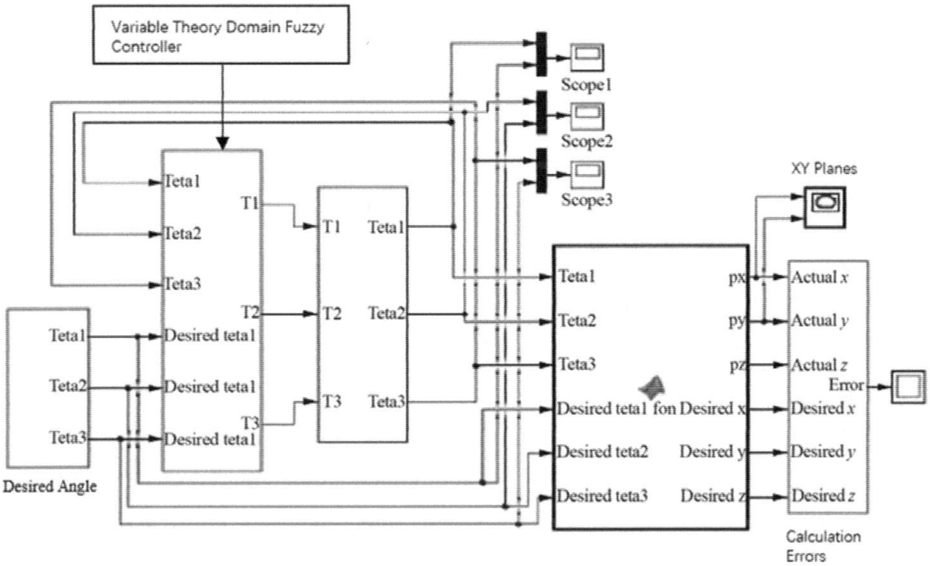

Figure 10. Simulation block diagram of fuzzy control system for the biped robot with variable theory domain.

The desired angle of each joint is obtained from motion capture experiments, and the desired trajectory is obtained in MATLAB from the gait planning in 2.3 above. Therefore, the parameters are set as follows: the quantization factor is taken as $K_e=K_{ec}=2.2$; the scaling factor is taken as $K_u=10$; the values of the relevant parameters of the scaling factor are taken as $\tau=0.98$, $\delta=0.002$, and $\rho=1$, respectively; the simulation step length is set as t=0.01 s; the total length of the simulation is set as T=4.0 s. The simulation data are obtained by adjusting the parameters of the torque angle of each joint in MATLAB's scope waveform diagrams.

5.2 Simulation result

The angular output angular displacement curves of the three joint angles of the hip, knee, and ankle joints can be obtained after the Simulink simulation is completed. Taking the ankle joint as an example, Figure 11(a) is the position tracking curve of the ankle joint using the fuzzy PID controller, and 11(b) is the position tracking curve based on the fuzzy control of the variable theory domain.

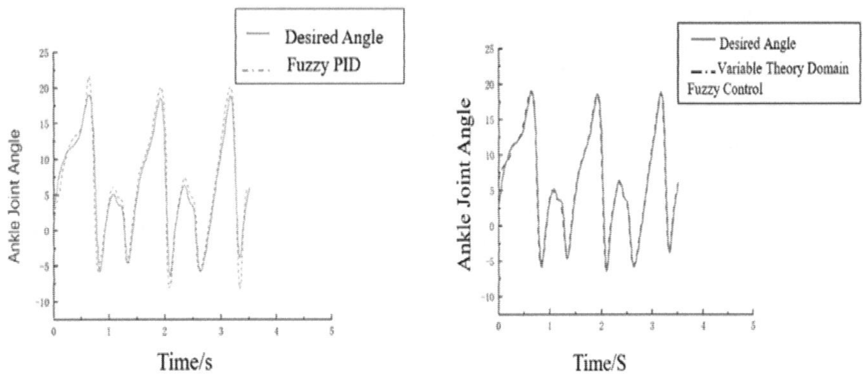

(a) (b)

Figure 11. Comparison of fuzzy PID and adaptive fuzzy PID simulation data.

The data results of joint angle values of 10 sampling points of the ankle joint in Figure 11 were selected for comparison using the sampling method, as shown in Table 2.

Table 2. Ankle angle value.

Time/s	*Desired Angle/rad*	*Fuzzy Controller/rad*	*Variable Theory Domain Fuzzy Control/rad*
0.1	7.52033	5.26571	8.45593
0.2	10.1308	9.37152	9.48214
0.3	11.4044	11.9834	11.5414
0.4	12.1873	13.292	12.3912
0.5	14.0662	14.3225	14.7933
0.6	18.0502	19.0287	18.6573
0.7	16.704	15.8295	16.5963
0.8	-2.75451	-5.69824	-3.45541
0.9	-2.30858	-1.22695	-2.463
1.0	3.36653	3.94748	3.91469

We take the first set of data as an example: at 0.1 s, the error between the fuzzy control and the desired angle is -2.25°, while in the variable theory domain, the error angle between the fuzzy control and the desired trajectory is +0.94°, with a difference of 1.32°.

6. Conclusion

In this paper, the kinematics model and Lagrangian dynamics equations of the bipedal robot are first established, and then based on the existing traditional fuzzy PID control algorithm, the theory of variable-theoretic-domain adaptive fuzzy control and the appropriate scaling factor are introduced. The Mamdani fuzzy reasoning method is adopted, a variable-theoretic-domain fuzzy PID control algorithm is designed for the movement of the joints of the lower limbs of the bipedal robot, which is a complex structure, and the data provided by the motion capture experiment are used as the basis. Based on the data provided by the experiment, the Simulink simulation environment is used to realize the simulation of the fuzzy PID and variable domain fuzzy PID of the bipedal robot and analyze it comparatively. The following conclusions can be drawn by comparing the expected data with the simulation data:

(1) In this paper, the design of a variable domain fuzzy control algorithm for the lower limb of a biped robot is realized by using the SIMULINK simulation testbed under MATLAB environment, and the corresponding desired tracking simulation curves are obtained.

(2) The simulation results show that the angular displacement of the lower limb joints of the robot in the use of fuzzy PID control system has a large error range of (-4.47°, 2.54°); the system which is prone to oscillation is unstable; variable-theoretic domain fuzzy PID controller error range changes in the initial oscillation of the initial 0.5 s within (-2°, 2°); the smaller the entire system overshooting is, the more stable the system is. Therefore, the selection of variable theory domain fuzzy control can accurately and stably track the expected trajectory, and accurately complete the control task under the expected path.

(3) The proposed algorithm effectively solves the contradiction between computational complexity and control accuracy, improves the response speed of the system, reduces the steady-state tracking error of the system, and improves the operation quality of the system. It shows the effectiveness and superiority of variable domain fuzzy control, which has strong engineering practice significance and reference value.

References

[1] Lyu J. L. (2013) Research and simulation of joint robot multi-motor control. Wuhan: Wuhan University of Technology

[2] Shi X. P. and Liu S. R. (2011) Research progress of manipulator trajectory tracking control. Control Engineering: 116-122.doi: 10.14107/j.cnki.kzgc.2011.01.031

[3] Li H. X. Interpolation mechanism of fuzzy control. Science in China Series E: Technical Science, 1998 (03): 259-267
[4] Wiput T. and Manukid P. (2017) Position control of a pneumatic surgical robot using PSO based 2-DOF H-infinity loop shaping structured controller. Mechatronics, 2017 (43(3)): 42-53.doi:10.1016/j.mechatronics.2017.03.001
[5] Wang T., Zhu A. D., Chen J. B., et al. (2019) Structural design and control of pneumatic parallel robot [J]. Hydraulic and Pneumatic 55-60.doi: 10.11832 /j.issn.1000-4858.2019.02.010
[6] Luo Y. W. (2007) Vicon 3D motion capture System's Application in human motion Analysis. Journal of Zhejiang Polytechnic of Industry and Trade, (2007 7(1)):40-42.
[7] Chen S. X., Shen H. P., Sun S. M., Cao K. and Chen B. (2014) Planning and simulation of humanoid robot walking gait on flat ground [J]. Mechanical Design:31-36. doi: 10.13841/j.cnki.jxsj.2014.12.007
[8] Zhong J. S. (2014). The PID control system theory. China Science and Technology Review, (18): 295-296
[9] Garrido R. and Trujano M. A. (2019) Stability analysis of a visual PID controller applied to a planar parallel robot [J]. International Journal of Control, Automation and Systems. (2019 17(6)):1589-1598. doi:10.1007/s12555-018-0178-7
[10] Sachin S., Aiswarya P. G., Monica G., et al. (2018) A comparative study of PID and neuro-fuzzy based control schemes for a 6-DOF robotic arm [J]. Journal of Intelligent & Fuzzy Systems. 35 (2018):5317-5327. doi: 10.3233/JIFS-169814
[11] Zhao J. and Yang H. Z. (2020) Intelligent Adaptive Fuzzy Control Strategy for Manipulator Motion [J]. Machinery Design & Manufacture, 8:192-196. doi: 10.19356/j.cnki.1001-3997.2020.08.045
[12] Tang L. W. and Yan H. Q. (2020) Fuzzy PID Control for Position Tracking of Serial Manipulator Based on PSO Algorithm [J]. Machine tool & hydraulics, 48:72-76.doi: 10/3969 /j.issn.1001-3881.2020.12.010
[13] Mohiuddine S. A. (2011) Some new results on approximation in fuzzy 2-normed spaces. Mathematical and computer modeling, 53:574-580.doi:/10.1016/j.mcm.2010.09.006
[14] Wen S. P., Zhao G. P. and Cai K. X. (2009) An adaptive algorithm of variable universe fuzzy control [J]. Control Theory and Applications, 26: 265-8
[15] Shao C., Dong X. W. and Wang X. F. (2010) The choice method of the variable universe fuzzy controller [J]. Information and Control, 39: 536-541. Doi:10.3724/SP.J.1219.2010.00536

Journal of Physics: Conference Series **2815** (2024) 012028 doi:10.1088/1742-6596/2815/1/012028

Numerical simulation study on the cold thread rolling forming of Ti-6Al-4V alloy fasteners with external threads and their defects.

Qiang Zhou* **and Jiawei Zhang**

School of Mechanical Engineering, Tiangong University, Tianjin, 300387, China

* Corresponding author's e-mail: pub090623@163.com

Abstract: A three-dimensional numerical finite element model for cold rolling forming of external threads on Ti-6Al-4V alloy fasteners for aerospace applications was established using DEFORM-3D. This model was employed to investigate the characteristics of the cold rolling forming process and analyze the defects associated with the rolling operation. The stability of the ratio between torque and tangential load at the radius value of the rolling wheel (72.5) validates the accuracy of the model. The research results indicate that due to the unevenness of the extrusion force, the metal flow velocity is higher adjacent to the tooth-shaped surface compared to the central position. This causes the metal on both sides to accumulate and squeeze towards the middle position at the tooth top, ultimately resulting in tooth top folding defects.

1. Introduction

Thread rolling stands as an efficient and cost-effective method for processing threaded fasteners, which involves extruding the workpiece with a thread rolling wheel to induce plastic deformation and form the threads. This process substantially improves the mechanical characteristics of the component, leading to notable enhancements in its structural integrity and performance. During the process of cold thread rolling, it is common for defects such as folding and cracking to occur on the surface of the threaded teeth. When the damage exceeds the specified limits, it can lead to a reduction in bolt strength and fatigue resistance. Hence, it is imperative to investigate the forming attributes of thread rolling and delve into the mechanisms underlying the generation of metallographic folds.

Scholars, both domestic and international, have delved into the realm of external thread rolling, employing methodologies like finite element analysis alongside experimental investigations to explore its diverse facets. Domblesky and Feng [1] formulated 2D and 3D finite element models for simulating the external thread rolling procedure. Their study demonstrated that the plane strain model effectively reflects the rolling behavior of the thread. Zhang et al. [2] utilized a combined approach of the Visco-Plastic Self-Consistent method and finite element method to establish a finite element model for thread rolling. Aktas and Kisioglu [3] analyzed the influence of forming parameters such as the number of passes on the material characteristics, microstructure, and microhardness. They verified the rationality of the processing parameters. Kramer and Groche [4] performed a comprehensive investigation into the impact of process parameters and billet preparation on defect formation. The results indicated that as the stroke speed increased, the forming force increased significantly, and crack defects caused by variations in the relative sliding velocity appeared at the root of the thread region. Zhang et al. [5] explored the effects of various lubrication conditions on the macroscopic and microscopic characteristics of the threaded surface. The study demonstrated that MoS_2 lubricating grease exhibited superior performance regarding surface roughness and defect reduction. Song et al. [6] employed finite element simulation to conduct process design and simulation studies on the thread-rolling process. Jiang et al. [7] integrated induction heating into the rolling process for thread processing. They examined the attributes of induction heating, alongside analyzing strain, displacement, and material flow behavior during hot rolling through finite element analysis. Yamanaka et al. [8] manufactured metal glass bolts via thread rolling. They utilized a 3D FEM to analyze the compressive stress pattern throughout the thread rolling procedure, identifying it as the primary facilitator in the formation of metal glass bolts. Wang et al. [9] characterized the morphology, microstructure changes, and fatigue life of the threaded surface and subsurface after rolling. Experimental results indicate that thread rolling improves the fatigue life of bolts, reduces surface roughness at the thread root, and results in phenomena such as grain refinement and increased hardness.

From the above discussion, it can be seen that most of the research on thread rolling has focused on aspects such as processing forming and process optimization. There is relatively less research on the forming characteristics and damage mechanism in cold thread rolling. Therefore, studying the formation mechanism of thread rolling defects is of great significance for improving the performance and lifespan of bolts.

2. Thread cold rolling experiment

The material used in the experiment is Ti-6Al-4V. The thread rolling system used for the rolling experiment consists of a platform, thread rolling wheel, and support block. The three-dimensional representation of the rolling system is depicted in Figure 1. After the thread rolling is completed, the bolts are cut. The cut bolts are then embedded and subsequently polished and finished to achieve smooth and polished surfaces. Afterward, the cross-section of the threads was observed using a microscope (Sigma 300, Zeiss).

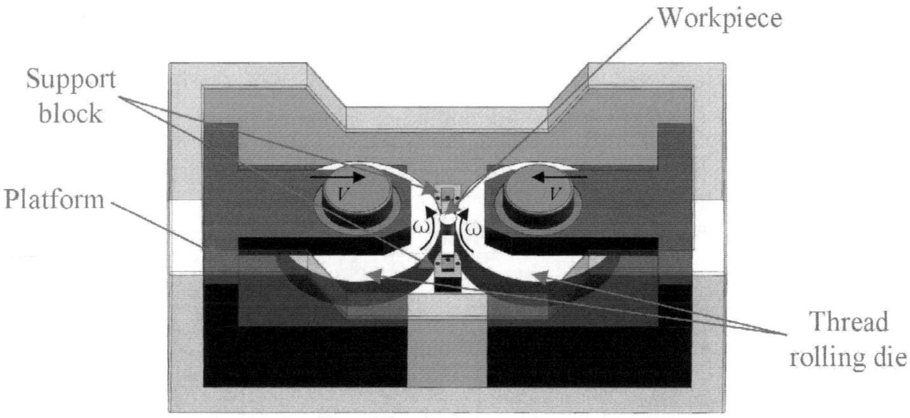

Figure 1. Three-dimensional schematic diagram of a thread rolling system.

3. Rolling process simulation modeling

Compared to the experimental methods for analyzing the thread-forming process and metallurgical folding damages, finite element simulation offers the advantages of being intuitive, cost-effective, and efficient. It can provide guidance for understanding the damage mechanism and forming process in thread rolling. Therefore, the thread rolling forming was studied using the finite element method.

3.1 Geometric model

Firstly, it is necessary to simplify the actual thread-rolling experimental model for the finite element model. Within the finite element analysis, only the bolt, rolling die, and supporting block need to be modeled, allowing for the omission of other components to enhance computational efficiency. The 3D model was created using UG (Unigraphics) software. The tooth profile of the rolling die was established with a half angle of 30° and a tooth height of 0.578 mm. Table 1 presents the specific parameters in detail. The rolling die model is shown in Figure 2(a) in the simulation software.

Table 1. The parameters of the thread rolling.

Parameters	Numerical value
Major diameter	144.956 mm
Pitch diameter	144.450 mm
Thread pitch	27
Thread pitch	1 mm
Width	30 mm

3.2 material model

The material model used in this study adopts the classical Johnson-Cook constitutive model, which describes the strain rate dependence of metallic materials. The model is concise and consists of three parts that describe strain hardening, strain rate strengthening, and temperature softening effects of the material. Due to the fact that thread rolling is a process performed at ambient temperature, the J-C formulation can be simplified by removing the parts related to temperature softening effects. The simplified model is as follows:

$$\sigma = (A + B\varepsilon^n)\left(1 + C\,ln\left(\frac{\dot{\varepsilon}}{\dot{\varepsilon}_0}\right)\right) \tag{1}$$

where σ represents stress and ε represents plastic strain; $\dot{\varepsilon}$ is the equivalent plastic strain rate; $\dot{\varepsilon}_0$ is the reference strain rate of the material; *A, B, C, n,* and *m* are constants that are often used in material models. This article uses material parameters *A*, *B*, *C*, and *n* with values of 782 MPa, 498 MPa, 0.028, and 0.28, respectively [10].

3.3 Boundary condition settings

The mesh is divided using DEFORM-3D. Since the deformation in rolling mainly occurs at the surface, while the strain is relatively small in the core of the workpiece, a refinement strategy is adopted to achieve a fine mesh on the surface and a sparse mesh in the core. The resulting workpiece model after partitioning is shown in Figure 2(b).

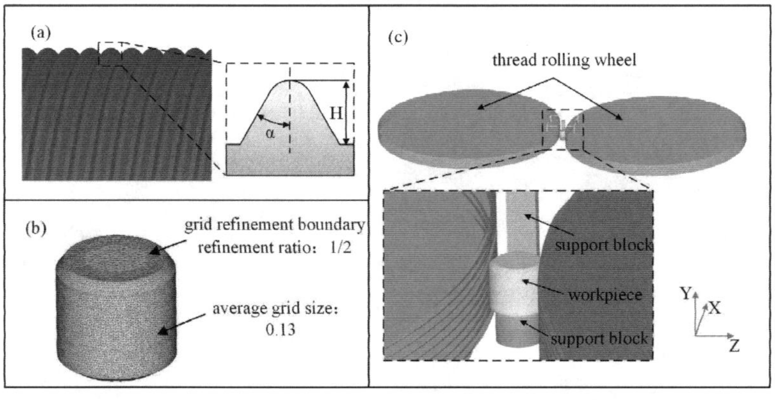

(a) Thread rolling wheel. (b) Mesh partitioning. (c) Thread rolling finite element model.

Figure 2. Schematic diagram of the finite element model for thread rolling.

The friction coefficient involving the workpiece and the thread rolling wheel is maintained at a steady value of 0.2. Similarly, the friction coefficient involving the workpiece and the support block is set to 0.1. During the rolling process, both thread rolling wheels rotate synchronously for feeding. The radial feed rate of the rolling wheels is established at 0.12 mm/s, while the rotational speed is configured at 30 RPM. The conclusive finite element model of the thread-rolling process is depicted in Figure 2(c).

4. Results and discussion

4.1 Torque characteristic analysis

During the cold rolling process, the thread rolling wheel rotates and feeds to drive the workpiece to rotate, thereby forming threads. Through the analysis of the torque applied by the thread rolling wheel, insights into the deformation characteristics of the workpiece throughout the rolling process can be garnered. The torque curve of the thread rolling wheel is shown in Figure 3. From the graph, it can be observed that the torque curve exhibits an overall increasing-then-decreasing trend. During the early stage of rolling, the slow increase in torque is due to the initial plastic deformation of the material. The material offers less resistance to deformation at this stage, being in a lower-strain hardening state. Hence, the torque increases at a slower rate. In the middle to later stages of rolling, the material starts to exhibit work-hardening effects, resulting in an increased resistance to further deformation. As a result, the torque experiences a rapid increase. Around 3.7 seconds into the processing, there is a sharp decrease in torque. This is because the threads have been mostly formed, and the processing enters the grinding stage. The advancement of the rolling wheel ceases, and the material undergoes only slight deformation, resulting in a decrease in torque.

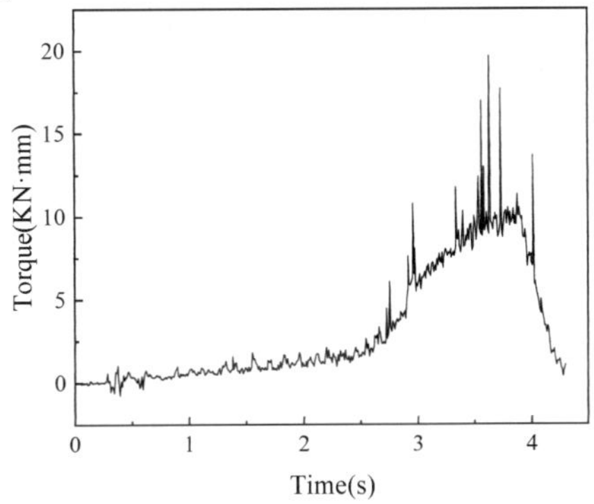

Figure 3. Torque curve graph.

Additionally, the tangential load on the rolling wheel was extracted, and the torque-to-tangential load ratio curve is shown in Figure 4. It can be observed that this value remains around 72.5, which is close to the diameter of the rolling wheel, confirming the reliability of the model.

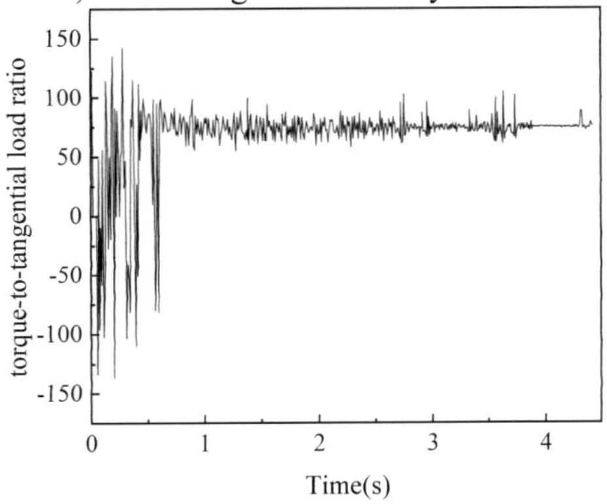

Figure 4. The variation curve of torque-to-tangential load ratio over time.

4.2 Analysis of thread crest defects

Due to the significance of bolt fasteners in the aerospace industry, higher demands have been placed on their inspection. Folding and cracking are common defects that occur during thread cold rolling processes. Observation of the formed threads was conducted using a scanning electron microscope (Sigma 300, Zeiss), as shown in Figure 5. From the figure, it can be observed that tooth-shaped damage is evident at the crest of the threads. When observing the defect morphology at the crest, it is evident that the defects exhibit irregular notches, primarily concentrated in the middle portion of the crest. Based on this observation, we can conclude that the defect type is folding. The folding at the crest of the tooth is attributed to the metal deformation induced by the pressure exerted during the tooth formation process. The metal on the sides of the thread experiences a faster flow rate, while the metal in the middle flows at a slower rate. This difference in metal flow ultimately causes the material on the sides of the thread to be squeezed towards the center, resulting in tooth crest folding.

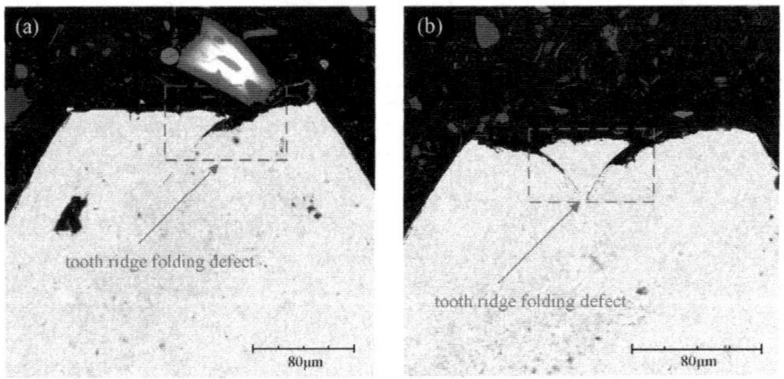

Figure 5. Image of tooth crest folding damage morphology.

To further explain the forming mechanism of tooth crest folding, the extracted tooth shapes from simulation results at different time steps are presented in Figure 6. When the thread rolling process starts, the material is subjected to compression from the tooth crest of the rolling wheel, causing it to accumulate towards the center and gradually form the thread root, as shown in Figure 6(a). As the rolling wheel continues to advance, the material undergoes further compression, causing it to pile up and form the thread flank. Nevertheless, throughout the cold rolling operation, the compression force from the rolling wheel on the workpiece is not uniformly distributed. The compression force on the sides of the thread is higher than the middle section. Due to the uneven distribution of compression force, the material flow rate increases on the sides, causing a depression to form at the middle position of the thread crest, as illustrated in Figures 6(b)-6(e). During the final stages of thread formation, due to the differences in material flow and the contraction of the tooth root space of the rolling wheel, the metal material on the sides of the thread flank starts to pile up towards the center, eventually forming a fold defect at the thread crest, as shown in Figure 6(f). Damage value is considered as the measure of the likelihood of damage occurrence, as observed in Figure 6. Observably, there is a gradual increase in damage value during the formation of the thread, reaching its peak at the thread crest.

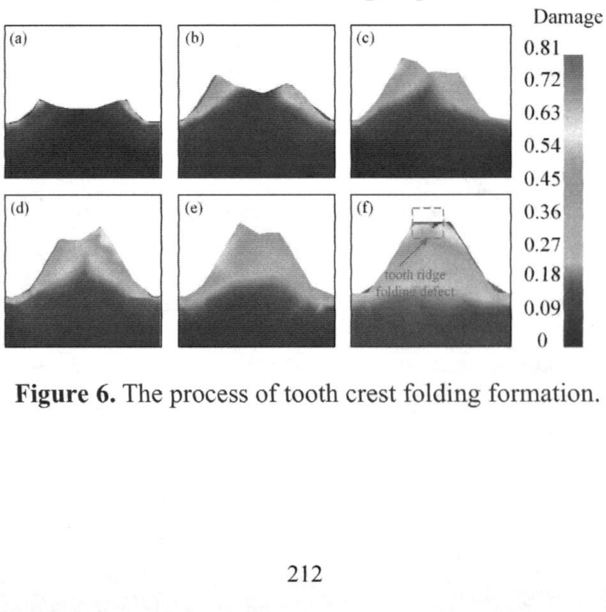

Figure 6. The process of tooth crest folding formation.

5. Conclusion

This paper establishes a finite element model for the cold thread rolling process and analyzes the deformation behavior of workpieces during the rolling process through torque curve analysis. The investigation into the workpiece's deformation behavior throughout the rolling process was conducted via torque curve analysis. The stability of the ratio between torque and tangential load at the radius value of the rolling wheel (72.5) validates the accuracy of the model. It further reveals the formation mechanism of tooth crowning defects. During the tooth-forming process, due to the gradient effect of extrusion force, there are differences in material flow. The metal flow rates on both sides are higher than the middle position. Eventually, the metal materials on both sides are squeezed and accumulated towards the center, forming tooth crowning defects.

References

[1] Domblesky J P and Feng F. (2002) Two-dimensional and three-dimensional finite element models of external thread rolling. Proceedings of the Institution of Mechanical Engineers Part B Journal of Engineering Manufacture, 216(4):507-517.

[2] Zhang S W, Zhang D W and Jiang H, et al. (2022) Numerical and experimental analysis of deformation behaviors and microstructure evolution in the thread rolling process. Journal of Materials Research and Technology, 19:230-242.

[3] Aktas S and Kisioglu Y. (2020) Effects of thread rolling processing parameters on mechanical properties and microstructures of high-strength bolts. Materials Testing, 62(10):1017-1024.

[4] Kramer P and Groche P. (2018) Defect detection in thread rolling processes – Experimental study and numerical investigation of driving parameters. International Journal of Machine Tools & Manufacture, 129:27-36.

[5] Zhang S, Zhang D and Zhao S, et al. (2023) Characteristics of surface and subsurface of formed thread parts by axial-infeed thread rolling process. Chinese Journal of Aeronautics, 36(3):471-481.

[6] Song J H, Lee H J and Lee J S, et al. (2012) Numerical and Experimental Approach for Incremental Thread Rolling Process of Micro-Sized Screws. Advanced Materials Research, 472-475:732-735.

[7] Jiang F, Zhao S D and Zhang S W, et al. (2022) Finite element analysis and experiment study on axial infeed hot rolling process of thread with Ti6Al4V. International Journal of Advanced Manufacturing Technology, 121(9-10):5873-5884.

[8] Yamanaka S, Amiya K and Saotome Y, et al. (2011) Plastic Working of Metallic Glass Bolts by Cold Thread Rolling. Materials Transactions, 52(2):243-249.

[9] Wang X M, Xiong X Y and Huang K H, et al. (2022) Effects of Deep Rolling on the Microstructure Modification and Fatigue Life of 35Cr2Ni4MoA Bolt Threads. Metals, 12(7):1224.

[10] Li A, Pang J and Zhao J, et al. (2017) FEM-simulation of machining induced surface plastic deformation and microstructural texture evolution of Ti-6Al-4V alloy. International Journal of Mechanical Sciences, 123:214-223.

Journal of Physics: Conference Series **2815** (2024) 012029 doi:10.1088/1742-6596/2815/1/012029

Research and application of self-diagnosis and automatic intervention functions for turbine regulating valve malfunctions in nuclear power plants

Changxian Gan [1,*], Xiaolong Jiang [1], Min Li [1], Huahuan Li [1], Xianhai Li [1] and Jiaxuan Li[1]

[1] Suzhou Nuclear Power Research Institute Co., Ltd., Shenzhen, Guangdong Province, 518000, China

[*] Corresponding author's e-mail: Ganchangxian@cgnpc.com.cn

Abstract: The inlet regulating valve of the turbine is very important to the control of the turbine. If the control is abnormal, it will create some uncontrollable safety risks, which may cause the turbine to trip, even to over-speed. The study of the structure and control mode of the turbine regulating valve shows some defects in the design of the turbine regulating valve in a nuclear power plant. In this paper, through the analysis of the fault mode, the method of fault self-diagnosis is studied, verified, and improved so that it has perfect self-diagnosis and intervention means, improves the reliability of turbine valve control, and eliminates the hidden danger of tripping.

1. Overview

The inlet regulating valve of the turbine is very important for controlling the turbine. When the control is abnormal, the corresponding alarm and automatic regulation should be set, and even the intervention function of the turbine protection should be triggered. These functions are used to ensure that the turbine speed control, pressure control, or load regulation is carried out without disturbance to ensure the safe and stable operation of the turbine. In terms of preventive maintenance management, during the operation of the unit, the power plant usually conducts regular valve activity tests on the turbine inlet steam regulating valve [1], strengthening the oil quality management of the oil supply system to improve the oil quality [2, 3]. When the engine is shut down for maintenance, the valve hydraulic servo-motor and its components are repaired, cleaned, or replaced. In terms of equipment operation and real-time monitoring, the routine faults of the turbine inlet control valve are usually monitored and alerted (such as command failure, feedback failure, large deviation of command and feedback, etc.). When the fault occurs, it is usually designed to maintain the state of the previous moment and exit the automatic operation, triggering the alarm and requiring manual intervention. Conventional thermal power plants and nuclear power plants have not proposed the concept of self-diagnosis of turbine control valve faults, and only basic faults are usually monitored for valve abnormalities. The turbine inlet regulating valve of a nuclear power unit with plate number CPR1000 is equipped with alarms to detect various failure modes. These include large deviations between the valve regulating command and feedback (known as valve jamming), valve position failure, command failure, deviation between the calculated flow rate and the actual valve position and system demand flow, unqualified valve opening and closing times, control card failure, and more. However, the valve hydraulic servo-motor body and its components are not designed, and there is no way to install monitoring sensors. Judging whether a device is faulty is impossible, and there is no warning for its adverse consequences. Under certain working conditions, if the equipment is abnormal, the turbine will be trip [3]. In summary, to further ensure the reliability and advancement of turbine-regulating valve control, it is necessary to improve the monitoring of its failure modes, form a relatively complete fault self-diagnosis function, and enable automatic intervention[4].

2. Research direction and content

The turbine inlet regulating valve is a hydraulic valve with an electric hydraulic regulating hydraulic servo-motor to overcome the disc-shaped spring force to lift the valve stem to realize the valve opening. The hydraulic servo-motor and its accessories (such as servo valve quick closing solenoid valve) are the most important core components of the steam turbine inlet control valve, which receives the signal of the turbine control system, and protection system, drives the turbine inlet valve, adjusts the steam intake volume, accurately controls the turbine speed and load, and makes quick closing action in an emergency to ensure the safe operation of the turbine. The control system of the turbine control valve controls the hydraulic servo motor and accessories. The control link, input signal, output signal, clip, actuator, and nuclear power unit have set up a relatively mature anomaly monitoring (as mentioned above). Still, they lack hydraulic servo-motor fault mode monitoring. By studying the fault mode of the hydraulic servo-motor (combined with operation and maintenance experience and reviewing a lot of literature) and analyzing the consequences[5], this paper collects practical cases of nuclear power and further supplements the monitoring elements required by the self-diagnosis function (Figure 1-① hydraulic servo-motor). Then, given the hydraulic servo-motor fault, the data is collected by simulating the fault test, and how to diagnose the anomaly through other parameters is studied on the premise that the measurement sensor cannot be added. The scheme of automatic intervention is studied, and experiments verify its effectiveness.

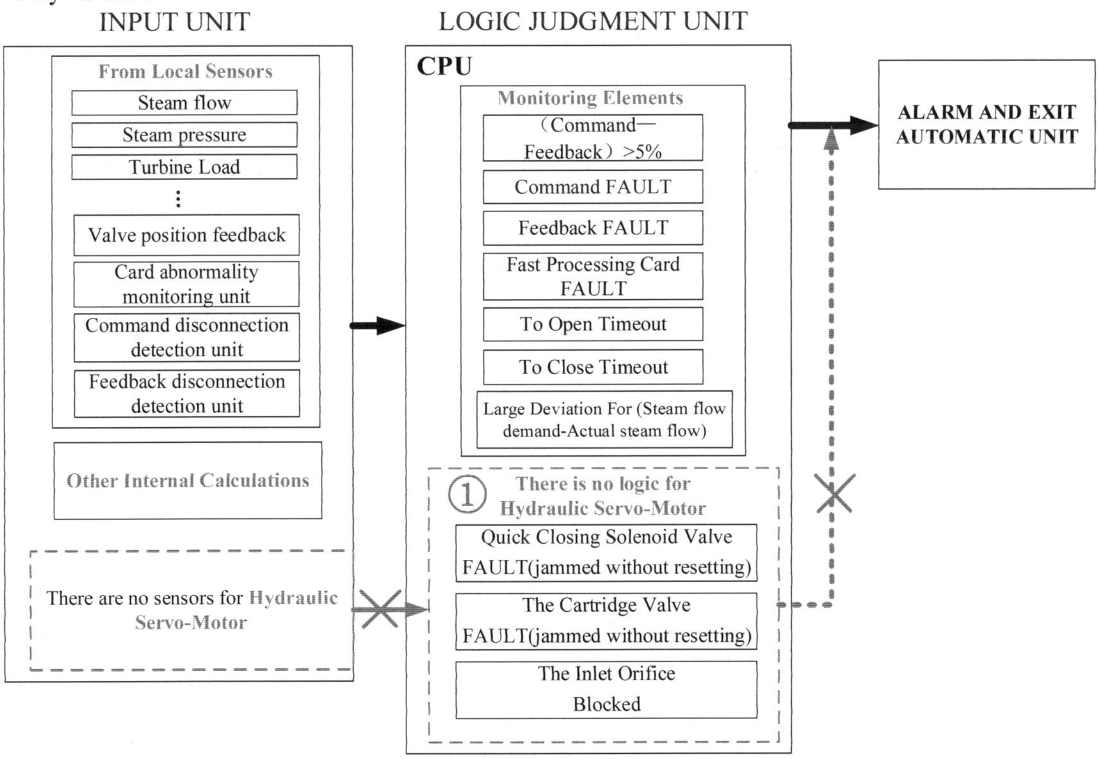

Figure 1 Schematic diagram of current design status monitoring.

3. Analysis of oil engine failure modes and their effects

In the process of turbine operation and maintenance in various power plants, the common fault phenomena of hydraulic servo-motor are poor oil quality, quick closing solenoid valve jammed, throttle hole blocked, servo valve jammed, etc. [2, 3]. According to statistics [6], only a nuclear power plant in the past ten years because of mud or impurity in oil caused abnormal valve or test failure times as many as 25 times. After research, we can conclude that the common failures and effects of oil engines are [6]:

① The oil quality is not good, and the oil whose particle size is not up to standard will block the servo valve, solenoid valve, and cartridge valve, which will cause abnormal valve adjustment or the hydraulic servo-motor cannot be closed in time when it needs to close quickly;

② The quick closing solenoid valve is jammed, resulting in the hydraulic servo-motor hydraulic oil working oil pressure cannot be established, or the hydraulic servo-motor needs to close quickly but cannot be closed in time;

③ The throttle hole before the quick close solenoid valve is blocked, resulting in the oil working pressure of the hydraulic servo-motor hydraulic oil cannot be established;

④ The cartridge valve is jammed, resulting in the oil working pressure of the hydraulic servo-motor not being established or the hydraulic servo-motor not being closed in time when it needs to be quickly closed;

⑤ The servo valve is jammed, resulting in the valve hydraulic servo motor not being adjusted normally because the rated flow rate of the servo valve is large, and it easily causes oil pressure fluctuation.

When the turbine is started for the first time after maintenance, the above faults usually do not occur. Even if the probability of occurrence is small, it can be repaired quickly and will not have major consequences on the unit.

In the normal operation of the unit, usually, the Fault Types ②③④ will not occur when there is a small probability of occurrence that will not affect the normal adjustment of the valve, and the valve needs to close quickly and may not be able to close in time. This situation cannot be found through other monitoring parameters, only through the valve activity test (ATT test) once every two months to find; And if the Fault Types (1) and (5) occur, the valve will not be able to adjust normally, and the existing command and feedback deviation logic will trigger the alarm and manually intervene.

In the unit's normal operation, we do the valve activity test (once every two months). If the Fault Types ① and ⑤ occur, the valve activity test will trigger the valve response timeout alarm that can be repaired. If the Fault Types ②, ③, and ④ occur, the timeout alarm of the ATT test will jump before it is triggered [6, 7] because of low oil pressure.

In most cases, when the hydraulic servo motor fails, the current design has the means and time to intervene, which will not cause great consequences. However, during the high-frequency valve activity test, the failure of the hydraulic servo motor may produce unacceptable consequences (turbine tripping). We will analyze, test, and study solutions to this problem further.

4. Cause analysis

The steam inlet regulating valve of the turbine is the only way to cut off the steam source when the turbine fails quickly. Therefore, during the unit's operation under load, these valves need to be tested for activity every two months to check the sensitivity of the valve actuator and verify whether the inlet steam cut-off valve (servo valve) and the inlet steam regulating valve are in a reliable state. During the valve mobility test, the automatic program will trigger the inlet steam regulating valve to close quickly, then we trigger the inlet steam regulating valve to open, and calculate the valve opening and closing times to ensure the requirements are met [7].

During the valve mobility test, there is a possibility that the quick-closing solenoid valve, cartridge valve, and other components on the valve actuator (oil actuator) will get stuck and cannot be reset normally after action. If these components are not reset normally, the valve will not open after receiving the command to open in the fully closed position, and the oil pressure of the EH oil supply system will rapidly decrease, triggering the turbine to trip.

The following description takes the turbine high-pressure cylinder inlet regulating valve as an example. The principle of the low-pressure cylinder inlet regulating valve is similar and will not be explained again. The turbine inlet regulating valve is a hydraulic valve with its electro-hydraulic adjustment hydraulic servo-motor, which is used to overcome the force of the disc spring to lift the valve stem to open the valve. The high-pressure anti-fuel station (which provides power oil for the electro-hydraulic regulating system, from now on referred to as EH oil) oil pump provides a certain hydraulic oil pressure for the hydraulic servo-motor. The EH oil system of each unit is equipped with two axial piston variable pumps, one main and one backup. The maximum flow rate of a single pump is 68 L/min, and the normal working pressure is 16MPa. The safety oil circuit (quick closing control circuit) of the oil engine is set up with 2*100% redundancy. The two quick-closing solenoid valves in column A/B control one safety oil circuit. Each safety oil circuit has a cartridge valve (used to establish safe oil pressure and cut off the rapid oil drain passage of the oil engine).

The oil circuit establishment process of valve opening is as follows, as shown in Figure 2:

① Safety oil pressure establishment: The quick closing solenoid valve is energized, the EH oil supply enters the cartridge valve → the cartridge valve controls the oil establishment and overcomes the internal spring force to close the cartridge valve → the cartridge valve cuts off the oil drain path

② Power oil establishment: When the servo valve is opened, the servo valve coil is charged, driving the servo valve internal slide valve displacement, and we open the oil inlet channel → EH oil is supplied through the hydraulic servo-motor into the valve cylinder → overcome the spring force to open the valve, the amount of oil into the cylinder determines the opening of the valve.

There are two ways to close the valve:

Method 1 (quick close): When power is lost, the safety oil pressure is released into the drainage through the quick closing solenoid valve. The cartridge valve is opened which allows the oil in the valve cylinder to quickly pressure into the drainage.

Method 2 (adjustment mode closed): The quick closing solenoid valve is still charged (the safety oil pressure is not lost), and the excess oil of the cylinder is discharged through the servo valve on the hydraulic servo-motor so that the valve is closed to the target opening.

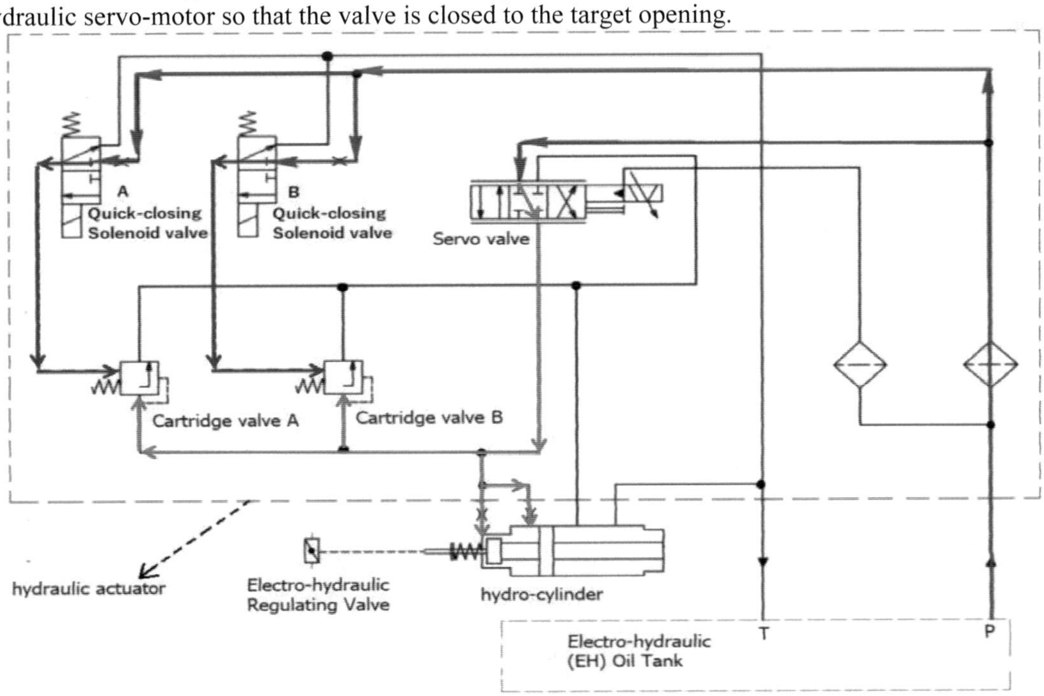

Figure 2 EH oil supply, oil circuit diagram when the valve is opened.

If the following three situations occur, there will be a safety hazard that causes the turbine to trip. If any one of the two quick closing solenoid valves on the hydraulic servo-motor does not return to the normal working state after being charged, or sludge impurities block its corresponding oil inlet orifice and make the solenoid valve fails, then the safety oil of the cartridge valve cannot be established, and the oil pressure in the valve cylinder will be completely discharged through the cartridge valve. The same is true if the two cartridge valves are jammed without resetting after action. If the valve receives the open command, the power oil will be continuously circulated through the servo valve, cartridge valve, and fuel tank (Figure 3).

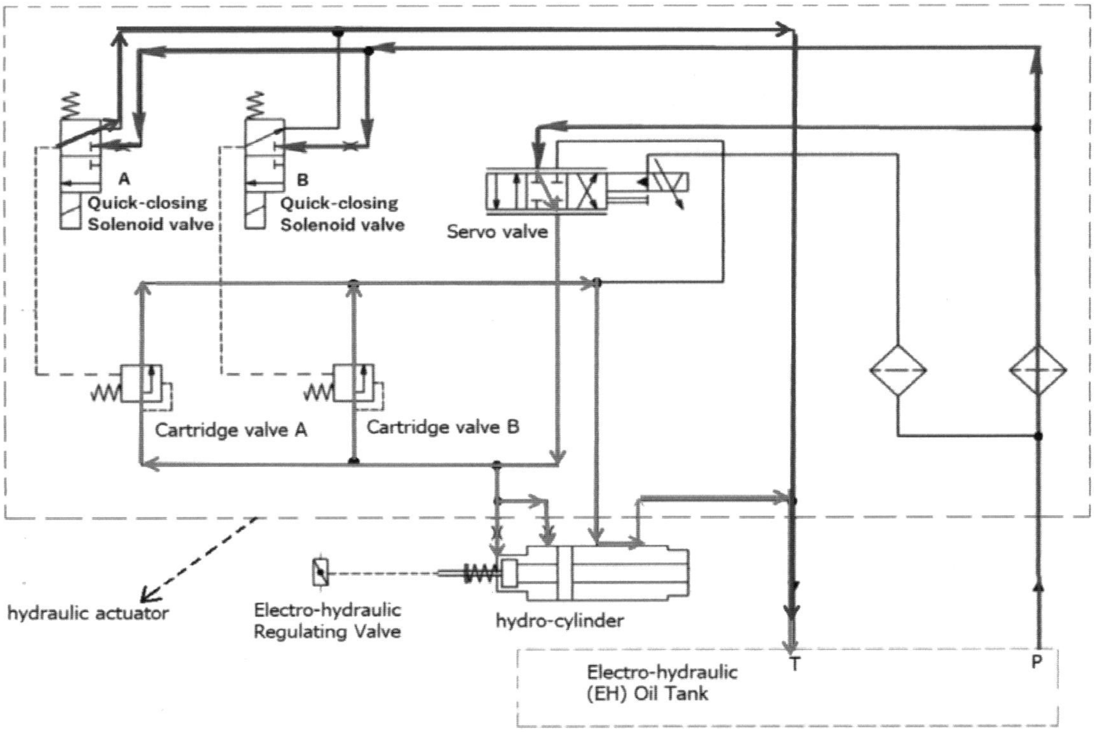

Figure 3 Schematic diagram of the oil circuit when the EH oil power oil and the oil drain are open.

The maximum flow rate of the EH oil pump is 68L/min, and the maximum flow rate of the servo valve is 280L/min. After the above oil circuit circulation, the EH system oil pressure will quickly drop to the trip, setting the value to trigger a shutdown.

Similar situations have occurred in thermal power plants [8], and nuclear power plants are no exception: In January 2023, Unit 1 of a nuclear power plant reduced the power to 89% FP to perform a door adjustment test (ATT test) according to the plan. During the high-profile door test of No. 2, when the servo valve receives a signal and opens, the shut-off valve (the principle is the same as that of the cartridge valve) gets stuck and does not reset, causing a large amount of power oil to be discharged from the unloading valve. The jointly started backup oil pump is still unable to maintain the oil pressure of the EH system, which ultimately leads to a trip.

According to the above research, it can be concluded that the current design of the turbine control valve lacks the fault self-diagnosis function and automatic intervention function of the hydraulic servo-motor valve, leaving a large safety hazard. In the following article, a fault simulation test will be conducted on the turbine regulating valve of a nuclear power plant, and a solution will be found. This method is also applicable to power plants with similar configurations.

5. Improvement scheme

5.1 Test and data acquisition

Relevant tests were conducted during unit maintenance to grasp the characteristics of the servo valve action under normal conditions during the ATT test and verify the impact of the above failure modes on the oil pressure of the EH system. We collect important unit transients of the unit envelope, such as no-load rejection, plant power transient test, and servo valve adjustment characteristic curve during short power grid failure. Combined with test data, etc., we analyze the impact of faults and provide data support for logical optimization plans. The following description takes the high-pressure cylinder regulating valve as an example.

5.1.1 Normal conditions of ATT test. During the ATT test, the regulating valve of the high-pressure cylinder is opened from 0% to 100%. In the process of fast opening, the maximum deviation between the command and feedback of the valve is < 5%. The maximum output current of the servo valve can reach 100%, and the minimum current is about 36%. The duration of the valve opening adjustment process of the servo valve is about 1.86s, and the oil pressure of the EH1 oil station is more than 14MPa. The characteristic curve is shown in Figure 4 below.

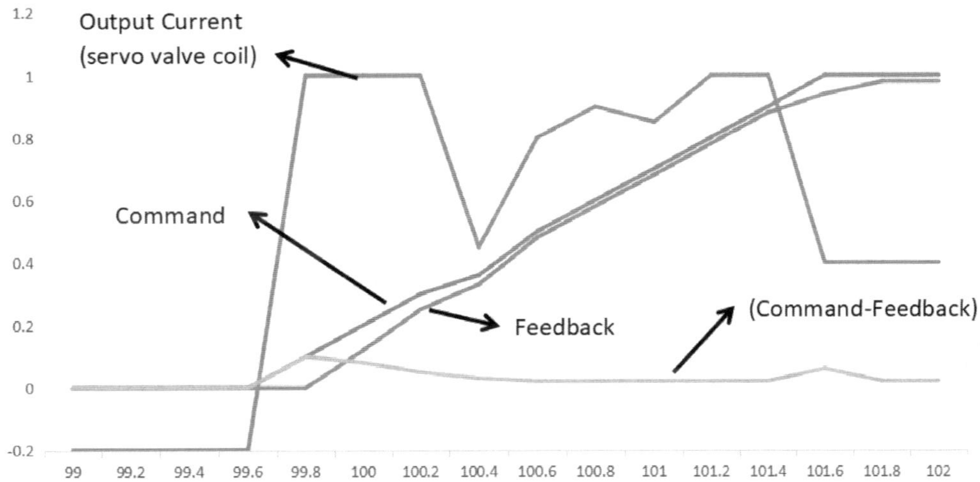

Figure 4 High-pressure cylinder valve opening process characteristic curve.

5.1.2 ATT test simulates abnormal conditions. For the high-pressure cylinder regulating valve, under the premise of opening the quick closing solenoid valve, the oil pressure change of the GFR master pipe of the EH1 oil station when the servo valve is different opening degree is simulated:

As the servo valve command increases to 35%, there is a drop in oil pressure. However, if the EH1 oil station pressure is gradually increased by 1% steps, the oil station pressure will still be able to maintain the system oil pressure despite a further decline trend. When the command increases to 41%, the EH1 oil station gradually drops; at about 6 minutes, EH1 triggers a low oil pressure interlock and low oil pressure station protection outage. The servo valve command is 75%, and the EH1 oil station triggers low oil pressure interlocking and low oil pressure protection outages for about 13s.

When the final output command of the servo valve reaches 100%, the oil circulation reaches the maximum, and the oil pressure of the system decreases the fastest (as shown in Figure 5 below). At 16:11:30, the command reaches the maximum and the oil pressure is 16.04MPa. After 4s, at 16:11:34, it is reduced to 13.1MPa. After 3s, at 16:11:37, it is reduced to 11.6MPa. The next moment reaches 11.5MPa backup pump will automatically interlock start. After another 3s and at 16: 11: 40, the oil pressure reaches the trip threshold of 10.5MPa, at which time the EH oil system automatically stops operation and the steam turbine triggers tripping.

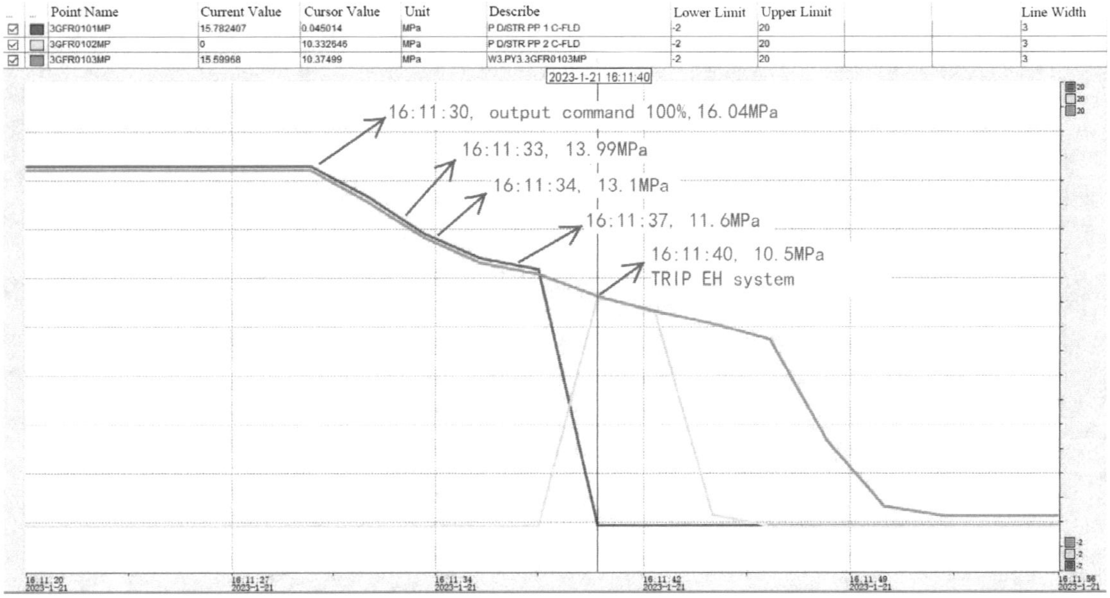

Figure 5 Simulated abnormal working condition of high-pressure cylinder regulating valve and system oil pressure trend diagram.

Combined with the above figure, the equipment action sequence is summarized in Figure 5. Since the controller output of the servo valve reaches 100%, the oil pressure continues to drop for about 7s, and the standby oil pump is interlocked to start. After another 3s, the oil pressure is reduced to the trip threshold, and the EH system and steam turbine are tripped. After multiple tests of multiple groups of valves, the test data are based on the action timing of Figure 6.

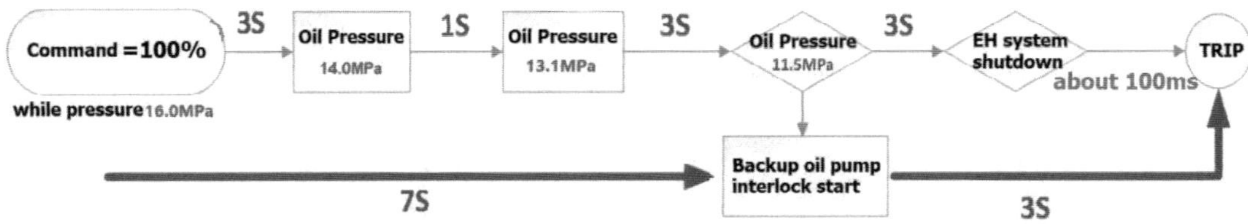

Figure 6 Timing diagram of verification results of high-pressure cylinder valve adjustment.

5.1.3 Historical transient data of a nuclear power plant. The transient test of the turbine mainly consists of six parts:

① 100%Pn platform variable load test ② trip-off test ③ reactor trip-off test ④100%Pn, 50%Pn dump plant power test ⑤100%Pn, 50%Pn dump no-load test ⑥RUNBACK test.

Selecting from the valve will close and open the extreme test and data analysis conditions, mainly for the 50%Pn dump no-load, dump power transient test, and the 100%Pn dump no-load, dump power transient test. The request is to analyze the historical trend of the oil pressure at the outlet of the EH master pipe of EH1/2/3 oil station. The analysis should consider the test scenarios where the load is dumped from 100%Pn to 50%Pn or from 100%Pn to factory power consumption and also when there is no load dumping. It is important to note that the oil pressure should never be lower than 14.9MPa.

5.1.4 Simulation of short-circuit power grid failure. When the short power grid failure occurs, the regulating valve of the high/low-pressure cylinder will recover after receiving the 1s quick shutdown signal, and the high and low-pressure cylinder section valves will be closed at the same time and then opened, resulting in the reduction of EH oil pressure. The duration of the servo valve command signal of the high-pressure cylinder adjusting valve reaching the maximum 100% (XD > 5%) is 1.8s, and the minimum pressure of the female pipe of the EH1 oil station is about 13MPa.

From the above test data, it can be concluded that when the deviation between the command and feedback is greater than 5% and lasts for more than 2 s. The oil pressure of the master pipe drops to about 13MPa. It is more appropriate to pay attention and intervene.

5.2 Design of improvement scheme

The fault self-diagnosis and automatic intervention scheme described in this paper is designed after obtaining many test data and obvious characteristic parameters. The improved scheme is realized in the DCS control system. The design idea is when the servo valve oil circuit is abnormal during the valve activity test, to prevent the oil pressure from continuing to drop, it is necessary to automatically cut off the servo valve in a reasonable time to prevent the oil circulation, leading to the oil pressure drop, to eliminate the hidden danger.

There is no oil pressure sensor on the valve hydraulic servo-motor, which cannot directly monitor the oil circuit of the hydraulic servo-motor; only the oil pressure of the master pipe of the EH oil system can be monitored. Because of the structure of the hydraulic servo-motor, it is not possible to design a new pressure measuring point. Therefore, the core of solving the problem is how to determine that the low pressure of the main pipe of the EH oil system is caused by the abnormal oil circuit of a certain valve, determine the failure of the valve, and trigger the automatic protection of the valve. We use the step-by-step inquiry and answer method to analyze this problem. Question 1: How to prevent oil pressure drop after EH oil circulation? Answer: We cut off abnormal oil circuit before EH oil pressure is too low. Question 2: How low is the oil pressure before intervention is necessary? Answer: This oil pressure value cannot be too low, and it needs to allow the automatic judgment and intervention protection logic to leave time through the test data and discussion with the turbine manufacturer, and finally by the system design and supporting manufacturers to give the dry predetermined value. Question ③: Under what conditions will the oil pressure of the EH oil system drop? Answer: The oil circuit of the valve hydraulic servo-motor is abnormal and needs to be monitored. This paper determines the abnormal oil circuit by monitoring the deviation of instruction and feedback. The oil pressure drop is caused by other reasons (such as pipe breaks); this situation has been considered in the system design to deal with measures; this paper does not need to provide plans. Question ④: How do you automatically cut off and terminate the oil circuit cycle? Answer: We switch the valve opening instruction to 0 or negative to make the servo valve close the oil inlet channel.

According to the above analysis, it is necessary to set two necessary logic conditions for automatic judgment: (1) the continuous deviation of the difference between the command and the feedback and (2) the intervention threshold after the oil pressure drops.

Combined with the test data, key characteristic parameters are selected and optimized based on the existing logic [9, 10]. As shown in Figure 7, the newly added fault self-diagnosis (①②) and automatic intervention (③) logic are in the box.

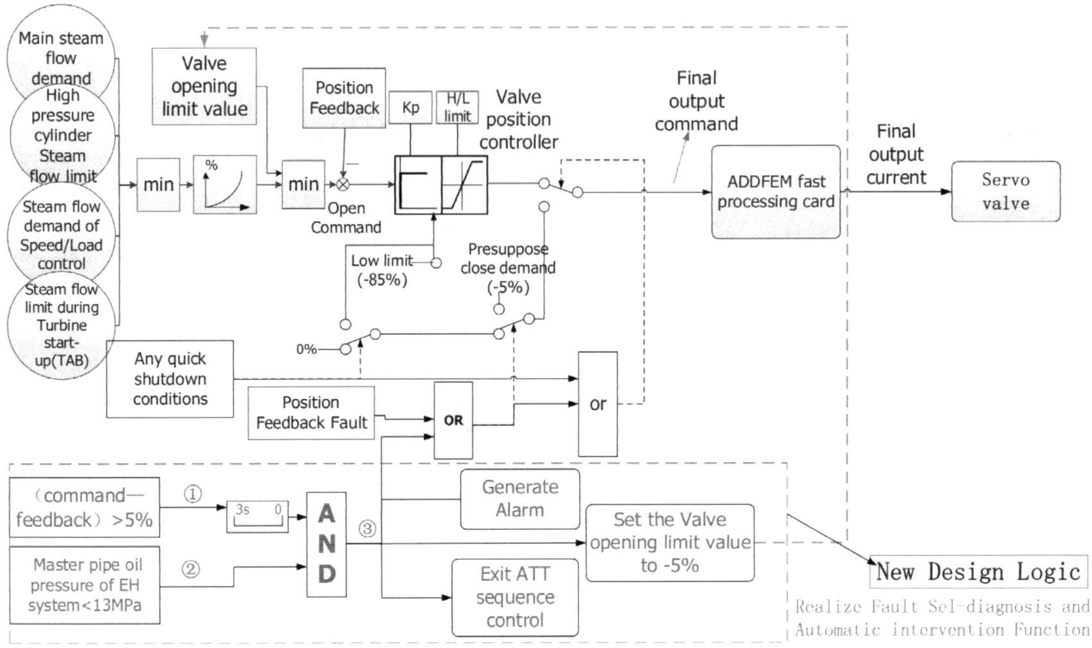

Figure 7 Protection logic diagram of newly added fault self-diagnosis and automatic intervention.

According to the valve operation manual, the start time of the high-pressure cylinder regulator must be less than 1.65s. Combined with the historical trend curve of the valve opening, if the valve is not faulty, there will be no more than 5% instruction and feedback deviation during the opening process; even if it is closed, it will be more than 1s time. Therefore, the valve instruction and feedback deviation can be greater than 5%, 8%, or 10% equivalent to determine whether the valve has a fault.

On the other hand, because the controller output of the servo valve = (instruction - feedback) *Kp, where Kp=20 (pure proportional control), that is, (instruction - feedback) > 5%, the servo valve can be fully opened, meaning that the oil flow of the servo valve reaches the maximum (the key point was selected during the test for observation and verification), the valve position still cannot follow the instruction change. It optimizes the program selection (instruction-feedback) by> 5% of this characteristic condition.

When the deviation between instruction and feedback is > 5% or more than 3s (the time is set according to the oil drainage velocity data of the previous simulation failure), the pressure-holding logic of automatic intervention is triggered when the pressure of the EH oil pipe is < 13MPa. At this time, the logic designed four automatic intervention instructions: ① It immediately closes the servo valve to cut off the hydraulic servo-motor; ② It sets the valve position limit value to -5%; ③ It automatically exits the valve activity test (prohibit receiving the new instruction of the automatic program of the test); ④ It is added a new alarm, requiring operation and maintenance personnel to check and deal with.

The logic triggers the exit of ATT compliance control, and the compliance control is completely withdrawn (rather than stopped at the current position), the valve command is completely taken over by the valve limit (-5%), and the valve remains in the fully closed position, which is a safe position for the system.

The above scheme is implemented in the Distributed Control System (DCS). The dashed box in Figure 8 shows the specific implementation method of this improvement plan for self-diagnosis and automatic intervention protection of hydraulic servo-motor faults. By introducing the existing EH system main pipe oil pressure signal through the acquisition card and performing the logic operation shown in Figure 7 in the controller, when the hydraulic motor is abnormal and affects the system oil pressure, the logic operation can automatically identify and locate which valve's hydraulic motor is abnormal. When the anomaly reaches the point where it will affect the operation of the system, it will automatically trigger the intervention protection logic (pressure holding logic). The alarm signal is transmitted to the operator workstation screen through the network for display. At the same time, the cut-off servo valve inlet oil pressure maintaining command is triggered, and the current signal is sent out through the high-speed analog output card, which acts on the coil of the on-site servo valve. The execution cycle of the protection signal link is about 50ms, which can quickly cut off the incoming oil and ultimately achieve the automatic intervention target of maintaining pressure.

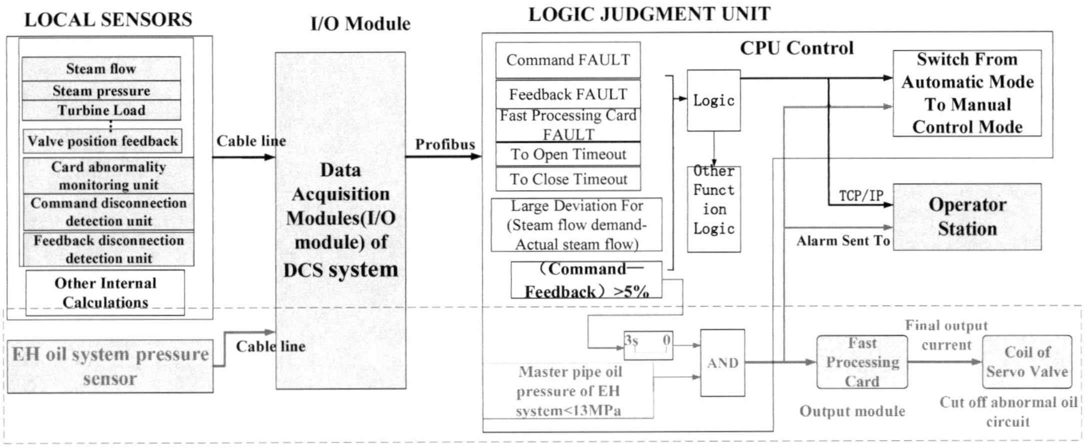

Figure 8 Hardware implementation diagram for fault self-diagnosis and automatic intervention protection.

The scheme was implemented in a nuclear power plant, and the fault of the fast-closing solenoid valve was verified during the on-site simulation of ATT. After the command and deviation of the servo valve were > 5%, the oil pressure of the GFR master pipe quickly dropped to about 12.95MPa and stopped falling. The oil pressure recovered to the normal working pressure of 16.0MPa (Figure 5), and the alarm of the highest response level was triggered. At this time, the valve remained in the safe position of full closure, and the turbine control was not affected, leaving sufficient inspection and reason time for operation and maintenance personnel. It shows that the valve successfully realizes the function of self-diagnosis and automatic intervention of the fault of the hydraulic servo-motor components. It should be further explained that after the improvement of 8 turbine inlet regulating valves, the control logic of one of the valves detected the abnormal oil circuit and triggered the pressure holding logic of automatic intervention during the simulation fault verification during the cold state. Although the oil pressure returned to the normal working pressure, contrary to expectations, the opening command of the valve was subsequently opened from -5% to 103% (see Figure 9), and the ATT sequence control did not exit.

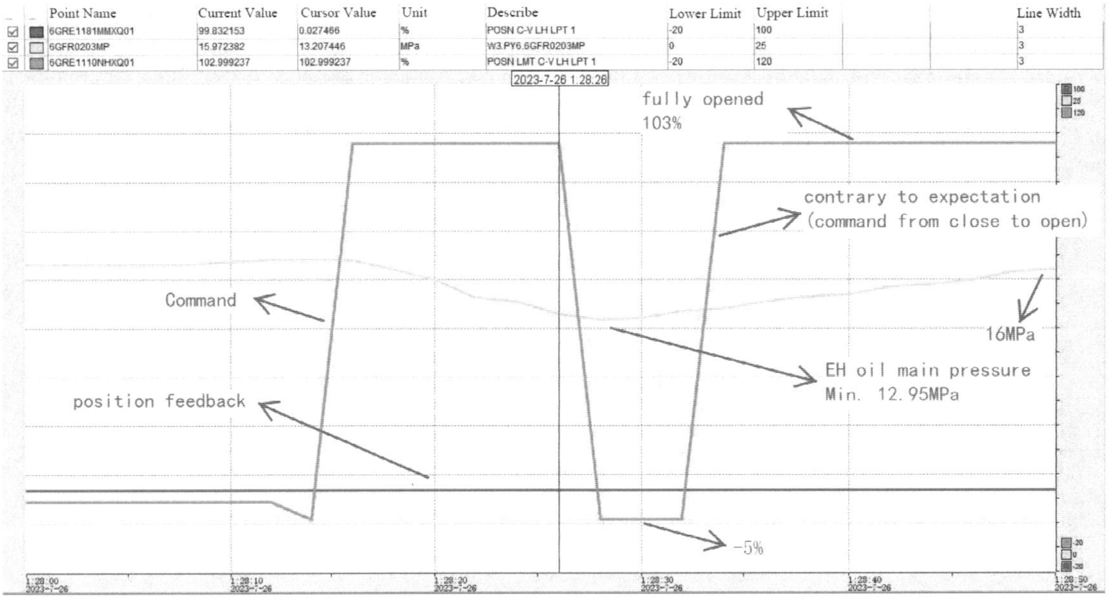

Figure 9 Does not conform to the expected curve.

After analysis, when the oil pressure logic of automatic intervention is triggered, the valve command immediately becomes 0. The valve is in the fully closed position, which does not meet the condition of "(instruction - feedback) > 5%", so the protection command will soon be reset (actually only triggered 203ms). The signal sent out is just not collected by the controller, which leads to the failure of sequence control to exit when the valve is verified. The effect is that the test compliance will continue to perform the stop sequence. If, during the daily operation of the unit with power, when the stop sequence goes to 55 steps, the test sequential control will issue an instruction to set the valve limit of the regulator to 103% and open the regulator. Because the fault of the solenoid valve jam has not been eliminated, the EH oil power oil and oil drain circuit are always in the on-state, and the valve cannot be opened normally; the test sequence control will stop to 55 steps. After receiving the opening command, the regulator activates the oil pressure reduction and protection logic, closing the valve which causes pressure oscillation in the oil system. This is not ideal for the system. It is necessary to add a delay module to the command (③ in Figure 6) to solve the problem of extending the signal triggering time.

6. Concluding remarks

By studying the failure mode of turbine regulating valves, this paper finds the hidden trouble of the current design of turbine regulating valves in a nuclear power plant. The hidden danger is unacceptable for the valve activity test with high frequency. Through many tests and studies, this paper puts forward an automatic diagnosis scheme for turbine-regulating valve hydraulic servo-motor failure, realizing the function of automatic intervention after failure and improving the system design defects of the turbine-regulating valve. By implementing the transformation in a nuclear power plant, the final test shows that the fault self-diagnosis and automatic intervention scheme successfully improves the reliability of turbine valve control and eliminates the hidden danger of turbine tripping.

References

[1] Quan Chen, Bo Zhang and Xiaodong Zhang. Pre-control measures for abnormality in Siemens ultra-supercritical steam turbine ATT test [J]. Henan Electric Power References, 2022(S1): 22-23
[2] Yingnan Wang, Liu Kecheng, Zhang Lijun, Shi Rongxue, Han Hesong. Causes and countermeasures for excessive particle size of electric power oil [J]. Hebei Electric Power (Volume 41), 2022(1):

Journal of Physics: Conference Series **2815** (2024) 012029 doi:10.1088/1742-6596/2815/1/012029

64-66
[3] Haihang Yu. Effect of anti-fuel oil quality on steam turbine speed control system and treatment method [J]. Technology and Market, 2020(6): 94-95
[4] M.M. Sultanov,Yu.M Chubko,A.V. Blagochinov,E.V. Kuryanova. Operational Control, Monitoring and Diagnostics of the Technical Condition of Steam Turbine Installations of Thermal Power Plants.2021 3rd International Youth Conference on Radio Electronics, Electrical and Power Engineering (REEPE) ,Electronic ISBN:978-1-7281-8398-5
[5] Priya Banerjee,Kamlesh Pandey.Implementation of Failure Modes and Effect Analysis on the electro-hydraulic servo valve for steam turbine.2016 IEEE 1st International Conference on Power Electronics, Intelligent Control and Energy Systems (ICPEICES).Electronic ISBN:978-1-4673-8587-9
[6] Ailin Zhou, Kailiang Zhu, Yuansheng Zhu, Jinhua Wang and Chengkai Jia. Development and Application Practice of Flushing Technology of Steam Turbine Control Oil System [J]. Mechanical Engineer, 2022(12): 74-79
[7] Xinwei Lian. Analysis of abnormal unit load fluctuations during the mobility test of steam turbine valves in nuclear power plants [J]. Electric Engineering, 2021 (14): 62-64
[8] Chunliang Wang, Mengzhong Ma and Xiaoyu Zhao. Analysis and processing of ATT experimental tripping [J]. Proceedings of the 2023 Electric Power Industry Technical Supervision Work Exchange Meeting and Professional Technical Forum (Volume 1), China Electric Power Technology Market Association, 2023.11
[9] Quan Chen, Bo Zhang and Xiaodong Zhang. Pre-control measures for abnormality in Siemens ultra-supercritical steam turbine ATT test [J]. Henan Electric Power References, 2022(S1): 22-23
[10] Donghui He, Zhenqi Ye and Yizhou Zhao. Improvement of ATT test control logic for 1000MW ultra-supercritical unit [J]. Heilongjiang Electric Power, 2019(41): 174-178

Journal of Physics: Conference Series **2815** (2024) 012030 doi:10.1088/1742-6596/2815/1/012030

Nonlinear adaptive robust control with continuous error reduction mapping

Zongliang Mou[1], Lei Xu[1,*], Jianhua Chen[1], Junbai Zhong[1], Nachuan He[1], Changdu Du[1] and Jun Chen[1]

[1]Sichuan University, Chengdu, Sichuan, 610000, China

[*]Corresponding author's e-mail: xulei@scu.edu.cn

Abstract: Achieving high stability and high-precision control by traditional control algorithms is difficult due to the strong nonlinearity, parameter uncertainty, and random disturbance characteristics of electro-hydraulic proportional servo systems. Herein, we developed a nonlinear adaptive control algorithm based on Lyapunov functions. By estimating the parameters and converging their errors, this algorithm solved the parameter adaptation problem. Additionally, based on convergence inference from the nonnegativity of the derivative of Lyapunov functions, the algorithm designed a disturbance mapping through the derived function. The random disturbance was reduced through the mapping, improving the stability of the system under random disturbance. Finally, we carried out continuous derivation and moved the mapping relationship out of the intermediate temporary controller. This reduced the complexity of the calculation during controller design, prevented performance degradation of the error convergence algorithm after complex calculation, and enhanced the robustness of the algorithm. The algorithm was verified through the simulation model established by Simulink and experience. Results showed that the nonlinear adaptive control algorithm demonstrates fast response, high accuracy, and strong anti-interference ability compared with the high-gain robust control algorithm. It realizes the adaptation of unknown parameters and reduces the control error introduced by random disturbance.

1. Introduction

Electro-hydraulic servo systems, which offer advantages such as a high power-to-volume ratio and high precision of displacement and force control [1], find extensive applications in industry. However, their models exhibit high nonlinearity[1] and parameter uncertainties, making it challenging to employ classical control theory for controller design. Although existing nonlinear controllers demonstrate high precision and stability in motion control[3], uncertainties in hydraulic systems, arising from factors including frictional forces, hydraulic parameter variations, and load fluctuations during operation, lead to parameter uncertainties. Additionally, nonlinear phenomena, such as leakage, external disturbances[1], and actuator vibrations, introduce random disturbances into the system. These drawbacks have impeded the development of high-precision controllers for electro-hydraulic systems. Therefore, mitigating the nonlinear effects caused by various factors and the impact of parameter uncertainties and disturbances on system stability has remained a focal point in controller research[4].

In the given motion trajectory, achieving a small working error is a fundamental requirement of modern electro-hydraulic control.[5]In recent years, research on electro-hydraulic servo control has been diverse, focusing mainly on strategies such as sliding mode control[6], unknown parameter estimation[7], observer design[8], and backstepping control[9]. The goal is to reduce the working error of electro-hydraulic servo systems and improve stability.

To enhance control, researchers shifted their focus to nonlinear controllers. For instance, Li et al. designed a fuzzy state observer to estimate unmeasurable quantities[10]. Khalil et al. used a high-gain observer to address the effects of model uncertainty and measurement noise[11]. These studies validated the superiority of nonlinear controllers over linear controllers to a certain extent. However, these approaches depend on the accuracy of observer output values, involve complex calculations for specific systems, and cannot adequately assess controller stability during the computation and design process. Lyapunov functions, known for their judgment of stability during the design process, became the foundation for numerous nonlinear controller designs. For example, Yao et al. resolved conflicts between adaptive and robust controllers with discontinuous projections[2] to meet the conditions for Lyapunov functions. These approaches required addressing relationships and conflicts among backstepping control, Lyapunov functions, adaptive controllers, and robust controllers. Moreover, they considered adaptive elements as unpredictable disturbances. Based on Lyapunov functions and backstepping control theories, this study developed a design scheme for a nonlinear adaptive controller. This scheme addressed conflicts between parameter adaptation and disturbance handling algorithms, reducing the disturbance threshold and its effects. Additionally, continuous derivation lowered the computational complexity of the controller in simulation and application scenarios.

2. Establishment of the system model

2.1 Description of the hydraulic system

Consider the typical hydraulic system as an example. The hydraulic pump delivers hydraulic fluid into the pipeline. The relief valve is used to prevent the pressure in the system from exceeding the rated value, ensuring the overall safety of the system, and it does not contribute to the establishment of the system model. The hydraulic fluid enters the hydraulic cylinder chamber through the main valve spool orifice, driving the displacement of the hydraulic rod and moving the load. The flow is influenced by the size of the orifice, which is directly affected by the displacement of the main valve spool. The main valve is electro-proportional, and its displacement is influenced by the input voltage u. The directional valve structure is illustrated in Figure 1.

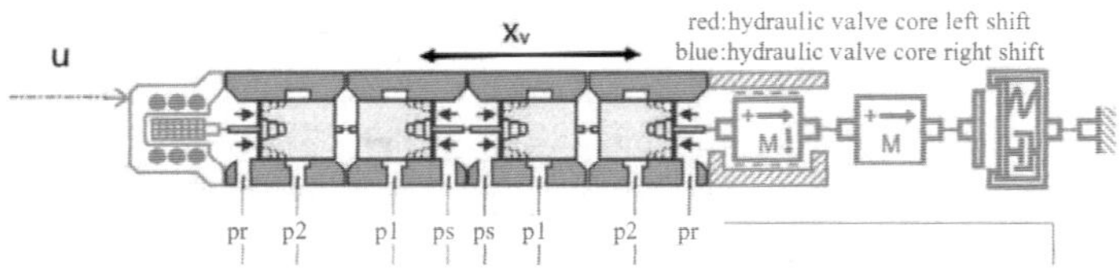

Figure 1 Directional Valve Schematic.

2.2 Model establishment

Consider the dynamic modeling of the actuator as follows:

$$mx_l'' = p_1 A_1 - p_2 A_2 - b_v x_l' - F_s + d\left(x_l, x_l', t\right) \tag{1}$$

where d is the bounded nonlinear disturbance term, $b_v x_l'$ accounts for viscous forces, and F_s characterizes various forms of frictional forces, which is expressed as

$$F_s = \operatorname{sgn}(\bullet) b_c \tag{2}$$

where b_c is the Coulomb friction coefficient and $\operatorname{sgn}(\bullet)$ is the sign function to determine the direction of the frictional force. Substituting Equation (2) into Equation (1) yields the dynamic model:

$$mx_l'' = p_1 A_1 - p_2 A_2 - b_v x_l' - b_c \operatorname{sgn}(\bullet) + d\left(x_l, x_l', t\right) \tag{3}$$

The continuity equation for the high-pressure chamber of the actuator is given by

$$Q_1 = A_1 x_l' + C_i(p_1 - p_2) + \frac{V_1}{\beta_e} p_1' \tag{4}$$

External leakage is not considered. On the other hand, internal leakage is considered solely in terms of the product of the fluid leakage coefficient and the pressure difference. Similarly, the continuity equation for the low-pressure chamber is expressed as

$$Q_2 = A_2 x_l' + C_i(p_1 - p_2) - \frac{V_2}{\beta_e} p_2' \tag{5}$$

The mathematical model for the electro-proportional servo valve area can be established as

$$\begin{cases} Q_1 = k_q x_v \left[S(x_v)\sqrt{p_s - p_1} + S(-x_v)\sqrt{p_1 - p_r} \right] \\ Q_2 = k_q x_v \left[S(x_v)\sqrt{p_2 - p_r} + S(-x_v)\sqrt{p_s - p_2} \right] \end{cases} \tag{6}$$

where S is a piecewise function related to the spool displacement direction, which is expressed as

$$S(*) = \begin{cases} 1, * \geq 0 \\ 0, * < 0 \end{cases} \tag{7}$$

where k_q is the displacement coefficient of the sliding valve, which is expressed as

$$k_q = C_d w \sqrt{\frac{2}{\rho}} \tag{8}$$

where C_d is the discharge coefficient, w is the area gradient of the sliding valve, and ρ is the fluid density. x_v is the displacement of the sliding valve, which is dependent on the voltage signal.

$$x_v = k_i u \tag{9}$$

We define the product of the two coefficients $k_q k_i$ as g, and define $\left[S(u)\sqrt{p_s - p_1} + S(-u)\sqrt{p_i - p_r} \right]$ and $\left[S(u)\sqrt{p_2 - p_r} + S(-u)\sqrt{p_s - p_2} \right]$ as R_1 and R_2. Then, the original definitions can be simplified to the following form:

$$\begin{cases} Q_1 = guR_1 \\ Q_2 = guR_2 \end{cases}. \tag{10}$$

Substituting Equation (10) into Equations (4) and (5) gives the following relationships:

$$p_1' = \frac{\beta_e}{V_1}(-A_1 x_l - C_i p_L + gR_1 u) \tag{11}$$

$$p_2' = \frac{\beta_e}{V_2}(A_2 x_l + C_i p_L - gR_2 u) \tag{12}$$

where $p_1 - p_2$ is defined as p_L, and subtracting Equations (11) and (12) results in

$$p_L' = g\beta_e\left(\frac{R_1}{V_1} + \frac{R_2}{V_2}\right)u - \beta_e\left(\frac{1}{V_1} + \frac{1}{V_2}\right)(Ax_2 + C_i x_3) \tag{13}$$

Considering equal force areas on both sides of the hydraulic cylinder, Equation (3) can be expressed as

$$mx_l'' = p_L A - b_v x_l' - b_c \operatorname{sgn}(\bullet) + d(x_l, x_l', t) \tag{14}$$

Defining $[x_1, x_2, x_3]^T = \left[x_l, x_l', p_L\right]^T$, Equations (13) and (14) lead to the following state equation: [12]

$$\begin{cases} x_1' = x_2 \\ x_2' = \dfrac{Ax_3 - b_v x_2 - b_c \operatorname{sgn}(\bullet) + d(x_l, x_l', t)}{m} \\ x_3' = g\beta_e\left(\dfrac{R_1}{V_1} + \dfrac{R_2}{V_2}\right)u - \beta_e\left(\dfrac{1}{V_1} + \dfrac{1}{V_2}\right)(Ax_2 + C_i x_3) \end{cases} \tag{15}$$

The above electro-hydraulic servo system structure is shown in Figure 2.

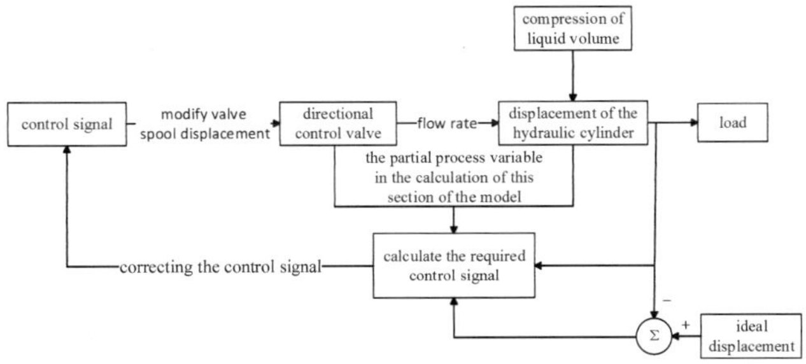

Figure 2 The structure of an electro-hydraulic servo system.

3 Improved controller design

The existing nonlinear controller design does not incorporate parameter adaptation and directly integrates unquantifiable values into the handling of random disturbances. This study proposes improvements to the controller based on Lyapunov theory from two perspectives: (1) developing Lyapunov functions to achieve parameter adaptation and (2) designing input signals to mitigate the impact of random disturbances.

3.1 Parameter adaptation

The expression of the control signal is derived based on a first-order state equation. Consider the first-order state equation of the system as follows:

$$x' = \theta\varphi(x,t) + \Delta(x,t) + u \tag{16}$$

where θ is a coefficient of the function of x, $\Delta(x,t)$ is an unpredictable disturbance, and u is the control signal. The derivation of the control signal is performed when θ is unknown.

$$z = x - x_d \tag{17}$$

where x is the desired value, x_d is the actual value, and z is the error. The Lyapunov function is set as follows:

$$V = \frac{1}{2}z^2 + \frac{1}{2}\tilde{\theta}^2 \tag{18}$$

$$V' = zz' + \tilde{\theta}\tilde{\theta}' \tag{19}$$

In Equation (19), $\tilde{\theta}$ is the error between θ and its estimated value $\hat{\theta}$, such as

$$\tilde{\theta} = \hat{\theta} - \theta \tag{20}$$

Substituting Equations (16), (17), and (20) into Equation (19) yields:

$$V' = z\left[\theta\varphi(x,t) + \Delta(x,t) + u - x'_d\right] + \tilde{\theta}\left(\hat{\theta}' - \theta'\right) \tag{21}$$

To ensure the negativity of the Lyapunov function derivative and avoid introducing the unknown θ in the design of the control signal u, we set the control signal as

$$u = -kz + x'_d - \hat{\theta}\varphi(x,t) - \Delta(x,t) \tag{22}$$

Substituting Equation (22) into Equation (21) gives

$$V' = -kz^2 - \tilde{\theta}\left[\varphi(x,t)z - \hat{\theta}'\right] - \tilde{\theta}\theta' \tag{23}$$

Consequently, we obtain

$$\hat{\theta}' = \varphi(x,t)z \tag{24}$$

3.2 Handling random disturbances

In the designed control signal u in Equation (22), $\Delta(x,t)$ is an unpredictable disturbance and cannot be directly output in the control signal. Considering the impact of $\Delta(x,t)$, we redesigned the control signal as

$$u=-kz+x'_d-\hat{\theta}\varphi(x,t)-hg(y) \tag{25}$$

Similarly, we have

$$V'=-kz^2-\tilde{\theta}\left[\varphi(x,t)z-\hat{\theta}'\right]-hzg(y)+\Delta(x,t)z-\tilde{\theta}\theta' \tag{26}$$

Given that θ is an unknown constant or slowly varying quantity, its derivative is zero. Equation (26) becomes

$$V'=-kz^2-k\tilde{\theta}^2-\tilde{\theta}\left[\varphi(x,t)z-\hat{\theta}'\right]+M_1+M_2 \tag{27}$$

Substituting Equations (18) and (24) into the above expression gives:

$$V'\le -kz^2-k\tilde{\theta}^2+\xi_1+\xi_2\le -2kV+\xi$$

Upon solving, we obtain $z^2+\tilde{\theta}^2\le\frac{\xi}{k}$, indicating error convergence. In the derivation above, the following conditions must be satisfied:

$$M_1=-hzg(y)+\Delta(x,t)z, M_2=k\tilde{\theta}^2$$

In condition (1), by setting $\Delta(x,t)$ and $\tilde{\theta}$ as bounded functions, we obtain condition (2):

$$h\ge\max\left|\Delta(x,t)\right|$$

At this point, we have $M\le -hzg(y)+hz$. Now, we define $y=hz$ and hope $G(x)=hz-hzg(x)\le\xi$. We take $G(x)=x-x\sqrt[\beta]{1-e^{\lambda x}}$, which is less than a certain value ξ for a given error, so this function has a smaller convergence value ξ in the case of a certain error value. Here, β and λ are adjustable variables, β should be an odd number to adjust the zero point and extreme value, and λ should be a negative number. Similarly, we have $\frac{1}{2}\tilde{\theta}^2\le\xi$, and theoretically, its value will continually converge. Figure 3 shows the behavior of the function under different variables. Therefore, we have the condition (3):

$$M_1\le\xi_1, M_2\le\xi_2,\ \xi=\xi_1+\xi_2$$

In this case, the control signal is given by

$$u=-kz+x'_d-\hat{\theta}\varphi(x,t)-h\sqrt[\beta]{1-e^{\lambda y}}. \tag{28}$$

Now, we verify the influence of different variables on the error values. Figure 4 depicts the impact of various variables on the error in a second-order system, while Figure 5 illustrates the theoretically allowable maximum error for each variable, i.e., $\frac{\xi}{k}$.

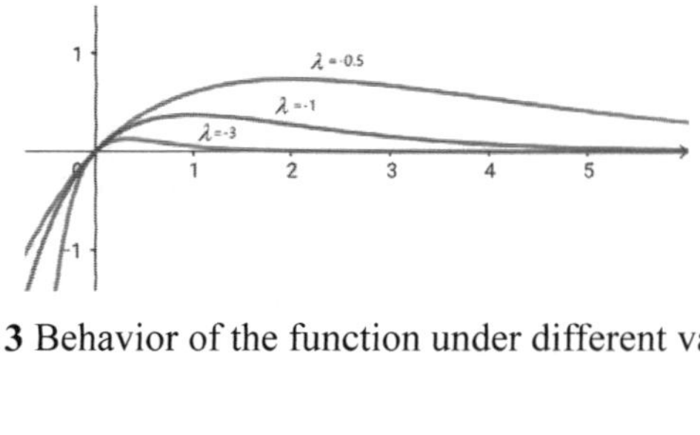

Figure 3 Behavior of the function under different variables.

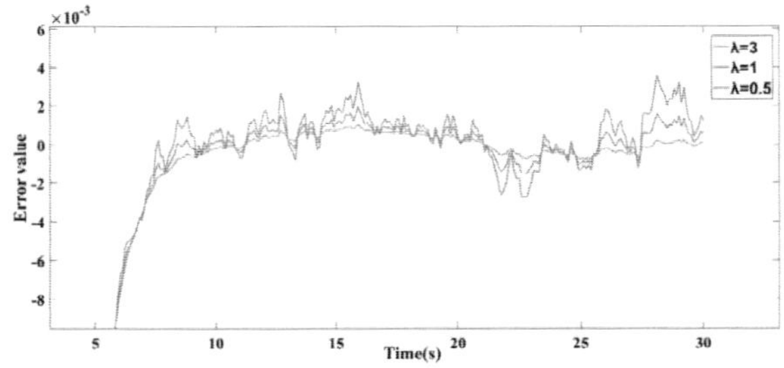

Figure 4 Changes in the error.

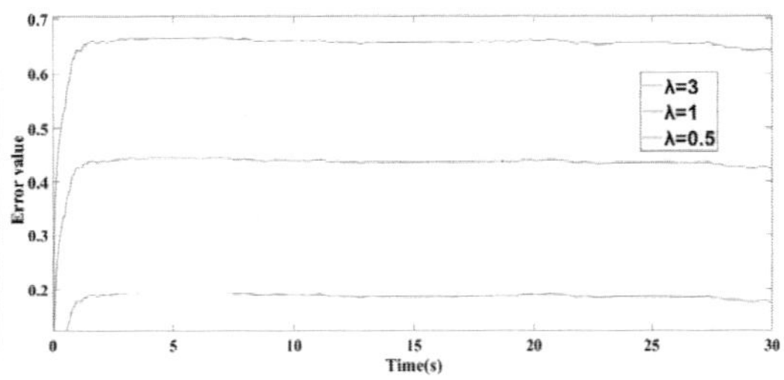

Figure 5 Permissible maximum error magnitude.

The results validate that this strategy can lead the error to converge within the expected maximum error range.

3.3 Derivation of a third-order system controller

The controller for the hydraulic system is designed based on Equation (15). The solution of the control signal in the first- and second-order system is as follows:

$$\alpha_1 = -k_1 z_1 + x'_{1d}. \tag{29}$$

Given that the second state equation involves unknowns and random disturbances, the Lyapunov function is obtained based on Equation (18).

$$V_2 = V_1 + \frac{1}{2} z_2^2 + \frac{1}{2} \tilde{\theta}_2^2 \tag{30}$$

Its derivative is given by:

$$V_2' = z_1 z_1' + z_2 z_2' + \tilde{\theta}_2 \tilde{\theta}_2'. \tag{31}$$

From Equation (17), we have $z = x_1 - x_{1d}$, and by substituting this into Equation (31), we obtain:

$$V_2' = z_1(x_1' - x_{1d}') + z_2 z_2' + \tilde{\theta}_2 \tilde{\theta}_2' = z_1(x_{2d} + z_2 - x_{1d}') + z_2 z_2' + \tilde{\theta}_2 \tilde{\theta}_2'. \tag{32}$$

In fact, $x_{2d} = \alpha_1$; thus, substituting it into Equation (29) yields

$$V_2' = -k_1 z_1 + z_2(z_1 + z_2') + \tilde{\theta}_2 \tilde{\theta}_2'. \tag{33}$$

Similarly, following the derivation approach outlined in Chapter 3, we have an intermediate control variable, which is expressed as

$$\alpha_2 = -z_1 - k_2 z_2 + (-k_1 z_1 + x_{1d}')' - \hat{\theta}_2 \varphi_2(x,t) - h_2 \sqrt[\beta_2]{1 - e^{\lambda_2 y_2}}. \tag{34}$$

In Equation (15), b_v is assumed to be an unknown constant or a slowly varying quantity with time (i.e., $\theta_2 = \frac{b_v}{m}$), and d is considered an unknown random fluctuation, while the rest of the parameters are assumed to be known. Combining Equations (15) and (31), we have the control variable, which is expressed as

$$\alpha_2 = -\frac{m}{A} z_1 - \frac{m}{A} k_2 z_2 + \frac{m}{A}(-k_1 z_1 + x_{1d}')' + \frac{b_c}{A} \operatorname{sgn}(\bullet) + \frac{\hat{b}_v}{A} x_2 - \frac{h_2}{A} \sqrt[\beta_2]{1 - e^{\lambda_2 y_2}} \tag{35}$$

where $\hat{b}_v' = -m x_2 z_2$.

The intermediate variable $\alpha_2 = x_{3d}$. Continuing with this approach, the final control signal is

$$u=\frac{-\frac{A}{m}z_2-k_3z_3+\beta_e\left(\frac{1}{V_1}+\frac{1}{V_2}\right)(Ax_2+C_ix_3)+(-\frac{m}{A}z_1-\frac{m}{A}k_2z_2+\frac{m}{A}(-k_1z_1+x_{1d}')'+\frac{b_c}{A}\operatorname{sgn}(\bullet)+\frac{\hat{b}_v}{A}x_2-\frac{h_2}{A}\sqrt[\beta_2]{1-e^{\lambda_2 y_2}})'}{g\beta_e\left(\frac{R_1}{V_1}+\frac{R_2}{V_2}\right)} \tag{36}$$

where

$$V_3=\frac{1}{2}z_1^{\ 2}+\frac{1}{2}z_2^{\ 2}+\frac{1}{2}\widetilde{\theta_2}^{\ 2}+\frac{1}{2}z_3^{\ 2}. \tag{37}$$

4. Continuous derivation of the controller

In the derivation presented in Part 2, the function limiting the impact of random disturbances is placed within the intermediate control variable α_2, that is, Equation (31). However, in the derivation of the third-order controller, the intermediate control variable α_2 appears in the form of its derivative. This is not conducive to the calculation and simulation of a third-order controller. Therefore, we considered handling the influence of random disturbances in the final controller. This section represents the continuous derivation, and the process is as follows:

Taking the derivative of Equation (34) and following the derivation approach in Equation (33), we obtain

$$V_3'=-k_1z_1^2+z_2\left(z_1+z_2'\right)+\tilde{\theta}\tilde{\theta}'+z_3z_3' \tag{38}$$

$$z_2=x_2-x_{2d} \tag{39}$$

According to Equations (15), (38), and (39), we obtain

$$V_3'=-k_1z_1^2+z_2[z_1+\frac{A}{m}(x_{3d}+z_3)-\frac{b_v}{m}x_2-\frac{b_c}{m}\operatorname{sgn}(\bullet)+\frac{1}{m}d-x_{2d}']+\tilde{\theta}\tilde{\theta}'+z_3z_3' \tag{40}$$

where $x_{3d}=\alpha_2$, and α_2 is defined as

$$\alpha_2=-z_1-k_2z_2+(-k_1z_1+x_{1d}')'-\hat{\theta}_2\varphi_2(x,t) \tag{41}$$

Substituting Equation (41) into Equation (40) and following the above approach, we have

$$V_3'=-k_1z_1^2-k_2z_2^2-z_3[\frac{A}{m}z_2+g\beta_e\left(\frac{R_1}{V_1}+\frac{R_2}{V_2}\right)u-\beta_e\left(\frac{1}{V_1}+\frac{1}{V_2}\right)(Ax_2+C_ix_3)-x_{3d}']+\frac{1}{m}dz_2 \tag{42}$$

Now, the designed control signal is

$$u=\frac{-\frac{A}{m}z_2-k_3z_3+\beta_e\left(\frac{1}{V_1}+\frac{1}{V_2}\right)(Ax_2+C_ix_3)+(-\frac{m}{A}z_1-\frac{m}{A}k_2z_2+\frac{m}{A}(-k_1z_1+x_{1d}')'+\frac{b_c}{A}\operatorname{sgn}(\bullet)+\frac{\hat{b}_v}{A}x_2)'+\frac{-h_2z_2\tanh(z_3)}{mz_3\gamma}}{g\beta_e\left(\frac{R_1}{V_1}+\frac{R_2}{V_2}\right)}. \tag{43}$$

where γ is a coefficient adjusting the convergence error. Changing its value directly affects the convergence error of the signal.

$$V_3'\le-k_1z_1^2-k_2z_2^2-k_3z_3^2+\frac{h_2z_2}{m}-\frac{h_2z_2\tanh(z_3)}{\gamma m} \tag{44}$$

Now, we analyze the performance of the function $\frac{h_2z_2}{m}-\frac{h_2z_2\tanh(z_3)}{\gamma m}$, which is also denoted as $\frac{h_2z_2}{m}\left(1-\frac{\tanh(z_3)}{\gamma m}\right)$. Without considering the introduction of the coefficient γ, its graph is shown in Figure 6.

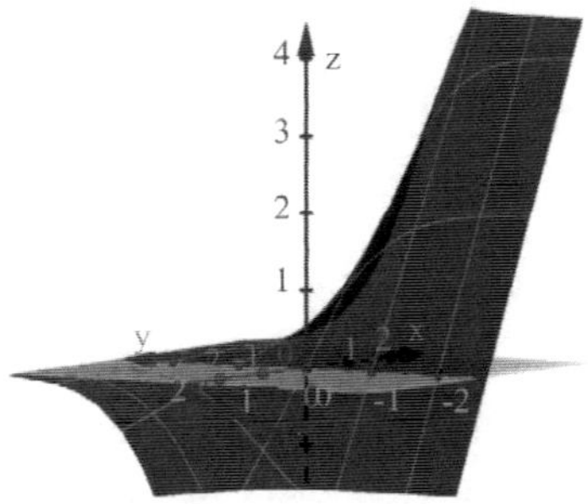

Figure6 Function plot.

Only when z_2 takes a larger positive value will the function characterize a larger value. However, considering that its initial value is negative, it can ensure signal convergence, so it is difficult to achieve larger values when exceeding the expected value. Even if a larger value appears, it can be adjusted by changing the coefficient. Therefore, we have a function that can converge to the numerical value of ξ. From Equation (44), we have

$$V' \leq -2kV + \xi \tag{45}$$

As per Section 3.2, there is error convergence.

5. Analysis of simulation and experimental results

To validate the derived results, simulations were conducted based on the calculations mentioned above. The algorithmic structure was implemented using Simulink. The numerical values for parameters are as follows:

Table 1. Numerical values selected for simulation.

parameter	m	β_e	C_i	$k_1 k_2 k_3$	b_c	g	R_1	R_2	A	$\beta_2 \lambda_2$
value	$10kg$	$1.5Pa$	0.05	3	$0.4N$	0.125	$5MPa$	$1MPa$	$5*10^{-3}m^2$	1

V_1 and V_2 are obtained from the area and displacement, that is, $V_1 = V_{01} + A_1 x_l$ and $V_2 = V_{02} - A_1 x_l$, where V_{01} and V_{02} are the initial volumes of the left and right chambers, respectively. $g = k_q k_i = 0.125$, and where k_q is derived from Equation (8). b_v is the parameter that must be adapted, representing either a constant or a slowly varying quantity (i.e., $b_v' = 0$). Based on Equation (8), d is the disturbance randomly fluctuating within [-100, 100], $h_2 = 200$, and the function constraining the random disturbance is $y = h_2 z_2 - h_2 z_2 \sqrt[\beta_2]{1 - e^{\lambda_2 h_2 z_2}}$, where parameters can be adjusted based on performance. The numerical values regarding random disturbance are shown in Figure 7. In the continuity derivation in section 3, an assumption was made that the value of z_2 is relatively small when positive. Figure 8 presents the verification results for this assumption. We compared the high-gain robust control algorithm with the improved control algorithm described in this study, along with the control algorithm yielded from the continuous derivation. The tracking of a setpoint signal and the corresponding error are depicted in Figures 9 and 10. The tracking of changing signals is shown in Figure 11, and the error variations are shown in Figure 12.

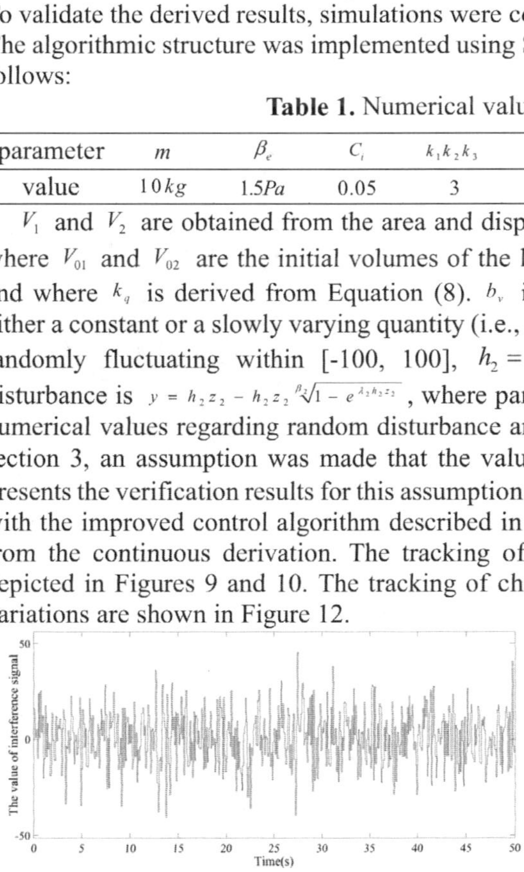

Figure 7 Magnitude of fluctuation values.

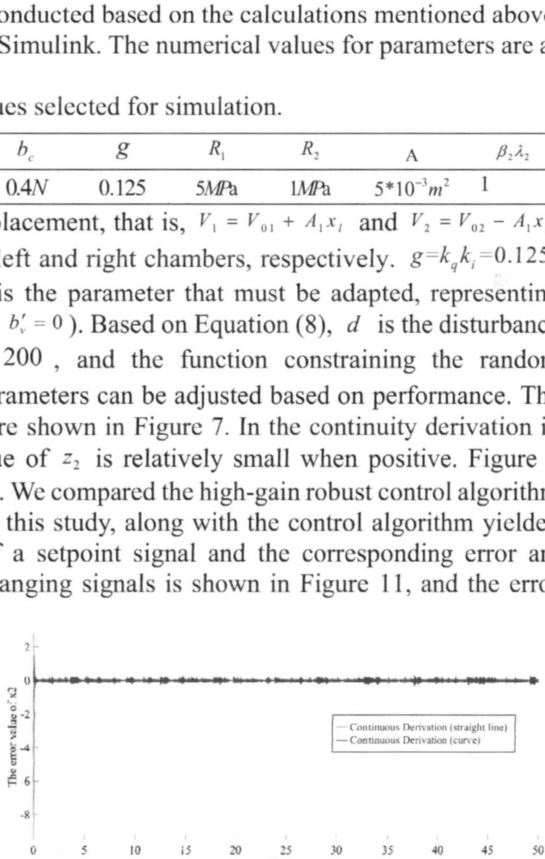

Figure 8 The range of values for x_2.

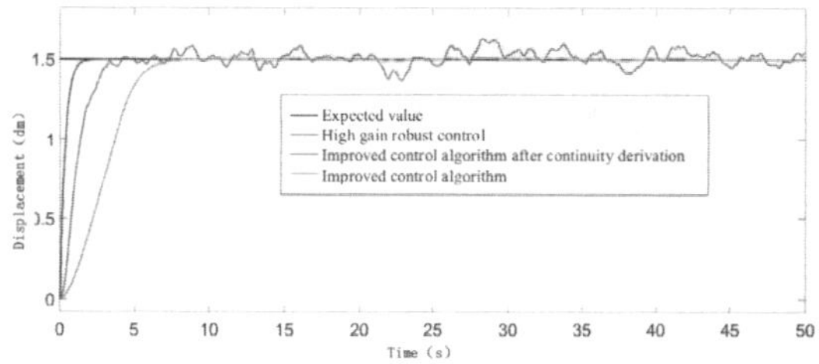

Figure 9 Setpoint signal tracking results.

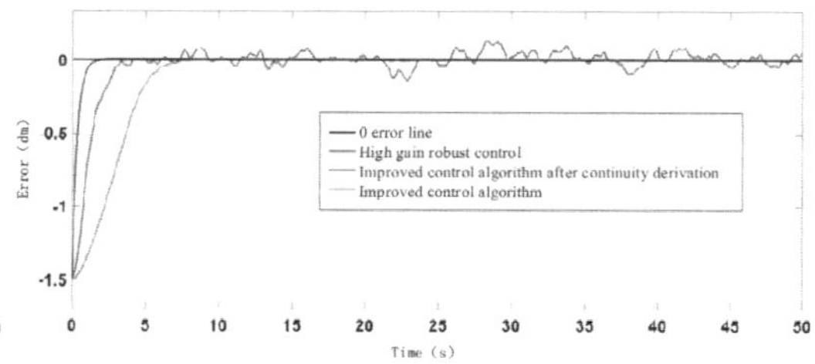

Figure 10 Setpoint signal tracking error.

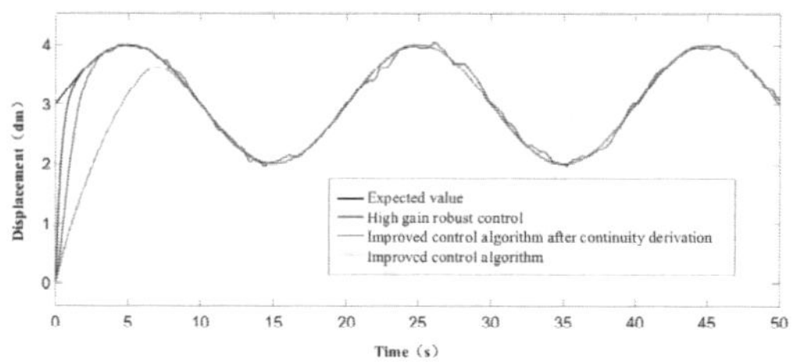

Figure 11 Changing signal tracking results.

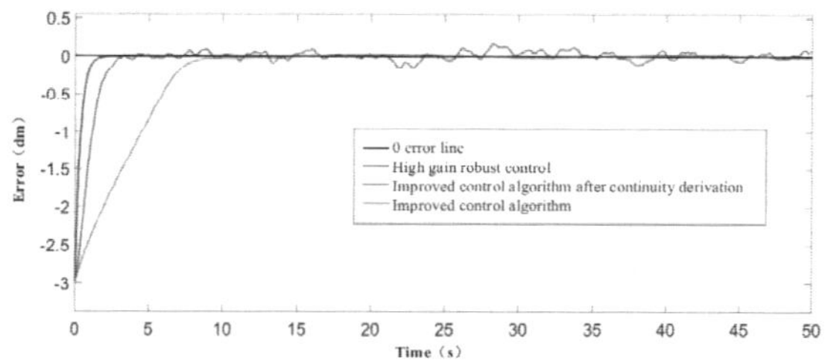

Figure 12 Changing tracking error.

Due to the superior performance of the improved control algorithm derived from continuity in the simulation results, it was selected for comparison with the high-gain robust control algorithm in the experiment. A hydraulic experimental platform is now established, with a structure as shown in Figure 13, and a physical object as shown in Figure 14. Its parameters are selected as commonly used parameters in hydraulic systems. The signal to be tracked is a sinusoidal displacement, and its tracking result is shown in Figure 15, with the tracking error shown in Figure 16. The error values between the expected and actual control quantities of x2 and x3 during the calculation process are shown in Figures 17 and 18.

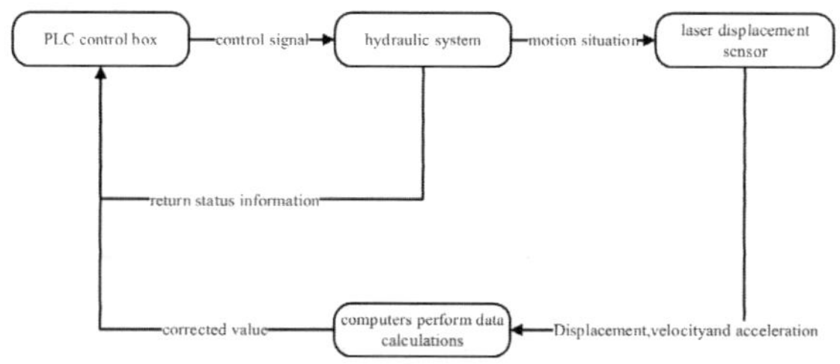

Figure 13 Hydraulic experimental platform structure.

Figure 14 Physical object (subject).

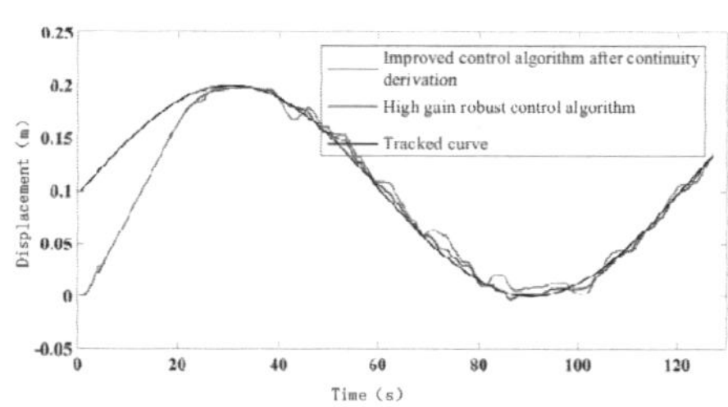

Figure 15 Displacement tracking results.

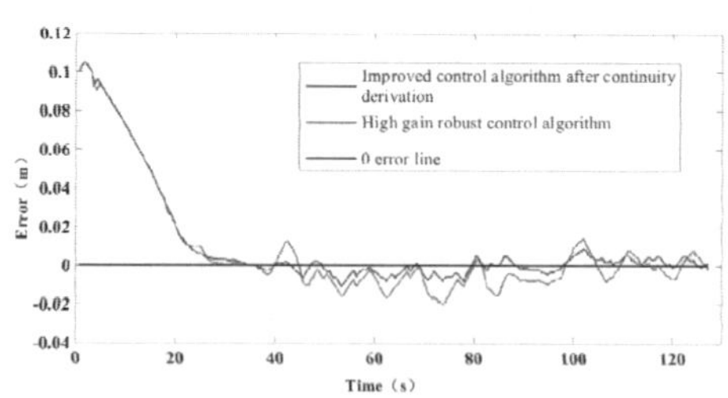

Figure 16 Displacement tracking error.

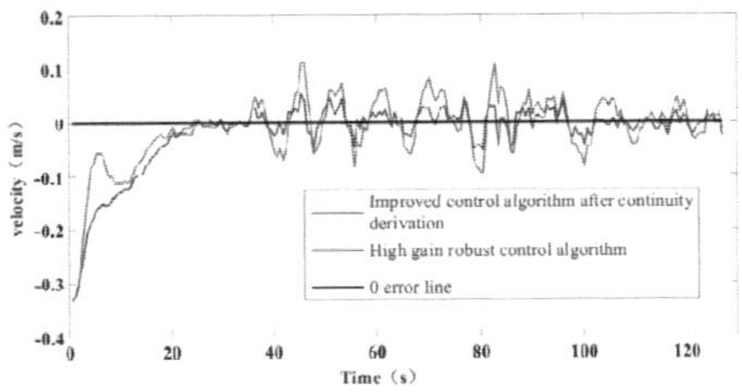

Figure 17 Tracking error of x_2.

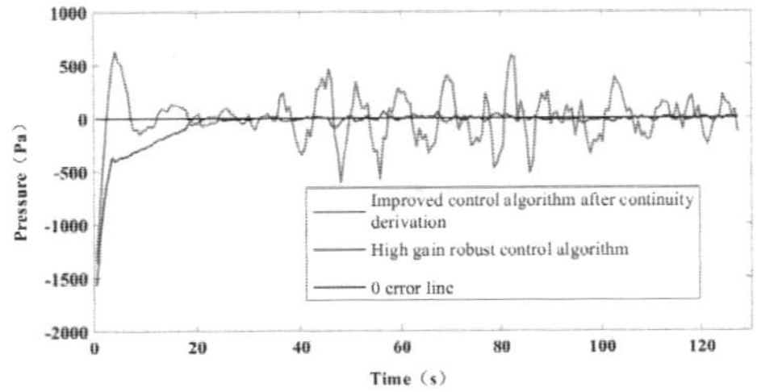

Figure 18 Tracking error of x_3.

Based on Figure 10, the improved control algorithm has a slightly lower response speed compared to the high-gain robust control algorithm. However, it effectively manages the impact of random disturbances on system stability. Based on simulation and experimental results, the improved control algorithm after continuous derivation not only enhances the system's response speed but also handles the impact of continuous random disturbances more effectively. Compared to traditional high-gain robust control algorithms, it significantly reduces steady-state errors and makes the system more stable.

6. Conclusion

Based on Lyapunov stability analysis and the method of reducing fluctuation mapping, this study established a nonlinear adaptive controller for robust control of electro-hydraulic servo systems. The designed controller addressed two main issues: 1) redesigning controllers due to variations in system parameters, adapting to unknown constant parameters; 2) the impact of disturbances, such as variations in frictional forces and unknown external disturbances, on system stability (i.e., managing random disturbances). Simulation verification was conducted using the Simulink model and experiment. The results indicate that this approach could effectively manage the influence of unknown disturbances. The adaptation of parameters could ensure the good performance of signal tracking and would not significantly negatively impact the strategy for handling unknown disturbances. Moreover, the system exhibited high robustness.

Acknowledgments

Fund project: This study was supported by the Sichuan Provincial Science and Technology Plan Project(2021YFG0060)

References

[1] Shen Y, Guo YQ, Zha X and Wang Y. Real-Time Hybrid Test Control Research Based on Improved Electro-Hydraulic Servo Displacement Algorithm. Sensors (Basel). 2023 May 15;23(10):4765. doi: 10.3390/s23104765. PMID: 37430679; PMCID: PMC10224155.

[2] Yao B, Bu FP and G. T. . -C. Chiu, "Nonlinear adaptive robust control of electro-hydraulic servo systems with discontinuous projections," Proceedings of the 37th IEEE Conference on Decision and Control (Cat. No.98CH36171), Tampa, FL, USA, 1998, pp. 2265-2270 vol.2, doi: 10.1109/CDC.1998.758681.

[3] Helian, B., Chen, Z., and Yao, B. (2021). Energy-saving and accurate motion control of a hydraulic actuator with uncertain negative loads. Chinese Journal of Aeronautics, 34, 253-264.

[4] Mingxing Y, Zhang Q and Xinliang L U, et al. Adaptive sliding mode control of a nonlinear electro-hydraulic servo system for position tracking[J]. Mechanics, 2019, 25(4): 283-290.

[5] Wos, Piotr and Dindorf, Ryszard. (2022). Synchronization of the movement for multi-cylinder electrohydraulic servo driver. EPJ Web of Conferences. 269. 01069. 10.1051/epjconf/202226901069.

[6] Chen G , Jia P andYan G ,et al.Research on Feedback-Linearized Sliding Mode Control of Direct-Drive Volume Control Electro-Hydraulic Servo System[J].Processes, 2021.DOI:10.3390/pr9091676.

[7] Woś, Piotr and Ryszard Dindorf. "Modeling and Identification of the Hydraulic Servo Drive." *EPJ Web of Conferences* (2019): n. pag.

[8] D. Won, W. Kim and M. Tomizuka, "Nonlinear Control With High-Gain Extended State Observer for Position Tracking of Electro-Hydraulic Systems," in IEEE/ASME Transactions on Mechatronics, vol. 25, no. 6, pp. 2610-2621, Dec. 2020, doi: 10.1109/TMECH.2020.2985619.

[9] H. Jia, H. Xie and G. Shen, "Position control of electro-hydraulic servo system based on disturbance observer and adaptive backstepping sliding mode controller," 2019 IEEE 8th International Conference on Fluid Power and Mechatronics (FPM), Wuhan, China, 2019, pp. 1261-1268, doi: 10.1109/FPM45753.2019.9035805.

[10] Y. Li, X. Min and S. Tong, "Observer-Based Fuzzy Adaptive Inverse Optimal Output Feedback Control for Uncertain Nonlinear Systems," in *IEEE Transactions on Fuzzy Systems*, vol. 29, no. 6, pp. 1484-1495, June 2021, doi: 10.1109/TFUZZ.2020.2979389.

[11] H. K. Khalil, "High-gain observers in nonlinear feedback control," 2008 International Conference on Control, Automation and Systems, Seoul, Korea (South), 2008, pp. xlvii-lvii, doi: 10.1109/ICCAS.2008.4694705.

[12] Yao, JY et al. "Friction compensation for low-velocity control of hydraulic flight motion simulator: A simple adaptive robust approach." Chinese Journal of Aeronautics 26 (2013): 814-822.

Journal of Physics: Conference Series **2815** (2024) 012031 doi:10.1088/1742-6596/2815/1/012031

Bushing fault diagnosis based on SVM and the improved sparrow search algorithm

Shengrong Li[1,*], Yefan Wang[1] and Zixi He[1]

[1]Jiangsu University of Science and Technology, Zhenjiang, Jiangsu, 212000, China

[*]Corresponding author's e-mail: 1810685513@qq.com

Abstracts. In order to address the issue of low precision in traditional bushing fault diagnosis, a bushing fault diagnosis method based on an improved sparrow search algorithm (ISSA) and support vector machine (SVM) is proposed in this paper. Firstly, the bushing vibration signals are extracted by wavelet packet, and the feature vectors are used as inputs for the SVM. In view of the impact of support vector machine parameters on the model, a sparrow search algorithm is proposed for intelligent optimization. To prevent reaching a local optimum, adaptive inertia weight is added based on the original approach. The final bushing fault diagnosis model is established by training. Comparison experiments with three fault diagnosis models, SSA-SVM, PSO-SVM, and SVM, found that the proposed method achieves complete diagnosis in a shorter time, and the diagnostic accuracy rate is 96.5%, which verifies the feasibility and effectiveness of the model.

1. Introduction

Bushings are an indispensable and important part of marine diesel engines. The prolonged exposure to intense heat and pressure, combined with the violent vibrations in the harsh environment, renders it particularly vulnerable to wear and breakage. Therefore, the detection of early defects in bushings is of great significance for the development of preventive maintenance programs to ensure the safe operation of diesel engines [1]. A solution to the issue of poor diagnostic precision in current methods is suggested, involving a support vector machine (SVM) and an improved sparrow search algorithm (ISSA) based diagnostic technique for bushing faults. This method produces efficient diagnostic results by enabling accurate fault identification in a short amount of time.

Experts, both domestically and internationally, have long been engaged in the development of bushing fault diagnosis methodologies. He et al. [2] proposed a bushing fault diagnosis using the wavelet domain moving beam formation (WMB) method. Deng et al. [3] proposed an SVD-based denoising and adaptive hybrid TMSST for bearing fault detection. Delanyo et al. [4] proposed a new analytic wavelet-based data augmentation (DAAW) method for the problem of classifying time series (TS) data in bearing fault diagnosis. Yan et al. [5] proposed a new approach based on sparse auxiliary parameter-adjustable variational modal decomposition (VMD) and whale optimal least squares support vector machine (WOA-LSSVM) for rolling bearings. Goto et al. [6] trained a support vector machine (SVM) model to predict bearing wear and damage and verified its effectiveness. However, the current fault diagnosis of diesel engine bushing still presents problems such as longer model training times and lower diagnostic accuracy.

Addressing the aforementioned issues, this paper introduces a method for diagnosing bushing faults, utilizing an improved sparrow search algorithm-optimized support vector machine (ISSA-SVM). The method not only combines the advantages of the SVM with small data size and strong generalization ability but also combines the ISSA, which can effectively avoid falling into the problem of local optimal solution. By comparing the SSA-SVM, PSO-SVM, and SVM fault diagnosis models, the proposed diagnosis model has higher precision in the diagnosis of bushing.

2. Research methodology

2.1 Support vector machine

The support vector machine algorithm (SVM) is proposed to solve binary classification problems [7]. The algorithm has great advantages in solving small sample, high dimensional, and nonlinear problems, and it can overcome dimensionality catastrophe and over-learning to a certain extent, so as to improve the computational efficiency [8]. The effectiveness of the SVM model in fault classification is heavily influenced by the algorithm's penalty factor C and kernel parameters σ. Therefore, a metaheuristic algorithm with powerful search capabilities is introduced to optimize these parameters quickly and efficiently.

2.2 Basic theory of sparrow search algorithm and its improvement

2.2.1 Basic theory of sparrow search algorithm. The sparrow search algorithm (SSA) is a novel approach to the optimization of population intelligence, which is based on the model of discovery and participant. The algorithm has good global search capability and convergence efficiency [9]. Although the sparrow algorithm is more efficient than other algorithms, it still suffers from the common problem of group intelligence algorithms: It's easy to get stuck in local optimum [10]. Consequently, this paper suggests an improved ISSA capable of effectively escaping local optimality in the search process.

2.2.2 Improvement of the sparrow search algorithm. Adaptive inertia weight factors have significant advantages in improving the global search capability and avoiding the results from falling into local optimal solutions. Different inertia weight factors have different effects on the search accuracy and the number of convergences. A larger weight factor can effectively improve the global search ability, and a smaller weight factor can effectively protect the sparrow's position. Therefore, this paper introduces an adaptive inertia weight factor into the SSA to control the search range and convergence speed, and the formula of the weight factor ω is shown below:

$$\omega_i^t = \begin{cases} \omega_{\min} + \dfrac{(\omega_{\max} - \omega_{\min})\left[f_{\text{avg}}^t - f(x_i^t)\right]}{f_{\text{avg}}^t}, & f(x_i^t) \leqslant f_{\text{avg}}^t \\ \omega_{\max} - \dfrac{(\omega_{\max} - \omega_{\min})\left[f(x_i^t) - f_{\text{avg}}^t\right]}{f_{\text{avg}}^t}, & f(x_i^t) > f_{\text{avg}}^t \end{cases} \tag{1}$$

where: $f(x_i^t)$ is the current sparrow's fitness value; f_{avg}^t and f_{min}^t are the average fitness value and minimum fitness value of all current sparrows, respectively; $w_{\max}$ and $w_{\min}$ are the preset maximum and minimum inertia coefficients; $w_{\min}$ is taken as 0.3 and $w_{\max}$ is taken as 0.8 to ensure that the weight factor has a reasonable adjustment range.

If the current fitness value is better than the average fitness value, this model should approach the global optimum solution and update the inertia weight value in a smaller direction. If today's sparrow adaptation value is lower than the average adaptation value, then the global optimum solution is far away. At this time, the inertia weight value should be updated in the larger direction, and the improved finder position should be updated as follows:

$$X_{i,j}^t = \begin{cases} X_{i,j}^t \cdot w & R_2 < ST \\ X_{i,j}^t + Q & R_2 \geq ST \end{cases} \tag{2}$$

Journal of Physics: Conference Series **2815** (2024) 012031 doi:10.1088/1742-6596/2815/1/012031

2.3 Wavelet packet feature extraction algorithm

Wavelet packet decomposition is a refined decomposition algorithm. It can efficiently decompose low-frequency signals and enhance its ability to decompose high-frequency signals [11]. This paper employs wavelet packet decomposition for the extraction of multi-scale spatial energy characteristics. The vibration signal of the faulty bushing is decomposed by a 3-layer wavelet packet to obtain the energy values of the vibration signal of the faulty bearing in different frequency bands, and the percentage of the energy of each node is taken. Finally, these normalized energy values are arranged into feature vectors in scale order as the basis of fault identification.

2.4 Algorithmic diagnostic process

The establishment of a bushing fault diagnosis model firstly requires the extraction of relevant feature vectors through the wavelet packet algorithm as the input of the SVM algorithm. Subsequently, the ISSA proposed in this paper serves to refine the relevant parameters of the vector machine. Finally, the corresponding fault diagnosis model is established, which is abbreviated as the ISSA-SVM fault diagnosis model, and the overall process structure is shown in Figure 1 with the following steps:

Step 1: The original vibration signal is extracted using the wavelet packet algorithm and decomposed into a finite number of feature vectors, which are used as inputs to the vector machine algorithm. The input samples are partitioned into training and test sets. Finally, parameters such as the number of sparrow populations are initialized.

Step 2: C and σ of SVM network are used as optimization objects for ISSA, while the positions of discoverer, follower, and alert are updated according to formulas. The optimal fitness (As shown in equation (3), a classification prediction error rate for fault diagnosis is adopted as the fitness function.) and the best position are updated.

$$Fitness = 1 - \frac{acc}{100} \tag{3}$$

where acc is the classification prediction accuracy.

Step 3: Whether the current iteration number reaches the maximum number of iterations or reaches a preset value; if satisfied, the iteration terminates; otherwise, proceed with optimization. The optimal parameters, C and σ, are input into SVM, and finally, the fault diagnosis model is constructed.

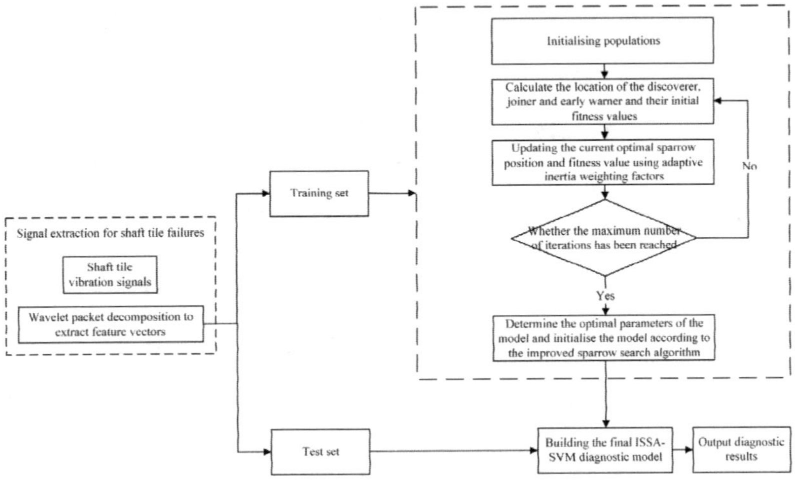

Figure 1. Flowchart of ISSA-SVM algorithm.

3. Experimental verification

3.1 Bushing troubleshooting test bench construction

The three-dimensional model of the bushing diagnosis experimental platform in this experiment is shown in Figure 2, which is mainly composed of the motor, lubrication pump, coupling, box, and other components. Most of the components in the experimental bench are mainly utilized in the KDF4000XE diesel generator parts to build. During the test, the acceleration vibration signal of the bearing cover is collected at 25.6 kHz sampling frequency, as shown in Figure 3. This experimental program categorized the abrasive wear into four levels according to its severity, which are normal, level 1, level 2, and level 3. Among them, level 1 is the lightest abrasive wear, while level 3 is the most serious abrasive wear. To prepare samples of these different levels of bushings failures, laser ablation was used to simulate live bushing abrasive wear failures. The plough furrow density for a level 2 fault is twice as high as for a level 1 fault, and for a level 3 fault, it is 3.5 times as high as for a level 1 fault. At an operating speed of 800 r/min, four failure tests have been conducted, as shown in Table 1.

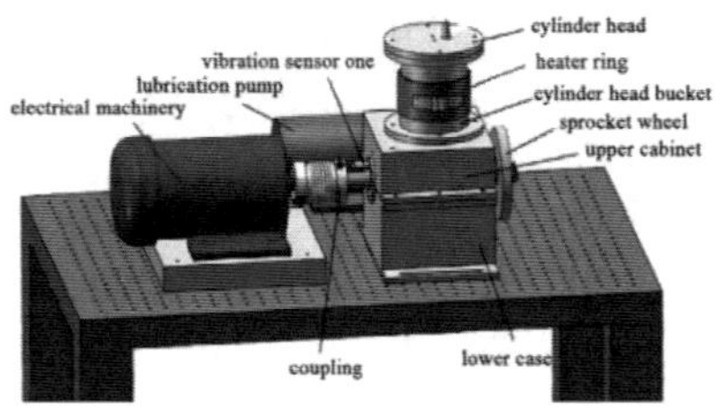

Figure 2. Test bench 3D model diagram.

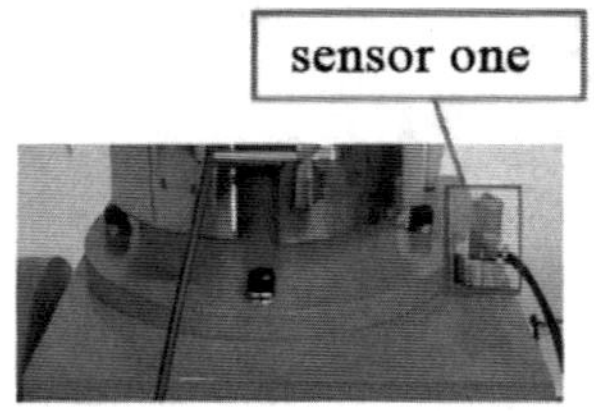

Figure 3. Sensor layout diagram.

Table 1. Experimental working conditions.

Experiment No.	Speed (r/min)	Bushing condition
1		Abrasive wear level 1
2	800	Abrasive wear level 2
3		Abrasive wear level 3
4		normalcy

Since SVM's performance is greatly influenced by parameters, an improved sparrow search algorithm is used to find optimal penalty factors C and kernel parameters σ of SVM. The initial parameter settings of the ISSA-SVM model is shown in Table 2.

Table 2. ISSA-SVM parameter settings.

Sparrow population	Maximum number of iterations	Optimization parameters	Search scope
50	100	C	[0.001,10000]
		σ	[0.01,1000]

3.2 Analysis of experimental results

In order to further demonstrate the effectiveness and accuracy of the proposed method, three fault diagnosis models, PSO-SVM, SSA-SVM, and SVM, were selected for comparison experiments. Experimentally, 500 sample data under specified working conditions are selected as the sample set to train and test the model, which contains 350 training sets and 150 test sets. Four models, ISSA-SVM, SSA-SVM, PSO-SVM, and SVM, are trained respectively, and after the completion of training, the test set is fault diagnosed, and the diagnostic results are shown in Figure 4. In the figure, "*" is the predicted test set; "○" is the actual test set. Vertical coordinates "1.0", "2.0", "3.0", "4.0" indicate normal, abrasive wear level 1, abrasive wear level 2, abrasive wear level 3, respectively. Through the final diagnostic results show that the diagnostic accuracies of the trained ISSA-SVM, SSA-SVM, PSO-SVM, and SVM fault diagnostic models are 96.5%, 95.2%, 93.2%, and 88.5%, respectively. The simulation results show that the diagnostic accuracies of ISSA-SVM, SSA-SVM, and PSO-SVM models are basically the same for normal and level 3 abrasive wear bushings, while the diagnostic accuracies of the SVM model are significantly lower. The ISSA-SVM model has a significant advantage over the other three models for fault diagnosis of level 1 and level 2 abrasive wear bushings.

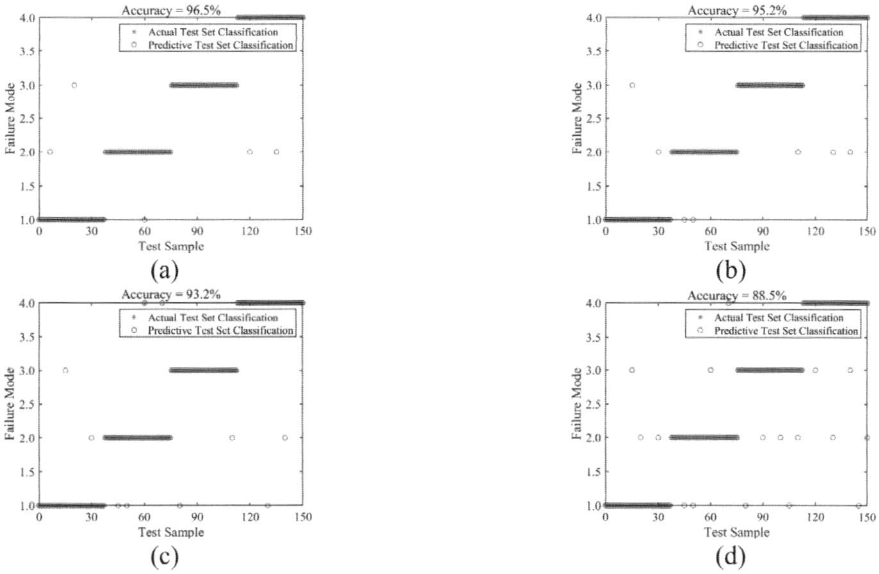

Figure 4. Diagnostic results of the four methods.

Table 3 lists the diagnostic results of ISSA-SVM, SSA-SVM, PSO-SVM, and SVM fault diagnosis models in terms of diagnostic time and diagnostic accuracy, respectively. As shown in Table 3, the accuracy of the four diagnostic models is 96.5%, 95.2%, 93.2%, and 88.5%. The diagnostic times are 1.4s, 2.5s, 4.8s and 5.2s. It is proved that the ISSA-SVM model has higher convergence precision and less error than other models, so it is able to accurately identify different types of faults in a short time.

Table 3. Diagnostic results of the sensor.

Speed(r/min)	Algorithms	Diagnostic time (s)	Diagnostic accuracy
800	ISSA-SVM	1.4	96.5%
	SSA-SVM	2.5	95.2%
	PSO-SVM	4.8	93.2%
	SVM	5.2	88.5%

4. Conclusion

In response to the issue of inadequate accuracy of traditional bushing fault diagnosis, this paper proposed a bushing fault diagnosis method based on ISSA-optimized SVM. In particular, in order to enhance the search capability of the SSA algorithm and prevent it from reaching a local optimum, adaptive inertia weight is included in this paper. The ISSA-SVM diagnostic model is ultimately created by combining the ISSA and SVM and training it using the feature vectors that the wavelet packet generated as training data. Compared with the other three fault diagnosis models, the diagnostic model has higher diagnostic accuracy and shorter diagnostic time, which truly realizes the rapid and accurate diagnosis of the bushing fault.

References

[1] Li B Q, Zhou H G, Liu J F and Kang C (2021) Research on weak fault feature extraction and detection of crankshaft bearing bushes used in marine diesel engines. Noise Vibrat. Control, 41, 134–138+168.

[2] He S Y, Hu D Y, Yu G, Liao A H and Shi W (2022) Trackside acoustic detection of bushing fault using wavelet domain moving beamforming method. Appl. Acoust., 195, 108851.

[3] Deng F Y, Liu C, Liu Y Q and Hao R J (2021) A hybrid SVD-based denoising and self-adaptive TMSST for high-speed train bushing fault detection. Sensors, 21, 6025.

[4] Kulevome D K B, Wang H, Cobbinah B M, Mawuli E S and Kumar R (2024) Effective time-series Data Augmentation with Analytic Wavelets for bearing fault diagnosis. Expert Syst. Appl., 249, 123536.

[5] Yan X A, Hua X, Jiang D and Xiang L (2024) A novel robust intelligent fault diagnosis method for rolling bearings based on SPAVMD and WOA-LSSVM under noisy conditions. Meas. Sci. Technol., 35(5): 056121.

[6] Daiki G, Tsuyoshi I, Takekiyo H, Shota Y, Keiichi K, Shigeyuki T and Akira H (2024) Failure diagnosis and physical interpretation of journal bearing for slurry liquid using long-term real vibration data. Struct. Health Monit., 23: 1201–1216.

[7] Ding S F, Sun Y T, Liang Z Z, Guo L L, Zhang J and Xu X (2024) An overview of support vector machine algorithms in weakly supervised scenarios. Chin. J. Comput., 1–25.

[8] Yin Y C, Zou Y J, Du T L and Zhang Y W (2024) Fault Diagnosis of Ship Low Freshwater Cooling System LFCS Based on SSA-SVM Algorithm. Comput. Simulat., 41(01): 548–553.

[9] Yu S W, Zhang X S, Lin D, Li Z Y and Xiong R (2024) Research on energy storage battery fault diagnosis by applying proba-bilistic neural network optimized by sparrow search algorithm.

[10] Li X R and Jin W Y (2023) Fault diagnosis of rolling bearings based on ISSA-SVM. J. Vibrat. Shock, 42: 106–114.

[11] Jiang H, Qiu L P and Jiang Q (2024) A Fault Diagnosis Method for Water Pump Bearings Based on Wavelet Packet Decomposition and MCKD Algorithm. J. ShenYang Ligong Univ., 43(02): 38–44.

Journal of Physics: Conference Series **2815** (2024) 012032 doi:10.1088/1742-6596/2815/1/012032

Research on calibration method for measurement error of multiple rate carrier energy meter

Fating Zhang *, **Xi Li, Yan Li and Zixiang Zhou**

Marketing Service Center of State Grid Qinghai, Xining Qinghai, 810000, China

*Corresponding author's e-mail: qwsd15690@163.com

Abstract: With the rapid development of smart grid technology, the multi-rate carrier watt-hour meter is an important measurement tool in power systems; its accuracy directly affects economic benefits and energy management efficiency. This paper proposes a calibration method of measurement error for a multi-rate carrier watt-hour meter to improve the measurement accuracy and reliability of the watt-hour meter under changing load conditions. The multi-rate carrier energy data is collected in real-time by high-precision data acquisition equipment to ensure the comprehensiveness and accuracy of the data. Using advanced data analysis technology to extract the amplitude characteristics of abnormal metering data, on this basis, the output deviation of equipment without input signal is eliminated, and the metering error calibration of the electric energy meter is completed. The experimental results show that this method can significantly improve the metering accuracy of multi-rate carrier watt-hour meters under different power consumption conditions and has good adaptability and practical value.

1. Introduction

With the rapid development of science and technology, as an important pillar of the national economy, the accuracy and reliability of its measurement technology are becoming more and more important. As an important measuring equipment in modern power systems, the accurate measurement of a multi-rate carrier watt-hour meter is directly related to the fairness of power transactions and the economic benefits of power enterprises[1]. The compound electricity price meter is divided into several periods every day. The electricity quantity of each period is calculated separately, and then the electricity price of each period is summarized. This kind of power charging method can effectively make users realize the active power consumption balance, promote the power consumption balance, reduce the system pressure during the peak period of power consumption, maximize volatilization, improve the capacity of power supply equipment, and improve economic benefits [2]. However, due to the influence of many factors, such as the performance of the equipment, the use environment, and human operation, the multi-rate carrier watt-hour meter will inevitably have measurement errors in practical application. Therefore, it is of great significance to study and develop an effective calibration method for metering error of multi-rate carrier watt-hour meter to improve the accuracy of power metering and ensure the normal operation of the power market. Through the study of the calibration method, the metering performance of an electric energy meter can be accurately evaluated, potential errors can be identified, and corresponding measures can be taken to correct them. At present, scholars at home and abroad have made some achievements in the research on the measurement error calibration method of multi-rate carrier watt-hour meters. Musaev et al. [3] proposed the load compensation method of an intelligent electric energy metering transformer. The parameters such as load size, power factor, and waveform distortion are analyzed using data analysis technology, and the capacitor is adjusted according to the numerical results of the parameters to complete the transformer load compensation. However, the data clustering effect of this method could be better. Magomedov et al. [4] proposed the energy consumption compensation method of the intelligent watt-hour meter. We collect smart watt-hour meters to collect real-time data on users' electricity consumption, identify users' electricity consumption patterns, peak hours, and peak energy consumption, and apply the correction values to the electric energy metering system for adjustment to realize dynamic compensation of energy consumption. However, this method could have higher efficiency in power transmission and distribution. Based on this, this paper studies the calibration method of measurement error of multi-rate carrier watt-hour meters.

2. Calibration method for the measurement error of electric energy meters

2.1. Collecting multi-rate carrier energy data

In actual use, the electric energy meter may face different conditions, such as voltage, current, and power factor, which will directly affect the measurement accuracy of the electric energy meter [5-7]. By collecting multi-rate carrier energy data, the metering characteristics of the watt-hour meter under various conditions can be understood more comprehensively to evaluate its error more accurately. A standard power source is used to supply power to the watt-hour meter to be calibrated. This standard source can provide stable and known voltage and current output:

$$a_t = \frac{\sum_{i=1}^{n} a_i}{2n} \tag{1}$$

where a_t represents the data set of electric energy metering current and n represents the total number of electric energy metering data. The load directly affects the current, which is one of the key factors of electric energy measurement. The carrier energy data under different current levels are obtained by adjusting the load, and then the metering accuracy of the watt-hour meter under different load conditions is analyzed. The power factor is an index to measure the effective utilization of electric energy, which reflects the metering characteristics of electric energy meters under different power factors. By changing the power factor setting of the power supply, the carrier energy data under different power factors are obtained to analyze the error of the watt-hour meter under different power factors. Therefore, by changing the load output, different power consumption conditions (including different power factors and frequencies) are simulated, thus generating multiple carrier energy data with different rates:

$$\begin{cases} I_{\max} = l_1 + (l_1 - l_2) \times 4 \\ I_{\min} = \max(l_2 - (l_1 - l_2), 0) \end{cases} \tag{2}$$

where l_1 represents the direction point position of the energy meter and l_2 represents the supplementary point position of the energy meter.

A high sampling rate is also an indispensable feature of data acquisition equipment. In the process of simulating different power consumption conditions, the measured value of the electric energy meter may change rapidly in a short time. If the sampling rate of data acquisition equipment is higher, these rapid changes may not be captured, resulting in data loss or distortion. Therefore, the high sampling rate can ensure that the data acquisition equipment can record the metering value of the electric energy meter in real time and accurately to obtain complete and continuous carrier energy data:

$$a_i' = a_t / d_{\max} \tag{3}$$

where $d_{\max}$ represents the maximum value of load capacitance.

Through the high sampling rate connected to the electric energy meter, the multi-rate carrier energy data is collected, which ensures the accuracy and reliability of the data and provides the data basis for the subsequent error analysis and calibration.

2.2. Extracting characteristics of abnormal measurement data

Based on collecting the metering data of the watt-hour meter, the multi-rate carrier energy data is preprocessed [8] to remove noise and invalid data:

$$W = \frac{\eta\alpha(t-1)}{a_i'} \tag{4}$$

where η represents the load loss and α represents the harmonic distortion.

Machine learning algorithm automatically identifies abnormal data by learning the inherent laws and patterns of data:

$$\begin{cases} \min f(Z), Z \in R^n \\ s.t. Z(i) = 1 or Z(i) = 0 \end{cases} \tag{5}$$

where $f(Z)$ represents the differential amplification of abnormal data, and $Z(i)$ represents the current flowing through the load.

Abnormal frequency refers to the ratio of abnormal data points in a time, reflecting the prevalence or severity of abnormal phenomena [9]. According to the characteristics of the data, we choose a time window to analyze the frequency change of abnormal events and find out whether there is a regular pattern:

$$R = \frac{W\sum_{k=1}^{n}|U(k)|e}{f(Z)} \tag{6}$$

where $U(k)$ represents the ambient temperature of the watt-hour meter in the running state of k, and e represents the calibration constant of the watt-hour meter [10].

We can better understand the abnormal data pattern by extracting the key features. When the system has an abnormal frequency and certain periodicity, it indicates that the system has problems, such as equipment failure.

2.3. Multi-rate carrier energy error calibration

Based on the characteristics of the extracted abnormal data, the source of the error is diagnosed. According to the characteristics of abnormal data and the metering principle of electric energy meters, the causes and types of errors are analyzed.

$$\varpi(A) = [\varpi(A_1), \varpi(A_2), \cdots, \varpi(A_x)]^T \tag{7}$$

where $\varpi(A)$ represents the peak factor.

According to the error analysis results, we set appropriate calibration parameters, improving its response-ability to small signals through gain adjustment:

$$K = \frac{R}{|S_K|} \varpi(A) \vartheta \tag{8}$$

where S_K represents the peak factor and ϑ represents the frequency.

We eliminate the output deviation of equipment when there is no input signal, that is, zero drift:

$$B = \frac{P\left(W_s + \sin^2 \varphi\right)}{K} \tag{9}$$

where P represents power factor, W_s represents zero drift rate, and φ represents phase angle. The calibration of the measurement error of the multi-rate carrier energy meter has been completed.

3. Experimental analysis

3.1. Experimental environment

This electric energy metering compensation and correction experiment is carried out under rated conditions. The rated voltage is set to 380 V, the rated current is set to 400 A, and the working frequency is set to 450 HZ in the laboratory. Under ambient temperature of 20°C±5°C and humidity of 75%, the pulse constant of the electric energy meter is set to 5500 IMP/KWh. After the experimental environment is set, the HBS1030 three-phase standard electric energy meter with an accuracy level of 0.05 is selected as the test object, and the error compensation and correction are carried out on the meter. The output voltage and current of the same standard source are measured by the HBS1030 three-phase standard electric energy meter and high-precision meter, and the measured electric energy is input into the meter calibration table in the form of the pulse signal. After receiving the pulse signal, the meter calibration table converts the pulse signal into electric energy. The experimental circuit includes eight complex rate carrier energy meters, corresponding to A1-A8, as shown in Figure 1.

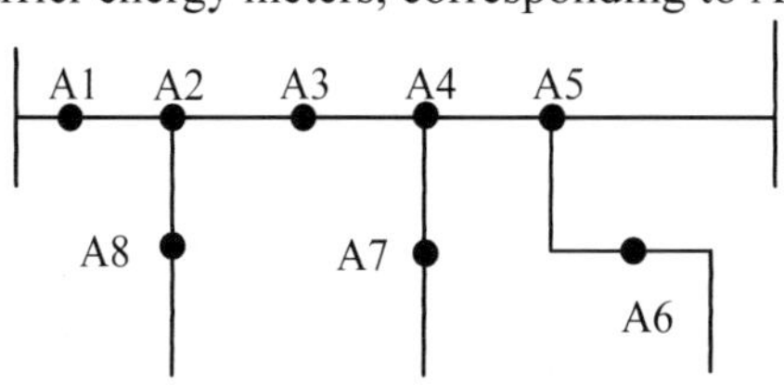

Figure 1. Experimental Object.

The specific content of single-phase line data is shown in Table 1.

Table 1. Data Collection Results.

Line Branch Number	Starting measurement point	Termination measurement point	Data output value
1	1	2	500
2	2	3	500
3	3	4	500
4	4	5	500
5	3	6	500
6	5	8	500
7	7	8	500

3.2. Analysis of experimental results

Experiments were carried out to verify the practical performance of this method. The method used by Musaev et al. [3] and the method used by Magomedov et al. [4] are selected as comparison methods, and different methods are used to compensate and correct the measurement error of electric energy metering equipment. The positioning results of abnormal metering points of electric meters by the proposed method, the method used by Musaev et al. [3] and the method used by Magomedov et al. [4] are shown in Figure 2.

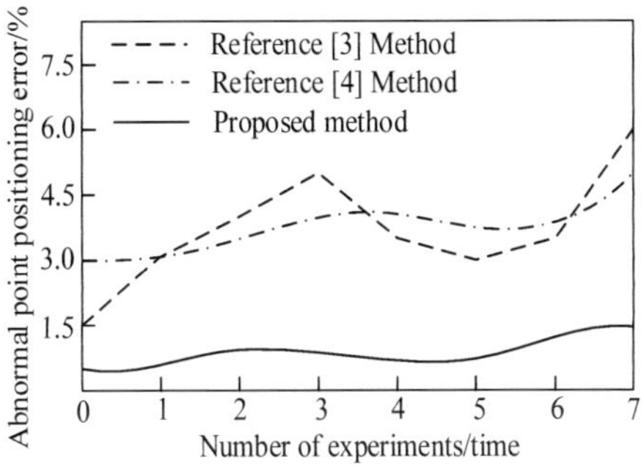

Figure 2. Location results of outliers by different methods.

According to Figure 2, the proposed method has high positioning accuracy for outlier measurement points and has shown stable application effects in many experiments. However, the accuracy of the abnormal point location by contrast method is reduced and fluctuates greatly.

Three methods were used to calibrate and compare the voltage measurement error of the complex rate carrier energy meter, and the calibration accuracy of the three methods was compared. The voltage diagnosis results of different methods are shown in Figure 3.

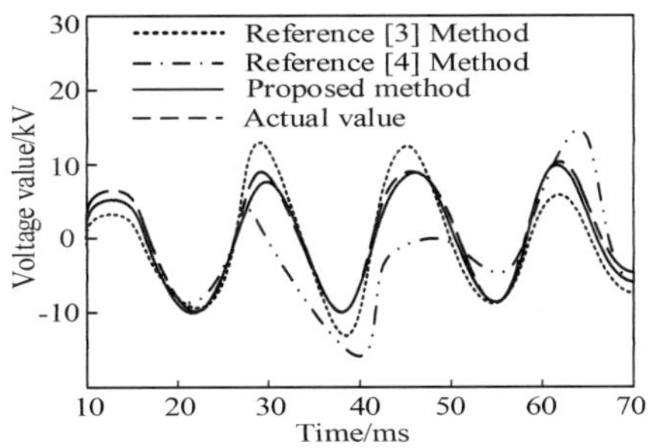

Figure 3. Voltage diagnosis results of different methods.

According to Figure 3, it can be seen that the difference between the voltage value and the actual value of the multi-rate carrier energy meter calibrated by the proposed method is small, and the measurement error calibration accuracy of the energy meter is relatively high. This indicates that the method can accurately evaluate and correct the errors of complex rate carrier energy meters in the measurement process, thereby improving the accuracy and reliability of electricity metering. The multi-rate carrier energy meter can measure different periods and electricity prices, which helps users use electricity reasonably and save electricity. By calibrating the error of the electricity meter, the effectiveness of energy conservation and emission reduction can be further improved, promoting sustainable development.

4. Conclusion

This paper studies the calibration method for measurement errors of complex rate carrier energy meters and proposes a systematic calibration method. This method fully considers the characteristics and working principles of complex rate carrier energy meters, adopts advanced calibration technology and equipment, and ensures the accuracy and reliability of calibration results. In future research, we will continue to improve and optimize this calibration method, continuously improving calibration accuracy and efficiency to meet the growing demand for electricity metering and management.

Acknowledgment
This paper was supported by the Research Projects of Marketing Service Center of State Grid Qinghai "Electric Energy Measurement Equipment Verification and Testing Technology Laboratory".

References

[1] Rind Y M, Raza M H, Zubair M, et al. Smart energy meters for smart grids, an internet of things perspective [J]. Energies, 2023, 16(4): 1974.
[2] Ma L, Meng Z, Teng Z, et al. A measurement error prediction framework for smart meters under extreme natural environment stresses [J]. Electric Power Systems Research, 2023, 218(1): 109192.
[3] Musaev T, Khabibullin M, and Fedorov O. Developing a Method Providing for the Accurate Assessment of Actual 6(10) kV Transformer Load Using Data from Smart Electric Energy Metering Systems [C]//2022 International Conference on Industrial Engineering, Applications and Manufacturing (ICIEAM), Sochi, Russian Federation. 2022.
[4] Magomedov I A, Khidivov M N, and Magomedov A I. Principles of construction of a multifunctional intelligent electric energy meter with advanced communication and functional capabilities [J]. Herald of Dagestan State Technical University. Technical Sciences, 2023, 50(1): 89-98.
[5] Viciana E, Arrabal-Campos F M, Alcayde A, et al. All-in-one three-phase smart meter and power quality analyzer with extended IoT capabilities [J]. Measurement, 2023, 206(1): 112309.
[6] Chen Z, Amani A M, Yu X, et al. Control and optimisation of power grids using smart meter data: A review [J]. Sensors, 2023, 23(4): 2118.
[7] Ghemari Z, Belkhiri S, and Saad S. A capacitive sensor with high measurement accuracy and low electrical energy consumption [J]. Applied Physics A, 2023, 129(5): 362-377.
[8] Williams D O, Li Z S, and Ghanavati A. Mitigating Electrical Losses Through a Programmable Smart Energy Advanced Metering Infrastructure System [C]//2023 IEEE International Conference on Prognostics and Health Management (ICPHM). IEEE, 2023: 153-157.
[9] Hafidz I, Priyadi A, Pujiantara M, et al. Development of IoT-Based Portable Power Quality Monitoring on Microgrids by Enhancing Protection Features [J]. IEEE Access, 2023, 11(1): 49481-49492.
[10] Xu W, Zhang P, Zeng W, et al. Research and Expectation of a New Generation of Smart Energy Meter based on IR46 Standard [C]//2023 12th International Conference of Information and Communication Technology (ICTech). IEEE, 2023: 455-463.

Journal of Physics: Conference Series **2815** (2024) 012033 doi:10.1088/1742-6596/2815/1/012033

A comprehensive investigation into the vibration fatigue characteristics of safety hangers for metro vehicles

Qihang Zeng[1,2], Fubing Zhang[1,*], Guanglin Tan[1], Fuxing Chen[3] and Qing Chen[1]

[1]School of Intelligent Manufacturing Engineering, Chongqing University of Arts and Sciences, Chongqing, 402160, China
[2]Chongqing City Vocational College, Engineering Research Center of Chongqing Education Commission of China in Running and Maintenance of Robots, Chongqing, 402160, China
[3]China University of Petroleum, Beijing

[*]Corresponding author's e-mail: zqh0823@foxmail.com

Abstract: To meticulously investigate the reasons behind the cracking of the safety crane in a subway vehicle, we conducted a comprehensive analysis of its modes and conducted mode tests accordingly. These tests revealed the natural frequencies of the crane to be approximately 340 Hz and 990 Hz. Additionally, to gain insights into the acceleration and dynamic stress levels experienced by the safety crane during operation, we conducted a line tracking test on the vehicle. Analysis of the test data revealed that the acceleration and dynamic stress of the crane exhibited significant energy levels around 340 Hz, indicating that the first-order mode was excited, resulting in resonance. A comparative evaluation of the fatigue data of dynamic stress, both before and after rail grinding, revealed that the fatigue damage caused by rail wave grinding significantly shortened the service life of the safety crane.

1. Introduction

Rail transit stands as the most efficient mode of transportation in alleviating urban congestion. By the conclusion of 2017, 32 cities in China had operational subway systems, with over 30 additional cities either under construction or having received approval. However, amidst this flourishing development, challenges such as the wheel polygon phenomenon and rail wave grinding have begun to surface. When such conditions trigger abnormal vibrations, they often lead to vibration-induced fatigue in components, ultimately resulting in damage [1-2]. Both domestically and internationally, vibration fatigue has been a contributing factor in numerous major accidents caused by the failure of critical components. Instances include the fatigue-related crashes of approximately 20 Wellington heavy bombers during the Second World War, structural cracks in a Chinese helicopter in 1970, and the sudden disintegration of a Taiwan, China airline's Boeing 747 in the Taiwan Strait airspace in 2002 due to fatigue failure. Clearly, the consequences of vibration and fatigue failures can be dire [3].

While the issue of vibration fatigue first garnered attention in the context of aircraft and carrier rockets, its significance has also become apparent in the rail transit industry, particularly in subway vehicles, as the sector experiences rapid growth. Scholars have extensively studied vibration fatigue, with Cui and Yao [4] assessing the vibration fatigue life of a connector using the nominal stress method and validating its accuracy through testing. Liu [5] conducted a fatigue life analysis that took into account the influence of dynamic structural responses under vibrational excitation on crack propagation. Additionally, Park et al. [6] employed a rigid-soft coupling approach to investigate the vibration fatigue life of container bodies. Ye et al. [7] conducted a thorough analysis of how residual stress and surface roughness impact the vibration fatigue performance of blades. Hu et al. [8] delved into the effects of vibration frequency on the vibration fatigue characteristics exhibited by aluminum alloy cantilever structures. Gona et al. [9] examined the structural response and fatigue properties when the gearbox body is coupled with axion stiffness. Zhu et al. [10] employed the frequency domain method to investigate the fatigue life of bogies. Moon et al. [11] not only explored the mechanisms and calculation methods of vibration fatigue but also proposed a methodology for conducting vibration fatigue tests. Drawing upon the optimization and comparative analysis of safety fracture issues, this paper offers measures aimed at preventing the recurrence of similar faults.

Based on the optimization and comparative analysis of the safety lifting fatigue fracture problem of a domestic subway vehicle, this paper puts forward measures to avoid the recurrence of similar faults.

2. Modality analysis

The structural dimensions and installation location of the subway wheel for the safety crane are clearly depicted in Figure 1. The wheel is securely fastened above the shaft box using bolts. During normal operation of the safety crane, the frame remains detached, ensuring that the wheel remains free-standing. However, when the vehicle is lifted, the frame and wheel pair align, effectively serving as a connector between the two components, facilitating the simultaneous lifting of the wheel pair. Consequently, during normal vehicle operation, the safety crane assumes a cantilever structure. Common instances of damage to the safety crane are illustrated in Figure 2.

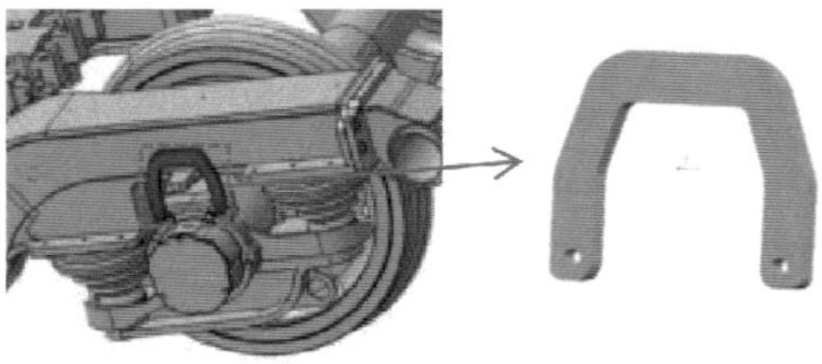

Figure 1. Safety hanger structure and installation position.

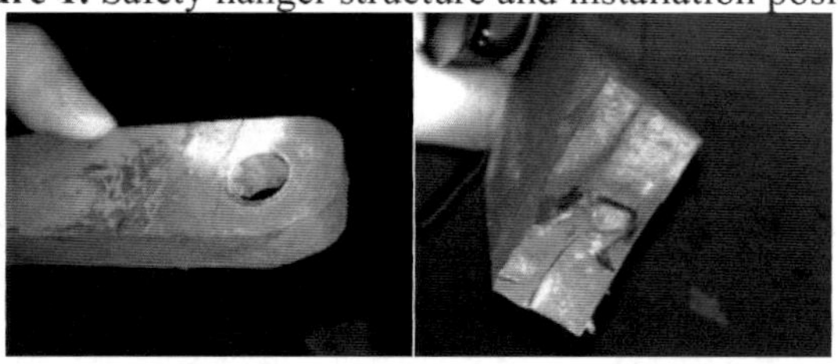

Figure 2. Picture of safety hanger break.

To analyze the vibration characteristics of the safety crane, both mode simulation calculations and mode experiments were conducted on its structure. The simulations were executed utilizing the Abaqus software. The finite element theory served as the basis for establishing the structural calculation model of the safety crane, assuming it to be comprised of homogeneous, isotropic elastic materials. The results revealed a first-order mode of 371 Hz and a second-order mode of 1,065 Hz. Experimentally, the measured first-order bending mode was 346 Hz, while the second-order torsion mode was 991 Hz. Both the test and simulation outcomes are presented in Table 1.

Table 1. Modes of safety hanger.

Model order	First-order mode		Second-order mode	
	Test	Simulation	Test	Simulation
Frequency f/Hz	346	371	991	1065
Vibration mode	-	Lateral bending	-	Twist

3. Line experiment

3.1 Introduction of the test

To accurately assess the vibration and stress levels experienced by the safety crane during actual operation, comprehensive testing was conducted on the vehicle throughout its entire journey from the starting station to the terminal. The measurement points encompassed both vertical and lateral acceleration, as well as the dynamic stress response of both the shaft box body and the safety crane.

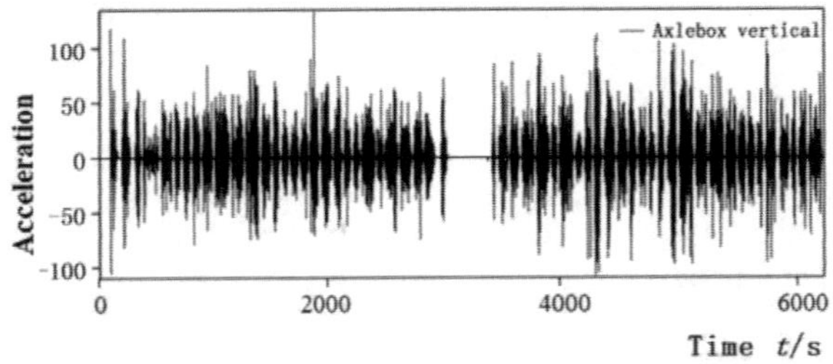

(a)

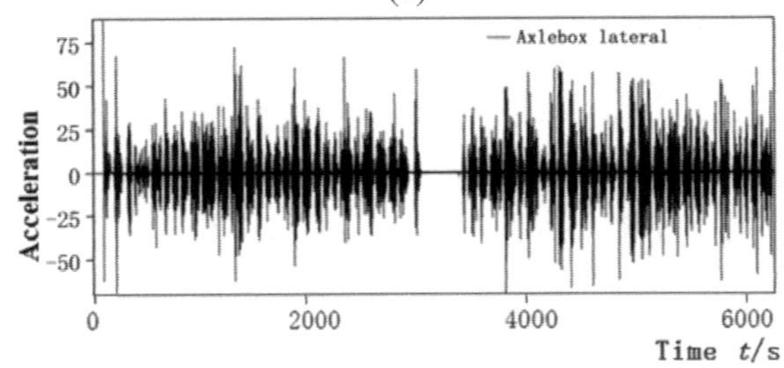

(b)

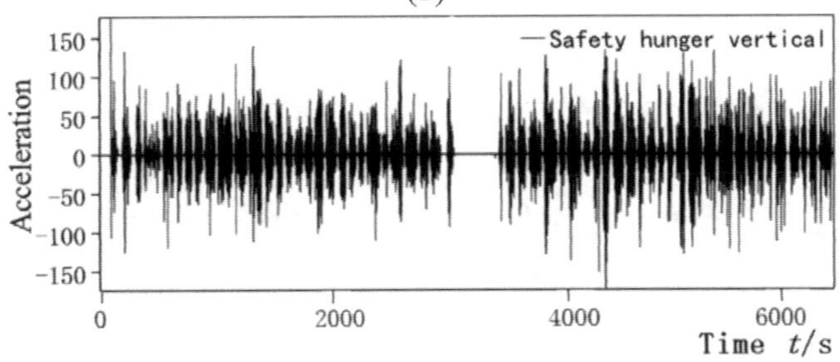

(c)

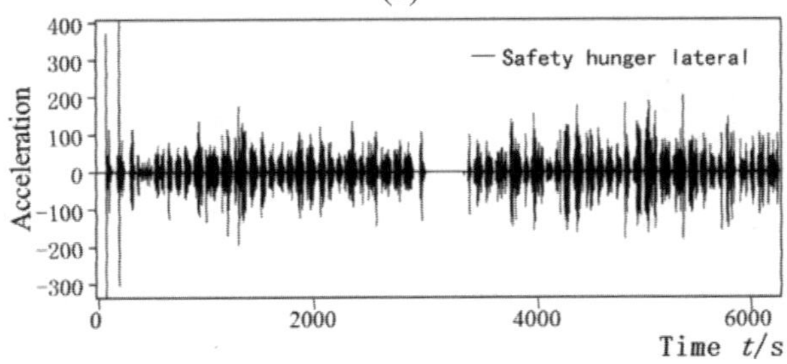

(d)

Figure 3. Overall acceleration data. (a)Vertical acceleration of the shaft box; (b) Lateral acceleration of the shaft box; (c) Safety hanging acceleration (d) Lateral acceleration of the safety suspension.

3.2 Data analysis

Based on the comprehensive acceleration time domain data collected throughout the journey from the initial station to the terminal, it is observed that both the lateral and vertical vibrations of the axle box remain relatively stable, without any significant spikes or increased sections as depicted in Figures 3 (a) and 3 (b). However, when it comes to the safe lifting position, Figure 3 (d) reveals a notable surge in vibration amplitude to approximately 400 g in specific sections (highlighted by the ellipse, representing the first two stations). In contrast, the majority of other sections exhibit a vibration amplitude of approximately 150 g. This preliminary analysis suggests the existence of rail grinding in those particular sections between the first two stations. It is hypothesized that, as the local iron vehicles traverse these sections at a certain speed, the excitation frequency generated by the grinding aligns closely with the first-order transverse bending frequency of the safety crane, leading to resonance. Field measurements indeed confirm the presence of a wave with a wavelength of approximately 0.056 m in this area. According to the vibration time frequency signal at the gear box can obtain the engagement frequency of the gear, and then infer that the running speed of the vehicle in the rail grinding segment is v = 57.60 km/h=16 m/s, and the excitation frequency f_c and the vehicle running speed v (m/s) and the grinding wavelength l (m) are as follows:

$$f_c = \frac{v}{l} = \frac{16\text{m/s}}{0.056\text{m}} = 285\ \text{Hz} \qquad (1)$$

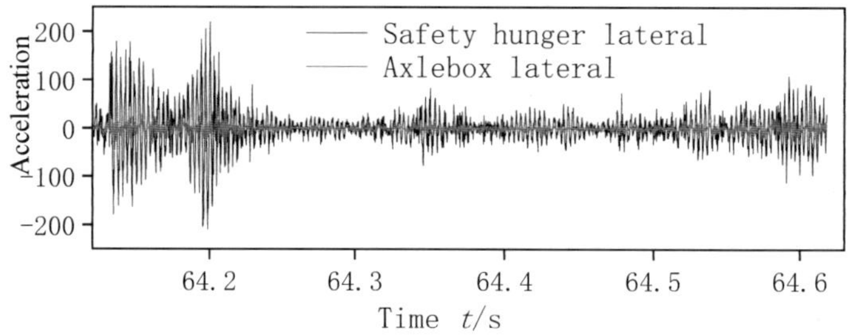

(a)

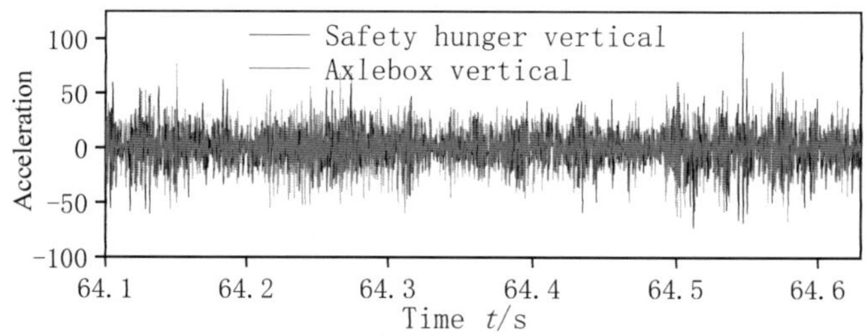

(b)

Figure 4. Vibration Transmission. (a) Lateral acceleration contrast; (b) Vertical acceleration contrast.

By comparing the vibration transfer in the time domain from the shaft box to the safety crane in Figure 4, several observations can be made. Firstly, when comparing the vertical acceleration amplitude of the safety crane with that of the shaft box, it is evident that there is no significant amplification, as depicted in Figure 4 (b). However, a notable difference is observed in the transverse vibration amplitude. Specifically, the amplitude of the safety crane increases by approximately 3 to 3.5 times when compared to the shaft box, as shown in Figure 4 (a). This finding indicates that there is a significant amplification of the transverse acceleration in the time domain.

3. 3 Comparative analysis of the fatigue lifespan

The dynamic stress data of the road is evaluated utilizing the total life method, which provides an estimation of the service life expectancy of the safety lifting measuring point under the current line conditions. The raw measured data often contains significant interference noise, necessitating data preprocessing procedures including zero drift correction, spike filtering, 50 Hz power frequency noise filtering, and common mode noise filtering [12]. The fatigue testing and life calculation theories rely on the material performance data associated with normal amplitude stress and strain cycles. The primary objective of load spectrum statistics is to transform the structural random stress-strain time history into a discrete graded constant stress-strain spectrum. This transformation is founded on the principles of Miner's damage accumulation and damage equivalence [12]. The rainflow counting method is a widely adopted technique for the fatigue life analysis of engineering structures. Consequently, this paper employs the rainflow counting method to categorize the measured dynamic stress data of the line. Structural fatigue is essentially a process of damage accumulation. In the railway industry, lifespan is typically characterized by either usage time or mileage, with the two metrics interchangeable based on the frequency of operation. For a single-cycle operation (from the starting station to the terminal), the damage incurred to a specific component of the safety crane is denoted as D. According to Miner's rule, the service life of the safety crane, expressed in terms of cycle times, is calculated as:

$$N = 1/D \quad (2).$$

3.4 Life-span assessment

The benchmark for evaluating lifespan is the FKM Analytical Strength Assessment, 5th Edition (hereinafter referred to as the Standard). The grade of material used for the safety lifting component is S355J2G3, which boasts a tensile strength of 510 MPa, a yield strength of 355 MPa, and a fatigue strength of 230 MPa after 1 million cycles, as indicated in the S-N curve. As per the Standard, the total allowable damage for structural steel without welding is 0.3. Once the accumulated damage surpasses this threshold, the structure is deemed to have failed. Considering the specific operating conditions of the safety crane, a safety factor of 1.35 has been applied. Utilizing the material parameters and the provisions outlined in the Standard, an S-N curve for the base material S355J2G3 can be constructed (Figure 5).

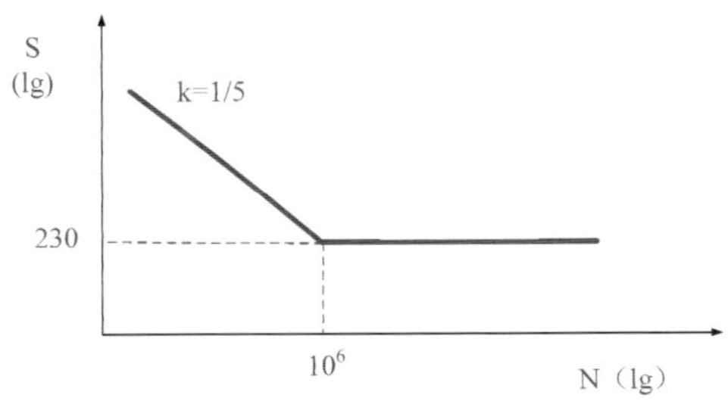

Figure 5. Safety hanger S-N curve.

The safety life of the subway round-trip test data was evaluated, but the actual sticking position of the strain gauge deviates from the fracture position. According to the finite element calculation, the relationship between the fracture position and the strain gauge paste position stress is 1.7 times the amplification of the strain gauge position. The results of the safety service life assessment are shown in Table 2.

Table 2. The service life of the measuring point.

Test point	Service life/running times	Service life/10^4 km
Point 1	5409	38.4
Point 2	47936	340.3

The data of test point 1 are evaluated as 1 755 752 199 round trips, $6.24*10^{10}$ km, for infinite life. It can be seen that the service life of the safety crane is greatly improved after removing the data from the wave grinding section.

4. Discussion and conclusion

After conducting a thorough analysis of the data obtained through simulation calculations and experimental tests on the line, several important observations and conclusions can be drawn.

Regarding the safety lifting mode analysis, our rigorous testing and finite element calculations have been validated by subsequent verification. The results indicate that the first-order transverse bending mode is 345 Hz, with a second-order natural frequency of 991 Hz. This information is crucial for understanding the dynamic behavior of the safety crane and ensuring its safe and effective operation. Also, our meticulous analysis of vibration data from the shaft box and safety crane has revealed the existence of track grinding between the first two stations. This grinding exhibits a wavelength of approximately 0.056 m, which closely resembles the wavelength typically observed in rail grinding on subway lines. This finding is significant as it highlights a potential source of wear and tear that could affect the performance and lifespan of the system. A comparison of dynamic stress data from both the grinding and non-grinding sections has shown a significant improvement in the service fatigue life of the system when grinding is absent. This suggests that reducing or eliminating track grinding can significantly enhance the durability and reliability of the railway infrastructure.

Based on the above analysis, it is advisable to conduct a comprehensive inspection of the line and subsequently undertake track polishing to mitigate potential damage and extend the lifespan of the safety crane. This recommendation is crucial for maintaining the integrity and safety of the system and ensuring its smooth operation. Finally, given the critical role of safety lifting in the overall integrity and safety of the system, it is strongly recommended that safety lifting be included in the lifecycle management of the railway infrastructure. This will ensure that all aspects of the system are taken into account in the planning, maintenance, and upgrading of the railway network.

In conclusion, our analysis has provided valuable insights into the vibration characteristics and performance of the safety crane and the railway infrastructure. These insights will help in improving the safety, reliability, and lifespan of the system, thereby enhancing the overall efficiency and performance of the railway network.

Acknowledgments

This work was supported by the Science and Technology Research Program of Chongqing Municipal Education Commission (Grant Nos. KJQN202001309, KJQN202301345), the Natural Science Foundation of Chongqing (No. cstc2021jcyj-msxmX0812), the Opening Project of Engineering Research Center of Chongqing Education Commission of China in Running and Maintenance of Robots (No. YWZX20220002).

References

[1] WANG QunSheng, ZENG Jing, LUO GuangBing, et al. Study on the vibration behavior of carbody underneath suspended systems under wheel profile wear [J]. Chinese Journal of Mechanical Engineering, 2016, 52(10):113-118.

[2] HE XiaoLong, ZHANG LiMing, LU LianTao. Effects of suspended equipment on the vertical vibration of high-speed vehicle [J]. Journal of Vibration and Shock, 2017, 36(19): 46-51.

[3] LIU WenGuang, CHEN GuoPing, HE HongLin, et al. Review of studying on vibration fatigue [J]. Chinese Journal of Engineering Design, 2012, 19(1): 1-8.

[4] CUI SiPeng, YAO WeiXing, XIA TianXiang. Nominal stress approach for fatigue life prediction of multi-fastener joints under vibration loading [J]. China Mechanical Engineering, 2014, 25(18): 2519-2522.

[5] LIU WenGuang. Calculation method of vibration fatigue life with structure crack growth [J]. Journal of Vibration Engineering, 2012, 25(1): 79-83.

[6] PIAO MingWei, FANG Ji, ZHAO QianXu, et al. Vibration fatigue analysis of container flatcar based on rigid-flexible coupling simulation [J]. Journal of Vibration and Shock, 1999, 69(2): 105-120.

[7] YE YongNeng, CHENG Ming, ZHANG ShIHong, et al. Influence of residual stress and roughness

on the vibration fatigue of blade [J]. Materials Science & Technology, 2015,23(5): 1-5.
[8] HU HaiTao, LI YuLong, SUO Tao, et al. Vibration fatigue and fracture performance of aluminum alloy 2024 [J]. Journal of Aeronautical Materials, 2013, 33(4):78-83.
[9] GAO Na, FENG HuiHua, YANG ShiQiu, et al. Dynamic analysis of the gear-box case structure based on stiffness coupling and its life estimation of vibration fatigue [J]. Journal of Mechanical Strength, 2017,39(1):221-225.
[10] ZHU Tao, XIAO ShouNe, YANG GuangWu, et al. Study on fatigue life in frequency domain for bogie frame [J]. Journal of Mechanical Strength, 2016, 38(1): 160-166.
[11] MOON S I, CHO I J, YOON D. Fatigue life evaluation of mechanical components using vibration fatigue analysis technique [J]. Journal of Mechanical Science and Technology, 2011, 25(3):631-637.
[12] LAN QingQun, LI ZhaoJun. Service life evaluation method for metro vehicle steering frame [J]. Modern Manufacturing Engineering, 2017(6): 79-82.

Journal of Physics: Conference Series **2815** (2024) 012034 doi:10.1088/1742-6596/2815/1/012034

Rated flow coefficient of geometric compensation control valve

Zhong Li*, Wei Wang, You Ming, Fengguang Cheng, Shengtao Geng, Hongbin Yu, Xiaojie Ye, Tingting Xu and Yulong Ni

Hefei General Machinery Research Institute Co., Ltd., Hefei Anhui 230031, China.

* Corresponding author's e-mail: lzrngcjq@126.com

Abstract: Aiming at the large design error of the rated flow coefficient of existing control valves, it was difficult to meet the requirements of the equipment system. This study adopts the principle of infinite approximation to the existing flow coefficient calculation equation for geometric shape correction compensation, based on which the rated flow coefficient design method was proposed, and simulation and test verification were carried out. The results show that according to the design method, the rated flow coefficient of the control valve can be accurately designed. The variation rule of the simulated value and the test value under the two kinds of spool inner parts is consistent, and the maximum error is not more than 10%, which meets the requirements of industrial standards. The experiment verifies the accuracy and applicability of the design method.

1. Introduction

With the rapid development of high-performance equipment technology and the improvement of performance parameters, there is a large error in the design of the rated flow coefficient of traditional control valves. In addition, the traditional control valve also has poor adjustment performance, low control accuracy, slow response time, limited adjustment interval, and serious erosion of valve internals under harsh conditions [1-2]. These have made it difficult to meet the system's requirements for high-precision control, fast response, lightweight and other indicators.

There were a lot of existing studies on control valves, mainly focusing on the regulation characteristics of traditional control valves in service [3-4], spool design [5-6], internal flow performance [7-8], test and engineering applications [9] and so on. However, there was little research on the modification of the rated flow coefficient of the control valve, and the relevant design methods were uneven. Therefore, based on the design of the rated flow coefficient of the traditional control valve, a revised design method of the rated flow coefficient of the jet control valve was proposed by using forward research and development. The simulation and experiment were iteratively verified, which had certain engineering practical significance to improve the accurate design of the rated flow coefficient of the control valve.

2. Structural principle

The structure principle of a high-performance jet control valve is shown in Figure 1, wherein jet internals are the core components of the control valve. The working principle of the control valve is similar to the louvre structure, and the multi-stage control window is designed on the moving and static slide. The high precision control is realised through the relative movement up and down.

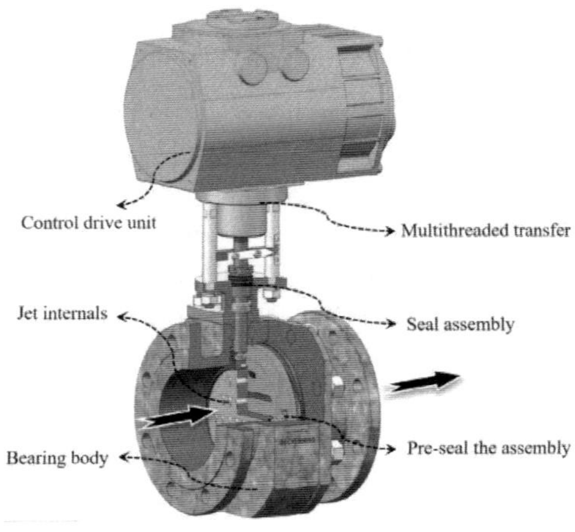

Figure 1. Structure principle of jet control valve.

3. Flow coefficient design

3.1 Derivation of flow coefficient of control valve based on bernoulli equation

The existing design theories of control valve flow coefficient are based on the evolution of the Bernoulli equation. As shown in Figure 2, the Bernoulli equation of the constant unit flow of the actual fluid is established according to the principle of conservation of energy, as shown in Equation (1).

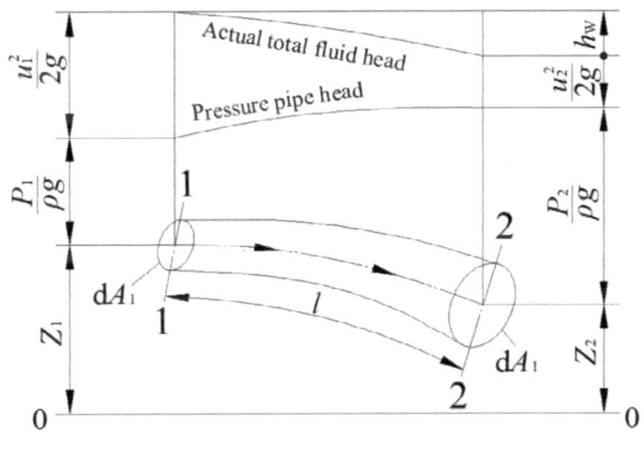

Figure 2. Geometric curve of fluid head change along the path.

$$Z_1+\frac{P_1}{\rho g}+\frac{u_1^2}{2g}=Z_2+\frac{P_2}{\rho g}+\frac{u_2^2}{2g}+h_W' \tag{1}$$

When Z1=Z2 (horizontal pipe), the capacity loss h_W' caused by friction is ignored, and Equation (1) becomes Equation (2):

$$P_1-P_2=\frac{u_2^2\rho}{2}-\frac{u_1^2\rho}{2} \tag{2}$$

According to the law of conservation of mass, the continuity equation of incompressible fluid per unit time is established, as shown in Equation (3):

$$Q=A_1u_1=A_2u_2 \tag{3}$$

where Q is the volume flow rate, A_1 and A_2 are the cross-sectional areas of the two flow sections, and u_1 and $u2$ are the flow rates through the two sections; $A_1 = \frac{\pi}{4} d_1^{\ 2}$, $A_2 = \frac{\pi}{4} d_2^{\ 2}$, d_1 and d_2 are the hydraulic diameter flowing through the two sections.

Combining Equation (2) and Equation (3) can obtain Equation (4):

$$Q = \sqrt{2g\left(\frac{P_1}{\rho g} - \frac{P_2}{\rho g}\right)} \times \frac{A_1}{\sqrt{\frac{d_1^{\ 4}}{d_2^{\ 4}} - 1}} \tag{4}$$

where the units of d_1 and d_2 are cm, $\Delta P = P_1 - P_2$, and the unit is kgf²/cm=1000 gf/cm²; g is the acceleration of gravity, which is 981 cm/s²; if $\zeta = \frac{d_1^{\ 4}}{d_2^{\ 4}} - 1 = \frac{A_1^{\ 2}}{A_2^{\ 2}} - 1$, then Equation (4) can be simplified to Equation (5):

$$Q = \frac{5.04 A_1}{\sqrt{\frac{d_1^{\ 4}}{d_2^{\ 4}} - 1}} \sqrt{\frac{\Delta P}{\rho g}} \tag{5}$$

$K_v = \frac{5.04 A_1}{\sqrt{\frac{d_1^{\ 4}}{d_2^{\ 4}} - 1}}$, and from the above, it is easy to know that $\zeta = \frac{d_1^{\ 4}}{d_2^{\ 4}} - 1 = \frac{A_1^{\ 2}}{A_2^{\ 2}} - 1$. Thus, $K_v = \frac{5.04 A_1}{\sqrt{\frac{A_1^{\ 2}}{A_2^{\ 2}} - 1}}$.

Based on the above principles, the Bernoulli equation of the actual pipeline system of the control valve is established. The cross-section of the pipeline before taking the valve is the I-I section, and the cross-section at the throttle window of the control valve is the II-II section. It is assumed that the cross-section at the throttle window is a circular cross-section.

According to IEC 60534-1, due to the inconsistency of units between K_v and C_v, the conversion relationship between them is C_v=1.156 K_v, then the flow coefficient of the control valve is expressed as Equation (6):

$$C_v = \frac{5.83 A}{\sqrt{\frac{A^2}{A_{100\%}^{\ 2}} - 1}} \tag{6}$$

where A is the area of the pipeline cross-section I-I section, $A_{100\%}$ is the area of the control valve throttling window II-II section, and the units of both are cm².

3.2 Geometric shape correction compensation

As mentioned above, two considerations have been neglected in the derivation of the flow coefficient of the existing control valve:

(a) The loss caused by medium friction is ignored.

In order to explain the principle and deduce the formula easily, many presuppositions were made during the establishment of the Bernoulli equation: (1) The mass force is only gravity; (2) Steady flow; (3) Incompressible fluid; (4) Slow variable flow; (5) The flow rate is constant. However, the actual flow inside the control valve is complex and changeable, and inevitably causes energy loss. In view of this aspect, some scholars have conducted studies and proposed an empirical correction coefficient a [10].

(b) The effect of geometry on the flow coefficient is ignored.

In the process of deriving the flow coefficient of the control valve, the Bernoulli equation of the control valve control system is established by assuming that both interrupt surfaces are ideally circular tubes to calculate (Figure 2), and the interrupt surfaces at the actual II-II regulating window of the control valve are different geometric shapes [11]. As shown in Figure 3, the red box lists the easy-e^{TM} control internals of Fisher, an internationally renowned control valve company. At the same time, for window control valves, the shape of a large number of control valve products at home and abroad has basically evolved from a regular quadrilateral and a regular trilateral. It can be seen that in the existing flow coefficient calculation equation, it is unreasonable to calculate the II-II cut-off surface as a circular section without considering the influence of geometric shape on the flow coefficient and divorced from the actual structure.

Based on this, a geometric shape correction compensation is proposed for the flow discontinuity surface at the II-II adjustment window. As shown in Figure 3, the Bernoulli equation model of the actual control valve control system was established. The pipeline cross-section in front of the valve was selected for sections I-I, and the projection plane of the control valve regulation window was selected for sections II-II.

Aiming at the projection plane, a control model with the adjustment window evolving step by step from the circle - regular12 - regular8 - regular6 - regular4 - regular3 is established by using the principle of infinite approximation under the precondition of equivalent geometric area. At the same time, η is the ratio of section II-II to section I-I, and $\eta = \frac{A_{\text{II-II}}}{A_{\text{I-I}}}$. The effects of nominal size and section ratio on flow coefficient under different geometric shapes were studied by numerical simulation method.

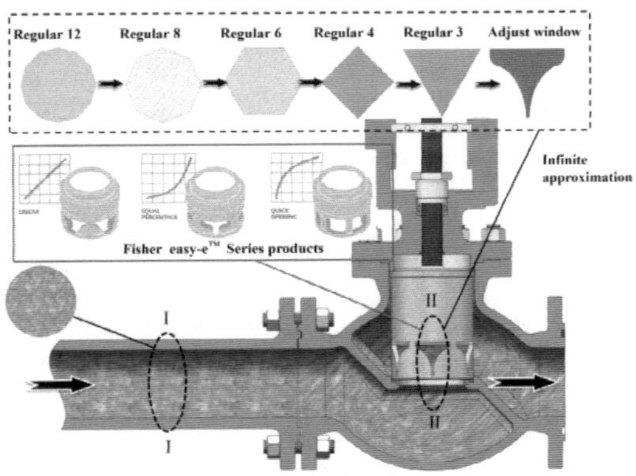

Figure 3. Bernoulli equation model of control valve control system.

Figure 4 shows the influence of nominal size and section ratio on flow coefficient under different geometric shapes. It can be seen that under different geometric shapes, the variation law of flow coefficient is basically the same with the change of nominal size, and there is no distortion due to the change of cross-sectional area ratio; that is, geometric shape is the key variable affecting the change of flow coefficient.

In order to quantitatively evaluate the degree of interference of geometric shapes on flow coefficient, based on Figure 5, the proportion of constant cross-section (take η=40%) and the flow coefficient corresponding to the circle under different nominal sizes are used as the basis to solve the proportion of flow coefficient corresponding to other geometric shapes and the relationship between the nominal size and the flow coefficient under different geometric shapes is established. It can be seen that under the same geometric shape, with the infinite approximation of the geometric shape, the proportion of the flow coefficient decreases in the lower parabola, and the smaller the nominal size is, the greater the reduction degree will be.

The data in Figure 6 were processed and analyzed, and the flow correction compensation function of the positive trilateral geometry was constructed, represented by β_T, to quantitatively characterize the degree of influence of the geometry on the flow coefficient. It is easy to know that β_T is a function related to the nominal size, and the geometric shape correction compensation function of the positive trilateral is obtained by fitting the curve, as shown in Equation (7).

$$\beta_T = 0.065\ln(d) + 0.66 \tag{7}$$

where d is the hydraulic diameter of the I-I section.

Combined with friction loss compensation, Equation (6) can be converted into Equation (8), that is, the revised flow coefficient calculation equation.

$$C_v' = \beta_T * \frac{5.83A}{\sqrt{\frac{A^2}{(A_{100\%} * \alpha)^2} - 1}} \tag{8}$$

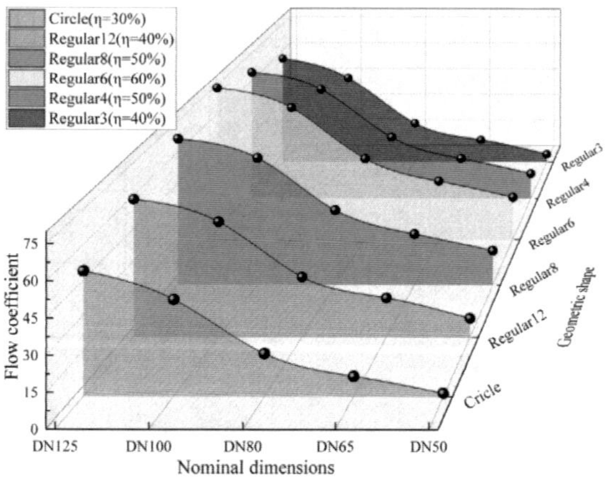

Figure 4. Effect of nominal size and cross-sectional area ratio on flow coefficient.

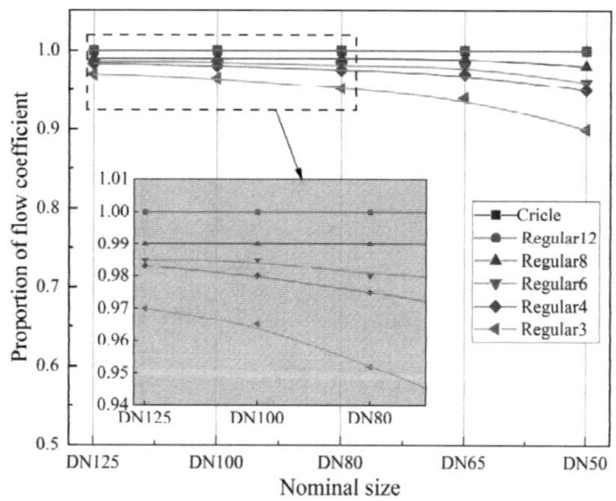

Figure 5. The relationship between the nominal size and the proportion of flow coefficient.

4. Numerical simulation

The simulation experiment is carried out based on the internal structure characteristics and working parameters of the control valve. According to the test standard regulations, the upstream and downstream of the control valve are set several times the nominal diameter of the through pipeline. To ensure the accuracy of the calculation data, the simulation analysis model is kept consistent with the real test, the impact of the valve inlet and outlet on the internal flow field is reduced, and the full development and balance of the internal flow is achieved. The control valve channel model grid is shown in Figure 7. For the convenience of comparative analysis, the numerical simulation results are shown in Section 5.2.

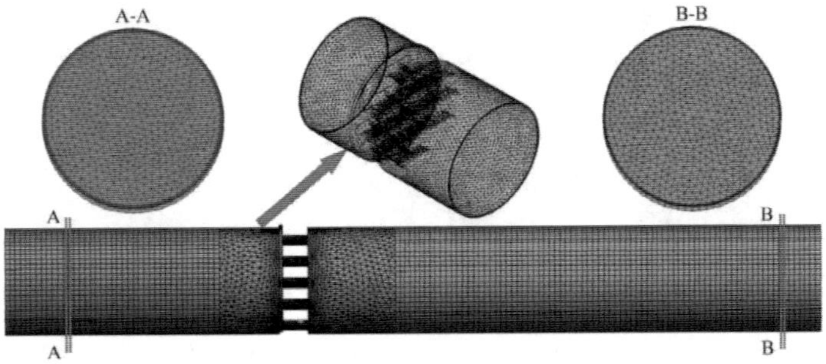

Figure 6. Grid model of control valve channel.

5. Test verification

5.1 Test principle

During the control valve test, the pressure difference before and after the test section of the valve under test is controlled and adjusted by the upper and downstream regulating valves. The test site is shown in Figure 7.

Figure 7. Rated flow coefficient test.

5.2 Analysis of rated flow coefficient

According to IEC 60534 series standards, the rated flow coefficient K_v of the control valve under the two spool structures is calculated and compared with the simulation results, as shown in Table 1:

Table1. Rated flow coefficient under different spool types.

Number of tests	Valve Spool 1#			Valve Spool 2#		
	Net differential pressure/(kPa)	Simulated value/(Kv)	Tested value/(Kv)	Net differential pressure/(kPa)	Simulated value/(Kv)	Tested value/(Kv)
Test 1	88.57	40.42	37.92	79.59	64.05	59.89
Test 2	112.45	41.28	38.42	108.44	63.78	61.24
Test 3	157.65	39.98	37.25	132.49	63.24	61.57
Test 4	188.96	40.21	38.29	164.86	64.11	59.04
Test 5	224.65	40.67	37.55	215.37	63.87	59.97
Average value	-	40.51	37.89	-	63.81	60.34

The data in Table 1 were quantitatively analyzed, and their relative errors were calculated, as shown in Figure 8 and Figure 9. Figure 8 shows the rated flow coefficient under the 1# spool structure. It is easy to know that under the five test conditions, the rated flow coefficient of the simulation and the test changes in the same law, showing a slight fluctuation trend, respectively, and the overall fluctuation is relatively stable. The simulated values are all larger than the test values, and the maximum error is 7.7%.

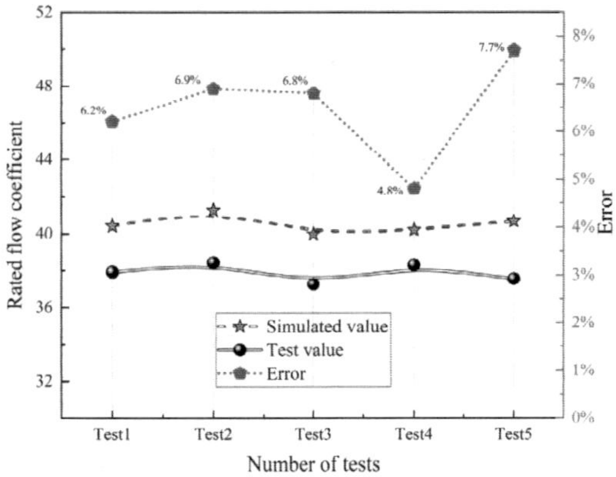

Figure 8. Comparison of rated flow coefficient under 1# spool inner.

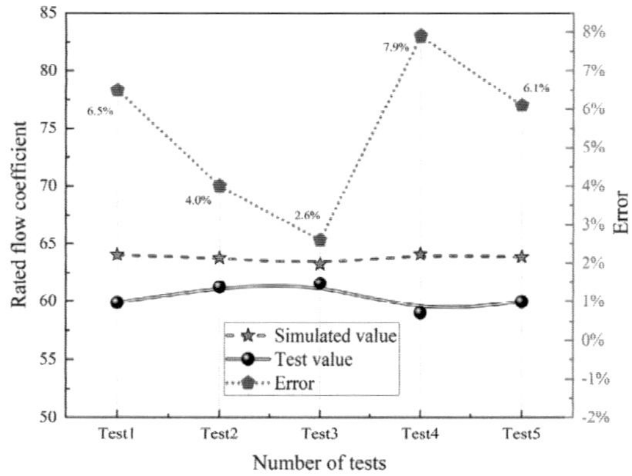

Figure 9. Comparison of rated flow coefficient under 2# spool inner.

Figure 9 shows the rated flow coefficient of the control valve under the 2# spool structure. It is easy to know that under the five test conditions, the rated flow coefficient of the simulation and the test changes in the same law, showing a slight fluctuation trend, respectively, and the overall fluctuation is relatively stable. The simulated values are all larger than the test values, and the maximum error is 7.9%.

6. Conclusion

(1) Using forward research and development, a calculation method for the rated flow coefficient of the control valve based on geometric shape compensation is proposed. The applicability of the design method is verified by numerical simulation and test, which reduces the cost of blind trial and iterative improvement in engineering and can realise the accurate design of the rated flow coefficient of the control valve.

(2) The rated flow coefficient of the output of the control valve developed according to the design method fluctuates slowly under the two spool internals, and the maximum error of the simulation and test is not more than 8%, which meets the requirements of 10% of the industry standard.

Acknowledgements

This work was financially supported by the Youth Science and Technology Foundation of GMRI (2022011501), the National Key R&D Program of China (2021YFB2500501), and the Anhui Province Key Key R&D Program(2022a05020001).

References

[1] Han X, WANG D D and WANG X W. Common problems and solutions of electric control valve selection in thermal station[C]//. Gas and heat, 2021:10.

[2] MING C D, WEN P, QIAO G C, et al. Design and patent structure analysis of cage type single-seat regulating valve[J]. Valve, 2023(06):660-668.

[3] Islam S U, Ahmed I and Din Z U. Energy Recovery Opportunity at Natural Gas Regulating Station by replacing Pressure Control Valve with Turbo Expander using Aspen HYSYS: A case study of WAH SMS (Sale Metering Station) [J]. Int. J. Innov. Sci. Eng. Technol, 2016, 3: 464-475.

[4] ZHANG H F, XU J Q, SUN Y, et al. Modeling and optimization of turbine control valve and control stage characteristics[J]. Journal of Mechanical Engineering, 2019,55(18):165-172.
[5] Li S X, Li C, Li Z, et al. Design optimization and experimental performance test of dynamic flow balance valve[J]. Engineering Applications of Computational Fluid Mechanics, 2020, 14(1): 700-712
[6] Hardy C S, Hudson C M, Harper M S, et al. Pressure is nothing without control: evolution of control valve design[M]//Fuel systems for IC engines. Woodhead Publishing, 2012: 115-128.
[7] Z Gao, Yue Y, Wu J, et al. The flow and cavitation characteristics of cage-type control valves[J]. Engineering Applications of Computational Fluid Mechanics, 2021, 15(1): 951-963.
[8] Ye J, Zhao Z, Cui J, et al. Transient flow behaviors of the check valve with different spool-head angle in high-pressure hydrogen storage systems[J]. Journal of Energy Storage, 2022, 46: 103761.
[9] LI Z, JIANG H L, FAN Y L, et al. Analysis and test on the flow capacity of a diaphragm solenoid valve used in Marine water system of a shi[J]. Mechanical design and research, 2020,36(04):57-61.
[10] WANG W B and HAO J S. Method for determining the equal percentage throttling window profile of sleeve regulating valve. Chinese patent:CN110705109A,2020-01-17
[11] Yu L J, Chen W P, Jin Z J, et al. Effect of sleeve orifices ratio on flow performance and hydrodynamic noise in the two-stage sleeve control valve[J]. Proceedings of the Institution of Mechanical Engineers, Part C: Journal of Mechanical Engineering Science, 2022, 236(14): 7705-7716.

Journal of Physics: Conference Series **2815** (2024) 012035 doi:10.1088/1742-6596/2815/1/012035

Study of self-actuated time-controlled tower microvalves

Hui Luo [1], Cuixue Ren [1] and Jingmin Li [1,*]

[1] Key Laboratory for Micro/Nano Technology and System of Liaoning Province, College of Mechanical Engineering, Dalian University of Technology, Dalian 116024, China

[*] Corresponding author's e-mail: jingminl@dlut.edu.cn

Abstract: Time-controlled microvalves are vital components of microfluidic systems, primarily used to regulate the sequential or precise introduction of reagents. However, existing time-controlled microvalves face certain limitations. For instance, those driven by magneto-thermal mechanisms require additional control elements, and fabricating some 3D time-controlled microvalves can be complex. To address these challenges, a self-driven time-controlled tower microvalve was designed and fabricated. This study investigates how structural parameters of the time-delay unit affect bubble generation and the time-delay effect. Experimental testing confirmed that the delay time of the time-controlled microvalve increases and then decreases with variations in the trapezoidal aspect ratio (a) and the upper and lower width ratios (b). Additionally, the delay time of a single unit extends as the width of the delay unit increases. The optimal parameters for the time-controlled microvalve were determined as follows: W_1=1, 000 μm, W_2=500 μm, and H=800 μm, achieving an overall delay time of 16.7±0.8 s.

1. Introduction

Microfluidics is a technology used to manipulate and process fluids at a micron scale, offering advantages such as compact size, rapid response speed, and efficient analysis [1]. It finds extensive applications in food safety [2], biomedicine, drug analysis [3], and environmental monitoring [4]. This technology enables precise fluid control through micro-devices like micro-valves and micro-pumps. Time-controlled micro-valves play a crucial role in microfluidic systems by regulating fluid movement to sequentially introduce reagents or maintain regular intervals.

Various control methods have been proposed for time-controlled microvalves, including magnetic actuation, thermal actuation, salt-based mechanisms, wax-based approaches, and capillary force actuation. Li et al. [5] introduced a novel magnetic timing valve for controlling fluid flow in paper-based microfluidic devices. Bazargan and Stoeber [6] demonstrated reversible gelation of a thermally responsive polymer solution in the control channel to deflect the membrane for flow control. However, time-controlled microvalves using magneto-thermal actuation require additional control units and pose challenges for integration into microfluidic systems. To address this, Tuccitto [7] proposed a salt-based autonomous time-delayed microfluidic valve utilizing NaCl and $CuSO_4$ crystals as actuator materials. Lai et al. [8] suggested using a surfactant to dissolve waxy hydrophobic barriers within fluidic channels. However, time-controlled microvalves employing soluble barriers for control are unstable [9], and the accompanying dissolved substances can alter fluid properties.

Capillary force-driven time-controlled microvalves, which retard fluid flow by altering the contact angle and channel structure, offer advantages such as no external energy requirement, good biocompatibility, and broad applicability. However, they have limitations related to structural complexity and stability. Ellinas et al. [10] used C_4F_8 plasma deposition to create superhydrophobic strips for microvalves. Nonetheless, inconsistent contact angles between delay regions and main channels often necessitate additional surface modifications, complicating fabrication. Xie et al. [11] proposed an efficient capillary microvalve based on microporous arrays, altering fluid-surface contact angles to slow flow. Yet, imprecise thermal embossing led to unstable flow rates. Mohammed and Desmulliez [12] developed a meandering delayed microvalve where fluid traverses a secondary channel to control timing, requiring extended secondary channels.

To overcome the complexities of existing time-controlled microvalves, a capillary-driven time-controlled tower microvalve based on capillary action is designed and fabricated. This valve delays fluid flow by inducing abrupt changes in contact angle through channel width adjustments. The paper establishes a bubble determination model, studies parameter effects on bubble generation and delay time, optimizes parameters, and designs and fabricates the microvalve. Finally, fluid experiments validate its performance. This time-delay valve has a simple structure, facilitating integration and enhancing microfluidic control system portability and practicality. This work serves as a blueprint for designing self-driven time-controlled microvalves, offering insights into enhanced microfluidic control system design.

2. Design and optimization

2.1. Structural design

Figure 1 illustrates the schematic of the time-controlled microvalve structure. Upon entry of the fluid into the microvalve from the inlet, a sudden change in contact angle induces capillary pressure from the sides, inhibiting fluid advancement. Despite this hindrance, fluid inertia and capillary forces between the top and bottom surfaces enable gradual progression until the gas-liquid interface breaches the microvalve inlet. Subsequently, the gas-liquid interface expands outward, and the fluid progressively adheres to the walls of the time-delay unit until reaching the subsequent time-delay valve unit.

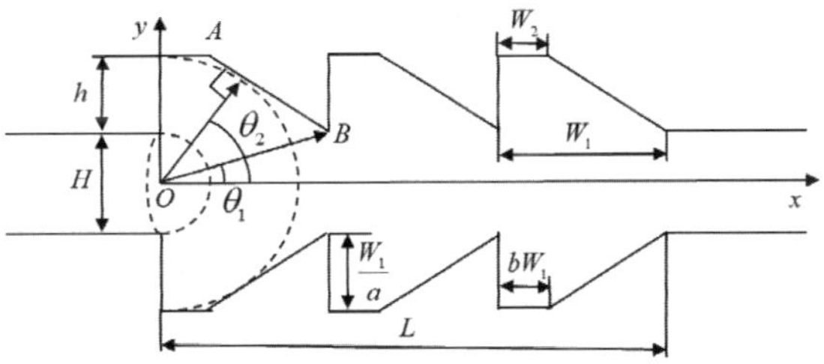

Figure 1. Schematic diagram of time-controlled microvalve structure.

To ensure that the time-controlled microvalve functions without generating bubbles, it is crucial to prevent the liquid-gas interface from coming into contact with the sloping wall of the trapezoidal structure. This contact can lead to bubble formation at the upper end of the trapezoid. For this purpose, a coordinate system is established within the time-controlled microvalve structure. A key determination is made based on this coordinate system: If the distance from Point O to line segment AB exceeds a threshold value R_1, air bubbles will not be generated. Mathematically, this condition can be expressed as:

$$\begin{cases} D_{O,l_{AB}} > R_1 = h + \dfrac{H}{2} & \theta_1 < \theta_2; \\ D_{OB} > R_1 = h + \dfrac{H}{2} & \theta_1 > \theta_2. \end{cases} \tag{1}$$

where $D_{O,l_{AB}}$ represents the distance from Point O to AB, θ_1 represents the angle between the vertical line and the x-axis, D_{OB} represents the distance between Points O and B, and θ_2 represents the

Journal of Physics: Conference Series **2815** (2024) 012035 doi:10.1088/1742-6596/2815/1/012035

angle between OB and the x-axis. By calculation, the expression for the line segment AB can be obtained as:

$$\frac{h}{(b-1)W_1}x - y + \frac{H}{2} - \frac{h}{(b-1)} = 0 \tag{2}$$

The distance from Point O to line segment AB is:

$$D_{O,l_{AB}} = \frac{\frac{(1-b)}{2}HW_1 + hW_1}{\sqrt{[(1-b)W_1]^2 + h^2}} \tag{3}$$

The distance from Point O to Point B is:

$$D_{OB} = \sqrt{W_1^2 + \frac{H^2}{4}} \tag{4}$$

2.2. Influence of structural parameters on bubble generation

The bubble-free selection of a time-controlled microvalve suitable for quantitative detection involved utilizing the orthogonal experimental method to systematically explore the relationships between bubble formation and key parameters. Three critical parameters, notch depth *h*, notch width W_1, and top-to-bottom width ratio *b*, were selected for analysis. The range of *h* was set from 100 to 300 μm, W_1 ranged from 100 to 500 μm, and *b* varied from 0.2 to 0.8. A three-factor five-level orthogonal table design was employed to explore various parameter combinations, recording the time *T* required for the liquid-gas interface to traverse the time-controlled microvalve. The simulation results are presented in Figure 2.

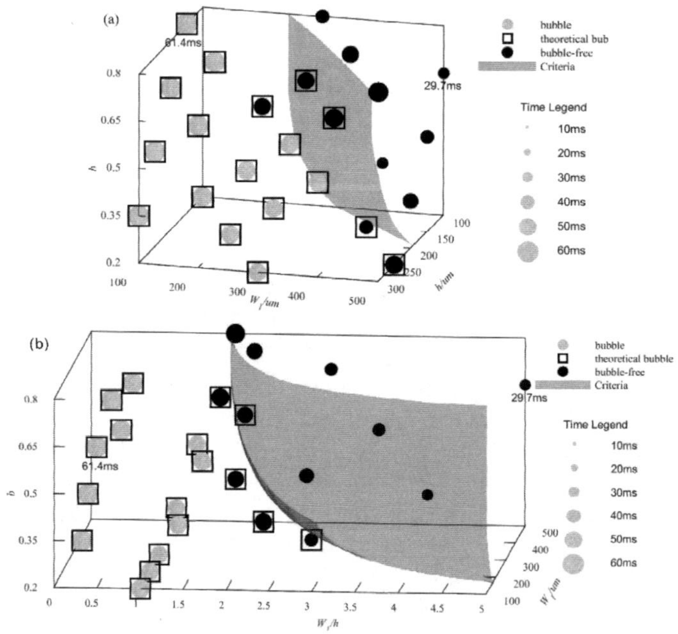

Figure 2. (a) Simulation result diagram; (b) Data diagram after changing the horizontal coordinate.

Figure 2 (a) illustrates that simulation data points resulting in bubbles are all located within the bubble determination condition, suggesting that this condition can effectively predict bubble formation during fluid flow. However, there are isolated data points without bubbles that are erroneously classified due to the fluid's tendency to follow the channel wall under capillary force influence, resulting in a half-moon-shaped liquid-gas interface rather than a semicircle. Compared to a semicircular interface, a half-moon shape allows more time for the liquid-gas interface to flow along the wall, reducing bubble formation probability and explaining discrepancies between theory and simulation.

From Figure 2 (a), data points with bubbles predominantly cluster on one side. To better investigate bubble appearance patterns, the horizontal coordinate h in Figure 2 (a) was transformed into the aspect ratio W_1/h, which gives Figure 2 (b). This figure shows that bubble regions in simulation results are primarily concentrated, where $W_1/h<1.5$. Analyzing the bubble determination conditions reveals that smaller trapezoidal widths and aspect ratios facilitate bubble formation. This phenomenon occurs because a longer h means the notch fluid takes more time to reach the upper end of the trapezoidal notch, while a smaller W_1 brings the gas-liquid interface closer to the trapezoidal beveled edge, promoting bubble formation.

2.3. Determination of structural parameters

To optimize the parameters affecting bubble formation and select the time-controlled microvalve with the longest delay time under a fixed length constraint, three key parameters were adjusted: the number of time-controlled microvalves N (calculated as $N= 3,000/W_1$), the trapezoidal width-to-height ratio a (defined as $a= W_1/h$), and the top-to-bottom width ratio b. These parameters were considered within a 3 mm length constraint. Initially, the top-to-bottom width ratio b was set to 0.5, which is associated with the minimal bubble formation based on previous relationships. All possible combinations from Table 1 were simulated, and the results are depicted in Figure 3.

Tabel 1. Simulation parameters.

Parameters	Value
Number of trapezoidal units (N)	3, 4, 5, 6, 7, 9, 12, 15, 18, 21, 24, 27, 30
Trapezoidal unit width (W_1)/μm	100, 111, 125, 142, 167, 200, 250, 333, 428, 500, 600, 750, 1, 000
Aspect ratio (a)	1, 2, 3, 4, 5
Top-to-bottom ratio (b)	0.5

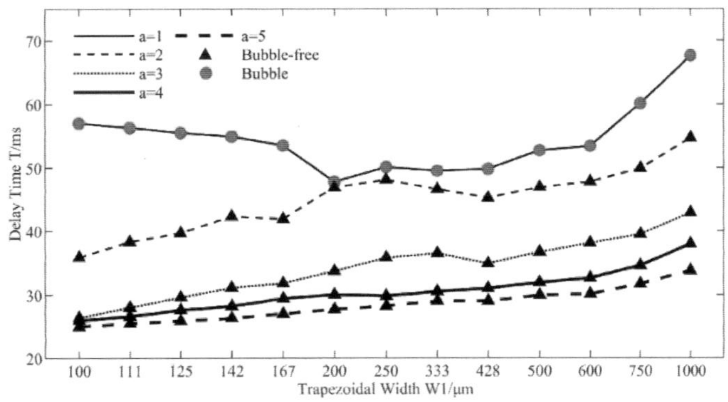

Figure 3. Effect of different a and W_1 on delay time.

As depicted in Figure 3, the combination exhibiting the poorest delay performance is (W_1=100 μm, a=5). As the trapezoidal aspect ratio a decreases, the delay time increases until a=1, at which point bubbles appear in all combinations. This trend occurs as the trapezoidal width-to-height ratio a decreases, and the trapezoidal height increases, requiring the gas-liquid interface to travel further along the trapezoid edge to return to the main channel, thus resulting in improved delay performance of the time-controlled microvalve. However, excessively high trapezoidal notches can cause the gas-liquid interface to travel too far along the main channel, leading to premature contact with the next time-controlled microvalve and bubble formation. Moreover, widening the trapezoidal notch enhances the delay effect of the time-controlled microvalve. Notably, simulation results identify the optimal combination for the time-controlled microvalve with the best delay performance and no bubbles as (W_1=1, 000 μm, a=2), achieving a delay time of 54.7 ms.

Given the rough search range for aspect ratio (a), further optimization is necessary. With the width of the delay cell (W_1) fixed at 1, 000 μm, all combinations of upper and lower width ratios (b) including 0.2, 0.35, 0.5, 0.65, and 0.8, along with trapezoidal aspect ratios (a) of 1, 1.25, 1.5, 1.75, 2, and 2.25, were selected for simulation, as illustrated in Figure 4. From Figure 4, it is evident that the combination yielding no bubbles and the best delay is (a=1.25, b=0.5), achieving a time-controlled micro-valve delay time of 77.2 ms. The dimensions of the time-delay valve cell in this optimal combination are (W_1=1, 000 μm, W_2=500 μm, H=800 μm). Decreasing the trapezoidal aspect ratio (b) weakens the delay effect of the time-controlled micro-valve. This weakening effect is attributed to the decreased value of (b), causing the trapezoidal hypotenuse to approach the origin (O-point), resulting in a shorter gas-liquid interface travel along the wall back to the main channel and thereby reducing the delay time.

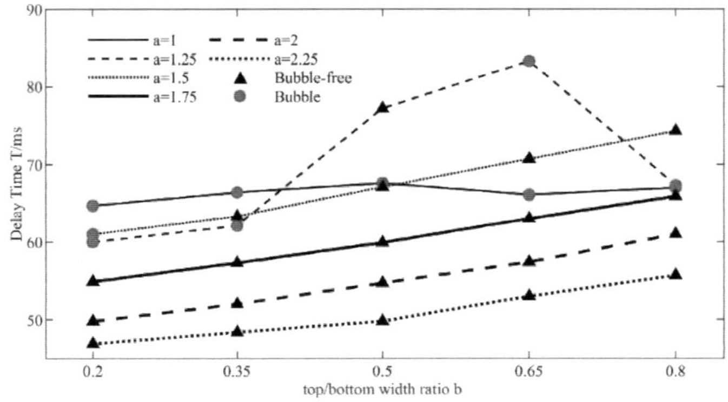

Figure 4. Effect of different a and b on delay time.

3. Experiment

3.1. Fabrication and bonding

The fabrication process of the time-controlled micro-valve is as follows: the glass substrate is wiped with an alcohol cotton ball, BN303 is spin-coated on it, and then SU8 molds are obtained through full exposure, spin-coating of SU8 adhesive, and pre-drying. According to photolithography, post-drying, and developing processes in sequence, the made SU8 molds are placed in an airtight container and silylated for 15 min. 20 ml of the PDMS pre-polymer solution are taken and injected into the SU8 molds by mixing with the curing agent in the ratio of 10:1. The PDMS was dried and cured and then de-molded to obtain the PDMS chip. The PDMS was placed in a plasma cleaner with a power of 100 w and an oxygen flow rate of 100 ml/min for 30 s. The PDMS was removed, bonded, and placed for some time to wait for the subsequent experiments, which are shown in Figure 5.

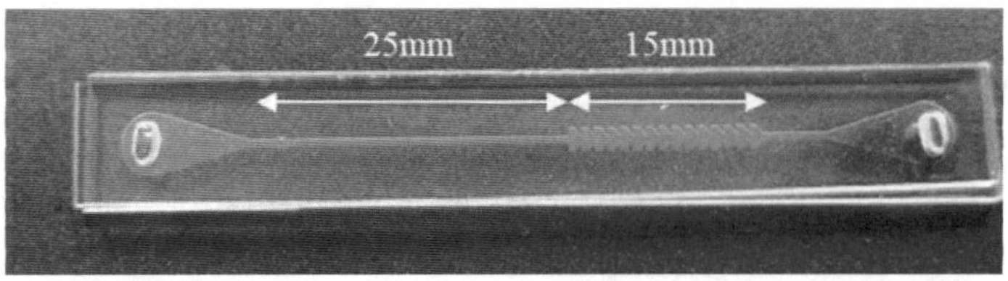

Figure 5. PDMS microvalve physical picture.

3.2. Experimental design of time-controlled microvalve fluids

To analyze the influence of each parameter on the performance of time-controlled microvalves, single-parameter fluidic experiments were designed. The main channel of the PDMS chip has a height of 100 μm and a width of 500 μm. The experiments focused on specific parameter combinations while keeping other variables constant. Firstly, when the delay cell width W_1 was set to 1, 000 μm and the upper and lower width ratio *b* was 0.5, different trapezoidal aspect ratios *a* (1, 1.25, 1.5, 1.75, 2, and 2.25) were investigated to study their effect on the delay performance of the time-controlled micro-valve. The experimental findings are depicted in Figure 6. Similarly, for a delay unit width W_1 of 1, 000 μm and a trapezoidal aspect ratio *a* of 1.25, various upper and lower width ratios *b* (0.2, 0.35, 0.5, 0.65, and 0.8) were studied to assess their impact on delay performance. The experimental outcomes are illustrated in Figure 7. Additionally, to analyze the effect of delay valve width W_1 on the delay time of the time-controlled micro-valve while maintaining a constant length, experiments were conducted with a delay valve length of 15 mm, trapezoidal height of 800 μm, and trapezoidal upper width of 500 μm. The delay valve width W_1 was varied (800, 900, 1, 000, 1, 100, and 1, 200 μm) to evaluate its influence on delay performance. The experimental data are presented in Figure 8.

4. Results and discussion

4.1. Influence of aspect ratio (a) on delay effect

The experimental results illustrating the effect of the trapezoidal aspect ratio of the time-controlled microvalve on the time-delay effect are presented in Figure 6. The figure indicates that as the aspect ratio increases, the delay time of the time-controlled microvalve initially increases and then decreases. This trend can be attributed to the relationship between the width-to-height ratio *a*: a smaller *a* results in a larger trapezoidal height, facilitating easier contact between the gas-liquid and the beveled edge, which leads to bubble formation and subsequently shorter delay times. As *a* increases to 1.25, the trapezoid height decreases, allowing the gas-liquid to reach the upper edge of the trapezoid and flow along the wall towards the next micro-valve unit, resulting in a longer delay time. At this aspect ratio, the delay time of the delay valve is measured at 16.7±0.8 s. Further increasing *a* moves the gas-liquid interface to the upper edge of the trapezoidal surface more quickly, leading to an overall reduction in delay time.

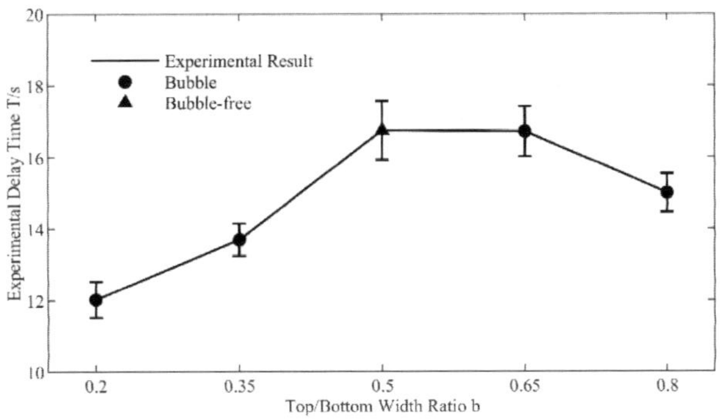

Figure 6. Influence of aspect ratio *a* on delay time.

4.2. Influence of upper and lower width ratios (b) on the delay effect

The effect of the upper and lower width ratios (*b*) of the time-controlled microvalve on the delay time is depicted in Figure 7. Upon examining the data in the figure, it is evident that as the upper and lower width ratios increase, the delay time of the time-controlled microvalve initially increases and then decreases. Notably, bubbles appear at all data points except for *b*=0.5. When *b* is small, the angle between the beveled surface and the main channel is also small, making it more likely for the expanding gas-liquid interface to contact the trapezoidal beveled wall first, leading to bubble formation. A smaller *b* value accelerates bubble formation, causing the fluid to move faster along the beveled surface towards the next time-controlled micro-valve, resulting in a shorter delay time. Conversely, larger values of *b* can also result in bubble formation. Although the gas-liquid interface flowing along the right-angled edge has sufficient time to reach the upper trapezoidal wall, an excessively long upper trapezoidal wall delays the fluid's arrival at the beveled wall. Consequently, when the gas-liquid interface in the middle of the channel reaches the beveled edge of the trapezoid, a large *b* value is also prone to bubble formation.

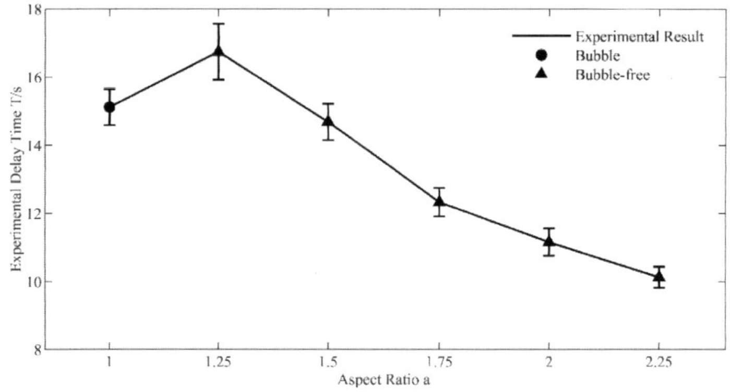

Figure 7. Influence of upper and lower width ratios *b* on delay time.

4.3. Effect of delay cell width on delay effect

The impact of time-controlled microvalve width (W_1) on delay time is illustrated in Figure 6. When examining the delay time of individual units, it is observed that the overall delay time increases as the width of the time-controlled microvalve unit increases. This increase is attributed to the fact that a wider delay valve requires more fluid to fill, resulting in an increased delay time.

However, upon analyzing the overall delay time of the microvalve with a length of 15 mm, it is noted that the delay time first increases and then decreases with increasing cell width. This behavior is explained by the proximity of the beveled edge of the trapezoid to the inlet of the delay valve. When the width of the trapezoidal cell is small, the distance between the beveled edge and the inlet is close, causing the fluid to inadequately fill the trapezoidal area and leading to bubble formation, thereby reducing the delay time of the valve. Conversely, with a larger delay valve width, although more area needs to be filled with fluid, the number of delay valves within the same length decreases. This reduction in the number of valves results in a smaller delay time for a larger valve width under the same length.

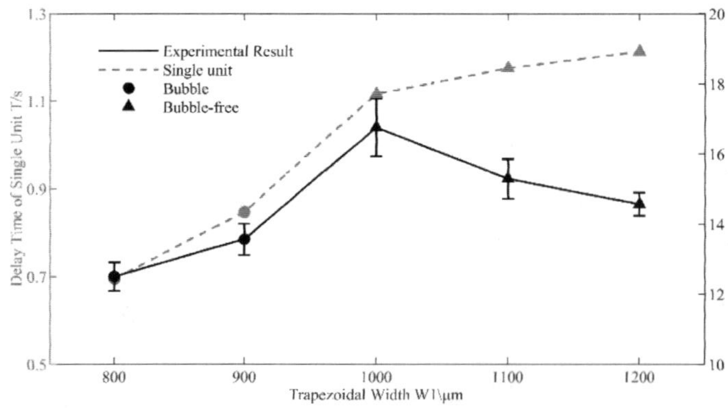

Figure 8. Effect of delay cell width on delay time.

5. Conclusion

In this study, a self-driven time-controlled tower microvalve was developed based on capillary theory to delay fluid flow by adjusting the channel width to alter the contact angle. A bubble determination model was established, and the effects of time-controlled microvalve parameters on bubble generation and delay time were investigated. The findings indicate that the smaller width-to-height ratio (a) and top-to-bottom width ratio (b) of the delayed valve unit result in increased bubble formation. As the trapezoidal aspect ratio (a) and upper and lower width ratio (b) increase, the delay time initially increases and then decreases. Moreover, the delay time of a single unit increases with the trapezoidal width (W_1), but the delay time of the valve at equal length follows a trend of increasing and then decreasing. For the delay valve parameters (W_1=1, 000 μm, W_2=500 μm, H=800 μm), the maximum delay time achieved was 16.7 ± 0.8 s. This time-controlled microvalve features a simple structure that is easy to integrate, and its delay time can be further adjusted by varying the number of delay units (N), notch width-to-height ratio (a), and top-to-bottom width ratio (b). Consequently, this design is suitable for microfluidic devices requiring sequential or timed addition of reagents, meeting the needs of quantitative detection on microfluidic chips. This study provides important insights for the design of microfluidic time-controlled microvalves.

Acknowledgments

The authors sincerely acknowledge the financial support from the National Key Research and Development Program of China (Grant No. 2020YFB2009000).

References

[1] Li Z., Liu C., and Sun J. (2023). Hydraulic – an electric analogy for design and operation of microfluidic systems. Lab on a Chip, 3, 311-3, 327. [CrossRef]

[2] Nunes J. K. and Stone H. A. (2022). Introduction: Microfluidics. Chemical Reviews, 122 (7): 6, 919-6, 920. [CrossRef]

[3] Cui P. and Wang S. (2019). Application of microfluidic chip technology in pharmaceutical analysis: A review. Journal of Pharmaceutical Analysis, 9 (4): 238-247. [CrossRef]

[4] Dhar B. C. and Lee N. Y. (2018). Lab-on-a-chip technology for environmental monitoring of microorganisms. Biochip Journal, 12: 173-183. [CrossRef]

[5] Li X., Zwanenburg P., and Liu X. (2013). Magnetic timing valves for fluid control in paper-based microfluidics. Lab on a Chip, 13 (13): 2, 609-2, 614. [CrossRef]

[6] Bazargan V. and Stoeber B. (2009). A thermally actuated microfluidic relay valve. TRANSDUCERS 2009-2009 International Solid-State Sensors, Actuators and Microsystems Conference, Denver, CO, USA 1, 289-1, 292. [CrossRef]

[7] Tuccitto N. (2016). A physicochemical timer: autonomous time-delayed microfluidic valves for environmental applications. Microfluidics and Nanofluidics, 20 (9): 129. [CrossRef]

[8] Lai Y., Tsai C., Hsu J., and YW Lu (2019). Microfluidic time-delay valve mechanism on paper-based devices for automated competitive ELISA. Micromachines, 10 (12): 837. [CrossRef]

[9] Chen Y., Wang J., Chen Y., Xiao C., and You H. (2023). A Sample-Triggered Microvalve Based on Material Property Regulation Mechanism for Reagent Storage and on-Demand Release. Advanced Engineering Materials, 25 (12): 2, 300, 127. [CrossRef]

[10] Ellinas K., Tserepi A., and Gogolides E. (2014). Superhydrophobic passive microvalves with controllable opening threshold: Exploiting plasma nanotextured microfluidics for a programmable flow switchboard. Microfluidics and Nanofluidics, 17: 489-498. [CrossRef]

[11] Xie Y., You H., Gao Z., Huang Z., and Yang M. (2018). An Effective Capillary Valve Based on Micro-hole Array for Microfluidic Systems. Analytical Sciences, 34 (11): 1, 323-1, 327. [CrossRef]

[12] Mohammed M. and Desmulliez M. P. Y. (2013). The manufacturing of packaged capillary action microfluidic systems using CO_2 laser processing. Microsystem Technologies, 19: 809-818. [CrossRef]

Journal of Physics: Conference Series **2815** (2024) 012036 doi:10.1088/1742-6596/2815/1/012036

Thermostatic interaction analysis of electric wheel brakes

Yan Yan*, Wenzhe Wang, Liangqiang Cao, Peng Hang, Xianlan Fu, Tao Huang

Department of Vehicle Technology, Air Force Early Warning Academy, Wuhan, Hubei, 430000, China

*Corresponding author's e-mail: 284528454@qq.com

Abstract. The electric wheel needs to integrate the hub motor and brake in the wheel hub, so the integration scheme and structural parameters of each component in the electric wheel determine whether the electric wheel can coordinate the operation and the quality of the running performance. In this paper, the thermosolid interaction analysis of the brakes of electric wheels is carried out through ABUQUS software, so as to judge that the performance of the brakes meets the requirements, which will play a decisive role in the safety performance and life limit of the automobile, so as to effectively improve the working stability of electric wheels.

1. Introduction

Electric wheeled vehicles can not only realize the perfect integration of the car's lightness of steering and the driver's road sense but also make the safety and flexibility of the car coordinated, boasting good application prospects and technical development potential. At present, most of the research focuses on the electric drive technology of electric wheel vehicles, and there are few studies specifically for the performance analysis of electric wheel brakes.

Liu [1] performed brake performance analysis for distributed electric vehicles and simulation analysis of regenerative braking simulations. Wan [2] established a braking dynamics model of multi-axle vehicles and studied the influence of multi-axle braking force distribution and brake performance on braking performance. Under the premise of regenerative braking, Jiang [3] proposed a method of lightweight braking and improved its performance. Xu [4] used finite element analysis to analyze the temperature and stress changes of the brake to confirm that its structure and performance met the requirements.

During the operation of the electric wheel, the temperature of the brake will rise sharply when the friction brake is used, which may cause thermal recession and even thermal fatigue and thermal cracking. In thta case, the brake will fail and ultimately affect the braking effect of the electric wheel. Therefore, when braking, it is necessary to take into account the change in temperature. It is also required to perform a thermo-solid interaction analysis of the brake in the electric wheel to verify the rationality of the structure.

2. Finite element analysis of braking

A 3D model of the motorized wheel structure is shown in Figure. 1.

Journal of Physics: Conference Series **2815** (2024) 012036 doi:10.1088/1742-6596/2815/1/012036

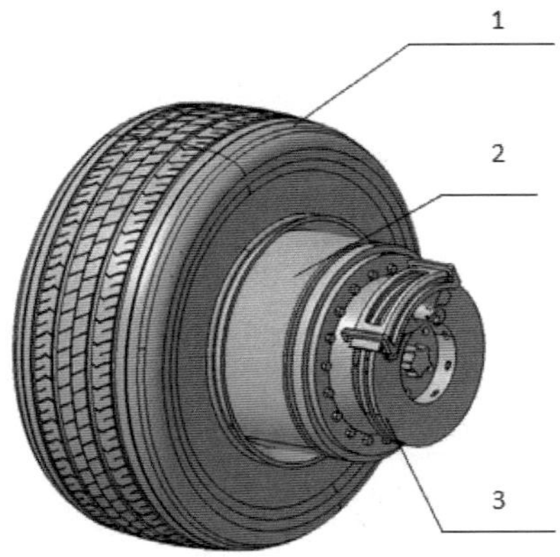

Figure. 1. Electric wheel assembly diagram.

1-tire; 2-hub motor, including an external rotor, inverter, stator, etc.; 3-brake.

2.1. Establishment of finite element modeling

When performing finite element analysis of brakes, the analysis of the model can be simplified. In the process of braking the automobile, the stress situation and constraints on both sides of the brake disc can be regarded as symmetrical, and the heat generated by friction braking is basically absorbed by the brake and the friction pad[5]. Therefore, in order to reduce the amount of calculation, the model is simplified, and the simplified geometric model is shown in Figure. 2.

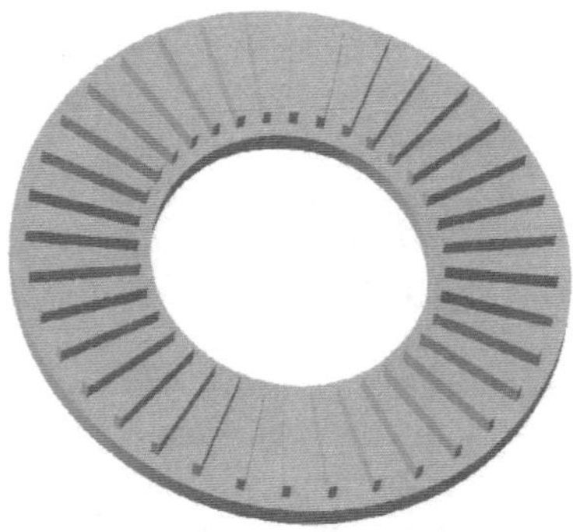

(1) Simplified model of brake disc

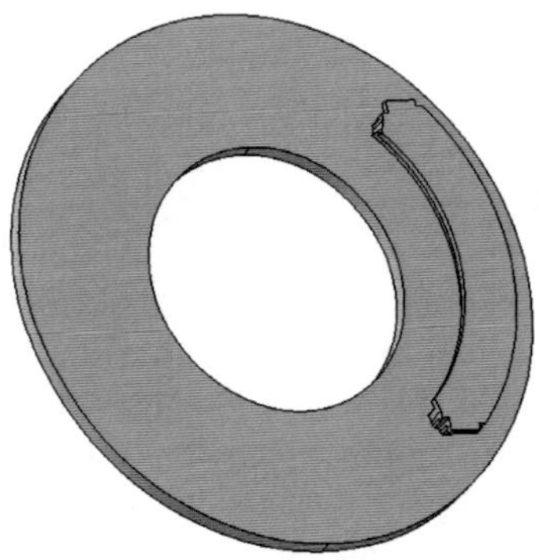

(2) Mesh model

Figure. 2. Geometry.

2.2. Diagram of the maximum temperature field distribution and the maximum stress moment

In finite element analysis, extreme values are an essential consideration[6], so firstly, the contour of the maximum temperature field distribution and the maximum stress contour are analyzed, as shown in Figure. 3.

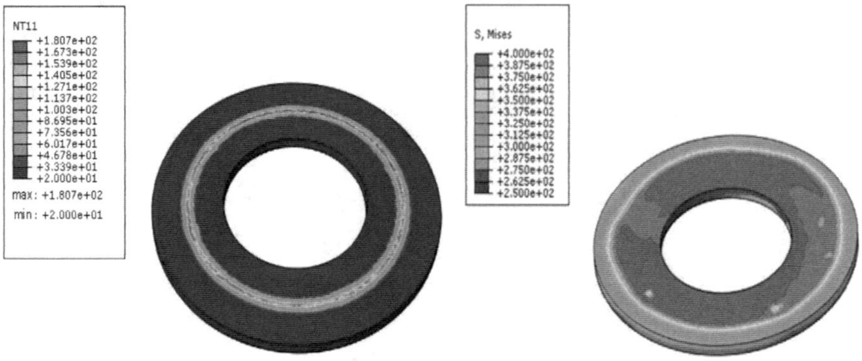

(1) Diagram of the time when the temperature is at its highest

(2) Contour diagram of the moment when the stress is at its highest

Figure. 3. Extreme distribution contours.

As can be seen in Figure. 3, both the maximum stress and the maximum temperature occur in the area where the brake disc is in contact with the friction block. The maximum stress is 400MPa, and the maximum temperature is 180.7°C. The material used in this article is HT250, and its tensile limit is 560MPa. The maximum temperature is also within the safety range of the material, which meets the requirements of the material.

3. Brake thermostatic interaction analysis

3.1. Characteristics of the temperature field distribution of the brake disc

In order to know more clearly the trend of temperature field distribution with time, six representative time points were selected[7], and their temperature contours are shown in Figure. 4. The temperature of the brake disc rises rapidly before 1.5s, from the initial temperature of 20°Cto 110.3°C, from 1.5s to 2.8s, the temperature increases from 110.3°C to 180.7°C, but from the 2.8s, the temperature has begun to show a downward trend, and by 3.5s, the maximum temperature of the brake disc is 163.7°C. When the car brakes, the reason why the temperature rises first at a lower level is that in the early stage of braking, the angular velocity of the brake disc is high, the heat flux density is high, the heat production speed is greater than the heat dissipation speed, and the heat begins to accumulate inside the brake disc, resulting in the temperature rise. In the later stage of braking, the vehicle speed decreases, the heat flux density is low, the heat production begins to be less than the heat dissipation, and the temperature of the brake disc begins to drop[8].

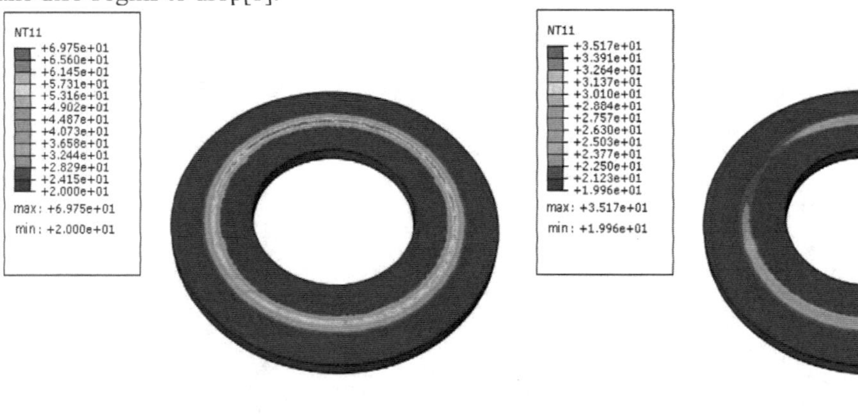

(1) t=0.5s

(2) t=1.0s

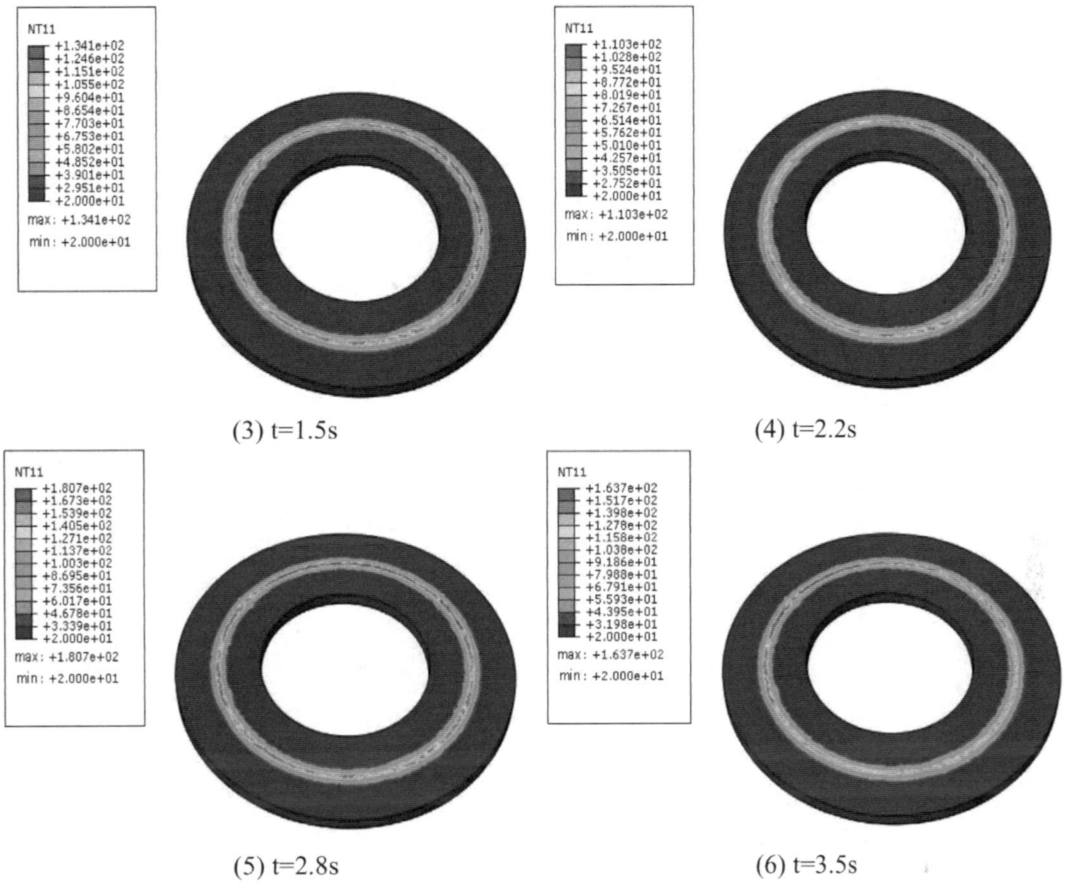

Figure. 4. Temperature field distribution contour map.

In order to describe the temperature change law with time more clearly, nine representative nodes were taken in the radial and axial directions of the brake disc. Node 1 is located on the innermost ring of the brake disc, node 2 has a radius of 115.8mm, node 3 is located at 130.5mm, node 4 is located at 162.7mm (near the friction radius), node 5 is located at 184.2mm, and node 6 is located on the outer ring of the brake disc. These six nodes are all located on the friction surface of the brake disc tangent on the same radius; nodes 9 and 6 are located at the axial node with a radius of 200mm, where node 7 is 2.1mm from the friction surface, node 8 is 4.7mm from the friction surface, and mode 9 is 8.3mm from the friction surface.

As can be seen from Figure. 5, the temperature curve of the five nodes 2, 3, 4, 5, and 6 in the braking process is obviously jagged. The trend of nodes 3, 4, and 5 is more pronounced, indicating that the temperature of the nodes fluctuates periodically during the braking process, and the period of shear fluctuation is getting longer and longer, which is caused by the continuous movement of the friction heat source and the slowing down of the moving speed with the time of braking. Among the six nodes, the temperature of node 4 is the highest, with the most apparent change because node 4 is located near the friction radius and is closest to the heat source. The temperature of node 6 of the brake is not obvious compared to nodes 3, 4, and 5, and the temperature is low because node 6 is on the outermost side, the friction isolation range is long, and node 6 is easy to dissipate heat. Node 1 is located on the inner ring of the brake disc, which is the farthest radial distance from the heat source. The temperature change is not obvious, and the temperature is the lowest[9].

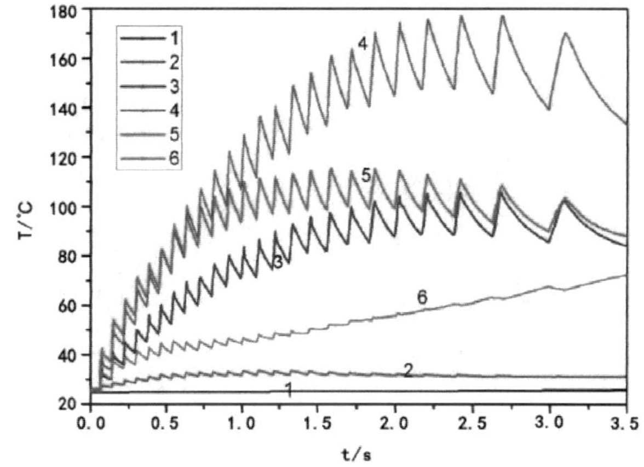

Figure. 5. Temperature curves of six radial nodes with time.

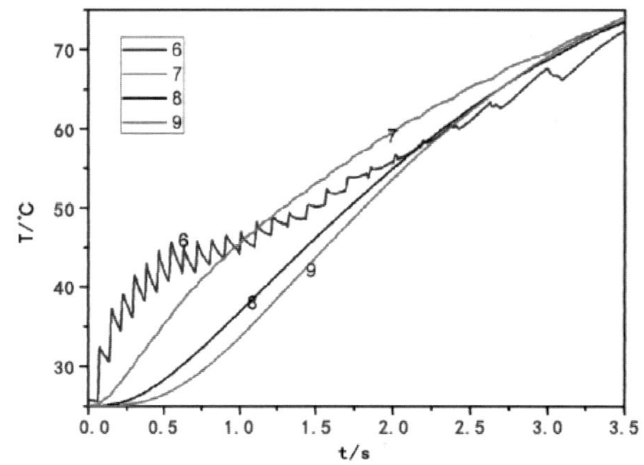

Figure. 6. Temperature curves of four axial nodes with time.

As can be seen from Figure. 6, there is a temperature difference between nodes 6, 7, 8, and 9 in the axial direction of the brake disc, and the temperature of nodes 7, 8, and 9 inside the contact surface is significantly lower than that of node 6, which is due to the fact that the speed of frictional heat generation is much greater than the speed of heat conduction. The curve of node 6 is "zigzag" and very obvious because node 6 is on the surface of the brake disc, which is not only subject to thermal shock but also forms thermal convection. On the other hand, the temperature curves of nodes 7, 8, and 9 are not zigzag because they are all inside the brake disc, are not affected by thermal shock and convection, and have a far separation area. After 2.5 seconds, the temperature of nodes 7, 8, and 9 is still rising, and the upward trend is greater than that of node 6, which is due to the slow heat conduction speed and delay.

3.2. Stress distribution characteristics of brake discs

3.2.1. Equivalent stress distribution characteristics of brake discs. In order to clarify the trend of the stress of the brake disc with time, six representative time points were selected, and the equivalent stress distribution was obtained, as shown in Figure. 7. It can be seen that the stress of the brake disc mainly appeared near the friction region, which was similar to the distribution of the temperature field, and the equivalent stress first increased and then decreased, and the maximum equivalent stress was 400MPa at doi:10.1088/1742-6596/2815/1/012036

t=3s. The friction zone attachment of the brake disc, the temperature is high, the radial temperature gradient is large, the deformation is uneven, and the pressure between the friction block and the brake disc exacerbates this unevenness, so that the maximum value of the equivalent stress appears in the friction range. By comparing the equivalent stress distribution and temperature change, it can be found that the maximum stress value appears in t=3s, the maximum temperature value appears in t=2.8s, and the maximum value of the two does not appear at the same time. The former appears a little later, and since the thermo-solid interaction process of the brake is very complex, it is affected by many factors, such as temperature, pressure, material, etc.

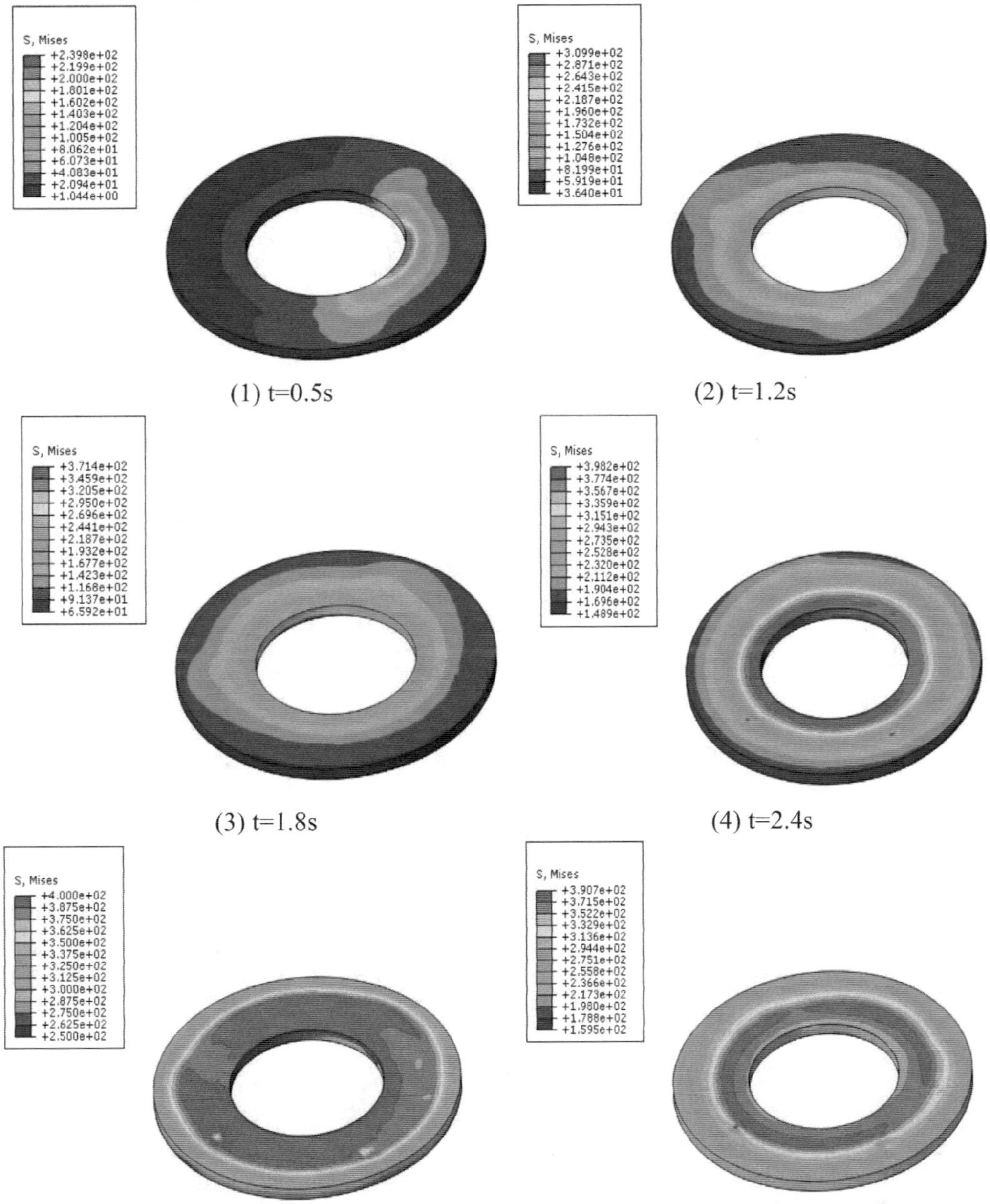

(1) t=0.5s (2) t=1.2s

(3) t=1.8s (4) t=2.4s

(5) t=3s (6) t=3.5s

Figure. 7. Equivalent stress distribution contour.

In order to further understand the relationship between stress and time, eight points used in the temperature field analysis were used for analysis. Figure. 8 is the equivalent force diagram of the six nodes radially on the friction surface on the brake disc. The equivalent stress is basically rising first and then decreasing, the stress curve fluctuates more violently with time, and the zigzag shape is irregular, which is affected by the moving heat source, and the temperature fluctuates, which will lead to the change of the contact state and the delay of heat conduction, making this zigzag shape irregular. As you can see from the diagram, the stress is the greatest at node 4, which is due to the fact that node 4 is near the friction region, the temperature is higher, the deformation is large, and the stress is high, followed by node 5. Node 1 is located in the inner ring of the brake disc, and when deformed, it is affected by the installation constraints, so the temperature and stress distribution of this node varies[10].

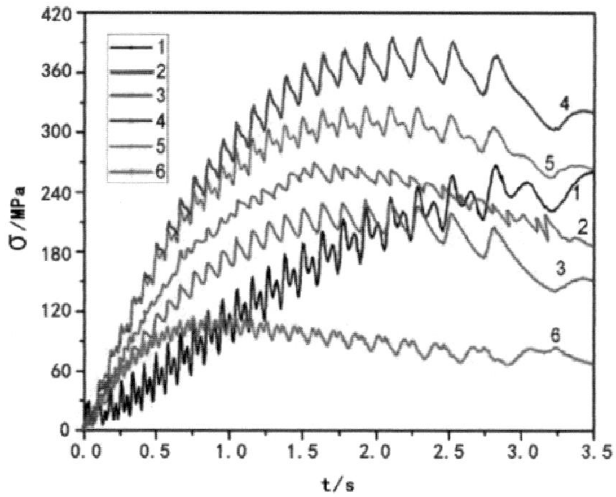

Figure. 8. Stress curves of six radial nodes with time.

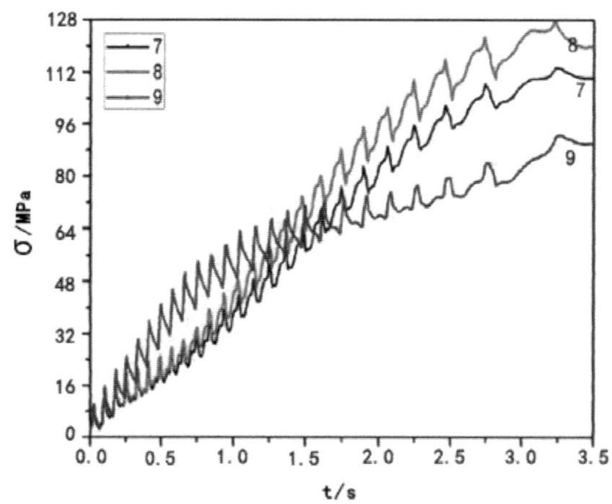

Figure. 9. Stress curves of three axial nodes with time.

The equivalent stresses of the three axial nodes 7, 8, and 9 are shown in Figure. 9. Due to the delay of heat transfer, the stress increases from the beginning of braking to the end of braking, and there is no phenomenon of first increasing and then decreasing. Node 9 is close to the vent hole, the heat dissipation is good, the temperature is low in the later stage of braking, and the stress is small.

3.2.2. Three-way stress distribution characteristics of brake discs. In order to know the stress distribution of the brake disc more clearly, the most representative four nodes were taken as the stress distribution curves of radial stress (S11), circumferential stress (S22), and axial stress (S33), as shown in Figure. 10. It is evident that the distribution of the three-way stress and the temperature field is similar, showing a jagged shape. The radial stress and axial stress are small, and the fluctuation amplitude is small, while the circumferential stress is much larger than the radial stress and axial stress, and the fluctuation amplitude is larger. High temperatures in the friction zone lead to temperature gradients in the radial and circumferential directions, resulting in large compressive stresses. In the axial direction, the temperature of the friction surface is higher than the temperature inside it, which also results in a large temperature gradient, resulting in a large compressive stress and a zigzag curve of the stress field[11].

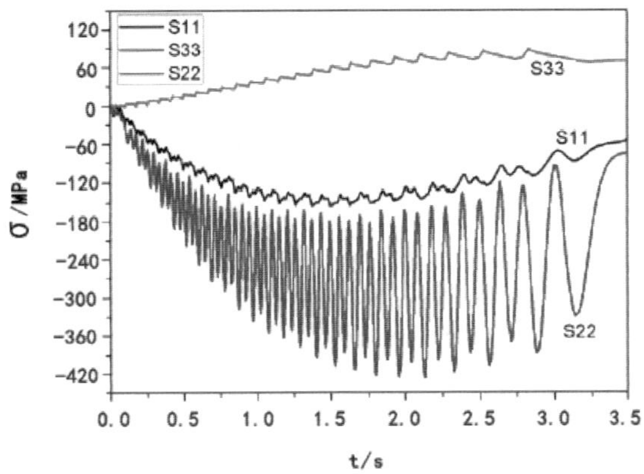

Figure. 10. Three-way stress curve of 4 nodes with time.

3.3. Thermal deformation characteristics of a brake disc

In the braking process of the brake, the distribution of the temperature field, the distribution of stress field, and the pressure distribution on the surface of the brake disc are uneven, and there is a large temperature gradient and stress gradient, so the deformation of the brake disc is uneven thermal deformation. As shown in Figure. 11, the thermal deformation of the brake disc has strong inhomogeneity in the radial direction. The deformation increases sequentially from the inner ring to the outer ring, because the outer side of the brake disc is in a state of free deformation[12]. In contrast, the inner ring is constrained, but the deformation is within a safe range to meet the use requirements.

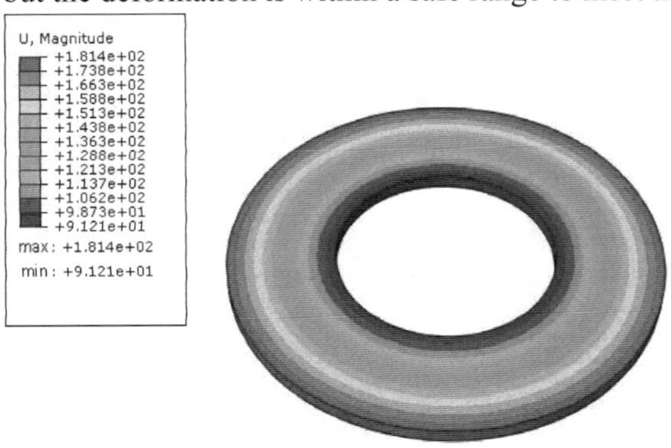

Figure. 11. Maximum displacement distribution.

4. Conclusion
In this paper, the performance of the brake under the condition of emergency braking of an electric wheel is studied by using ABUQUS software. The temperature change and stress distribution characteristics of the brake disc are obtained according to the thermosolid interaction analysis of the brake disc. The results show that the strength requirements and high-temperature resistance of the brake meet the performance requirements of electric wheel braking. The research in this paper can effectively determine the performance requirements of electric wheel brakes and improve the working stability of electric wheels.

References
[1] Liu Xinwen. Research on regenerative braking of distributed drive electric vehicles[D]. Chongqing: Chongqing University, 2019.
[2] Wan Zhen, Gao Feng, et al. Dynamic model and braking performance analysis of multi-axle vehicle braking[J]. China Mechanical Engineering, 2008, 19
[3] JIANG Xun. Lightweight design of electric vehicle braking system based on regenerative braking[D]. Wuhan: Wuhan University of Technology, 2018.
[4] Xu Junbo. Thermal structure coupling analysis and optimization of disc brakes for electric vehicles[D]. Hefei: Anhui University of Technology, 2020.
[5] XIE Dan. Design and analysis of high-speed electric wheel structure[D]. Chongqing: Chongqing University, 2021.
[6] Wang fang. Development of new energy vehicles[EB/OL]. New Energy Vehicle Network, 2022. https://gss3.bdstatic.com.
[7] Yang Weihua, Li Yourong, Fang Zifan, et al. Design of new multifunctional electric wheels and simulation of vehicle dynamic performance[J]. Science Technology and Engineering, 2016, 14(12): 113-117.
[8] Lv Fengyang. Research on coordinated control strategy of regenerative braking and pneumatic braking for pure electric buses[D]. Changchun: Jilin University, 2018.
[9] Wang Zhengxiang. Static and dynamic strength analysis and experimental study of front axle of heavy-duty truck[D]. Hefei: Anhui University of Science and Technology, 2016.
[10] Ultra Light Steel Auto Suspension Final Report. Lotus Engineering Services, Inc. 2013.
[11] Matsumoto, Masahide, Abe.Structural Optimization for Configuration Design of the motorcycle body. Proceedings of the 6th International Pacific Conference On Automotive Engineering, 2009, 12: 1399-1406.
[12] Matsumoto, Masahide, Abe.Structural Optimization for Configuration Design of the motorcycle body. Proceedings of the 6th International Pacific Conference On Automotive Engineering, 2016, 8: 1399-1406.

Journal of Physics: Conference Series **2815** (2024) 012037 doi:10.1088/1742-6596/2815/1/012037

Fault diagnosis and analysis of automobile capacitive touch display false triggering

Tingting Liu [1], Ling Mao [1], Xiaoling Chen [1], Zhifeng Li [2] and Jing Nie [1,*]

[1] Department of Mechanical and Electrical Engineering, Wenhua College, Wuhan 430074, Hubei, China
[2] Technology Center, Valeo Automotive Internal Control (Shenzhen) Co., LTD., Shenzhen 518103, Guangdong, China

*Corresponding author's e-mail: 137154911@qq.com

Abstract. Nowadays, automobile handling is mainly concentrated in the central control display. The use of high-definition and large-size capacitive touch displays has become the mainstream trend. Driving information and entertainment information are displayed on it, and the operation of the electronic control function of the car is generally realized by touching the touch panel of the central control display. As the integration of the central control display becomes higher, the fault of the display needs to be paid more attention to. The common faults include display faults and touch faults. The display fault will affect the display of vehicle information, and the driver cannot obtain vehicle status information; the touch fault will affect the car operation and can not carry out human-machine interaction. Therefore, once the central control display fails, it will affect the safety driving of the car. This paper studies the false triggering failure which is a kind of touch fault of the vehicle's 9-inch capacitive touch display. Firstly, we test the parasitic capacitance of faulty parts and repeat the touch failure. After testing and analyzing, it has been found that the mechanical structure design of the display and the air gap bonding process can change the value of parasitic capacitance which leads to false triggering. This paper proposes solutions to solve this failure which are changing the wideness of glue to avoid glue overflow and strengthening size management to improve the spacing between the touch panel and the TFT display screen. After the implementation of the solutions, the false triggering failure no longer occurs, which proves that the solutions are effective and feasible.

1. Introduction

The automobile central control display is an important human-machine interaction window between the car and the driver. With the rapid development of touch and LCD technology and the continuous update in the field of automotive electronics, more intelligent applications and functions of modern cars are integrated into the central control display, such as speed, navigation, and other driving information, as well as weather and entertainment information [1]. The onboard display system is gradually developing in the direction of intelligence, multimedia, and networking [2]. Currently, a 7/8 inch display is the standard of entry-level models, a 10-inch screen is the mainstream configuration of many OEMs, 12-inch and above size display is the configuration of high-equipped models, and even manufacturers have launched a triple screen display, large size central control display is popular to consumers.

Common faults in the central control display include black screens, flash screens, false triggering, and touch failures. Once the display fails, the human-machine interaction will not be performed, which will also bring great hidden dangers to safe driving [3]. The fault of the central control display of modern cars needs to be resolved in time and eliminate the occurrence of known problems. Therefore, the structural design and software security of the automotive central control display needs to attract more attention, and the manufacturing process also needs to be continuously upgraded to ensure its reliability and stability. This paper will take the false triggering failure of a 9-inch capacitive touch display on a certain model as an example, analyze the root cause, and propose solutions.

2. Touch technology and structure design of automotive capacitive touch display

2.1. Projective mutual capacitive touch technology

At present, automobile central control display generally adopts projective mutual capacitive touch technology. The projective mutual capacitive touch display adopts a double-layer ITO structure to form a matrix-distributed electrode with X-axis and Y-axis crossed vertically [4], as shown in Figure 1. The upper X axis is the induction electrode RX, the lower Y axis is the driving electrode TX, and mutual capacitance Cm is generated between the two electrode matrices [5].

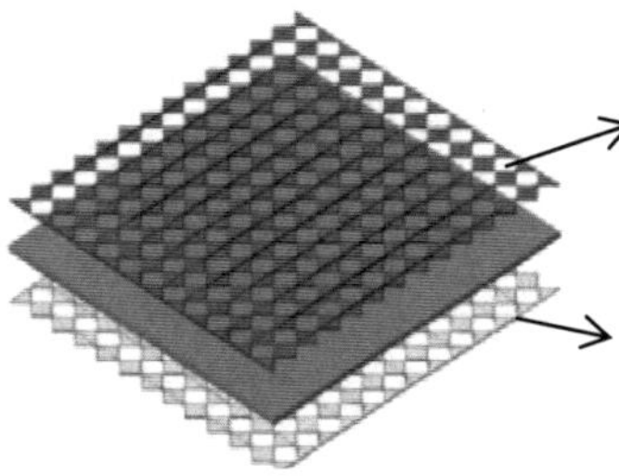

Figure 1. Double-layer electrode structure design of capacitive touch display.

When the finger touches the touch panel, the finger is equivalent to a conductor, resulting in a part of the power line from the driving electrode TX to the induction electrode RX transferred to the finger, the field strength between the two electrodes is weakened, and the mutual capacitance is reduced [6], as shown in Figure 2.

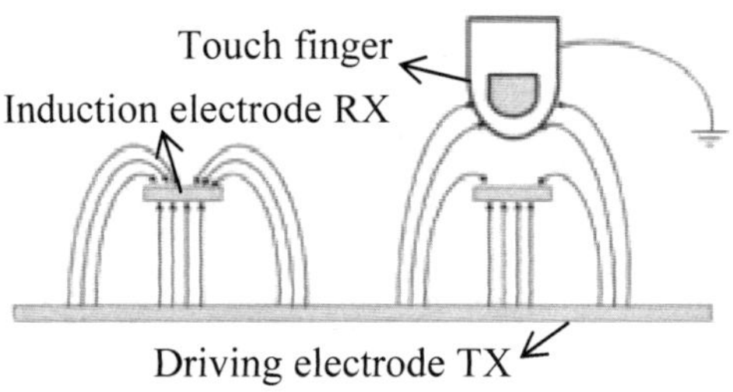

Figure 2. Power line distribution diagram of electrodes with or without finger touch.

The principle of mutual capacitance detection is shown in Figure 3. When no fingers touch the capacitive touch panel, the mutual capacitance between TX and RX is Cm; when the capacitive touch panel has been touched by a finger, the mutual capacitance is Cm'. Because the human body is a conductor, a part of the current will be absorbed when touched by the finger, thus making the mutual capacitance between the electrodes smaller [7]:

$$\Delta C = C_m - C_m^{'} \tag{1}$$

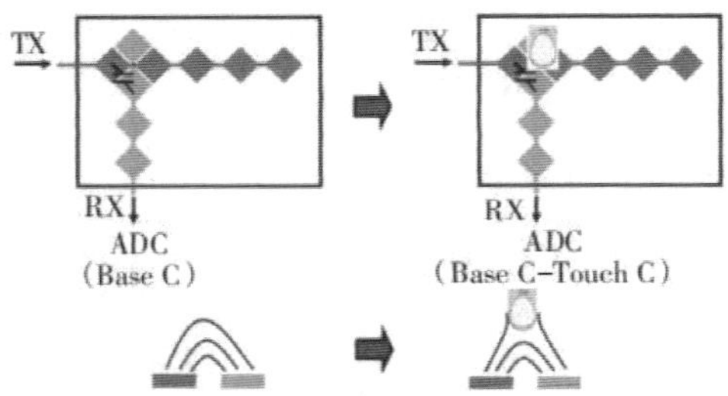

Figure 3. Principle of mutual capacitance detection.

In a cross-matrix electrode, each induction electrode Rx and each drive electrode Tx cross, and there is mutual capacitance at the intersection. When detecting mutual capacitance, the touch drive chip provides drive signals to the drive electrode in turn, and the induction electrode receives the signals simultaneously through multiple channels. Therefore, after completing a drive cycle, the mutual capacitance value at the intersection of all Tx and Rx electrodes in the entire two-dimensional plane can be obtained.

When the finger touches the capacitive touch panel, the coupling capacitance of the electrode near the contact point will be changed, and the touch chip can obtain the coordinates of each contact point according to the data of the capacitance change of the touch panel on the two-dimensional plane. Through this measurement method, the real coordinates of each contact point in multi-finger touch can be accurately calculated [8].

According to the parallel plate capacitance formula $C=\frac{\varepsilon s}{d}$, the plate capacitance value is affected by three parameters: medium parameter ε, cross-sectional area s, and plate spacing d. For the 9-inch automobile capacitive touch display, the parameters ε and s area are fixed, so it is important to consider the influence factors of plate spacing d on capacitance.

The change of plate spacing d will affect the sensitivity of the capacitive touch display. We assume that the initial value of the distance between the two plates is d_0, the initial capacitance of the capacitor is c_0, and when the plate distance has a certain increment, the capacitance is Δd:

$$C=\frac{\varepsilon s}{d+\Delta d}=C_0+\Delta C \tag{2}$$

$$\begin{aligned}\Delta C&=C-C_0\\&=\frac{\varepsilon s}{d_0+\Delta d}-\frac{\varepsilon s}{d_0}\\&=\frac{\varepsilon s(-\Delta d)}{d_0+\Delta d}\\&=-\frac{\varepsilon s\bullet\Delta d}{d_0^{\ 2}}\bullet(1+\frac{\Delta d}{d_0})^{-1}\end{aligned} \tag{3}$$

When $\frac{\Delta \mathrm{d}}{d_0}<<1$, the above formula can be expanded into a series form:

$$\Delta C=\frac{\varepsilon s}{d_0}\bullet(-\frac{\Delta d}{d_0})\bullet[1-\frac{\Delta d}{d_0}+(\frac{\Delta d}{d_0})^2-(\frac{\Delta d}{d_0})^3+...]$$

$$= C_0 \bullet (-\frac{\Delta d}{d_0}) \bullet [1 - \frac{\Delta d}{d_0} + (\frac{\Delta d}{d_0})^2 - (\frac{\Delta d}{d_0})^3 + ...] \quad (4)$$

The resulting sensitivity K is:

$$K = \frac{\Delta C}{\Delta \mathrm{d}} = \frac{\varepsilon s}{{d_0}^2} \bullet [1 - \frac{\Delta d}{d_0} + (\frac{\Delta d}{d_0})^2 - (\frac{\Delta d}{d_0})^3 + ...] \quad (5)$$

From Formula (5), it can be seen that the effect of the change of plate spacing d on the sensitivity is not a constant, and only when the ratio $\frac{\Delta \mathrm{d}}{d_0}$ is very small, it can be considered as an approximately linear relationship. When ε and s are fixed, the sensitivity K is inversely proportional to the square of the plate spacing d, and the smaller the plate spacing d is, the higher the sensitivity is [9]. Because the sensitivity varies with the plate spacing, it is crucial for the selection of the plate spacing d [10]. For capacitive touch display, the key factors affecting the plate spacing d are the display's bonding process and structural design.

2.2. Display bonding process and structure design

2.2.1. Display air gap bonding technology. The structure of the display is composed of three main parts, from top to bottom: the protective glass, the touch panel, and the TFT display. The three main components need to be sequentially bonded, and generally need two bonding processes to complete, firstly bonding the protective glass and the touch panel, and then boding the touch panel and the TFT display. Depending on the screen size and resolution, there are two processes: air gap bonding and optical bonding. The protective glass and the touch screen are generally optical bonding, and the air gap bonding is generally used between the touch panel and the TFT display.

For air gap bonding technology, as shown in Figure 4, the touch panel and the TFT display are fitted along the edges of the four sides with optical glue, and there is an air gap layer in the middle. Air gap bonding technology is used by most small and medium-sized displays.

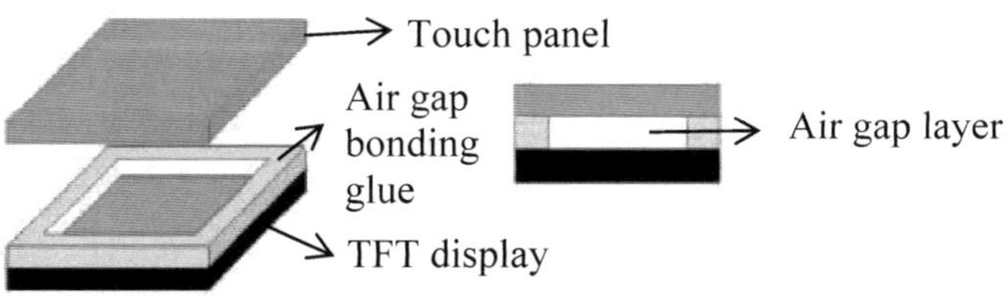

Figure 4. Air gap bonding section.

Display which is laminated through the air gap bonding process will create an air layer between the touch panel and the display [11]. The thickness of the air gap layer, that is, the spacing d between the touch panel and the TFT display screen, is an important parameter affecting the capacitance value. Plate spacing d must be set to a reasonable value, and too much spacing will cause the display water vapor, fog, excessive reflectivity, and other problems; too small spacing can cause false triggering problems. Therefore, in the display design, it is especially important to consider the gap parameters.

2.2.2. Display structure design. At the same time, in addition to the air gap bonding process, the structural design of the display is also an important factor affecting the plate spacing d. A metal bracket is designed for fixing the TFT display, as shown in Figure 5. Since the metal bracket has a certain thickness, the spacing between the touch panel and the display screen is increased. At the same time, the size deviation and installation error of the bracket will lead to a change in the gap, which will affect the capacitance and touch sensitivity.

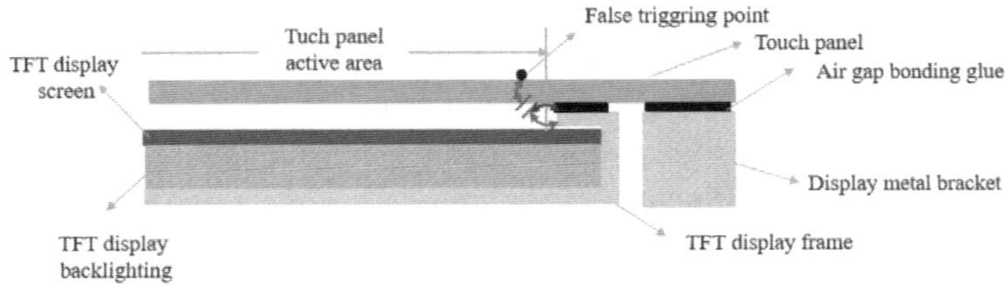

Figure 5. Display structure design profile.

3. False triggering fault test of the touch display

3.1. Fault phenomenon

The automobile capacitive touch display of a certain model has 13 cases of false triggering fault in 3 consecutive months, and the display is supplied by the same supplier. The fault description is as follows: when the vehicle is driving on a bumpy road, the button in the upper right corner of the display is false triggering and blinking.

3.2. Fault testing

After the faulty part is returned, the appearance check is normal, and the bench function test is carried out. When shaking or pressing the display border, the top right edge of the screen will be in a triggered state, and the fault can be repeated.

When the TFT module is powered off, we connect the TFT screen/bracket/support/PCBA with the GND of the test tool and test the parasitic capacitance of the touch panel. The value of the parasitic capacitor of the faulty part in the upper right corner is greater than 600, and the figure has a sharp angle. The parasitic capacitor of the good part is less than 200, as shown in Figure 6.

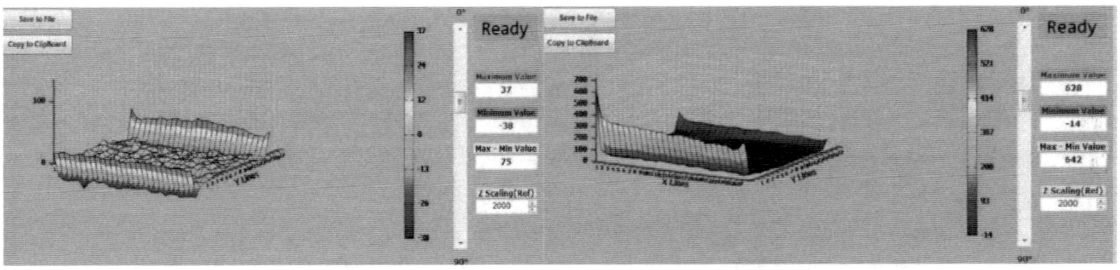

Figure 6. Measuring the capacitance of a faulty part and a good part.

It can be seen that the test result of the faulty part shows that the parasitic capacitance change value is greater than 600, causing the upper right corner to be mistakenly triggered. The capacitance test of faulty parts and good parts is shown in Table 1.

Table 1. Capacitance test of faulty parts and good parts.

Parts	Capacitance value test			
Faulty parts	#1:603 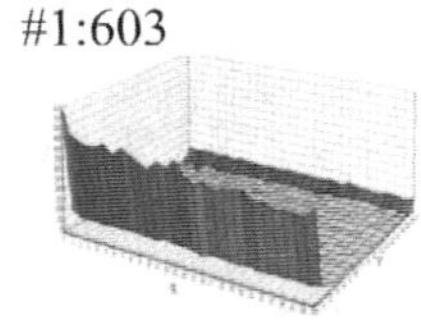	#2:1180 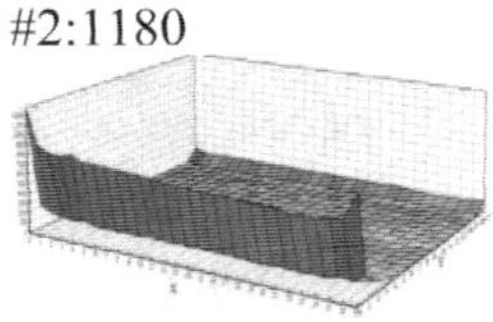	#3:1272 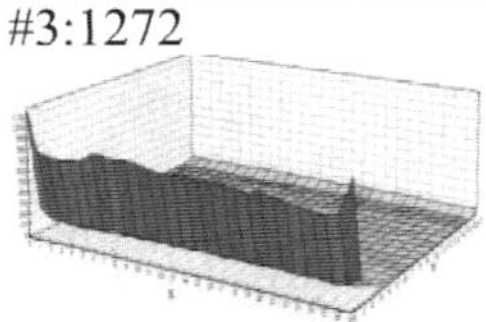	#4:1071 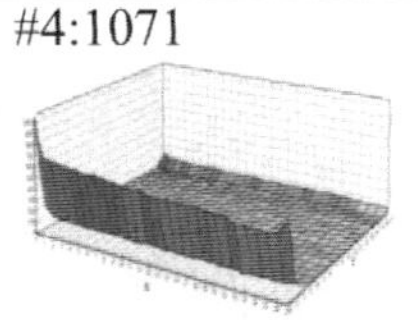
Good parts	#1:183 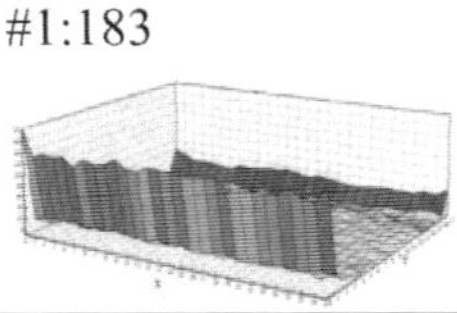	#2:124 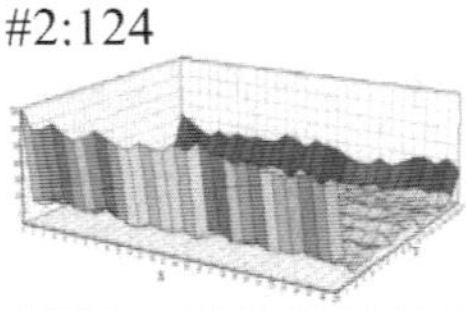	#3:162 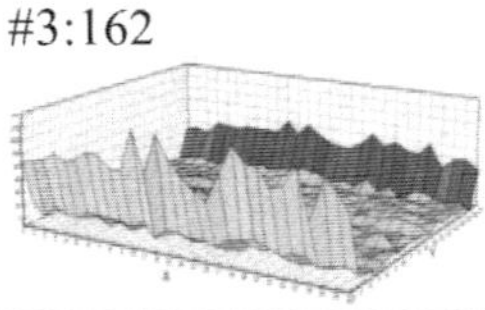	#4:198 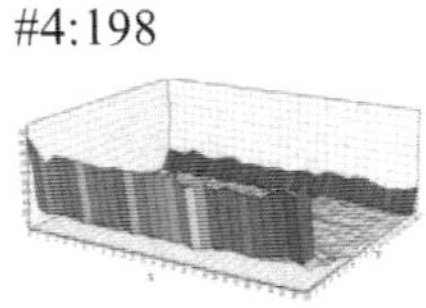

Journal of Physics: Conference Series **2815** (2024) 012037 doi:10.1088/1742-6596/2815/1/012037

4. Fault cause analysis

The upper right corner of the display is falsely triggered and is in the same position. Preliminary analysis of the possible reasons are (1) glue overflow caused by false triggering and (2) the air gap between the touch screen or TFT display is too small, resulting in a false trigger. For the above possible reasons, we check one by one.

4.1. Cause analysis of air gap bonding

Further disassembly analysis of the faulty parts, in the process of squeezing the TFT frame or shaking, the display is triggered. Abnormal points are found which confirm that there is a slight glue overflow in the upper right corner of the screen, as shown in Figure 7. After inquiring about the production records of the faulty parts and testing, it was found that the glue adhesive width was too large which led to the overflow of glue.

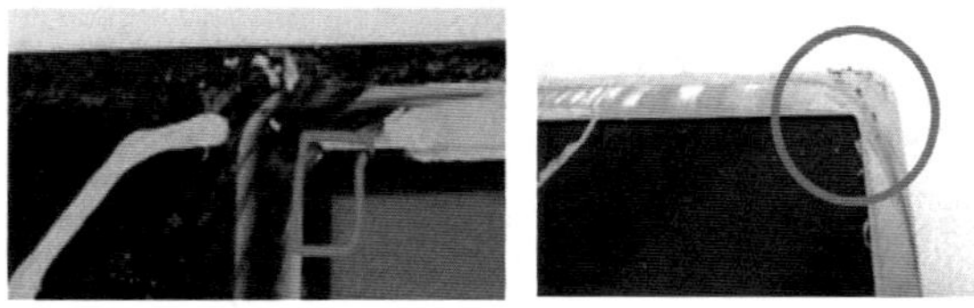

Figure 7. Measuring the capacitance of a faulty component and a normal component.

4.2. Touch panel and display air gap analysis

Another key factor affecting the air gap between the touch panel and the display is the thickness of the metal bracket that holds the display in place. The normal metal bracket thickness is 1.35 mm and is used to fix the 4 points of the TFT module. The size requirements are shown in Figure 8.

The metal bracket thickness of the faulty part is shown in Table 2. The thickness at points 1, 2, and 4 is smaller than the normal size, which will squeeze the display close to the touch panel, resulting in glue overflow and small spacing. The point 1 position is the upper right corner position where the display is triggered by mistake. This size exceeds the lower limit, causing the whole position to sag. Finally, after the TFT display is installed, the TFT display will sink. The gap between the TFT display and the touch panel is reduced in this position. After the TFT display is installed, the strength of the point 1 position is high, and the strength of the point 4 position is low, so the point 4 position will also be pulled to the lower limit, and the size will be reduced.

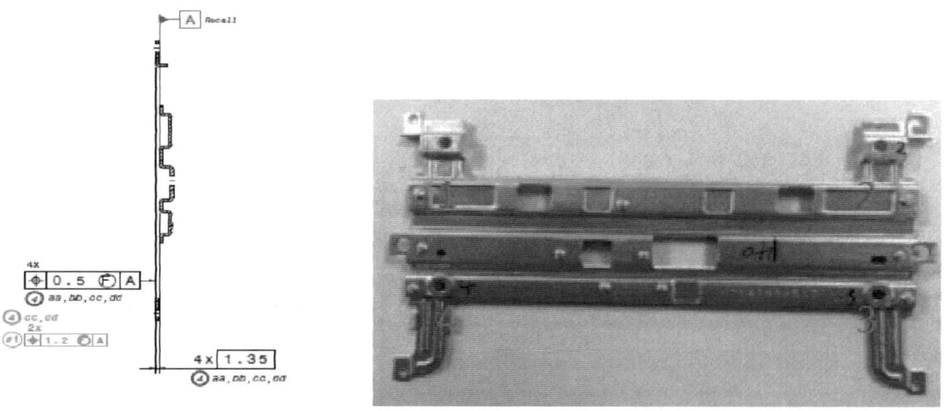

Figure 8. Metal bracket thickness requirements.

Table 2. Thickness test data of metal bracket.

Faulty parts	Point 1 (mm)	Point 2 (mm)	Point 3 (mm)	Point 4 (mm)
#1	1.13	1.05	1.22	1.23
#2	0.86	1.02	1.12	1.27
#3	0.84	1.01	1.18	1.05
#4	0.89	1.03	1.19	1.09

Table 3. Distance of touch panel and display.

Parts	Point 1 (mm)	Point 2 (mm)	Point 3 (mm)	Point 4 (mm)
Good parts#1	9.496	9.459	9.720	9.554
Good parts#2	9.463	9.570	9.788	9.323
Faulty parts #1	9.159	9.254	9.760	9.107
Faulty parts #2	9.114	9.158	9.634	9.263
Faulty parts #3	9.074	9.208	9.697	9.031
Faulty parts #4	9.135	9.203	9.728	9.081

The distance between the touch panel and the TFT display is mainly controlled by dimension A as shown in Figure 9. The dimension A of the faulty parts is measured as shown in Table 3. The result shows that the size of the touch screen and display screen of the faulty part is too small at the two positions of point 1 and point 4.

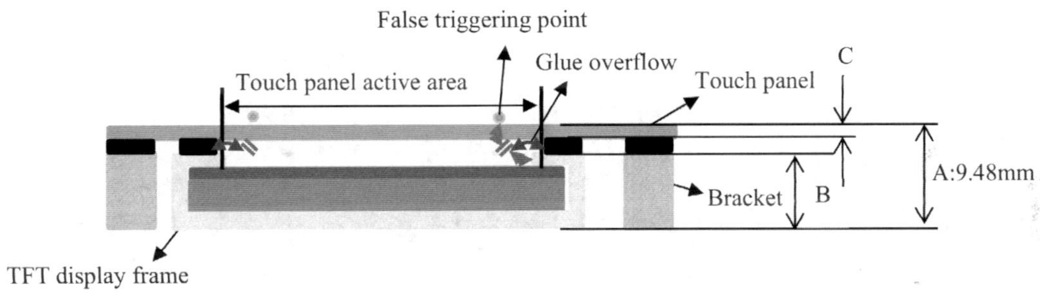

Figure 9. Distance of touch panel and display.

In summary, the test results show that the root cause of the fault is that the spacing between the touch panel and the display in the upper right corner is too small, resulting in incorrect triggering. The edge of the TFT display is far away from the touch screen activation area, so there is no effective capacitance change and false positive point. However, in an abnormal state, if the glue layer is thinner, the spacing between the TFT display and the touch panel is reduced, or the glue layer partially overflows, being close to the activation area of the touch screen. The above two factors will form an effective trigger parasitic capacitor, resulting in a false trigger.

According to the above analysis, the simulation of glue overflow and the small spacing between the touch screen and TFT display of the 2pcs of product replicas showed that the two replicas had the same fault phenomenon as the after-sales parts, and the upper right corner mistakenly triggered the capacitance value of the upper right part of the display to be greater than 600. After the after-sales faulty parts are disassembled without damage, the upper right corner capacitance is 230, which is normally less than 600, as shown in Figure 10.

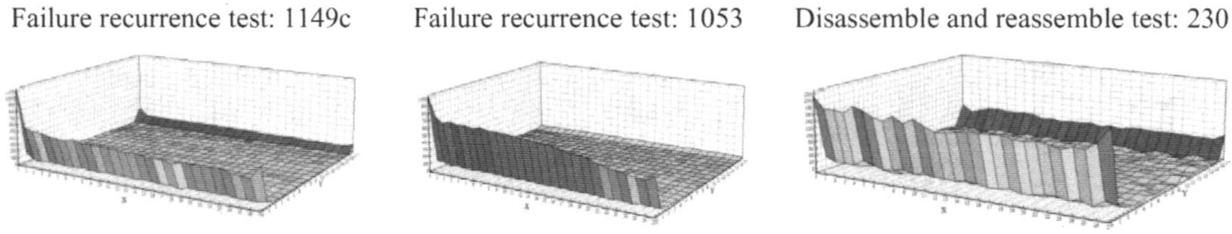

Figure 10. Failure recurrence test.

To sum up the cause analysis, it can be concluded that there are two root causes for the display false triggering: (1) there are glue overflow problem in the upper right corner between the touch panel and TFT display; (2) the spacing at the top right corner between the touch panel and TFT display is too small.

4.3. Solutions

Long-term solution: (1) the air gap bonding process should be improved to avoid glue overflow. We strictly control the amount of glue applied, adjust the glue wideness from 3.9 mm to 2.8 mm, and adjust the upper and lower limits from 1.5~5.3 mm to 2.0~3.2 mm; (2) we adjust tolerances to ensure the spacing between the touch panel and TFT display, modify the screw height of the metal support, and adjust the height of the blue side to the dimension upper tolerance, as shown in Figure 11. We adjust the spacing between the TP touch screen and TFT display to 1.7 mm to avoid glue overflow and small clearance resulting in false triggering; (3) during the production of metal brackets, the supplier shall do SPC control on the height of the screws; (4) experimental verification of the above measures. We detect the initial parasitic capacitance, sample hot and cold shock test, sample vibration test, and sample after testing parasitic capacitance. After the implementation of the long-term solution, the problem of after-sales false triggers no longer occurs, and the solution is effective.

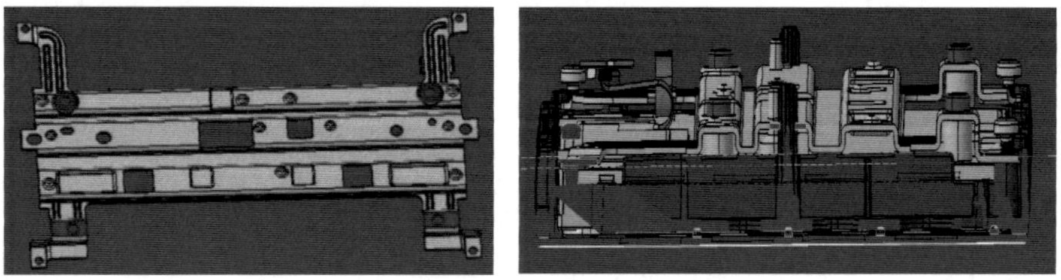

Figure 11. Size control of metal bracket.

5.Conclusion

This paper analyzes and studies the root causes of the false triggering problem of the 9-inch central control display of a car. Due to the air gap bonding process, an effective solution is proposed with minimal changes. If the optical bonding process is used on this 9-inch display, the trigger problem caused by the air layer can be avoided. With the improvement of cost and product yield, it is believed that optical bonding technology will gradually be implemented on automobile central control displays.

Acknowledgments

This work was supported by Hubei Provincial Colleges and Universities Outstanding Young and Middle-aged Science and Technology Innovation Team Program Project: Intelligent Logistics Equipment Digital Twin(No.T2021043), and the Provincial first-class undergraduate major construction project of Electrical Engineering and Automation of Wenhua College (No. J09006904).

References

[1] Lu Chao, 2016, Research on key technologies of capacitive touch liquid crystal display applied to the in-vehicle multimedia system. Shanghai Jiao Tong University.

[2] Zhao Xiaozhen, 2021, Design and research of touch screen in airborne display system[J]. Electronic Design Engineering, pp 46-50+55.
[3] Pan Xinli, 2018, Design method of small and medium size capacitive touch screen. Modern Information Technology, pp 56-57+60.
[4] Zhao Haizeng and Liu Tengfei, 2018, A review of capacitive touch screen technology. Henan Science and Technology, pp 10-12.
[5] Lee C ,Park K J ,Piao C , et al, 2018, Mutual Capacitive Sensing Touch Screen Controller for Ultrathin Display with Extended Signal Passband Using Negative Capacitance.Sensors,pp 3637-3637.
[6] Xie Jiangrong, Pan Fengming and Wu Zhengnan, 2016, Projected capacitive touch screen electrode design analysis. Journal of Optoelectronic Technology, pp 200-204.
[7] Liu Yudong, 2023, Analysis of optimization strategies for sensitivity and anti-noise performance of in-vehicle capacitive touch screens. Electronic Technology, pp 31-33.
[8] Nathan A ,Sharma P ,Tsangarides P C , et al, 2019, Invited Paper: Multi-Force Touch Technology Augmenting Capacitive Touch in Displays.SID Symposium Digest of Technical Papers, pp472-475.
[9] Long Lingli, Zhang Yingfan and He Yiyang, 2014, Capacitive touch screen in the parasitic capacitance principle and measuring method. Mechanical and Electrical Information, pp 137-139.
[10] Piljoong K ,Sanghyun H ,Yunho J , et al, 2021, A PAPR Reduction Technique for Fast Touch Sensors Adopting a Multiple Frequency Driving Method on Large Display Panels.Sensors, pp 429-429.
[11] Zhang S.Q., Zhang S.G. and Shu C.Y., 2023, Lamination process and application of double-layer LCD touch panel. Shanghai Light Industry, pp 137-139.

Journal of Physics: Conference Series **2815** (2024) 012038 doi:10.1088/1742-6596/2815/1/012038

FPGA-based cesium optical pump magnetometer data recording system design

Da Teng*, **Guoyu Hei, Yang Wu, Min Chen and Yue Zhao**

School of Electronic Information and Communication Engineering, Chongqing Aerospace Polytechnic, Chongqing, 400021, China.

* Corresponding author's e-mail: tim21927@163.com

Abstract. A cesium optical pump magnetometer is a kind of magnetic detection instrument commonly used in geophysical and military fields, which has significant advantages such as high sensitivity, insensitivity to attitude change, and no zero drift. Compared with manned aircraft, rotary-wing UAVs have the advantages of flexible takeoff and landing, low cost, and suitability for small measurement areas, so the aeromagnetic system based on the integration of rotary-wing UAV platforms and cesium optical pump magnetometers has a strong practical value. This paper investigates an FPGA-based CS-3 cesium optical pump magnetometer with a high-precision magnetic field data recording system for building a miniaturized aeromagnetic measurement system based on a rotary-wing unmanned aerial vehicle (UAV) platform, and the system achieves a magnetic field measurement accuracy of 3.4 pT.

1. Introduction

With the booming economy, the role of mineral resources has become increasingly prominent. Therefore, it is of considerable social and engineering significance to develop a miniaturized aeromagnetic measuring system founded on an unmanned aerial vehicle (UAV) with a rotary wing platform that can perform magnetic field measurements with high efficiency and low costs [1].

A cesium optical pump magnetometer is a kind of atomic magnetometer with high sensitivity and precision, which is widely used in the field of magnetic field measurement. In this paper, we use the CS-3 model cesium optical pump magnetometer, whose magnetic field measurement range is 15000~105000 nT, sensitivity is 0.0006 nT/√Hz(rms), precision is 0.01 nT, noise level is 0.002 nT, and its performance parameters can fulfill the needs of high-precision aeromagnetic measurements [2]. The frequency of the Larmor frequency signal, which is proportional to the magnetic field information and is the output signal of the CS-3 cesium optical pump magnetometer, must be measured. The most popular techniques for measuring frequency are direct frequency measurement, equal-precision frequency measurement (EPFM), and others. To improve the accuracy of the frequency measurement of the Larmor signal, the conventional equal-precision frequency measurement (EPFM) method is modified in this study [3].

2. Equal-precision frequency measurement (EPFM)

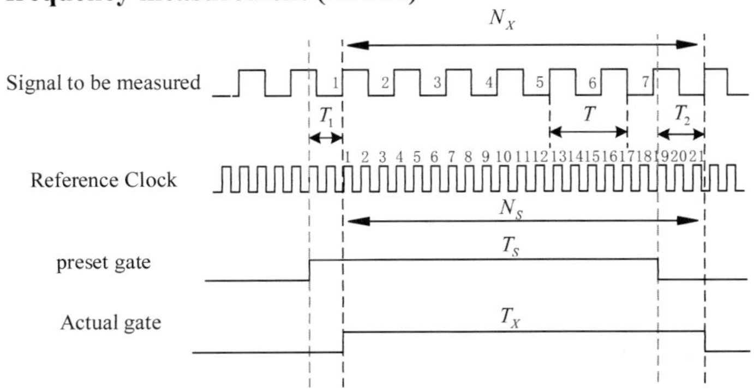

Figure 1. The timing diagram of the equal-precision frequency measurement method.

The timing diagram of the equal-precision frequency measurement method is shown in Figure 1.The following equation can be used to determine the frequency of the signal that the equal-precision frequency measurement method needs to measure:

$$F_X = \frac{N_X}{N_S} * F_S \tag{1}$$

where N_s is the standard clock signal's count value at the actual gate time, F_s is the frequency value of the reference clock signal, and N_x is the count value of the signal to be measured during the actual gate time. The error calculation equation of the equal-precision frequency measurement method is:

$$\frac{\Delta F_X}{F_X} = \frac{\Delta N_X}{N_X} - \frac{\Delta N_S}{N_S} + \frac{\Delta F_S}{F_S} \tag{2}$$

The source of error of the equal-precision frequency measurement method is improved from the ± 1 word count error of the gate signal to the signal that needs to be measured in the direct frequency measurement method to the ± 1 word count error of the gate signal to the reference clock signal. Since the reference clock signal is constant, for both high-frequency and low-frequency signals, the error of the equal-precision frequency measurement method is likewise constant [4]. Since the signal to be measured by the equal-precision frequency measurement method is strictly aligned with the gate signal, $\Delta N_X = 0$, Equation (2) can be further written as:

$$\frac{\Delta F_X}{F_X} = \pm \frac{1}{N_S} + \frac{\Delta F_S}{F_S} \tag{3}$$

Since the reference clock error ΔF_S is several orders of magnitude smaller than the signal error ΔN_S to be measured, Equation (2) can be further written in the case of neglecting the reference clock signal error ΔF_S:

$$\frac{\Delta F_X}{F_X} = \pm \frac{1}{N_S} = \pm \frac{1}{F_S * T} \tag{4}$$

As can be seen from Equation (4), the source of error in equal-precision frequency measurement is mainly the value of N_S, i.e., the count value of the standard frequency signal during the gate time. So there are two main ways to improve the accuracy of the equal-precision frequency measurement method: one is to increase the gate time T ; the other is to increase the frequency value of the system clock F_S. The first way is difficult to realize due to the limitation of the sampling rate, so the main way to improve the accuracy of the equal-precision frequency measurement method is to increase the frequency value of the system clock F_S.

3. Improved equal-precision frequency measurement method

The precision of the cesium optical pump magnetometer in measuring magnetic fields is 0.01 nT, and there is a linear relationship between the external magnetic field value B_0 and the Larmor frequency f_0, as shown in Equation (5).

$$B_0 = f_0 / 3.49857 \tag{5}$$

From Equation (5), the corresponding frequency measurement accuracy is 0.035 Hz, so the frequency measurement accuracy of the system should be better than 0.035 Hz. The project's aeronautical magnetic measurement system has a 10 Hz sample rate, and the corresponding gate time is 100 ms. Assuming that the frequency of the system clock adopts a crystal of 50 MHz, the relative inaccuracy of the equal-precision frequency measurement method can be calculated by using Equation (4) as follows:

$$\frac{\Delta F_X}{F_X} = \pm\frac{1}{F_S * T} = \pm\frac{1}{100\times10^{-3} * 50\times10^{6}} = \pm0.2\times10^{-6} \tag{6}$$

The frequency measurement range of the cesium optical pump magnetometer is 50 kHz ~ 350 kHz. Taking the maximum value of 350 kHz, the value of the error ΔF_X can be calculated as:

$$\Delta F_X = (350\times10^{3})\times(\pm0.2\times10^{-6}) = \pm0.07 \tag{7}$$

As can be seen from Equation (7), when the system clock is 50 MHz, the maximum error within the range is 0.07 Hz, which can not meet the minimum accuracy requirements of the cesium pump magnetometer. To safeguard the system's overall measurement accuracy, the frequency measurement accuracy needs to be one order of magnitude higher than the sensor accuracy. Therefore, the equal-precision frequency measurement (EPFM) method should be enhanced accordingly, and the core is to improve the frequency of the system clock. The PLL phase-locked loop is used to multiply the system clock of the FPGA, and the corresponding error equation when multiplying the system clock frequency to 400 MHz is as follows:

$$\frac{\Delta F_X}{F_X} = \pm\frac{1}{F_S * T} = \pm\frac{1}{100\times10^{-3} * 400\times10^{6}} = \pm0.25\times10^{-7} \tag{8}$$

The system error is calculated by taking the signal's frequency that needs to be measured as 350 kHz:

$$\Delta F_X = (350\times10^{3})\times(\pm0.25\times10^{-7}) = \pm0.0087 \tag{9}$$

From Equation (9), when the system clock frequency is 400 MHz, the corresponding maximum system error is 0.0087 Hz, which is one order of magnitude smaller than the sensor error of 0.035 Hz. Therefore, the enhanced equal-precision frequency measurement method can theoretically meet the frequency measurement accuracy standards of the cesium optical pump magnetometer at a sampling rate of 10 Hz [5].

4. Improved equal-precision frequency measurement method FPGA programming

This design uses Quartus software to develop and design the FPGA program based on Verilog HDL hardware description language. Designing circuits using hardware description language has the advantages of high design complexity, short design cycles, and strong design portability. Firstly, the program input is carried out, and the software synthesizes, compiles, and routes the circuits described by the hardware description language. Then functional and timing simulations are performed. Finally, the compiled program is downloaded into the FPGA. The overall top-down design flow is adopted.

The register-level concept of the enhanced equal-precision frequency measurement method is displayed in Figure 2. The system clock is a 100 MHz constant temperature crystal. In the figure, sys_clk is the 400 MHz reference clock signal multiplied by the PLL phase-locked loop to the system clock; clk_fx is the TTL square wave signal to be measured; sys_rst_n is the system reset signal; fs_cnt is the count value of the reference clock signal output by the 32-bit counter in the gate time; The count value of the signal that needs to be measured in the gate time by the 32-bit counter is represented by fx_cnt.

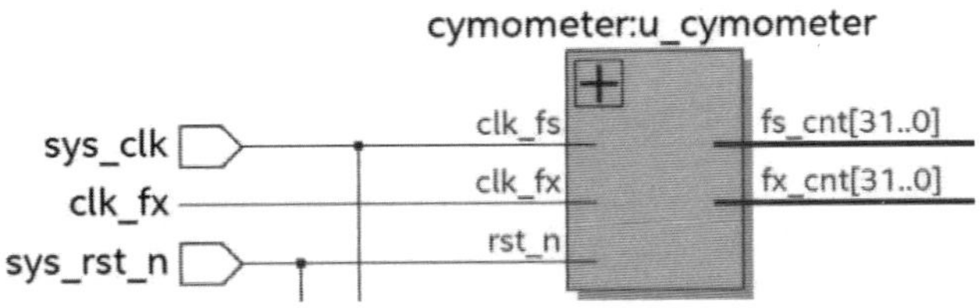

Figure 2. Quartus Software RTL Schematic.

The pin configuration interface of Quartus software is shown in Figure 3, and a reasonable pin configuration can make the FPGA run more stably. The system input pins include the signal to be tested, clk_fx, the system clock signal, sys_clk, the reset signal, sys_rst_n, and the serial port receive signal, uart_rxd. The output pins are the serial port transmit signal, uart_txd. The pin configurations include the configuration of the pin direction, pin position, pin I/O BANK, and the pin level protocols to complete the linkage between the internal program of the FPGA and the external hardware circuits to realize the preset logic functions.

Node Name	Direction	Location	I/O Bank	VREF Group
in altera_reserved_tms	Input			
in clk_fx	Input	PIN_127	7	B7_N0
in sys_clk	Input	PIN_24	2	B2_N0
in sys_rst_n	Input			
in uart_rxd	Input	PIN_50	3	B3_N0
out uart_txd	Output	PIN_43	3	B3_N0

Figure 3. Quartus Software FPGA Pinout.

The frequency measurement module is measured by using a SignalTap logic analyzer, and the function signal generator output is set to a 5 V TTL square wave signal at 173197 Hz based on the common frequency of the cesium optical pump magnetometer to measure the geomagnetic field. The function signal generator output setting is shown in Figure 4.

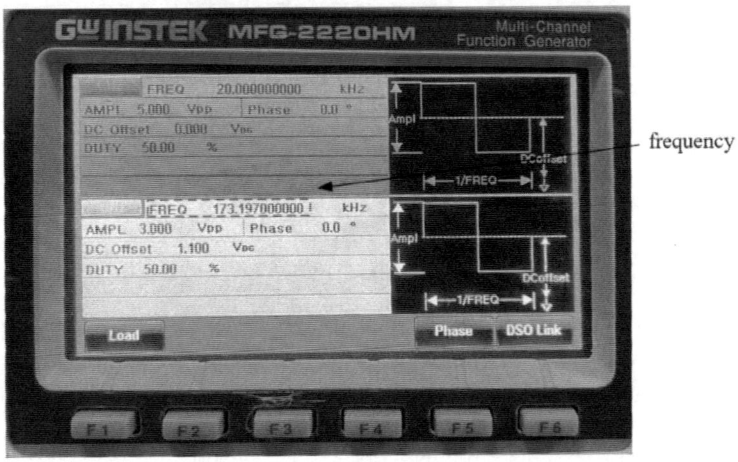

Figure 4. Function Signal Generator Output Settings.

Journal of Physics: Conference Series **2815** (2024) 012038 doi:10.1088/1742-6596/2815/1/012038

The set TTL square wave signal is input to the PIN_127 pin of the FPGA for frequency measurement, and the values of each register are observed with the SignalTap Logic Signal Analyzer. The actual measurement is shown in Figure 5. Among these, the measurement signal, clk_fx, is observed to see if it is entered normally; the functionality of the frequency measurement method is confirmed by observing the counting values of the two counters, fx_cnt and fs_cnt.

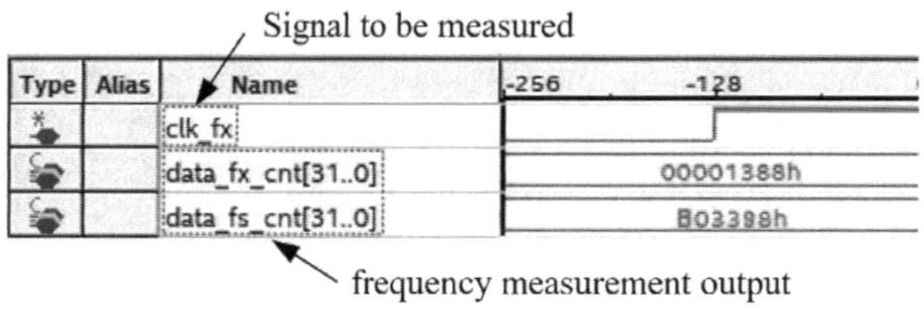

Figure 5. SignalTap Logic Analyzer Measurements.

The count value of fx_cnt is the hexadecimal number 1388 and the count value of fs_cnt is the hexadecimal number B03398. This can be resolved by using Equation (10) to determine the signal's frequency value for measurement. The actual project is derived by sending the data from the serial port to the STM32 terminal for floating point operation [6].

$$F_X = \frac{N_X}{N_S} \times F_S = \frac{f_X\_cnt}{f_S\_cnt} \times clk\_f_X = \frac{5000}{11547544} \times 400 \times 10^6 = 173197.002 \quad (10)$$

From Equation (10), the measured frequency value is 173197.002 Hz, which is consistent with the preset value.

5. Frequency measurement accuracy test

In order to verify whether the data acquisition function of the data collection system meets the preset performance index, a test platform is built for testing [7]. The specific scheme is that the input signal adopts a function signal generator to produce a TTL signal, which is used to simulate the output signal of the CS-3 cesium optical pump magnetometer. The actual test diagram is shown in Figure 6.

Figure 6. Test Environment Setup.

Using a function signal generator to generate a TTL square wave signal in the frequency range of 50 kHz ~ 350 kHz, the full frequency band of the cesium pump magnetometer is simulated to calibrate the frequency measurement error of the program [8]. The step size is selected as 50 kHz, each sampling point is measured for at least 2 minutes, and the maximum value of the error is taken as the absolute error value. The input frequency is increased by 0.001 Hz and the output change is observed. The output change corresponds to the minimum frequency value for the resolution of the current measurement value. Table 1 displays the results of the measurements [9].

Table 1. Frequency Measurement Results.

Frequency (Hz)	Test Results (Hz)	Absolute Error	Relative Error	Resolution (Hz)
50000	50000.001	0.001	0.19×10^{-7}	0.001
100000	99999.999	0.001	0.1×10^{-7}	0.002
150000	150000.002	0.002	0.13×10^{-7}	0.003
200000	199999.998	0.002	0.1×10^{-7}	0.003
250000	249999.996	0.004	0.16×10^{-7}	0.007
300000	300000.006	0.006	0.19×10^{-7}	0.009
350000	350000.007	0.007	0.19×10^{-7}	0.012

Throughout the whole frequency range, the enhanced equal-precision frequency measurement method achieves a relative error measurement accuracy of 0.19×10^{-7} and an absolute error that ranges from 0.001 Hz to 0.007 Hz. The maximum resolution in the frequency band range is 0.012 Hz, and the corresponding magnetic field value is 3.4 pT.

The measured results in the range of 50 KHz~350 KHz are taken to compare with the theoretically calculated values in order to further examine the experimental results by comparing the difference between the measured and theoretical calculations of the method. The results are displayed in Table 2:

Table 2 Comparison of Measured Results and Theoretical Calculations.

Frequency (Hz)	Measured Absolute Error (Hz)	Maximum Theoretically Calculated Absolute Error (Hz)	Difference Between Theoretical and Actual Values (Hz)
50000	0.001	0.0012	0.0002
100000	0.001	0.0025	0.0015
150000	0.002	0.0037	0.0017
200000	0.002	0.005	0.003
250000	0.004	0.0062	0.0022
300000	0.006	0.0075	0.0015
350000	0.007	0.0087	0.0017

Comparing the two, it is evident that the method's measured absolute error values are less than the theory's greatest absolute error value, indicating that the method can meet the requirements of the cesium optical pump magnetometer for frequency measurement accuracy.

6. Conclusion

This research presents an FPGA-based high-precision frequency measurement technique that is investigated to realize high-precision magnetic field data inclusion. Based on the above analysis, the main conclusions can be summarized as follows:

(1) The equal-precision frequency measurement method is theoretically analyzed and calculated, and an improved method is proposed on this basis. The enhanced method's theoretical calculation confirms that it can, in principle, match the cesium optical pump magnetometer's requirements for precision in frequency measurement.

(2) To measure the improved equal-precision frequency measurement method, an FPGA-based experimental platform is constructed. The measured results are compared with theoretically calculated values, demonstrating that the method can meet the frequency measurement accuracy requirements of the cesium optical pump magnetometer.

The above conclusions provide a basis for further building FPGA-based aeromagnetic measurement systems in the future.

Acknowledgments

This study was supported by a grant from the Scientific Research Program of Chongqing Municipal Education Commission (KJQN202303004, KJQN202303002, KJQN202303003, KJQN202103003, KJQN202203004, KJQN202203006, and KJQN202303008).

Journal of Physics: Conference Series **2815** (2024) 012038 doi:10.1088/1742-6596/2815/1/012038

References

[1] Sui Y, et al. Compact fluxgate magnetic full-tensor gradiometer with spherical feedback coil [J]. Review of Scientific Instruments,2014,85(1)01470

[2] S. Sun, W. Yan, Q. Zhang, et al. "High-precision frequency acquisition method based on FPGA," CSAA/IET International Conference (AUS 2022), Nanchang, China, 2022, pp. 183-186, doi: 10.1049/icp.2022.1521.

[3] P. H. Prajapati, A. R. Patel, A. D. Darji, et al. "FPGA-Based Data Acquisition System With Preprocessing of Plasma Signal," IEEE Transactions Plasma Science, vol. 49, no. 11, pp. 3597-3614, Nov. 2021.

[4] Tang Y, Liu Z, Pan M, et al. Detection of Magnetic Anomaly Signal Based on Information Entropy of Differential Signal [J]. IEEE Geoscience and Remote Sensing Letters, 2018,15(4):512-516.

[5] D. Li, et al. "Design of security detection digital sensor system based on FPGA," 2022 IEEE 16th International Conference on Solid-State & Integrated Circuit Technology (ICSICT), Nanjing, China, 2022, pp. 1-3.

[6] S. Sivanathan, M. A. Roula, K. Li, et al. "Design of an FPGA-Based High-Speed Data Acquisition System for Frequency Scanning Interferometry Long-Range Measurement," in IEEE Open Journal of Instrumentation and Measurement, vol. 3, pp. 1-10, 2024, Art no. 7000110, doi: 10.1109/OJIM.2023.3347268.

[7] L. Fu, Q. Li and B. Qu, "Miniaturized High Precision Time Interval Measurement Technology Based on FPGA," 2023 IEEE 16th International Conference on Electronic Measurement & Instruments (ICEMI), Harbin, China, 2023, pp. 7-11, doi: 10.1109/ICEMI59194.2023.10270350.

[8] X. Cen, J. Chen, Z. Zhu, et al." Design and Implementation of Ultrashort Baseline Transmission and Data Acquisition Module Based on FPGA," 2023 IEEE International Conference on Signal Processing, Communications and Computing (ICSPCC), ZHENGZHOU, China, 2023, pp. 1-5.

[9] M. Hamel and H. Mohellebi. A LabVIEW-based real-time acquisition system for crack detection in conductive materials [J]. Mathematics and Computers in Simulation, 2020,167(1):381-388.

Journal of Physics: Conference Series **2815** (2024) 012039 doi:10.1088/1742-6596/2815/1/012039

Estimation of ultrasonic flight time based on SHADE algorithm and extended kalman filtering

Peng Mao, Jinglin Xiong and Xisheng Li*

School of Automation and Electronic Engineering, Beijing University of Science and Technology, Beijing, 100083, China

*Corresponding author's e-mail: lxs@ustb.edu.cn

Abstract: Traditional methods of using ultrasound to measure oxygen concentration have the issue of inaccuracy and susceptibility to noise interference when measuring the time of flight (TOF) of ultrasonic echo signals. A proposed algorithm for ultrasonic parameter estimation, utilizing an asymmetric Gaussian model, aims to enhance the precision and reliability of TOF estimation, aiming to reduce the influence of noise interference. The algorithm combines an improved Self-Adaptive Differential Evolution (SHADE) algorithm based on successful historical records with an Extended Kalman Filter (EKF). The improved SHADE algorithm controls the degree of mutation strategy greediness during the mutation operation based on the number of iterations, thereby enhancing the algorithm's iteration speed. Then, the obtained solution is used as the initial value for the EKF to estimate the model parameters, thus addressing the issue of EKF's dependency on accurate initial values to obtain precise outputs. Through simulation and experimental measurements of ultrasonic oxygen concentration, several commonly used parameter estimation algorithms are compared with the proposed SHADE-EKF algorithm. Results demonstrate that the SHADE-EKF algorithm exhibits superior performance in estimating the TOF of ultrasonic waves.

1. Introduction

Ultrasonic waves are widely used in the field of detection, including non-destructive testing [1], positioning [2], ranging [3], medical applications [4], flow measurement [5], concentration measurement [6], and more. The principle of using ultrasonic waves to detect oxygen concentration is based on the different propagation speeds of ultrasonic waves in different oxygen concentrations, which allows for the calculation of oxygen concentration based on the gas state equation. The primary method for detecting oxygen concentration using ultrasonic waves is the time of flight (TOF) method. This involves measuring the time of arrival of the ultrasonic echo signal at a fixed distance.

The commonly used methods to measure TOF include the threshold method [7], the cross-correlation method, and the curve fitting method [9]. The curve fitting method obtains the TOF by fitting the actual echo signal with a mathematical model. Common mathematical models used for fitting ultrasonic echo signals include the Gaussian model and the mixed exponential model. Compared to the threshold method and the cross-correlation method, the curve fitting method has stronger anti-interference capabilities and a wider range of applications, including different variants of ultrasonic transducers and various operating conditions. Following the development of the ultrasonic echo model, the curve fitting technique employs a suitable optimization algorithm to accurately estimate the parameters and achieve the TOF. Traditional approaches for parameter estimation of ultrasonic echo signals consist of Least Squares Estimation

(LSE), Artificial Bee Colony algorithm (ABC), Particle Swarm Optimization (PSO) algorithm, Differential Evolution (DE) algorithm, and Extended Kalman Filter algorithm (EKF). However, these algorithms have their limitations. LSE and EKF depend on given initial values. At the same time, ABC, DE, and PSO require a sufficient number of iterations to achieve accurate results and have slow iteration rates in the later stages, making them prone to convergence to local optima. In the context of real-time detection of oxygen concentration using ultrasonic waves, these algorithms require refinement to attain rapid, stable, and precise outcomes within a restricted number of iterations.

Based on the aforementioned challenges, this study introduces a novel approach that integrates the adaptive parameter-based Differential Evolution (SHADE) algorithm with the Extended Kalman Filter (EKF) for parameter estimation of ultrasonic echo signals. Initially, an asymmetric Gaussian model is utilized to construct the ultrasonic echo signal model. Subsequently, during the mutation phase of the SHADE algorithm, the iteration count is harnessed to regulate the exploration-exploitation balance, enabling SHADE to swiftly converge towards an approximate optimal solution. This solution serves as the starting point for the EKF algorithm, which refines the estimate in a minimal number of iterations. Through extensive simulation and experimentation, the findings corroborate that this hybrid algorithm yields precise parameter estimates within a short duration, thereby fulfilling the exigencies for real-time monitoring of oxygen concentration.

2. Model of ultrasonic echo signals

Utilizing the echo signal model of the ultrasonic transducer, an amplitude spectrum of the transducer's echo signal is constructed by applying a Gaussian model spectrum. The central point of the spectrum aligns with the transducer's center frequency, with the bandwidth remaining consistent [9].

$$A(\theta\boldsymbol{\theta}t) = env \cdot \cos(2\pi f_c(t-\tau) + \psi(t-\tau)^2 + \varphi) \tag{1}$$

where $\boldsymbol{\theta}' = [\beta, \alpha, \tau, f_c, \varphi]$, $\beta, \alpha, \tau, f_c, \varphi$ are the amplitude factor, bandwidth, arrival time, centre frequency, and phase, respectively. Figure 1 shows the simulated ultrasonic callback signal using the Gaussian model when $\boldsymbol{\theta}' = [1, 10, 0.9, 2, 1]$ and the envelope after performing the Hilbert transform.

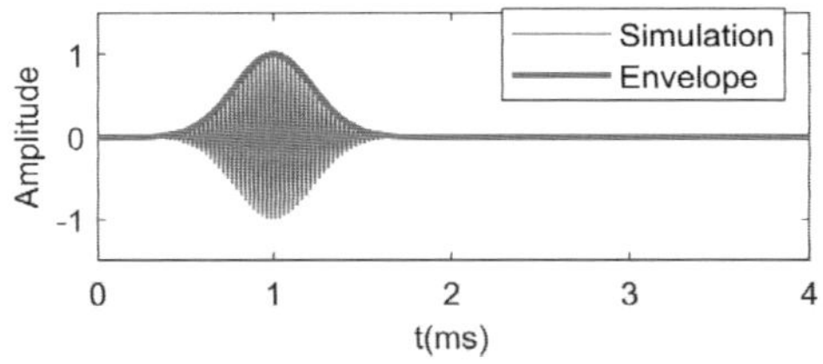

Figure 1. Gaussian Model and Envelope Graph.

However, the actual ultrasonic echo signal is asymmetric. To more accurately simulate the ultrasonic echo signal, two additional parameters need to be added to the above model: one called the asymmetry factor that controls the echo skewness, and the other called the chirp rate. The optimized model is known as the Asymmetric Gaussian Model(ACGM):

$$A(\boldsymbol{\theta}, t) = env \cdot \cos(2\pi f_c(t-\tau) + \psi(t-\tau)^2 + \varphi) \tag{2}$$

$$env(\theta; t) = \beta e^{-a(-r\tanh(k(t-\tau)))(t-\tau)^2} \tag{3}$$

where $\theta = [\beta, a, r, k, \tau]$, r is an asymmetry coefficient ranging from $[-1, 1]$, ψ denotes the linear chirp rate, and k is a positive integer constant, $\tanh(mt)$ represents the hyperbolic tangent function of the k-th order.

Figure 2 depicts the envelope graph of the ultrasonic echo signal simulated using the ACGM model with $\boldsymbol{\theta} = [1, 10, 0.9, 2, 1]$, along with the result after applying the Hilbert transform. The amplitude of the ultrasonic echo signal is $1V$, with a bandwidth of $10KHz$, an asymmetry coefficient of 0.9, $k = 2$, an arrival time of $1ms$, and a central frequency set to $40KHz$.

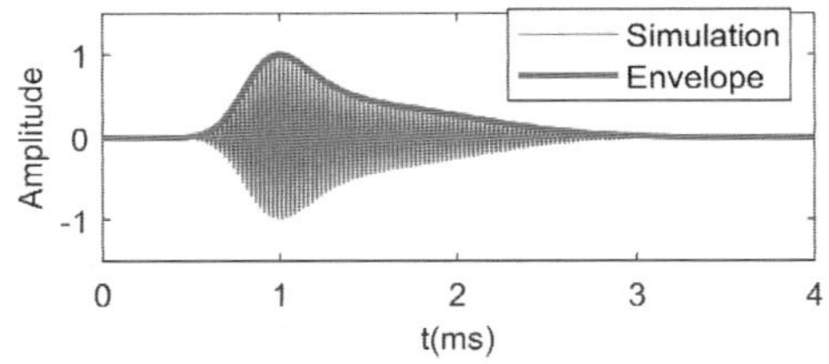

Figure 2. ACGM and Envelope Graph.

3. Optimization algorithm

In this section, a novel algorithm merging the benefits of the SHADE and EKF algorithms is introduced. This innovative approach enables swift and accurate parameter estimation for nonlinear models, showcasing enhanced precision and accelerated convergence rates.

3.1. SHADE algorithm

SHADE algorithm is an enhanced version of the DE algorithm, which performed impressively by securing the third position in the CEC 2014 competition. It leverages the successful control parameter settings stored in its historical memory to inform the selection of future values. The enhancements in the SHADE algorithm facilitate rapid and accurate convergence to the global optimum when tackling nonlinear optimization challenges. The fundamental stages of the SHADE algorithm encompass initialization, assignment of control parameters based on historical memory, mutation, crossover, selection, and update of historical memory.

3.2. Extended kalman filter algorithm

EKF is a commonly employed nonlinear state estimation filter that utilizes system measurements and dynamic models to estimate the system's state. It is an extension of the Kalman Filter, allowing for the handling of some nonlinear systems.

3.3. SHADE-EKF algorithm

While the SHADE algorithm boasts robust global search capabilities, the attained solution may not always be the optimal one, leaving room for enhancement in terms of convergence speed. On the other hand, the EKF algorithm demonstrates rapid iteration towards the optimum with appropriate initial values. However, deviations of initial values from the optimum can lead the EKF algorithm to converge to local optima or even diverge, resulting in suboptimal or inaccurate outcomes. To mitigate these challenges and harness the strengths of both algorithms, this study introduces a hybrid approach. Initially, the refined SHADE algorithm explores a broad range to identify promising solutions, which are subsequently utilized as initial values for the EKF algorithm. This strategy enables the EKF algorithm to swiftly iterate towards the optimal solution. The flowchart illustrating the SHADE-EKF algorithm is depicted in Figure 3.

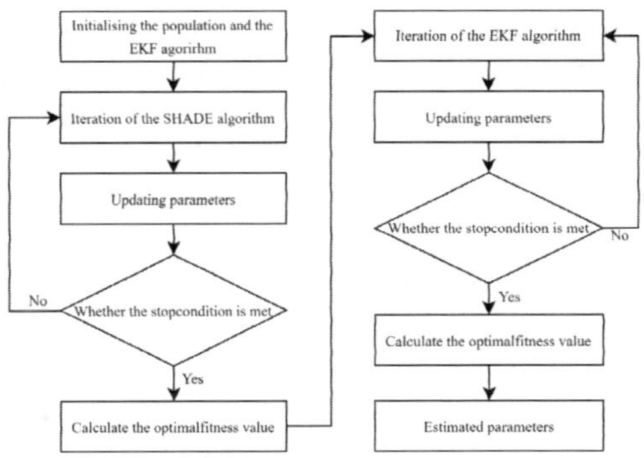

Figure 3. Diagram Illustrating the SHADE-EKF Algorithm.

Journal of Physics: Conference Series **2815** (2024) 012039 doi:10.1088/1742-6596/2815/1/012039

4. Estimation of ultrasonic echo signal parameters

4.1. Construct the objectiveffunction

In this paper, for the ultrasonic echo signal model, to achieve a more accurate estimation of the time of ultrasonic echo signal arrival, the following fitness function is constructed to minimize the mean squared difference between the actual echo parameters and the estimated echo parameters [10].

$$f(\theta) = \sum_{n=0}^{N-1} (env(\boldsymbol{\theta}, t) - y)^2 \tag{4}$$

where y represents the envelope of the sampled actual echo signal.

4.2. Parameter settings

To test the performance of the SHADE-EKF algorithm in estimating parameters of ultrasonic echo signals, simulation experiments were conducted using the asymmetric Gaussian model introduced in Chapter 2. The parameters of the ultrasonic model were set as: $\boldsymbol{\theta} = [1, 10, 0.9, 2, 1]$, and the center frequency of the ultrasonic signal was set to $40KHz$.

In the simulation experiments, the population size for all intelligent algorithms was configured to 50, with a maximum iteration count of 500. Each algorithm was run independently 100 times to obtain its average performance, ensuring a fair comparison.

4.3. Simulation

In this section, we compare the SHADE-EKF algorithm with several other advanced optimization algorithms, including PSO, ABC, SO, ACO, EKF, SHADE, and L-SHADE, using the criteria of Best Fitness (BF), Mean Best Fitness (MBF), Standard Deviation (SD), and Running Time.

Table 1. Estimation Results of Different Algorithms for ACGM.

	PSO	ABC	SO	ACO	EKF	SHADE	L-SHADE	SHADE-EKF
BF	50.05	10.86	14.61	2.37	0.2737	1.7708E-08	1.7708E-08	1.7708E-08
MBF	485.8	95.51	80.21	18.45	12.62	0.2144	0.6757	0.1193
SD	141.1	41.10	33.43	5.62	39.50	1.508	2.6881	0.5377
Running Time	0.121	0.156	1.556	1.75	0.21	2.2351	0.70317	1.5438

From Table 1, it can be observed that in 100 experiments, the PSO, ABC, SO, ACO, and EKF algorithms did not reach the optimal fitness value after 500 iterations and required more iterations to discover the optimal solution. In contrast, the SHADE algorithm, L-SHADE algorithm, and SHADE-EKF algorithm have all achieved the optimal solution in estimating the ACGM parameters, outperforming the other algorithms. Additionally, while the running time of the SHADE-EKF algorithm is slightly slower than that of the L-SHADE algorithm, it ranks second among these three algorithms. Moreover, the SHADE-EKF algorithm exhibits the smallest standard deviation, indicating its higher stability and robustness. For a more comprehensive comparison, the convergence speed of the successful three algorithms can be observed in the following figure:

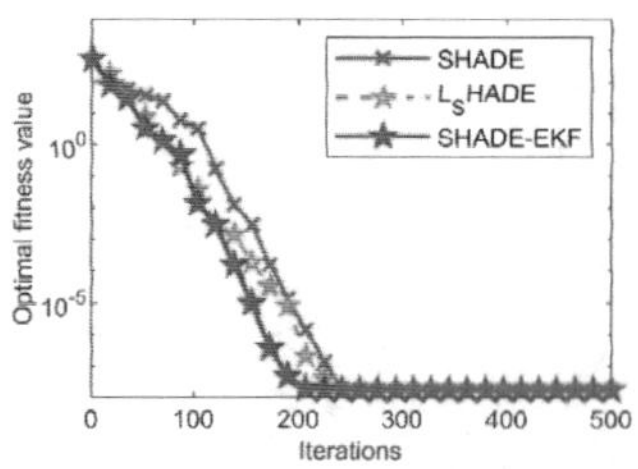

Figure 4. Convergence Plots of the Three Algorithms.

From Figure 4, it is evident that during the 500 iterations, the SHADE-EKF algorithm shows the fastest convergence, followed by L-SHADE. The SHADE-EKF algorithm outperforms in terms of convergence speed.

Lastly, to test the algorithm's resistance to noise interference, Gaussian white noise with signal-to-noise ratios of 20 dB, 15 dB, and 10 dB was introduced to the initial signal, depicted in Figure 5:

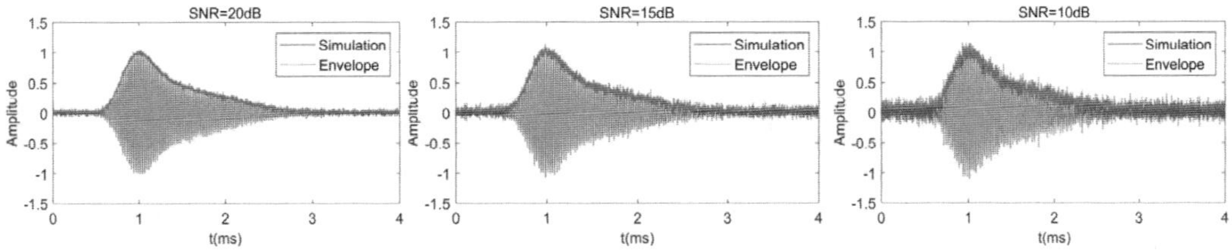

Figure 5. Signal with Noise Interference.

Figure 6 displays the envelopes of the original signal, the envelope with noise interference, and the envelope estimated by the SHADE-EKF algorithm.

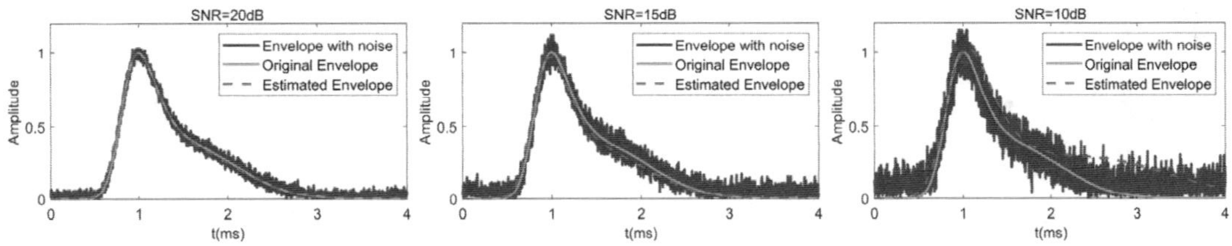

Figure 6. Fitted Envelope Plots.

The findings suggest that the SHADE-EKF algorithm is minimally impacted by noise interference at signal-to-noise ratios of 20 dB and 15 dB, with only minor errors occurring at 10 dB. Overall, the SHADE-EKF algorithm exhibits minimal sensitivity to noise interference.

5. Experiment

To showcase the superior performance of the SHADE-EKF algorithm in real-world measurements, we conducted comparisons utilizing the threshold method, L-SHADE, and SAHDE-EKF algorithms. The results are outlined in the table 2. The results for each of these cases are averaged over 100 measurements to avoid chance errors.

Table 2. Test Results Table.

Oxygen Concentration	Temp	Threshold Method	L-SHADE	SHADE-EKF
21%	25°C	22.8%	21.7%	21.5%
21%	27°C	22.6%	22%	21.4%
21%	29°C	22.7%	21.9%	21.6%
93.4%	25°C	94.7%	93%	93.5%
93.4%	27°C	92.8%	93.8%	93.6%
93.4%	29°C	94%	93.8%	93.6%

Experimental findings reveal that the precision of measurements obtained through the parameter estimation approach surpasses that of the conventional threshold method. Furthermore, when compared to the L-SHADE algorithm, the SHADE-EKF algorithm demonstrates a reduced error and superior performance in problem-solving.

6. Conclusion

This paper presents a combined algorithm integrating SHADE and EKF, which utilizes an asymmetric Gaussian model to characterize ultrasonic echo signals and measure Time of Flight (TOF). In this combined algorithm, the improved SHADE algorithm is used for initial iteration to obtain a reasonable initial value, which is then fed into the EKF algorithm for parameter estimation. This algorithm combines the advantages of both approaches, addressing issues such as the SHADE algorithm getting trapped in local optima, the EKF algorithm's dependence on initial values, and the need for a large number of iterations to obtain accurate results. Through simulation and experimentation, it has been demonstrated that this algorithm exhibits superior performance in terms of noise immunity, accuracy, robustness, and stability. The parameter settings of this algorithm are tailored specifically for the asymmetric Gaussian model in the context of TOF measurement. When applied to solve other problems, parameter adjustments according to specific scenarios are necessary to evaluate and optimize the algorithm's performance in those contexts.

References

[1] Lu Z, Yang C, Qin D, et al. (2016) Estimating ultrasonic time-of-flight through echo signal envelope and modified Gauss Newton method[J]. Measurement, 94: 355-363.
[2] Saad M M, Bleakley C J, Ballal T, et al. (2012) High-accuracy reference-free ultrasonic location estimation[J]. IEEE Transactions on Instrumentation and Measurement, 61(6): 1561-1570.
[3] Khyam M O, Ge S S, Li X, et al. (2016) Highly accurate time-of-flight measurement technique based on phase-correlation for ultrasonic ranging[J]. IEEE Sensors Journal, 17(2): 434-443.
[4] Mu L B, Xu K J, Liu B, et al. (2019) Echo signal envelope fitting based signal processing methods for ultrasonic gas flow-meter[J]. ISA transactions, 89: 233-244.
[5] Gao W, Liu W, Hu Y, et al. (2021) A novel NaCl concentration detection method based on ultrasonic impedance method[J]. IEEE Transactions on Ultrasonics, Ferroelectrics, and Frequency Control, 68(10): 3226-3233.
[6] Gao W, Liu W, Hu Y, et al. (2021) A novel NaCl concentration detection method based on ultrasonic impedance method[J]. IEEE Transactions on Ultrasonics, Ferroelectrics, and Frequency Control, 68(10): 3226-3233.
[7] Fang Z, Su R, Hu L, et al. (2021) A simple and easy-implemented time-of-flight determination method for liquid ultrasonic flow meters based on ultrasonic signal onset detection and multiple-zero-crossing technique[J]. Measurement, 168: 108398.
[8] Yacef N, Bouden T, Grimes M (2019) Accurate ultrasonic measurement technique for crack sizing using envelope detection and differential evolution[J]. NDT & E International, 102: 161-168.
[9] Demirli R, Saniie J (2001) Model-based estimation of ultrasonic echoes. Part I: Analysis and algorithms[J]. IEEE Transactions on Ultrasonics, Ferroelectrics, and Frequency Control, 48(3): 787-802.
[10] J. Xiong, P. Mao and X. Li (2023) Enhancing ultrasonic time-of-flight estimation using adaptive differential evolution and Levenberg-Marquardt algorithm[J]. IEEE Sensors Journal, 24(2): 1224-1232.

Optimization and performance verification of track vehicle multi-friction pair liquid-cooled full-disc brake system

Dianzhao Yang[1], Zhipeng Li[2], Mingxing Qiu[2], Lijin Han[1, *] and Quanfu Han[2]

[1] School of Mechanical Engineering, Beijing Institute of Technology, Beijing, China
[2] Inner Mongolia First Machinery Group Co., LTD, Inner Mongolia, China

[*] Corresponding author's e-mail: lj.han@163.com

Abstract. The brake system is a crucial component for enhancing the reliability and maneuverability potential of tracked armored vehicles, with a particular emphasis on improving braking performance and reliability. This paper proposes a novel multi-friction pair liquid-cooled full-disc brake system by extending the friction pair and introducing a liquid-cooling structure. Simulation and experimental results demonstrate that the proposed solution significantly accelerates the heat dissipation rate of the full-disc brake system compared to its initial version. Moreover, the presented approach increases the maximum average braking torque by 24.7% and reduces the braking time by 15.2% compared to the original version of the brake system.

1. Introduction

With the increasing demands on vehicle performance, the brake system has become a critical core component to ensuring high mobility and reliability of tracked vehicles [1-3]. Due to the complex service conditions and growing mobility requirements of tracked vehicles, improving the braking performance and reducing the operating temperature of the full-disc brake system has become a crucial research direction in the braking system of tracked armored vehicles.

During the braking process, the friction generated by asymmetric and repeated cyclic braking leads to high temperatures in the brake system, resulting in a decrease in performance. The majority of existing full-disc brake systems for tracked armored vehicles are air-cooled. Due to the low airflow velocity and poor thermal conductivity inside the vehicle body, the brake system operates in a high-temperature environment for extended periods. Studies by Mackin et al. [4], Zhou et al. [5], and others have shown that the rapid thermal stress generated during the braking process is closely related to the service life of the brake disc. Wang et al. [6] investigated temperature evolution of the train brake disc during high-speed braking by experiments, theoretical analysis, and finite element modeling. And results show that the temperature variation caused the corresponding fluctuation of instantaneous frictional coefficient and thermal stress distribution in the disc, which results in the thermal damages. Therefore, the design of high-performance brake systems has become increasingly important [7]-[11].

In this paper, a new type of multiple friction pairs liquid-cooled full-disc brake system is proposed through the extension of the friction pair and the design of a liquid cooling structure, resulting in improved performance. Finally, combining simulation and experimental results, it is demonstrated that the proposed new type of multiple friction pairs liquid-cooled full-disc brake system has better operating temperatures and braking performance compared to its initial version.

2. Design of the multi-friction pairs liquid-cooled full-disc brake system

The full-disc brake system adopts a structure with multiple alternating dynamic and static friction pads, demonstrating excellent braking performance and serving as a high-performance mechanical braking device for high-speed tracked vehicles [12]. To enhance vehicle braking performance and brake system reliability, as shown in Figure 1, the number of friction pairs has been increased compared to the original brake system.

Figure 1. Original brake system.

The specific pressure and the unit sliding power of the friction pair are critical indicators affecting the performance of the full-disc brake. When the vehicle reaches its maximum deceleration, the maximum braking force on the vehicle's drive wheel is:

$$F_{\max} = a(1 + \delta)M \tag{1}$$

where δ is the mass increase factor, taken as $\delta=0.16$; a is the maximum braking deceleration; M is the tank mass. The corresponding maximum braking torque for a single-side brake is $M_{zg} = \frac{F_{\max} r}{2i_c}$, and r and i_c are the active wheel radius and the side drive ratio, respectively.

The specific pressure of the friction pair is:

$$p = \frac{M_{zg}}{n\mu A r_e} \tag{2}$$

where n is the number of friction pairs; μ is the dynamic friction coefficient for iron-based powder metallurgy on iron-based powder metallurgy (μ=0.4); A is the friction area of the friction disc; r_e is the equivalent friction radius for braking torque. Here, $r_e = (r_b + r_s)/2$. Due to the requirements of porosity and processing technology, the actual friction area of brake discs in this order of magnitude generally differs from the ideal friction area by 60 to 90 cm^2, assuming it is 75 cm^2. And then:

$$A = A_r - 75 = \pi(r_b^2 - r_s^2) - 0.0075 \tag{3}$$

where r_b and r_s are the inner radius and the outer radius of the brake disc respectively, and A_r is the ideal area of the brake disc.

The unit sliding power of the friction pair is:

$$\rho = \frac{W}{nA} \tag{4}$$

where W is the unit slip power of the brake, and assuming that the vehicle's kinetic energy is entirely dissipated in the brake, it can be expressed as:

$$W = \frac{\delta_1 G v^2 \eta_c \eta_x}{4g} \tag{5}$$

where δ_1 is the mass increase factor after the passive part of the main clutch, taken as 1.16; G is the tank gravity (N); v is the tank travel speed (m/s); η_c and η_x are the efficiencies of the lateral decelerator and the drive unit, respectively; g is the acceleration due to gravity.

The basic parameters of the vehicle equipped with the brake are shown in Table 1. Due to the influence of actual processing conditions and operating conditions, the relevant parameters of the brake are set, as shown in Table 2.

Table 1. Vehicle basic parameters.

Input parameter	Parameter
Combat weight	45 t
Side drive ratio	4.42
Drive wheel radius	0.318 m
Maximum braking deceleration	5 m/s^2

Table 2. Brake parameters comparison.

Project	Multi-friction brake	Dual-friction brake
Number of friction pairs	4	2
Outer radius	0.193 m	0.227 m
Inner radius	0.133 m	0.155 m
Equivalent effective radius	0.163 m	0.191 m
Brake disc area	0.054 m2	0.079 m2

Based on the design parameters, various indicators are presented in Table 3. It can be observed that the proposed multi-friction pair liquid-cooled full-disc brake outperforms the dual-friction pair brake.

Table 3. Performance comparison of two brakes.

Indicators and related parameters	Multi-friction brake		Dual-friction brake	
	40 km/h	85 km/h	40 km/h	85 km/h
Maximum braking force F_{max}	261 kN		261 kN	
Maximum braking torque M_{zg}	9391 N · m		9391 N · m	
Efficiency of the side reducer and drive device η_c、η_c	0.95		0.95	
One-sided brake sliding power W	1.45 MJ	6.57 MJ	1.45 MJ	6.57 MJ
The specific pressure of the friction pair p	0.664 Mpa		0.782 Mpa	
The unit sliding power of the friction pair ρ	0.67 kJ/cm^2	3.04 kJ/cm^2	0.92 kJ/cm^2	4.16 kJ/cm^2

3. Structure of multi-friction-pair liquid-cooled full-track brake

According to the design parameters, a three-dimensional model of the multi-friction-pair liquid-cooled full-track brake is established. As shown in Figure 2, its structure mainly includes the connecting gear (1), brake disc assembly with inner friction lining (2), connecting seat assembly (3), rotating cam disc assembly (4), moving cam disc assembly (5), ventilated disc assembly (6), moving disc assembly (7), and brake disc assembly with outer friction lining (8).

On each side of the individual brake, there is a brake disc assembly with a friction lining ((2) and (8)), and the two brake disc assemblies are fixed on both sides of the connecting gear by bolts. The ventilated disc assembly is fixed on the gear by a spline and can move axially. Several steel balls (14) are placed between the wedge-shaped grooves of the rotating cam disc assembly and the moving cam disc assembly to produce radial tightening effects during rotational motion. The moving disc assembly is positioned by double-headed bolts (12) and can only slide axially.

The brake disc assembly with the friction lining is riveted with a circle of friction lining (10) on the inner side, and the connecting seat assembly is equipped with a circle of friction lining (9), which is positioned by pins (15). The moving cam disc assembly is equipped with several friction pads, also positioned by pins. The two friction surfaces of the ventilated disc assembly are riveted with several friction linings. The two friction surfaces of the moving disc assembly are fitted with several friction pads positioned by cylindrical pins.

The brake cylinder drives the rotating cam disc, causing it to rotate. Simultaneously, it pulls the rotating return spring (shown as (19) in Figure 3). After the rotating cam disc rotates, it generates amplified axial thrust, propelling the moving cam disc in the axial direction while pulling the axial return spring (13). After the braking process is completed, the brake returns to its original position under the influence of the brake cylinder's return and the return spring.

The sharp increase in temperature during the brake operation can significantly affect its performance. Therefore, the average temperature rise of the proposed multi-friction pair full-disc brake was calculated. According to its 3D model, the measured masses of the rotating and non-rotating discs are 59.56 kg and 83.89 kg, respectively. Assuming that initially, the majority of the heat is concentrated on the brake disc surface [13], and then rapidly cooled by surrounding components and airflow, the average temperature rise of the brake (after a single braking until the vehicle reaches a stationary state) is calculated as follows:

$$\Delta t = \frac{W}{427Cm} \tag{6}$$

where, 427 is the heat equivalent, C is the specific heat capacity (assuming C=0.50 kJ/kg· ℃ for cast iron), and m is the weight of the heat-absorbing component (Kg). It is assumed that the heat generated by the brake is distributed in a 1:1 ratio between the rotating and non-rotating discs. The calculation results are shown in Table 4.

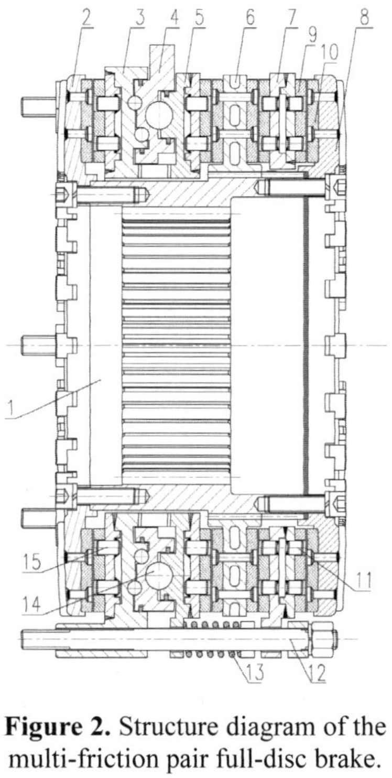

Figure 2. Structure diagram of the multi-friction pair full-disc brake.

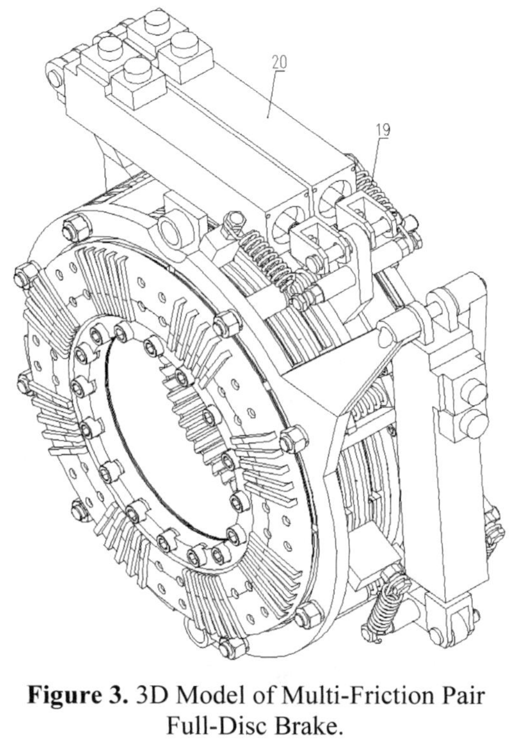

Figure 3. 3D Model of Multi-Friction Pair Full-Disc Brake.

Table 4. The average temperature rise of the multi-friction pair brake.

Indicators and related parameters	Multi-friction brake	
	40 km/h	85 km/h
Average temperature rise of non-rotating disc	40.5°C	183.4°C
Average temperature rise of rotating disc	57.0°C	258.3°C

According to Table 4, it can be observed that the brake has a significant temperature rise in the air-cooling mode, which would compromise the performance and reliability of the brake. Therefore, cooling channels are arranged in the moving disc assembly, moving shoe disc assembly, and connecting seat assembly to enhance the brake's heat dissipation conditions, allowing the brake to work with liquid circulation cooling to reduce operating temperatures.

4. Simulation analysis of cooling effect for multi-friction-pair liquid-cooled full brake

To verify the superiority of the proposed multi-friction-pair liquid-cooled full brake, Ansys was applied to simulate the brake temperature when the vehicle decelerated rapidly from 85 km/h to 0 km/h. Taking friction pair 1 as an example, finite element analysis was conducted on its temperature variation. The calculation results are shown in Figure 4 to Figure 6.

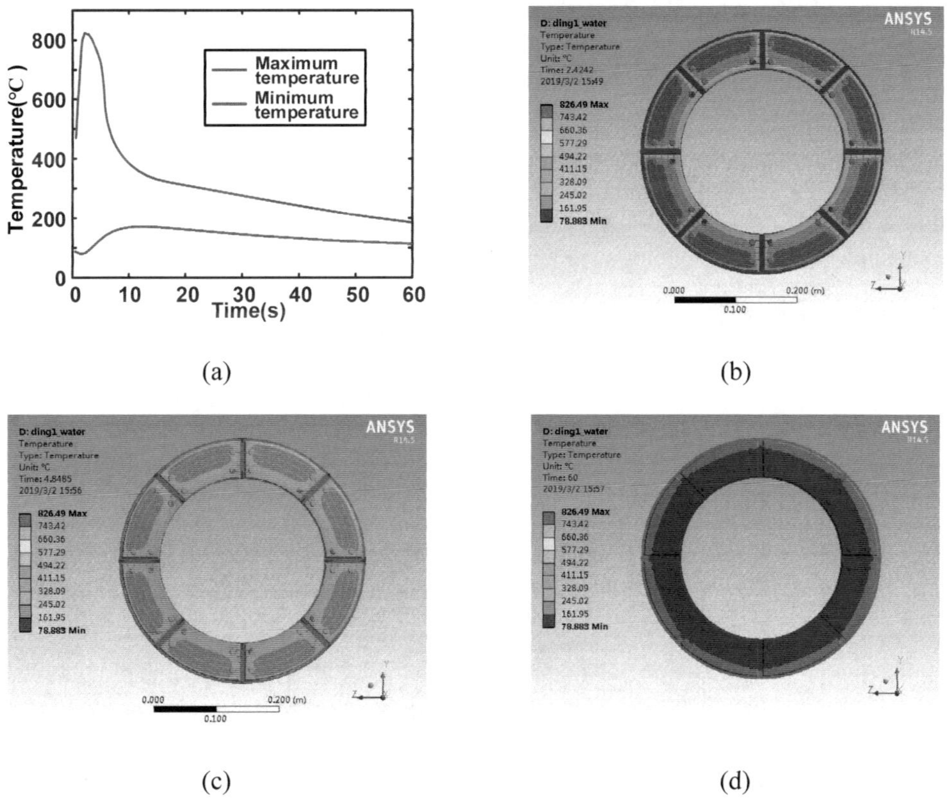

Figure 4. Friction pair 1 fixed plate water cooling analysis results: (a) Temperature variation curve, (b) Maximum temperature contour map, (c) Temperature contour map at the end of braking and (d) Analysis-end temperature contour map.

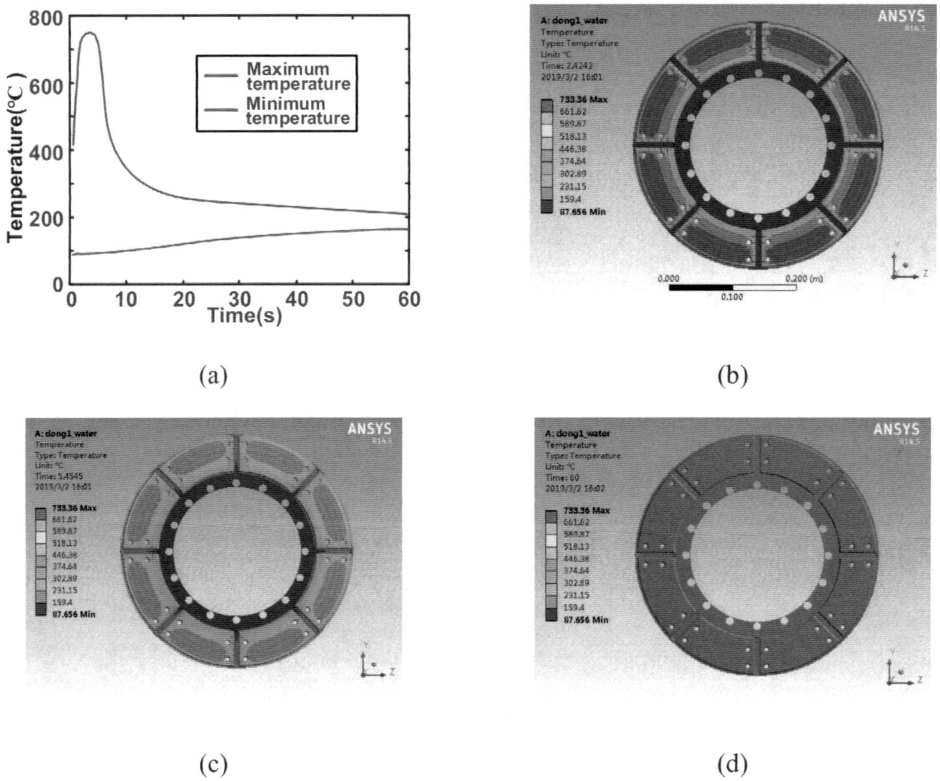

Figure 5. Friction pair 1 moving plate water cooling analysis: (a) Temperature variation curve, (b) Maximum temperature contour map, (c) Temperature contour map at the end of braking and (d) Analysis-end temperature contour map.

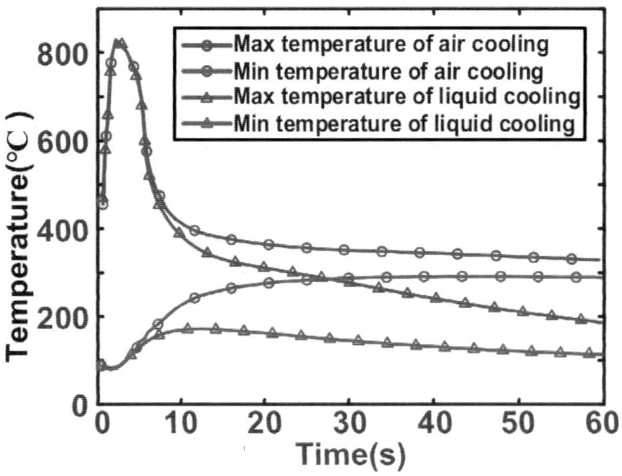

Figure 6. Friction pair 1 fixed disk temperature change curve.

According to the simulation results, the temperature of the brake increases sharply during the rapid deceleration of the vehicle. At this point, the temperature difference between the two cooling methods is relatively small. However, as time progresses, the rate of temperature decrease in the water-cooled brake is much higher than that in the air-cooled brake. This indicates that the multi-friction pair liquid-cooled full-disc brake has better heat dissipation performance.

5. Experimental testing on the bench of multi-friction pair liquid-cooled full disk brakes

In this section, experimental comparisons of the braking performance between the multi-friction pair liquid-cooled full disk brake and the twin-friction pair full disk brake were conducted. The experimental parameters and conditions are detailed in Table 5 and Table 6, respectively.

Table 5. Experimental project.

Project	Experimental content
Apply emergency braking to the brake under the maximum actuating force.	(1) The experiment for the single-side brake was conducted with a configured inertia of 22.5 tons. Prior to the experiment, the test subject should undergo sufficient running-in. (2) Emergency braking is performed according to the conditions specified in Table 6.

Table 6. Experimental conditions.

Initial speed before braking n_p	Vehicle speed	Brake oil pressure	Kinetic energy
2200 r/min	60 km/h	6.4 Mpa	3.13 MJ

The structure of the experimental platform is shown in Figure 7. The platform is primarily composed of a 750 kW drive motor, high-power clutch, inertia flywheel, speed, and torque measuring instrument, liquid-cooled brake, brake cylinder and its control system, temperature sensors, and supply system. The actual layout of the experimental platform is depicted in Figure 8.

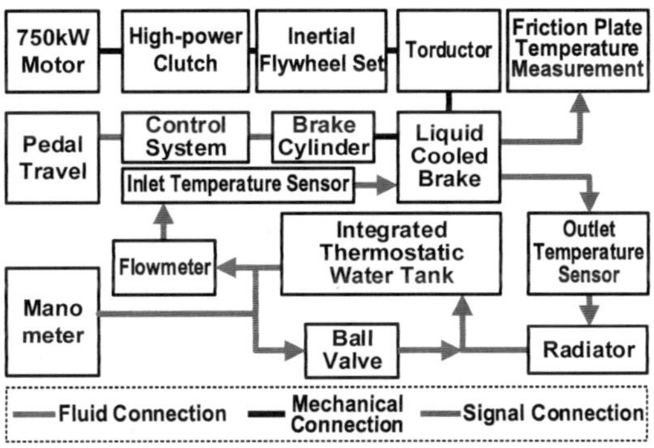

Figure 7. Schematic diagram of the experimental platform structure.

Figure 8. Experimental platform.

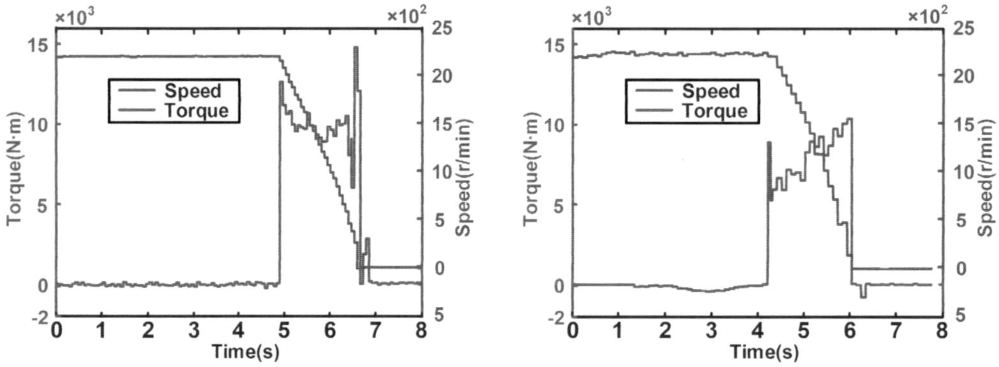

Figure 9. Brake performance comparison: (a) Multi-friction-liquid-cooled full-brake disc and (b) Double-friction full-brake disc.

As shown in Figure 9, by collecting torque-time and speed-time curves of the two brakes in the braking process and comparing their performance, it can be seen that the multi-friction pair liquid-cooled brake proposed in this paper can increase the maximum average braking torque by 24.7%, and the maximum braking torque can exceed 12 kN·m. The braking time can be reduced by 15.2%. This indicates that, without increasing the size of the brake and the thrust of the brake cylinder, the braking efficiency of the full-disc brake can be effectively improved by increasing the number of friction pairs.

6. Layout of multi-friction pair liquid-cooled full-disc brake on vehicle

When integrating the multi-friction pair liquid-cooled full-disc brake into tracked armored vehicles, as shown in Figure 10, an option is to parallelize its cooling water circuit with the engine's water circuit, eliminating the need for an additional independent cooling system. If space permits, it is also possible to consider equipping the multi-friction pair liquid-cooled full-disc brake with a separate cooling system to achieve independent control of the cooling system for the liquid-cooled full-disc brake. This helps reduce unnecessary circulation in the cooling water circuit of the liquid-cooled full-disc brake, although such a configuration will increase the overall complexity of the cooling system and occupy more space.

The connection between the multi-friction pair liquid-cooled full-disc brake and the vehicle body is relatively simple, providing ample space for brake cylinder layout. When fixed together with the transmission box, they can be lifted together, facilitating the subsequent inspection, replacement, and maintenance of the brake.

Figure 10. Multi-friction pair liquid cooled overall brake loading situation.

7. Conclusion

The objective of this study is to enhance the reliability and maneuverability of tracked armored vehicles, with a specific focus on improving braking performance. By extending the friction pairs and introducing a liquid-cooling structure, an innovative multi-friction pair liquid-cooled full-disc brake is proposed. Simulation and experimental results indicate: (1) Following rapid deceleration of the vehicle, the temperature reduction rate of the proposed multi-friction pair liquid-cooled full-disc brake is significantly higher than that of the twin-friction pair air-cooled brake, highlighting the excellent heat dissipation performance of the multi-friction pair liquid-cooled brake. (2) The multi-friction pair liquid-cooled brake can increase the maximum average braking torque by 24.7%, the maximum braking torque can exceed 12 kN·m, and the braking time can be shortened by 15.2%.

References

[1] Ning K, Li H and Zhang H 2004 Research and development of dry disk brake *Veh. & Pow. Tech.* **1** 16-22
[2] Yang X, Yang Z and Zhang Z 2017 Regenerative braking strategy research based on multi-factor input fuzzy control *J. Hunan Univ.: Nat. Sci.* **10** 17-24
[3] Peng X and Wen K 2018 Control strategy of optimal braking force distribution for vehicle with brake-by-wire system after single wheel brake failure *J. Hunan Univ.: Nat. Sci.* **45** 44-53
[4] Mackin T J, Noe S C, Ball K J and et al 2002 Thermal cracking in disc brakes *Eng. Fail. Anal.* **9** 63-76
[5] Zhou Z, Zhang H and Fan Q 1998 Investigation in failure mechanism and service life of brake disc *China Petr. Mach.* **26** 29-31
[6] Wang Z, Han J, Liu X, Li Z, Yang Z and Chen E 2019 Temperature evolution of the train brake disc during high-speed braking Adv. Mech. Eng. **11**.
[7] Zhang S, Zhang Y, Ning K, Yan N and Fang Q 2019 Research on forecasting methods of braking energy/force distribution for high speed tracked vehicles *Acta Arma.* **40** 904-14
[8] Xia L and Deng Z 2018 Calculation and analysis of friction torque and energy dissipation of electromechanical brake actuator *J. Hunan Univ.: Nat. Sci.* **45** 48-56
[9] Adamowicz A and Grzes P 2011 Analysis of disc brake temperature distribution during single braking under non-axisymmetric load *Appl. Therm. Eng.* **31** 1003-12
[10] Teng J, Li W, Cheng D and Chen Z 2008 Calculation of the temperature on Al-Si/SiCp composite material brake disc during braking *J. Hunan Univ.: Nat. Sci.* **35** 71-74
[11] Adamowicz A and Grzes P 2011 Influence of convective cooling on a disc brake temperature distribution during repetitive braking *Appl. Therm. Eng.* **31** 2177-85
[12] Lin Z and Chen J 1994 Design of the clearance compensating adjuster in hydaulie multiple-friction-disc brake *China Mech. Eng.* **5** 48-49
[13] Yan A and Xu H 2003 Analysis of temperature and stress of electro-rheological(ER) brake using ANSYS *Mech. Sci. Tech.* **22** 556-557,652

Journal of Physics: Conference Series **2815** (2024) 012041 doi:10.1088/1742-6596/2815/1/012041

Consistency and repeatability verification for vehicle in the loop test platform based on digital twin of proving ground

Zhichen Huang[1,2], Qiangsheng Liu[3,4], Liang Chen[1,2], Yingjie Ke[3,4], Xiongdong Huang[3,4] and Xinghui Zhang[1,2,*]

[1]Xiamen Product Quality Supervision and Inspection Institute, Fujian, Xiamen, 361023, China

[2]Fujian Intelligent Connected Vehicle Product Quality Inspection Center, Fujian, Xiamen, 361023, China

[3]Xiamen Kinglong United Automobile Industry Co., Ltd, Fujian, Xiamen, 361023, China

[4]Key Laboratory of Intelligent Connected Commercial Vehicle (Information Discipline) of Fujian Province, Fujian, Xiamen, 361023, China

[*]Corresponding author's e-mail:zhangxinghui@xmzjy.org

Abstract. A high-confidence vehicle-in-the-loop test method based on the digital twin of proving ground (DTPG-VIL) is proposed for the test of autonomous driving vehicles. The overall architectural design of the DTPG-VIL and the corresponding components are introduced, with a focus on describing the scene simulation system based on the digital twin of the proving ground. A King Long intelligent driving sightseeing bus for test verification is introduced, and its positioning performance is verified by the high-definition positioning base station and GPS/INS combined navigation system to verify the accuracy of its positioning precision and meet the requirements of the digital twin test so that the DTPG-VIL can obtain more information from the bus. Taking the automatic emergency avoidance function as an example, the method, process, and experimental results of the control test between the DTPG-VIL test and the real vehicle test in a proving ground are designed. Through the correlation coefficient analysis and transient response analysis, consistency and repeatability comparisons are carried out to verify the high confidence of the digital twin test.

1. Introduction

According to the testing process, the testing content of autonomous driving (intelligent connected) vehicles can be divided into model-in-the-loop testing, software-in-the-loop testing, hardware-in-the-loop testing, driver-in-the-loop testing, vehicle-in-the-loop (VIL) testing, closed proving ground testing and open road testing [1]. Vehicle-in-the-loop test is to embed the whole vehicle into the virtual environment for testing, mainly based on the rotary hub (chassis dynamometer) and the closed proving ground [2].

The overall architecture of the chassis dynamometer-based VIL test platform studied by Wang et al. includes a road simulation subsystem, a virtual scenario subsystem, a sensor simulation subsystem, a platform measurement and control system, and a test management platform [3]. The VIL co-simulation framework studied by Li et al. includes a driving robot, a rotary hub pedestal, a scenario virtual subsystem, a sensor simulation subsystem, etc.[4]. The research focuses on the chassis dynamometer-based VIL test platform, including the steering follower system of the rotating hub stage [5], the virtual test scene reconstruction method, and the sensor mode data simulation and injection method [6] etc. Diewald et al. use the radar target simulator embedded in the chassis dynamometer-based VIL test platform to solve the problem of the lack of real radar echoes[7]. Chen et al. utilized an in-vehicle video dark box for virtual scene injection [8]. Chassis dynamometer-based VIL has been applied to ADAS function tests such as autonomous emergency braking tests [9], longitudinal control tests [10], as well as road limit condition tests [11]. The chassis dynamometer-based VIL, combined with environmental simulation, allows flexible and repeatable indoor testing. However, chassis dynamometer and sensor simulation equipment are expensive, and restricted steering angles lead to limited test scenarios.

The closed proving ground-based VIL (PG-VIL) test platform studied by Butenuth et al. consists of a scenario simulation subsystem. This visualization system adds virtual targets and scenarios to the real-world test environment, aiming to ensure the maximum possible number of real states of the components under test [12]. The application of the PG-VIL focuses on testing and calibrating the autonomous driving functionality, usually ignoring the sensors used. Selim et al. designed a VIL test platform that is capable of simulating static environments, traffic flows, and the corresponding static environment sensor signals and dynamic environment sensor signals for ADAS functional validation [13-14]. The PG-VI has been applied to fully automatic control testing at intersections [15], right-turn collision avoidance system development testing [16], and automatic emergency braking scenario testing for a stationary vehicle in front of the vehicle [17]. Drechsler et al. also have validated repeatability by SIL vs. VIL [17]. Son et al. have also carried out validation of the consistency of real vehicle tests and VIL test results by means of a scenario injection method [18]. Besides, research on the PG-VIL test platform tends to focus on VIL integration-related methods as well as the effectiveness verification of automated driving function tests.

Therefore, this study designs a VIL test platform based on the digital twin of a proving ground (DTPG-VIL), which especially has convenient comparison test conditions to vs. real vehicle tests. Taking the 3rd-level King Long automated driving sightseeing bus as the test object, the positioning performance of the test object is verified to meet the data accuracy requirements through the high-definition positioning base station and the GPS/INS combined navigation system. Taking the automatic emergency avoidance function as an example, the method and process of the control experiment are designed. The high confidence level of the test platform is verified by correlation coefficient analysis and transient response analysis of the control experiment results.

2. DTPG-VIL architecture

The overall architecture of the DTPG-VIL is shown in Figure 1 below.

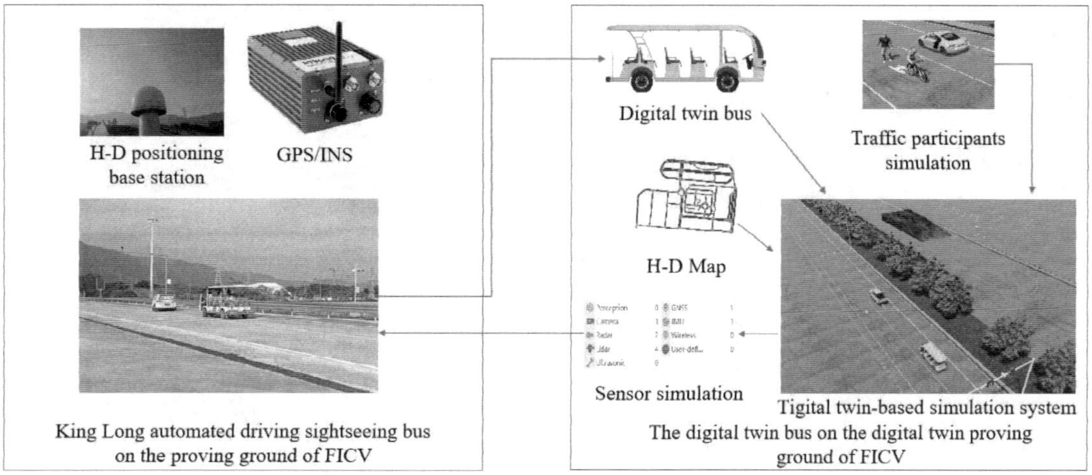

Figure 1. The overall architecture of the DTPG-VIL.

The digital twin-based simulation system takes the high-definition map of the closed proving ground as the blueprint and takes the road and its ancillary facilities as the objects to establish a one-to-one digital twin test platform. The system has components such as traffic participant simulation, traffic flow simulation, environment simulation, and sensor simulation. The Vehicle Under Test (VUT) receives the simulation scenario data and executes it with automatic control according to the driving task. The execution results are fed back to the simulation system, and the feedback data include but are not limited to, information such as position, speed, acceleration, elevation, braking status, and steering status. Before the digital twin test, the high-definition positioning base station and GPS/INS combined navigation system in the closed proving ground are used to verify the positioning function of the VUT to ensure the accuracy of test data.

The closed proving ground of Fujian Intelligent Connected Vehicle Product Quality and Inspection Center (FICV) can carry out automatic driving function tests such as traffic signal recognition and response, road traffic infrastructure and obstacle recognition and response, pedestrian and non-motorized vehicle recognition and response, peripheral vehicle driving state recognition and response, automatic emergency risk avoidance, vehicle positioning and so on.

Taking the closed proving ground of FICV as the object, the high-definition map, as shown in Figure 2, is collected and produced, containing attribute data such as road network, lane network, road markings, and the geometry, attributes, and relationships of road facilities. The scene simulation system supports the docking of high-definition map data with the engine, supports high-definition map rendering, and is capable of digital scene modeling, as shown in Figure 3.

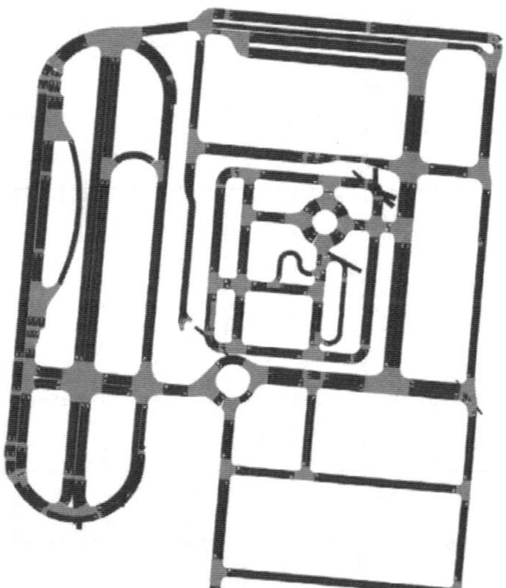

Figure 2. High-definition map of FICV.

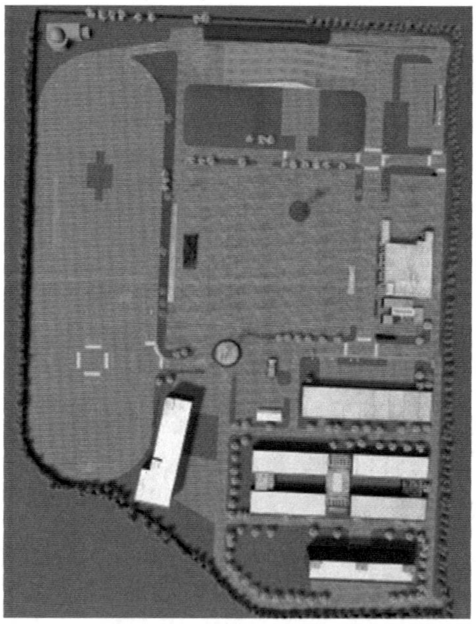

Figure 3. Digital twin proving ground of FICV.

3. Vehicle under test

3.1. The 3rd-level king long automated driving sightseeing bus

Based on the multi-sensor fusion sensing capability of the camera, LIDAR, and millimeter-wave radar, King Long automated driving sightseeing bus, as shown in Figure 4, can realize the recognition, prediction, and tracking of pedestrians, vehicles, and non-motorized vehicles.

Figure 4. King Long automated driving sightseeing bus off product line.

Based on the multi-sensor fusion positioning and map-building capabilities of BeiDou/GPS satellite positioning, inertial guidance, and LiDAR, the bus can create a high-definition map of the FICV closed proving ground, as shown in Figure 5, including intersections, roundabouts, tunnels, ramps, and other scenarios have been established. The vehicle positioning can be achieved in the constructed map.

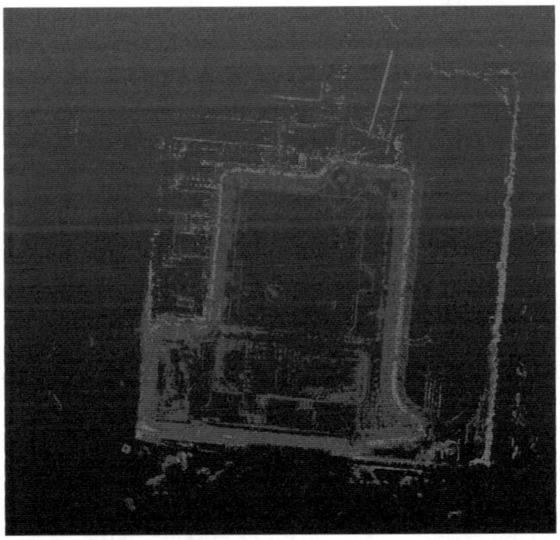

Figure 5. Lidar point cloud map of FICV proving ground created by the bus.

Based on the capabilities of fusion sensing, fusion positioning, and high-performance intelligent driving platform, chassis domain controller, the bus can perform automatic driving functions such as lane keeping, low-speed following, automatic emergency avoidance, fixed-point stopping, intersection passing, tunnel passing as shown in Figure 6, and so on.

Figure 6. Simulated tunnel passing test.

The King Long automated driving sightseeing bus is designed to have a maximum speed of 30 km/h in manual mode and 20 km/h in automatic mode, with a maximum sensing distance of 30 m.

3.2. H-D positioning base station and GPS/INS combined navigation system

GPS/BeiDou RTK high-definition positioning base station corrects the error of the satellite positioning system and gives the positioning target a differential signal through the network to achieve accurate positioning, which has the advantages of high accuracy, high speed, high efficiency, etc. The GPS/INS combined navigation system can detect the position, speed, altitude, and other parameters of the vehicle, and combined with the high-definition positioning base station; the positioning accuracy can reach the centimeter level. The key performance indicators of the positioning function are shown in Table 1 below.

Table 1. Key performance indicators of positioning function.

H-D positioning base station		GPS/INS combined navigation system	
Performance	Indicators	Performance	Indicators
Plane accuracy	<1 cm	Positioning accuracy	1 cm
Elevation accuracy	<2 cm	Speed accuracy	0.5 km/h

3.3. Positioning function test of VUT

The test cases are designed as follows:

- (1) The bus accelerates from a standstill to 27 km/h, then travels at a constant speed in a straight line for 30 s.
- (2) The bus accelerates from a standstill to 5 km/h, and then travels at a constant speed in a straight line for 60 s.
- (3) The bus accelerates from a standstill to travel in a straight line and then randomly accelerates and decelerates during the traveling process, with a traveling time of 120 s.

The position (x_1, y_1) and distance d relative starting point, speed v detected by GPS/BeiDou RTK high-definition positioning base station + GPS/INS combined navigation system positioning are the reference values, and the position (x_2, y_2) relative starting point outputted by the bus is the object to be measured. The positional error is adopted by Equation (1):

$$e = \sqrt{(x_2 - x_1)^2 + (y_2 - y_1)^2} \tag{1}$$

The test results are shown in Figure 7~ Figure 12, in which the horizontal coordinate is distance *d(m)*, and the vertical coordinate is speed *v(km/h) or* positional error *e(m)*:

(1) Experiment 1, high-speed driving condition.

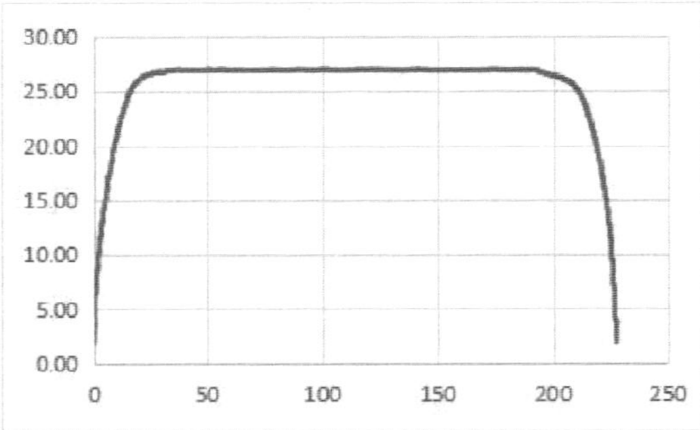

Figure 7. Distance and speed curve of experiment 1.

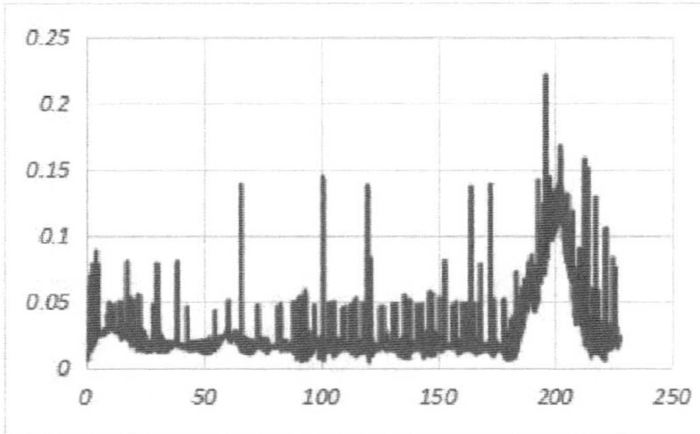

Figure 8. Distance and error curve of experiment 1.

(2) Experiment 2, low-speed driving condition.

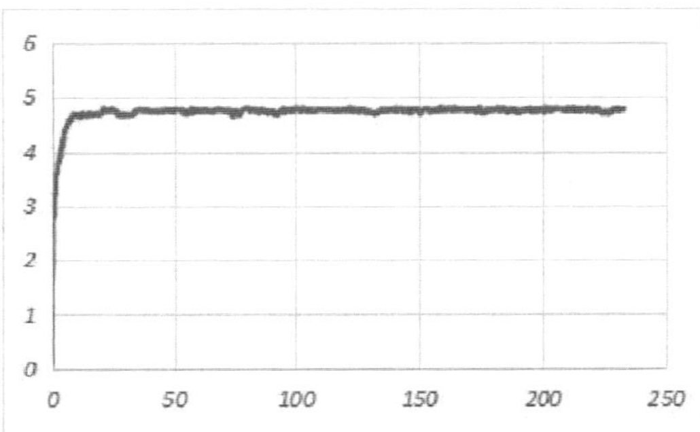

Figure 9. Distance and speed curve of experiment 2.

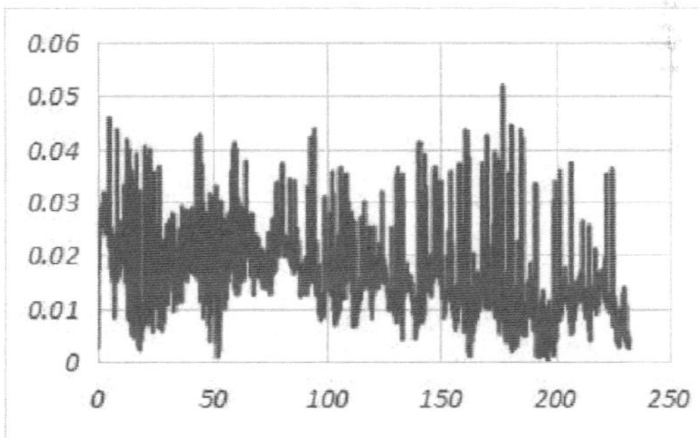

Figure 10. Distance and error curve of experiment 2.

(3) Experiment 3, random acceleration/deceleration driving condition.

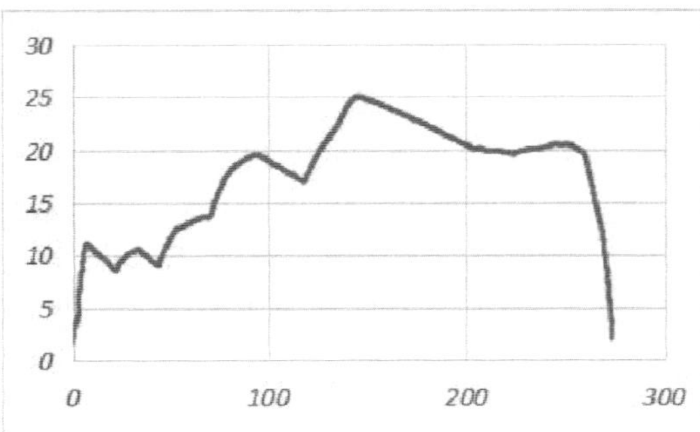

Figure 11. Distance and speed curve of experiment 3.

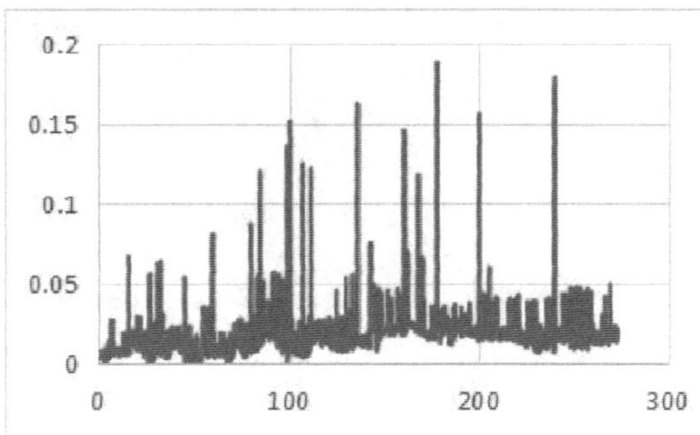

Figure 12. Distance and error curve of experiment 3.

The root mean square adopted by Equation (2) is used as an indicator for the statistical analysis of errors. The test results are shown in Table 2.

$$rms = \sqrt{\frac{\sum_{i=1}^{n} e_i^2}{n}} \tag{2}$$

Table 2. Root mean square of distance error.

Experiment	rms
1	0.04 m
2	0.02 m
3	0.02 m
Average	0.03 m

This distance error is accurate and valid compared to many regulatory standards that require test equipment position acquisition accuracy of no more than 0.1 m. Therefore, vehicle position data can be obtained directly from the CAN bus signal of the VUT and combined with speed, acceleration, braking, turn indicator information, etc., can be sent to the simulation system to realize the vehicle's digital twin, which makes the digital twin test more realistic.

4. Controlled test method of DTPG-VIL vs. real vehicle test

4.1. Controlled test method

An automatic emergency risk avoidance function (customized by King Long) is taken as an example of the controlled test method. In the long straight two-way two lanes, there is a King Long automated driving sightseeing bus that is stationary, and a stationary target object is 100 meters ahead of this lane. The bus automatically accelerates from a stationary state to 18 km/h (5 m/s) and then moves forward at a constant speed until the automatic braking stops. The test group uses a digital simulation car as the target object, and the control group uses a semi-truncated balloon car as the target object. The rest of the test conditions are basically the same, and two tests are conducted, respectively, as shown in Figures 13 and 14. During the autopilot task, the driver acts as a safety officer and does not perform driving operations.

Figure 13. The test group using a digital simulation car.

Figure 14. the control group using a semi-truncated balloon car.

4.2. Test results

The test results are shown in Figures 15-22 below. In the time-velocity curve, the horizontal coordinate is time *t* (*ms*), the sampling period is 10 ms. The vertical coordinate is vehicle fusion localization perceived velocity v_{vut} (*m/s*). In the time-distance curve, the horizontal coordinate is time *t* (ms), the sampling period 10 ms, and the vertical coordinate is the distance v_{object} *(m)* between the vehicle fusion localization perceived position and the position of the digitally simulated target or the semi-truncated balloon vehicle.

(1) Experiment A. A digital simulation car.

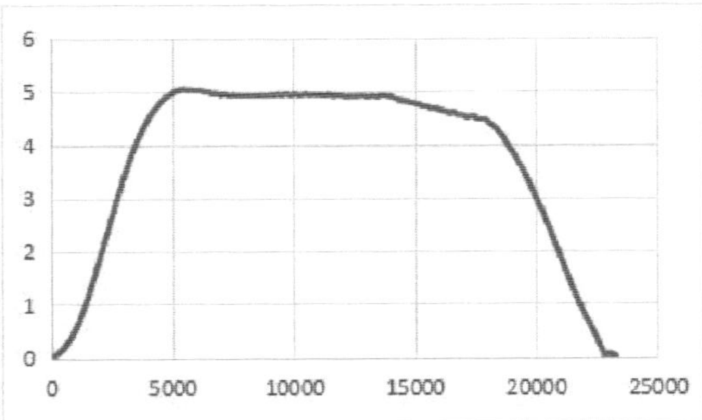

Figure 15. Time-velocity curve of experiment A.

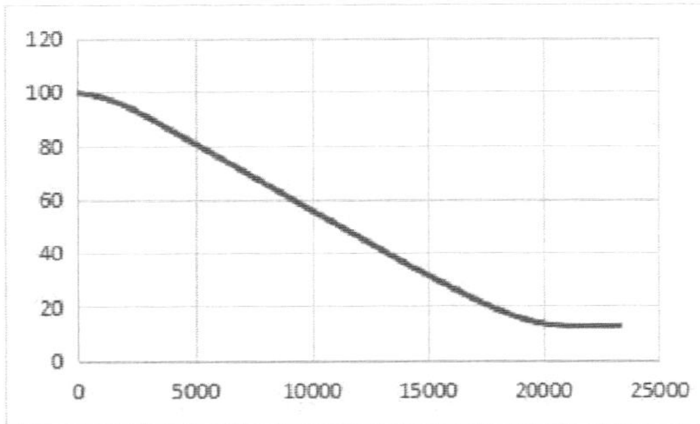

Figure 16. Time-distance curve of experiment A.

(2) Experiment B. A digital simulation car.

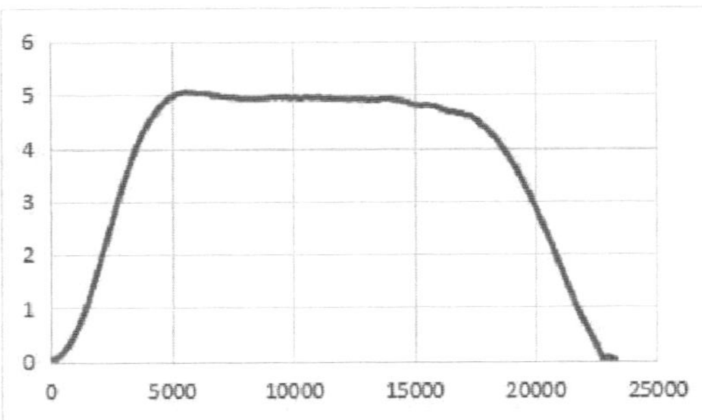

Figure 17. Time-velocity curve of experiment B.

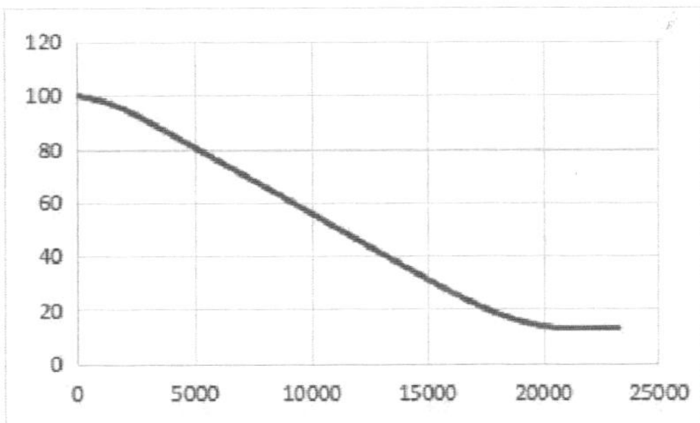

Figure 18. Time-distance curve of experiment B.

(3) Experiment C. A semi-truncated balloon car.

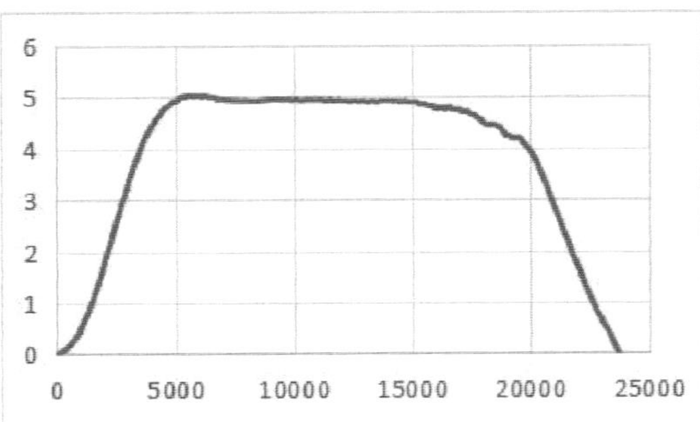

Figure 19. Time-velocity curve of experiment C.

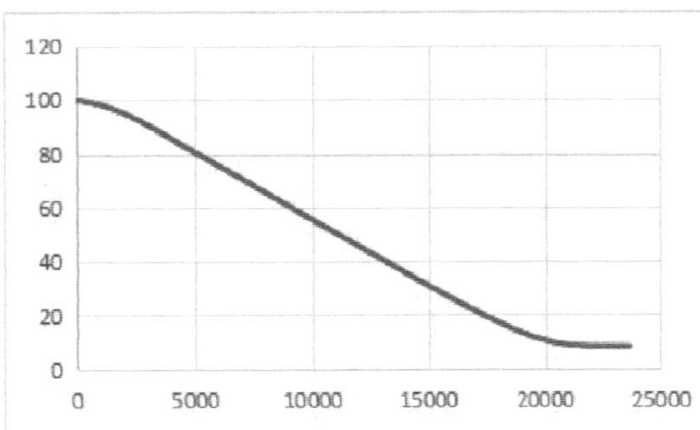

Figure 20. Time-distance curve of experiment C.

(4) Experiment D. A semi-truncated balloon car.

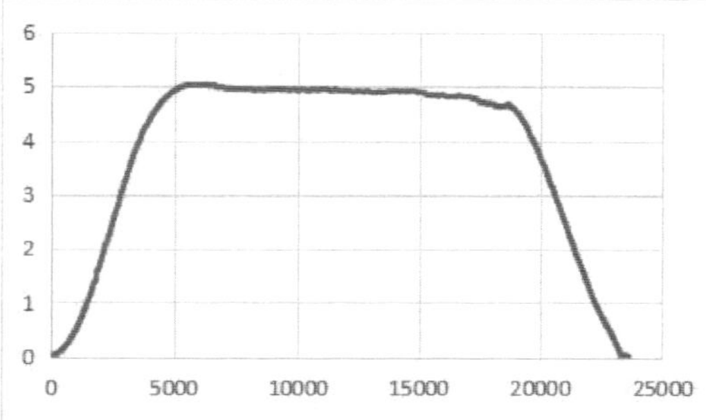

Figure 21. Time-velocity curve of experiment D.

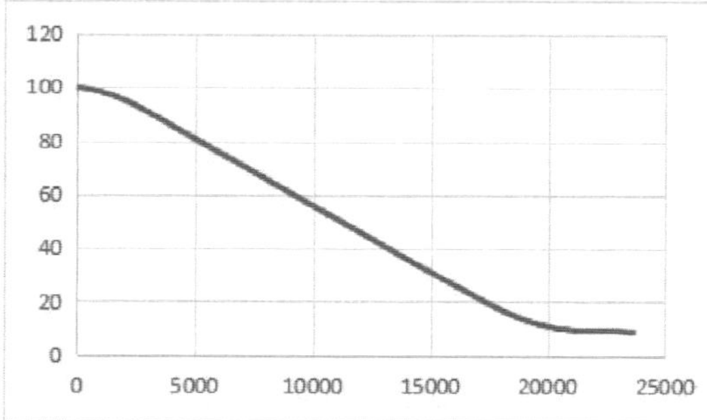

Figure 22. Time-distance curve of experiment D.

5. Consistency and repeatability analysis

5.1. Consistency analysis

The time series comparison method was used to quantitatively verify the consistency of the test results [18]. The correlation coefficient is a quantity used to study the degree of linear correlation between the variables. The Pearson correlation coefficients of variable X and variable Y are shown by equation (3):

$$r = \frac{\sum_{i=1}^{n}(x_i - \bar{x})(y_i - \bar{y})}{\sqrt{\sum_{i=1}^{n}(x_i - \bar{x})^2 (y_i - \bar{y})^2}} \tag{3}$$

The same length time of v_{vut} and d_{object} of the bus from start to stop were selected for correlation analysis. The length of time of 23.73 s of experiment C is taken as the standard, and the remaining three sets of experimental tail data are supplemented with the values of v_{vut}=0 m/s and d_{object}=$d_{object\text{-}stop}$. The comparison results are shown as table 3 and table 4:

Table 3. Correlation analysis of cohort tests.

Experiment	Correlation between speeds	Correlation between distances
A and B	99.95%	100.00%
C and D	98.71%	100.00%

The above datas show that, performed under identical conditions, the test results are highly correlated between the test groups and between the control groups, but there are also random errors.

Table 4. Correlation analysis of heterogeneous group tests.

Experiment	**Correlation between speeds**	**Correlation between distances**
A and C	98.43%	99.96%
A and D	99.05%	99.97%
B and C	98.16%	99.93%
B and D	98.80%	99.95%

The above data show that, even with target object differences, the results of the digital car twin simulation test group and the semi-truncated balloon car control group are highly correlated under otherwise identical conditions and can be considered to be in general agreement, taking into account the presence of random errors.

5.2. Repeatability analysis

The vehicle longitudinal dynamics can be simplified and modeled as shown by equation (4) for a typical second-order system [19].

$$\mathrm{F} - \mathrm{F_f} = m\ddot{x} \tag{4}$$

Referenced to the relevant transient response metrics of the second-order system [19], combined with the time-velocity curve characteristics (small amount of overshooting, steady state value closing to the desired value, etc.), the following metrics are chosen for the analyses and comparisons:

- Delay time t_d: the time required for the time-velocity curve to reach half of the desired value for the first time.
- Rise time t_r: the time required for the time-velocity curve to rise from 0% to 100% of the desired value.
- Peak time t_p: the time required for the time-velocity curve to reach the first peak of the overshoot.
- Maximum overshoot M_p: the maximum peak value of the time-velocity curve.
- Emergency avoidance time t_b: the time required for the time-velocity curve to fall from 98% of the desired value (with an allowable error of 2% on the steady state value) to 10%.
- Emergency avoidance distance d_b: the distance traveled by the bus during emergency avoidance time t_b.
- Relative error e: the difference between the test value and the mean value as a percentage of the mean value.

Using the average value and relative error as the statistical analysis index, the results of the test groups are analyzed and compared as table 5:

Table 5. Transient response analysis of the test group.

Indicator	t_d	t_r	t_p	M_p	t_b	d_b
Experiment A	2.46	5.09	5.33	5.04	8.30	23.46
Experiment B	2.50	5.17	5.48	5.05	8.13	22.29
Average 1	2.48	5.13	5.41	5.05	8.22	22.88
e_A	-0.81%	-0.78%	-1.39%	-0.10%	1.03%	2.56%
e_B	0.81%	0.78%	1.39%	0.10%	-1.03%	-2.56%

The above data show that for the two trials of the digital twin simulation test group, the maximum relative error index is $|t_p|$=1.39% in the acceleration phase and $|t_b|$=1.03%, $|d_b|$=2.56% in the emergency avoidance phase. Considering the existence of random errors, it can be assumed that the two experiments are basically the same, which means the high repeatability of DTPG-VIL.

Using the average value and relative error as the statistical analysis index, the results of the control group test are analyzed and compared as table 6:

Table 6. Transient response analysis of the control group.

Indicator	t_d	t_r	t_p	M_p	t_b	d_b
Experiment C	2.5	5.22	5.52	5.03	8.99	26.44
Experiment D	2.52	5.31	5.61	5.03	8.64	25.67
Average 2	2.51	5.27	5.57	5.03	8.82	26.06
e_C	-0.40%	-0.85%	-0.81%	0.00%	1.99%	1.48%
e_D	0.40%	0.85%	0.81%	0.00%	-1.99%	-1.48%

The above datas show that for the two trials of the semi-balloon car control group, the maximum relative error indicators were $|t_r|$=0.85% in the acceleration phase and $|t_b|$=1.99%, $|d_b|$=1.48% in the emergency avoidance phase. Considering the presence of random errors, it can be assumed that the two experiments are basically the same, which means the high repeatability of the real vehicle test.

Assuming that the control group test Average 2 is an estimate of the true value, the test group Average 1 is analyzed in comparison with Average 2 as table 7:

Table 7. Comparative analysis of transient response between test and control groups.

Indicator	t_d	t_r	t_p	M_p	t_b	d_b
Average 1	2.48	5.13	5.41	5.05	8.22	22.88
Average 2	2.51	5.27	5.57	5.03	8.82	26.06
e_{12}	-1.20%	-2.56%	-2.88%	0.30%	-6.81%	-12.20%

The above data show that the maximum relative error indicators for the test group and control group tests are |tp|=2.88% in the acceleration phase and |tb|=6.81%, |db|=12.20% in the emergency avoidance phase. In the acceleration phase, the error is within 3%, which is still considered to be the effect of random error. In the emergency avoidance phase, the error is greater than 5%, but the error between the test group and the control group is smaller (as shown in Tables 6 and 7), and the existence of systematic error should be considered. Further analysis of acceleration shows that the test group decelerated earlier than the control group, which is explained by the fact that the vehicle is designed for a maximum sensing distance of 30 m. However, the test group injected the VUT by sensing the target and its distance from the beginning stage of the test and started to decelerate at a distance of 50 m.

6. Conclusions

In this paper, a vehicle-in-the-loop test platform based on the digital twin of the proving ground was designed, which makes it easy to realize the control experiment with the closed proving ground real vehicle test. The high-definition positioning base station and GPS/INS combined navigation system were used to verify that the VUT has high-definition positioning performance and meets the requirements of test data accuracy, which means more data can be combined feedback to the simulation system. Using Pearson correlation coefficient analysis and transient response analysis to compare the results of the digital twin test and the control test of the real vehicle test, the consistency and repeatability of the results were analyzed, which confirms the high confidence of the digital twin test method. In particular, the transient response analysis method was used to find the systematic errors between the digital twin test and the real vehicle test, which confirms the possible impact of ignoring the sensor sensing capability on the DTPG-VIL. In the follow-up, DTPG-VIL will be extended to other autonomous driving function tests, and the virtual sensors will be corrected before the experiment.

Acknowledgments

This work was carried out under the funding of the Science and Technology Project of Fujian Provincial Market Supervision Administration (No. FJMS2023050) and the Science and Technology Program Project of Xiamen Municipal Science and Technology Bureau (in the field of future industries).

References

[1] Wu Q, Sun Y F, Dang G L, et al. (2019) Testing and evaluation techniques. In: Keqiang Li and Mingyuan Bian (Eds.), *Electric Vehicle Engineering Handbook Volume VI Intelligent Connected.* Machinery industry press, Beijing. pp.293-294.

[2] Zhu B, Zhang P X, Zhao J, et al. (2019) Research progress on virtual testing of self-driving cars based on scenarios.J. *China Journal of Highway and Transport*, 32:1-19.

[3] WANG R M, ZHAO X M, XU Z G, et al. 2022 Design of an in-the-loop virtual simulation test platform for autonomous driving vehicle.J. *Automobile Technology*, 4:1-7.

[4] Li, H, Makkapati, V P, Wan L, et al. (2023) A. Validation of Automated Driving Function Based on the Apollo Platform: A Milestone for Simulation with Vehicle-in-the-Loop Testbed. J. *Vehicles* 5:718–731.Vehicles | Free Full-Text | Validation of Automated Driving Function Based on the Apollo Platform: A Milestone for Simulation with Vehicle-in-the-Loop Testbed (mdpi.com)

[5] Wang W W, Zhao X M, Zhen W, et al. (2020) Research on Steering-Following System of Intelligent Vehicle-in-the-Loop Testbed. J. *IEEE Access*, 8:31684-31692.Research on Steering-Following System of Intelligent Vehicle-in-the-Loop Testbed | IEEE Journals & Magazine | IEEE Xplore

[6] ZHAO X M, CHENG J J, XU Z G, et al. (2019) Indoor rapid test platform for self-driving cars based on whole-vehicle-in-the-loop simulation. J. *China Journal of Highway and Transport*, 32:124-136. doi: 10.19721/j.cnki.1001-7372.2019.06.013

[7] Diewald A, Kurz, C, Kannan, P V, et. Al. (2021) Radar Target Simulation for Vehicle-in-the-Loop Testing. J. *Vehicles,* 3: 257–271. Vehicles | Free Full-Text | Radar Target Simulation for Vehicle-in-the-Loop Testing (mdpi.com)

[8] CHEN Y F, QIU X Y, ZHOU Y, et al. (2023) A simulation test method for vehicle-in-the-loop smart driving based on visual camera scheme. J. *Shanghai Auto*. 1:8-12. doi: 10.3969/j.issn. 1007-4554.2023.01.02

[9] Gao Y, Zhao X M, Xu Z G, et al. (2021) An Indoor Vehicle-in-the-Loop Simulation Platform Testing Method for Autonomous Emergency Braking. J. *Journal of Advanced Transportation*, 4:1-12.An Indoor Vehicle-in-the-Loop Simulation Platform Testing Method for Autonomous Emergency Braking (hindawi.com)

[10] Miriam D R, Simeon I, Kevin M S, et al. (2020) Experimental Evaluation of Longitudinal Control for Automated Vehicles through Vehicle-in-the-Loop Testing. J. *SAE Technical Paper* 2020-01-0714. doi:10.4271/2020-01-0714

[11] Guo R L, Yang C, Liang, D. (2022) Applied study on functional safety limit scenario test of a steering system based on whole vehicle in the loop. In: 2022 6th CAA International Conference on Vehicular Control and Intelligence, Nanjing, China. doi:10.1109/CVCI56766.2022.9964976.

[12] Matthias B, Roland Kt, Paul P. (2017) Vehicle-in-the-Loop Real-world Vehicle Tests Combined with Virtual Scenarios. J. *ATZ worldwide*, 119:52-55.

[13] Selim S, Martin R, Marlies M. (2020) A Vehicle-in-the-Loop Methodology for Evaluating Automated Driving Functions in Virtual Traffic. In: 2020 IEEE Intelligent Vehicles Symposium (IV), Las Vegas, NV, USA. doi:10.1109/IV47402.2020.9304811

[14] Selim S, Martin R, Marlies M, et al. (2021) Hybrid Testing: A Vehicle-in-the-Loop Testing Method for the Development of Automated Driving Functions. J. *SAE International Journal of Connected and Automated Vehicles*,4:133-148. doi: 10.4271/12-04-01-0011

[15] Fayazia S A, Vahidia A, Luckowb A (2019) A Vehicle-in-the-Loop (VIL) verification of an all-autonomous intersection control scheme. J. *Transportation Research*: *Part C*, 107:193-210. doi.org/10.1016/j.trc.2019.07.027.

[16] Pongsathorn R, Yuta A. (2015) Development of Collision Avoidance System in Right Turn Maneuver Using Vehicle-in-the-Loop Simulation. J. *Journal of Robotics and Mechatronics*, 27:627-635.JRM Vol.27 p.627 (2015) | Fuji Technology Press: academic journal publisher (fujipress.jp)

[17] Maikol F D, Varun S, Fabio R, et al. (2022) Dynamic Vehicle-in-the-Loop: A Novel Method for Testing Automated Driving Functions. J. *SAE International Journal of Connected and Automated Vehicles*,5:367-380. doi:10.4271/12-05-04-0029.

Journal of Physics: Conference Series **2815** (2024) 012041 doi:10.1088/1742-6596/2815/1/012041

[18] Son W, Ha Y, Oh T, et al. (2022) PG-Based Vehicle-In-the-Loop Simulation for System Development and Consistency Validation. J. *Electronics*, 11,4073.Electronics | Free Full-Text | PG-Based Vehicle-In-the-Loop Simulation for System Development and Consistency Validation (mdpi.com).
[19] He Y H, Yang C, Liu Y H, et al. (2019) Automotive test technology. Machinery industry press, Beijing.

Journal of Physics: Conference Series **2815** (2024) 012042 doi:10.1088/1742-6596/2815/1/012042

Design and research of fuzzy PID-based liquid pump performance testing platform

Yuntao Hou[1,*], Zequan Wu[1], Xiaohua Cai[1] and Zhancheng Li[1]

[1]Heilongjiang Academy of Agricultural Machinery Sciences, Heilongjiang Academy of Agriculture Sciences, Harbin, Heilongjiang, 150081, China

[*]Corresponding author's e-mail: hyt19840324@163.com

Abstract. The pump is a crucial component of agricultural sprayer machinery, directly impacting its operational quality. This paper designs a performance testing platform for pumps. Utilizing a brushless DC motor as the power source for the pump, a closed-loop mathematical model for the brushless DC motor is established. Simulation experiments and indoor experimental tests are conducted. Simulation results indicate that the motor control system exhibits fast response, no overshooting, and excellent dynamic and static performance. Experimental tests on the platform show that the error range of motor speed is less than 0.3%, achieving excellent control performance. The constructed pump performance testing system achieves precise measurement of pump flow rates at different speeds, demonstrating outstanding performance.

1. Introduction

The liquid pump is a vital component of agricultural sprayer machinery, serving the purpose of generating high pressure for the liquid to overcome flow resistance during its passage through the pipeline. It supplies the energy required for liquid atomization, spray, and the mixing and stirring of the liquid [1-2]. The performance of the liquid pump directly influences the operational quality of the sprayer machinery. Therefore, in experimental trials of sprayer machinery, the detection of liquid pump performance is essential [3-4].

Brushless DC motors are known for their advantages, such as simple structure, absence of sparking during commutation, excellent speed regulation, and reliable operation [5-7]. They can achieve smooth speed control over a wide range [8-9]. To ensure the accuracy and stability of liquid pump performance testing, this paper employs a brushless DC motor to provide power for the pump. The brushless DC motor control system adopts a dual closed-loop control of Speed and current. The system employs a fuzzy PID control algorithm, which can effectively respond to sudden changes in load and achieve smooth speed control of the motor. By continuously adjusting the motor speed, precise control of the liquid pump flow rate is achieved. This allows for precise speed-flow characteristic curves to be obtained for testing pump performance.

2. System structure and operating principle

The liquid pump performance testing system is composed of a liquid pump, a brushless DC motor, a pressure sensor transmitter, a flow sensor transmitter, a torque-speed sensor transmitter, and other components. Its operating principle is as follows: start the motor, transmit power to the pump through a coupling, and the plunger of the pump undergoes reciprocating motion. The liquid from the medication tank enters the pump through the water inlet pipeline and filter, creating a pressurized liquid flow.

Journal of Physics: Conference Series **2815** (2024) 012042 doi:10.1088/1742-6596/2815/1/012042

During testing, the system primarily controls and measures parameters such as pressure, Speed, and flow of the liquid pump, obtaining the relationship between Speed and flow under different pressures. This allows for the evaluation of the pump's operational performance.

3. Motor control system model

To ensure excellent speed regulation, outstanding resistance to load disturbances, and good starting performance of the electronic control system, the brushless DC motor control system adopts a dual closed-loop control strategy for speed and current [10-11]. The control structure diagram is depicted in Figure 1, where the current loop and speed loop are connected in a cascading manner. The system employs a fuzzy PID control algorithm to achieve precise control. Brushless DC motors are nonlinear and coupled systems. Although traditional PID controllers have advantages such as ease of implementation and simplicity of structure, when faced with changes in load in nonlinear and coupled systems like brushless DC motors, the motor speed may suddenly decrease or increase, leading to transient instability in the system. In such cases, the advantage of fuzzy control, which exhibits strong anti-disturbance capability, becomes apparent. Fuzzy control can effectively adjust PID parameters to cope with load changes, enabling precise motor speed control. Consequently, accurate characteristic curves of pump flow rate and motor speed can be obtained to evaluate pump performance.

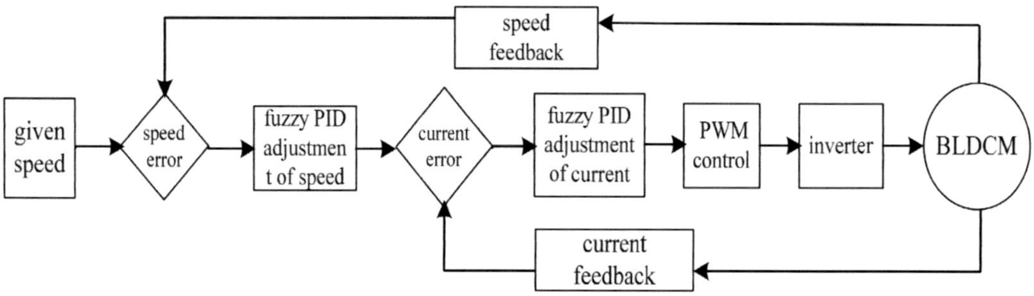

Figure 1. Double closed-loop control block diagram.

3.1. Current loop structure design

The current loop is designed to control the torque of the electric motor, emphasizing excellent tracking performance. Therefore, in this paper, the current loop is constructed based on the design principles of a Type I system. The open-loop transfer function of this current loop is expressed as follows:

$$W_I(s)=\frac{K_I}{R(T_l s+1)+K_I} \tag{1}$$

In the equation, K_I represents the amplification coefficient of the current regulator, T_l is the time constant of the current regulator, and R is the phase resistance of the motor.

The structure of the current loop is illustrated in Figure 2. In this diagram, U^*_i represents the given voltage for the current, U_i represents the feedback voltage of the current, and I is the armature current.

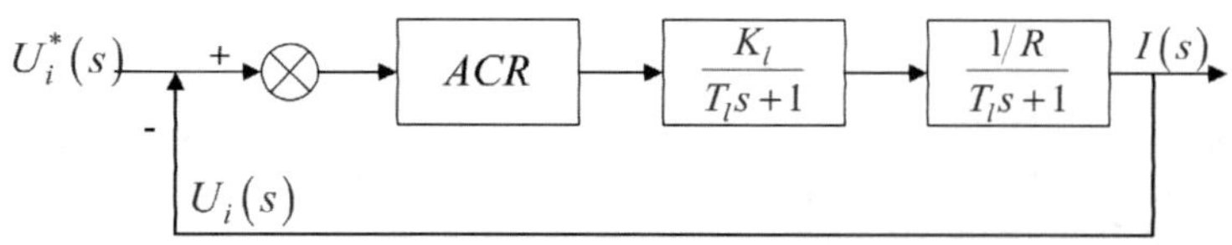

Figure 2. Current loop of structure.

3.2. Loop structure design

To achieve dynamic speed tracking and steady-state zero error, an integral component in the speed loop must be introduced. Therefore, in the design process, a structure based on a Type II system is adopted. The transfer function of the speed loop is expressed as Equation (2), ensuring that the system possesses the desired performance characteristics:

$$W_n(s)=\frac{K_n KR(\tau_n s+1)}{\tau_n C_e T_m s^2(K_l s+1)} \tag{2}$$

In the above expression, τ_n represents the lead time constant of the speed regulator, T_m is the electromechanical time constant, C_e is the back electromotive force coefficient of the motor, K_l is the amplification coefficient of the current regulator, K_n is the amplification coefficient of the speed regulator, and K represents the open-loop gain.

Figure 3 illustrates the structure of the speed loop. I_L represents the current under load, U^*_n denotes the prescribed voltage for Speed, U_n corresponds to the speed feedback voltage, and n signifies the motor speed.

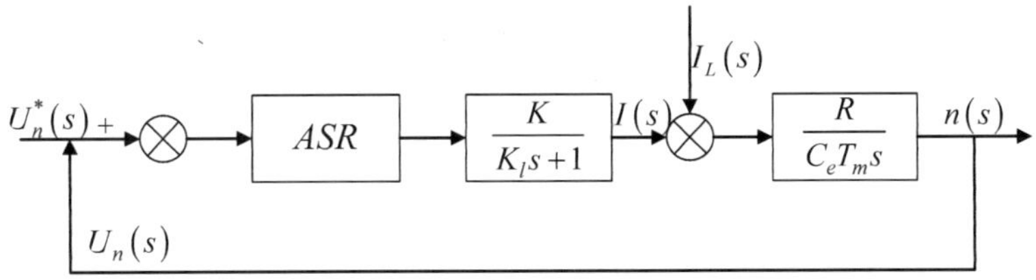

Figure 3. Speed loop of structure.

3.3. Implementation of the control algorithm

3.3.1. Incremental PID control. The PID (Proportional-Integral-Derivative) control principle involves forming a control output by linearly combining the proportional, integral, and derivative components of the error to regulate the controlled object. There are two modes of PID control: positional PID and incremental PID. The general formula for PID control is:

$$u(t)=K_p\left(e(t)+\frac{1}{T_i}\int_0^t e(t)dt+T_d\frac{de(t)}{dt}\right)+u_0 \tag{3}$$

In the above formula, K_p represents the proportional gain; T_i is the integral time constant, T_d is the derivative time constant, e(t) denotes the deviation between the setpoint and the actual output, and u_0 is the initial value of the control variable.

When applying the PID control algorithm to a control system, it is necessary to discretize the formula (3), resulting in:

$$u(k)=K_p e(k)+K_i\sum_{j=0}^{k} e_j+K_d\left(e(k)-e(k-1)\right)+u_0 \tag{4}$$

In the above formula, $K_i=K_p(T/T_i)$ and $K_d=K_p(T_d/T)$ are the discretization coefficients for integral and derivative terms, where T is the sampling period, and k is the current sampling instant.

Based on the discretization process in Formula (4), the output value at the (k-1) th sampling instant is obtained as:

$$u(k-1)=K_p e(k-1)+K_i\sum_{j=0}^{k-1} e_j+K_d\left(e(k-1)-e(k-2)\right)+u_0 \tag{5}$$

Subtracting Formula (5) from Formula (4) yields the formula for incremental PID control:

$$\Delta u(k) = K_p(e(k) - e(k-1)) + K_i e_k + K_d(e(k) - 2e(k-1) + e(k-2)) \quad (6)$$

The incremental PID control algorithm is an optimized form of PID control that has the advantage of not requiring cumulative calculations, thereby reducing computation time. When the system experiences disturbances leading to significant errors, the incremental PID control algorithm can employ logical judgment for error detection and elimination. This approach enhances the flexibility of the system, enabling faster response to changes, and concurrently reduces computational complexity.

3.3.2. Fuzzy PID control. In the process of testing the performance of a liquid pump, the real-time variation of the pump's Speed, i.e., the motor speed, is crucial. Using a fixed set of PID parameters has drawbacks, such as poor adaptability and insufficient control stability. To enhance stability and precision in testing, the paper integrates fuzzy control theory on top of traditional PID, achieving adaptive tuning of PID parameters.

The basic process of fuzzy control is to first fuzzify the input variable, then perform fuzzy inference, and finally defuzzify to obtain the control variable. This process requires the use of a knowledge base (including a database and a rule base). The principle of fuzzy PID control is to use a fuzzy controller to complete the tuning of PID parameters, and the control strategy structure is shown in Figure 4.

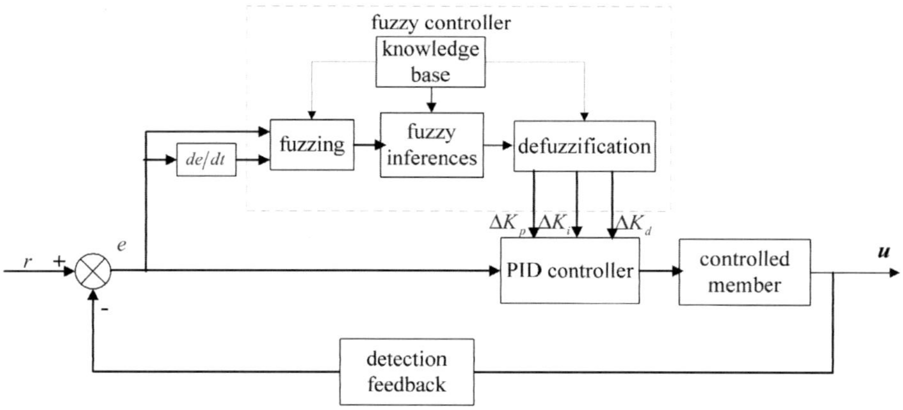

Figure 4. Structure diagram of fuzzy PID controller.

The inputs of the Fuzzy PID controller include the error *e* and the rate of change of error, *etc*. In the fuzzification process, it is necessary to quantize these two input variables using quantization factors and map them to their respective variable domains. Based on the quantization results and fuzzy subsets, the membership degrees of the input variables to various fuzzy subsets can be obtained. The error *e* and the rate of change of error, as well as the fuzzy subsets of the output variables ΔK_p, ΔK_i, and ΔK_d, are all {NB, NM, MS, Z0, PS, PM, PB}. The domain range for each variable is [-3, 3]. The quantization factors are:

$$K_e = \frac{n_e}{e_{\max}}, K_{e_c} = \frac{n_{ec}}{ec_{\max}} \quad (7)$$

Here, *e* and *ec* represent the fuzzy levels, e_{max} and ec_{max} stand for the maximum values of the variables *e* and *ec*, and during the defuzzification process, a proportional factor is applied to convert the fuzzy quantity into a precise quantity. The proportional factors are:

$$K_{K_P} = \frac{K_{P\max}}{L}, K_{K_i} = \frac{K_{i\max}}{L}, K_{K_d} = \frac{K_{d\max}}{L} \quad (8)$$

$K_{p\max}$, $K_{i\max}$, and $K_{d\max}$ represent the maximum values of PID parameters, and *L* represents the fuzzy level of the output variables. Once the membership degree of the variables in the fuzzy subsets is determined and the fuzzy rules for K_p, K_i, and K_d are applied, the defuzzification process follows. This process converts the fuzzy quantity into a precise quantity, leading to the derivation of fuzzy control rules for ΔK_p, ΔK_i, *and* ΔK_d. Consequently, the input-output relationship of the fuzzy controller is established, and the 3D effects of K_p, K_i, and K_d are illustrated from left to right in Figure 5.

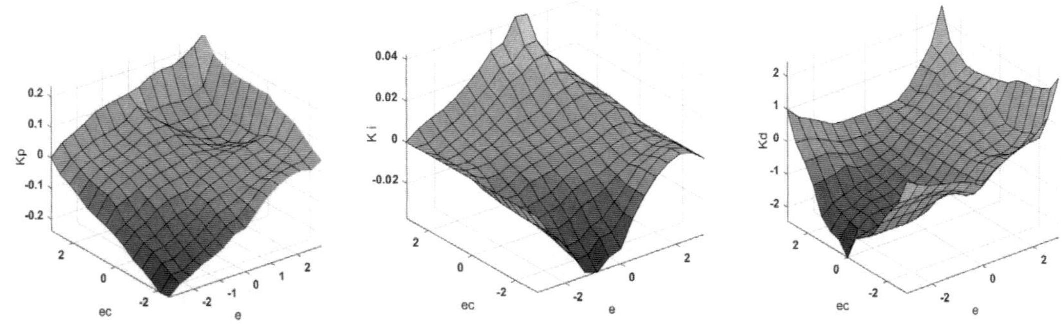

Figure 5. K_p, K_i, and K_d.

4. Liquid pump performance testing platform performance experiments

4.1. Simulation experiment.

In the Simulink environment of Matlab software, a double-closed-loop control system model for a brushless DC motor is constructed, as shown in Figure 6. The parameters of the brushless DC motor are as follows: the stator phase winding resistance R is 0.24 Ω, mutual inductance M is 48.5μH, moment of inertia J is 33.5 kg/cm^2, pole pairs np is 4, rated torque T_C is 2.5 Nm, and rated speed W_C is 3520 rpm.

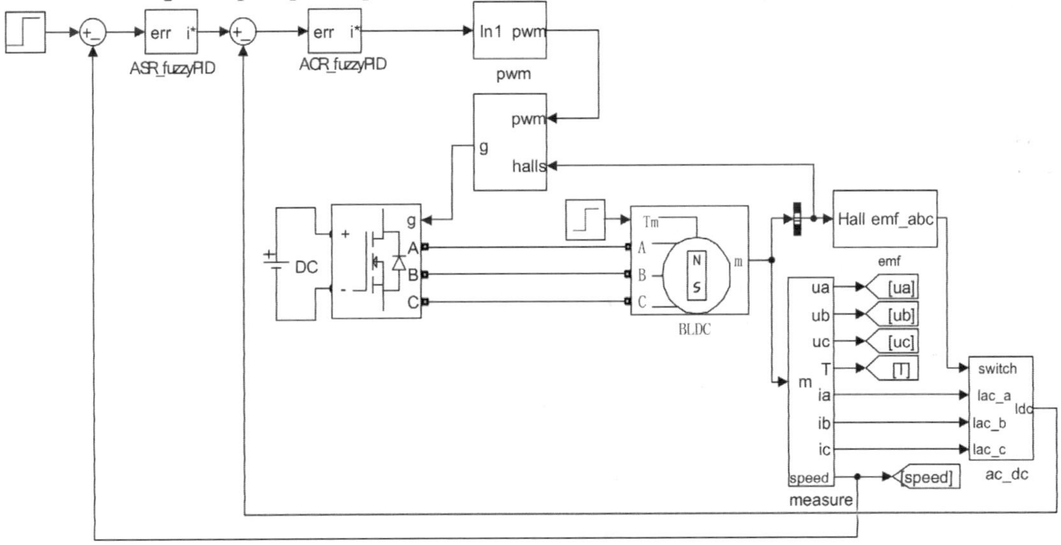

Figure 6. Simulation model of brushless DC motor double closed-loop control system.

The waveforms of motor speed and electromagnetic torque are depicted in Figures 7 and 8. In the simulation experiment, the simulation time is set to 0.5s. The initial load torque is set to 1N·m, and the initial set speed is 1500 rpm. At 0.15s, the load torque is increased to 4N·m, and at 0.25s, the set Speed is increased to 3500 rpm. From the simulation waveform, it can be observed that the system responds quickly after startup, and the motor quickly stabilizes and operates at the initial Speed and initial load torque. After the sudden increase in load, it can quickly reach a steady state, and the speed curve of the motor can effectively track the set Speed. During the sudden change in Speed, the torque value also has a corresponding jump, but the system exhibits excellent disturbance rejection performance, and the torque can quickly return to 4N·m and stabilize operation.

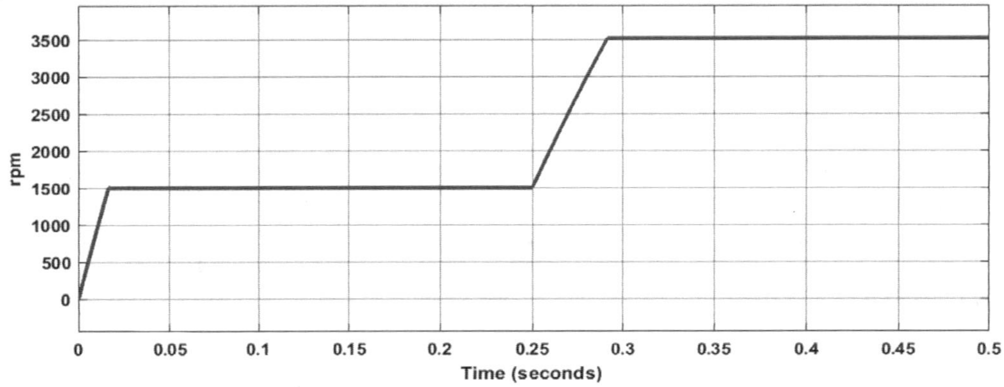

Figure 7. Speed waveform.

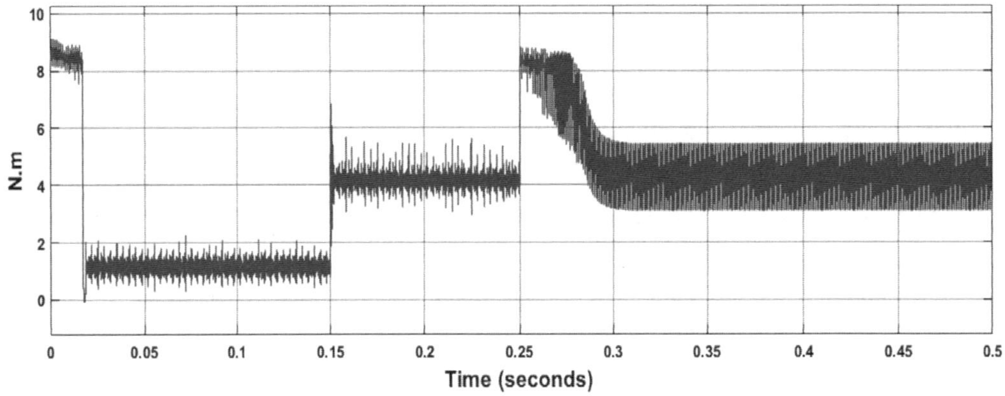

Figure 8. Electromagnetic torque waveform.

4.2. Experimental testing and analysis.

To assess the performance of the brushless DC motor control system and the spray stability of the pump driven by the brushless DC motor, an experimental platform has been set up, as shown in Figure 9. The experimental platform consists of a liquid circulation system and a sensor testing system. The liquid circulation system includes a motor, water tank, pump, filter, pressure regulator, three-way valve, ball valve, hoses, stainless steel pipes, and nozzles. The sensor testing system includes inlet and outlet pressure transmitters, flow transmitters, torque transmitters, and speed transmitters, among others.

Figure 9. Liquid Pump Performance Testing Platform.

Journal of Physics: Conference Series **2815** (2024) 012042 doi:10.1088/1742-6596/2815/1/012042

The brushless DC motor is controlled to gradually increase its Speed from 500 r/min to 3500 r/min. After setting the Speed, each speed test is repeated three times, and the experimental results are shown in Table 1.

According to Table 1, the error range of the motor speed is less than 0.3%, achieving excellent control performance and meeting the requirements of the actual system.

Table 1. Motor Speed Test Results.

Test Run	Set Speed (rpm)	Actual Speed (r/min)			MAPE（%）
1	500	498	501	500	0.20%
2	1000	1000	1003	1002	0.17%
3	1500	1498	1497	1501	0.14%
4	2000	2003	2001	2000	0.07%
5	2500	2502	2501	2498	0.07%
6	3000	3003	3002	3001	0.06%
7	3500	3502	3502	3501	0.05%

When controlling the pump outlet pressure to remain constant (P=2 MPa), the characteristic curve of pump flow rate Q versus motor speed n is obtained, as shown in Figure 10. It is evident that the pump's output flow rate Q is directly proportional to the feedback signal of the motor speed n. That is, as the speed signal increases, the pump flow rate increases almost linearly with the motor speed. It can be concluded that the flow rate-speed characteristic of the pump is relatively stable and meets the operational requirements.

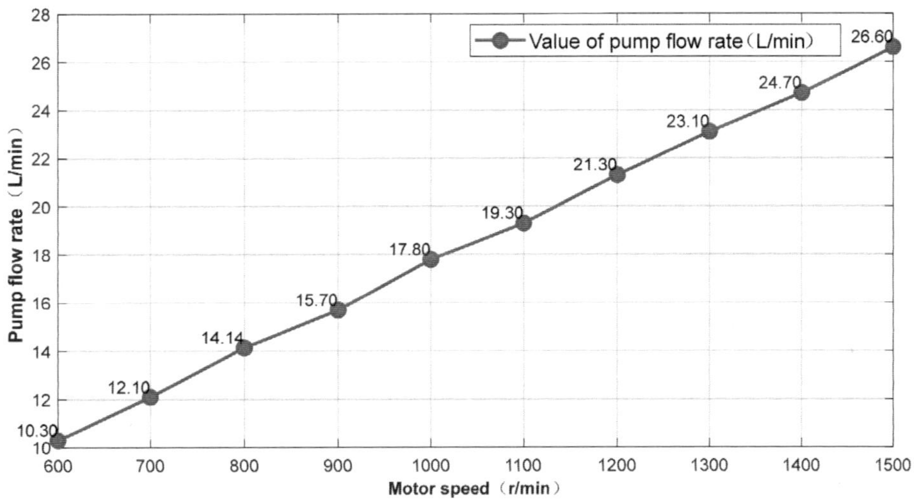

Figure 10. Pump Flow Rate vs. Motor Speed Characteristic Curve.

5. Conclusions

(1) A Liquid pump performance testing platform was designed, utilizing a brushless DC motor to provide power to the pump. A dual-closed-loop DC brushless motor control system based on the fuzzy PID method was designed. A Simulink model was established and simulated, and the results indicate that the waveforms align with theoretical analysis. The system exhibits fast response, no overshoot, and possesses excellent dynamic and static performance.

(2) The experimental results from the test bench indicate that the error range of the motor speed is less than 0.3%, demonstrating excellent performance of the motor control system. During the process of testing the pump's performance, real-time parameters such as pump flow rate, pressure, and motor speed were collected. Combined with pump pressure control and through seamless adjustment of the motor speed, the pump performance testing system constructed in this paper achieved precise detection

of pump flow rates at different speeds. Consequently, accurate speed-flow characteristic curves were obtained for precise evaluation of pump performance.

(3) The fuzzy rules of the fuzzy controller in the motor speed control system designed in this paper are obtained based on expert experience and continuous adjustments of speed simulation waveforms. Although this method is effective and highly targeted, it also requires extensive experience and a significant amount of time. Therefore, in subsequent research on fuzzy PID control strategies, attempts can be made to combine them with single-neuron and neural network algorithms for further investigation.

Acknowledgments
This work was financially supported by the Scientific Research Project of Heilongjiang Academy of Agricultural Sciences (CX23GG17) and the Major Projects of Key R&D plans in Heilongjiang Province (2022ZX05B02-03).

References

[1] Yang Xuejun, Yan Herong and Zhou Haiyan. (2005). The design and realization of the technique of multi-channel workflow in the NGOSS system. *Journal of Agricultural Mechanization Research* 27(01),161-162+165.
[2] Xu Jun, Xu Jianyong, Qi Jibao, Wang Bin and Chen Jian. (2013). Design of performance testing system for the positive displacement pump used in crop protection equipment. Acta Agriculturae Zhejiangensis 25(05),1114-1118.
[3] He Xiongkui. (2019). Research and development of crop protection machinery and chemical application technology in China. Chinese Journal of Pesticide Science 21(5-6), 921-930.
[4] Jiang Saike, He Xiongkui, Liu Yajia, Wang Changling, Song Jianli and Zeng Aijun. (2023).Development Status and Prospect of Intelligent Plant Protection Equipment and Technology.Plant Health and Medicine 2(04),1-11.
[5] Yang Yang, Zhang Gang, Zha Jiayi, Li Yankai, Zhang Tie and Chen Liqing. (2020). Design of Automatic Steering System Based on Direct Connection of DC Motor and Full Hydraulic Steering Gear. Transactions of the Chinese Society for Agricultural Machinery 51(08), 44-54+61.
[6] Ding Youchun, Liu Weipeng, Dong Wanjing, Chen Liyuan, Liu Xiaodong and Jin Wei.(2021). Design and Experiment of Seed Metering Device for Rapeseed Based on Seeding Frequency Feedback. Transactions of the Chinese Society for Agricultural Machinery 52(10),175-185+232.
[7] Al Mashhadany, Y. I. M., Abbas, A. K., & Algburi, S. S. (2022). Modeling and analysis of brushless DC motor system based on intelligent controllers. Bulletin of Electrical Engineering and Informatics, 11(6), 2995-3003.
[8] Kanungo, A., Choubey, C., Gupta, V., Kumar, P., & Kumar, N. (2023). Design of an intelligent wavelet-based fuzzy adaptive PID control for brushless motor. Multimedia Tools and Applications, 82(21), 33203-33223.
[9] Gobinath, S., & Madheswaran, M. (2020). Deep perceptron neural network with fuzzy PID controller for speed control and stability analysis of BLDC motor. Soft Computing, 24(13), 10161-10180.
[10] Arof, S., M. Noor, N., Mohamad, R., Noorsal, E., Mawby, P. A., & Arof, H. (2020). Close loop feedback direct current control in driving mode of a four quadarnts drive direct current chopper for electric vehicle traction controlled using fuzzy logic. Progress in Engineering Technology II, 57-71.
[11] Hammoodi, S. J., Flayyih, K. S., & Hamad, A. R. (2020). Design and implementation speed control system of DC motor based on PID control and matlab simulink. International Journal of Power Electronics and Drive Systems, 11(1), 127-134.

Research on agricultural monitoring and progressive map creation in the context of ground-air coordination

Xiaoyong Li [1,*]

[1] Kashi University, Kashgar, Xinjiang, 844000, China

[*] Corresponding author's e-mail: 630769090@qq.com

Abstract: By combining network RTK technology, 2D LiDAR, and UAV, this study investigates how to achieve agricultural monitoring effectively. A collaborative agricultural monitoring scheme was designed to accurately and comprehensively gather agricultural information in response to the demand for targeted sequential monitoring at specified agricultural locations. To establish a low-cost and low computational power-dependent 3D map of obstacle point clouds, a progressive 3D mapping scheme based on 2D LiDAR was designed and implemented. The effectiveness of the design was validated through simulation experiments and partly through real-world testing.

1. Introduction

In agriculture, drones primarily use low-altitude remote sensing to monitor crop growth, crop pests and diseases, crop lodging status, yield estimation, and other aspects. Radoglou-Grammatikis et al. [1] extensively discussed the application of drones in low-altitude remote sensing in agriculture. In precision agriculture, in-depth research has been conducted by equipping drones with monitoring devices such as RGB cameras, multispectral sensors, thermal imagers, and others [2-9]. However, relying solely on onboard monitoring devices for agricultural monitoring still does not provide a comprehensive and accurate acquisition of agricultural information [10]. Xiangpeng et al. [10] discussed the limitations of using low-altitude remote sensing alone for agricultural monitoring and highlighted some shortcomings. For instance, in the current research stage, if only low-altitude drone remote sensing is used for agricultural monitoring, its flight stability and level of intelligence during the monitoring process are somewhat limited when facing complex operational environments. To obtain richer information, it is usually necessary to equip drones with multiple onboard sensors. This, on the one hand, makes it challenging for drones to bear large payload capacities, and factors such as sensor accuracy, battery life, and communication distance carried by the drone will affect the practicality of drone remote sensing.

Regarding utilizing Unmanned Aerial Vehicles for coordinated aerial-ground data collection, Li et al. [11] assumed limited node energy and employs a strategy to minimize data collection time. Tsai et al. [12] investigated the problem of generating optimal flight paths considering minimum flight time during UAVs' coordinated data collection process.

Building upon previous research and considering practical applications, a collaborative agricultural monitoring scheme was designed in this study, utilizing drones as aerial mobile monitoring platforms to conduct coordinated aerial-ground agricultural monitoring, thereby achieving a more comprehensive collection of agricultural information.

To achieve efficient and safe monitoring by repeatedly implementing agricultural monitoring under low-cost design, this paper deeply studies the continuous construction of obstacle maps based on two-dimensional LiDAR. A progressive mapping scheme based on 2D LiDAR is designed, where, with increased monitoring frequency, the drone's dependency on onboard computer processing power during flight gradually decreases. Optimized obstacle avoidance flight paths can be obtained as the obstacle map construction becomes more refined. Adopting this design approach can significantly enhance the safety and efficiency of agricultural monitoring under existing hardware conditions.

2. Overall concept

2.1. A collaborative agricultural monitoring scheme

Integrating ground collaborative agricultural data collection with a UAV autonomous navigation system has obvious advantages, as the UAV can autonomously avoid obstacles, fly to the nodes, and complete monitoring tasks. Firstly, it is necessary to clarify the visiting sequence of all ground nodes to be collected [12]. Additionally, the integration of the UAV autonomous navigation system with the collaborative data collection system needs to be considered to obtain the state transition relationships, as shown in Figure 1.

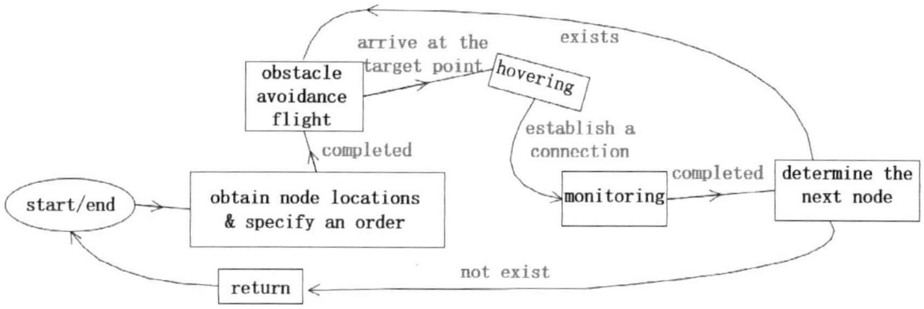

Figure 1. State transition relationships of collaborative monitoring scheme for agricultural data collection.

2.2. Airborne onboard monitoring

Using drones as aerial monitoring platforms, various equipment such as RGB cameras, multispectral sensors, thermal imaging cameras, etc., can be mounted on drones for agricultural monitoring [1]. For instance, in this study, an RGB camera is mounted on drones to monitor the area surrounding data collection points. Low-altitude remote sensing images are captured using a non-distorted monocular RGB camera, as shown in Figure 2. The onboard computer utilizes NVIDIA JETSON NX and runs ROS nodes. Collaborative monitoring activities can be carried out when the UAV flies above the ground collection nodes. The onboard terminal can simultaneously capture high-definition images below the drone and obtain the drone's current pose, as depicted in Figure 3. By subscribing to both the control messages corresponding to the topic (start_ACQ) and the image messages corresponding to the topic (usb_cam/raw_image), the node can perform image, position, and attitude acquisition and storage.

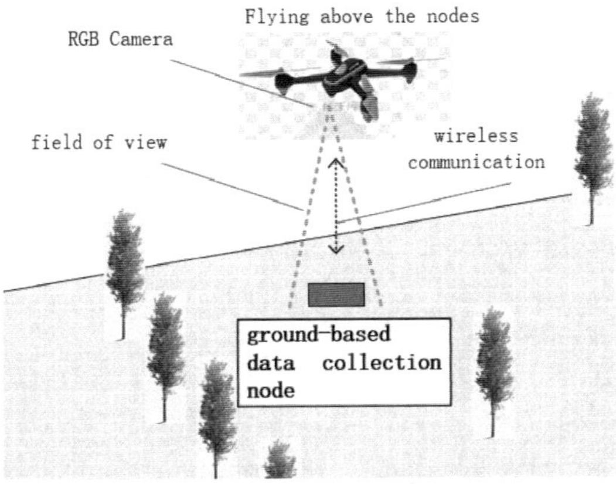

Figure 2. Airborne onboard monitoring.

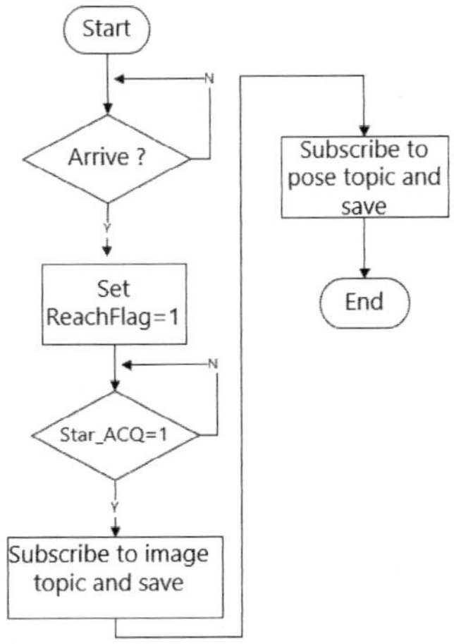

Figure 3. Onboard monitoring process.

2.3. Ground data collection node

According to the actual needs of this research, the ground data collection node is mainly used to monitor data that is difficult for the drone to obtain accurately, such as soil temperature, humidity, pH value, nutrient information, etc. When the drone flies over the ground data collection node, the ground node establishes a TCP connection with the UAV's communication node via WiFi. Then, the ground node uploads the monitored data to the drone. This process is illustrated in Figure 4(1).

When deploying nodes, we first obtain positioning data through a high-precision GNSS module, transfer it to a laptop via a USB port for processing, and finally send the position data (latitude, longitude, altitude) to the node via a TCP connection, as shown in Figure 4.

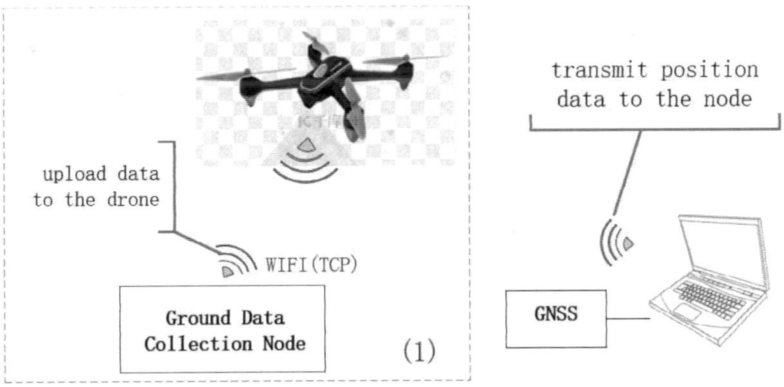

Figure 4. Node data transmission and deployment.

3. Processing and analysis of node positions and point cloud maps

3.1. Processing of global position data

In this study, the global positions of all ground nodes are represented in the WGS-84 coordinate system and form a set of locations to be visited. To facilitate autonomous navigation of UAVs in a local coordinate system, it is necessary to transform the global positions of the nodes into a local coordinate system with the UAV's takeoff point as the origin.

Generally, this transformation is done in two steps [14, 15]. The first step is to convert the points represented in the geodetic coordinate system (latitude, longitude, altitude) to the ECEF (Earth-Centered,

Earth-Fixed) coordinate system. The second step is to convert the points in the ECEF coordinate system to the local coordinate system with the UAV's takeoff point as the origin.

3.2. From global coordinate system to local coordinate system

The first step is to convert the point P from the WGS-84 coordinate system to the ECEF coordinate system from a geometric perspective, as shown in Figure 5. Given the position representation of point P in the WGS-84 system as $P_{WGS-84}(\lambda, \varphi, h)$, the three-dimensional Cartesian coordinates of point P in the ECEF system as $P_{ECEF}(x_0, y_0, z_0)$ can be calculated, as expressed in Equation (1).

$$P_{ECEF}(x_0, y_0, z_0) = \begin{cases} x_0 = (N+h)\cos(\varphi)\cos(\lambda) \\ y_0 = (N+h)\cos(\varphi)\sin(\lambda) \\ z_0 = (N(1-e^2)+h)\sin(\varphi) \end{cases} \tag{1}$$

In Equation (1), φ represents the latitude, λ represents the longitude, h represents the height above the reference ellipsoid (in meters), N represents the ellipsoidal radius of curvature, and e^2 represents the square of the ellipsoid's eccentricity.

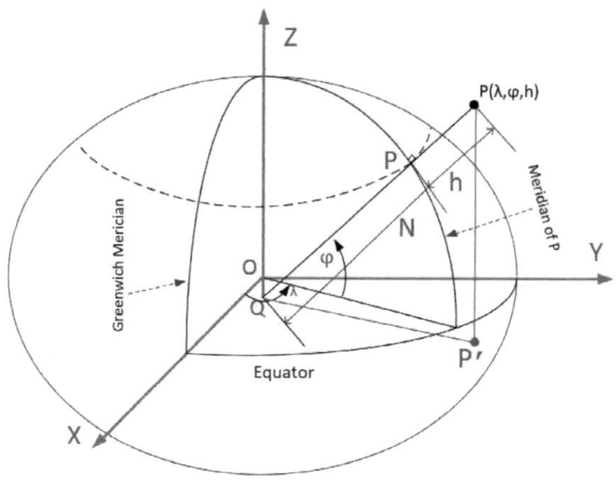

Figure 5. Transforming point P from the WGS-84 coordinate system to the ECEF coordinate system.

Point P in the ENU coordinate system is represented by transforming it from the ECEF coordinate system. In the ENU coordinate system, the coordinate origin is defined as the UAV's takeoff point.

Typically, for the representation of a point in different coordinate systems, merely obtaining the positional relationship between the two coordinate system origins is not enough, but the attitude relationship between the two coordinate systems should also be known.

In this study, the ENU coordinate system is adopted as the local coordinate system, and the directions of its three axes are clear. The ECEF coordinate system is adopted as the global coordinate system, and the directions of its three axes are also clear. For the calculation of coordinate transformation, this, to a certain extent, reduces the complexity of representing a point in the ECEF coordinate system in the ENU coordinate system.

We explain the process of transforming from the ECEF coordinate system to the ENU coordinate system using a geometric approach, as shown in Figure 6, where point P is used as the origin to construct the ENU coordinate system. This process requires a rotation operation and a translation operation to be performed separately.

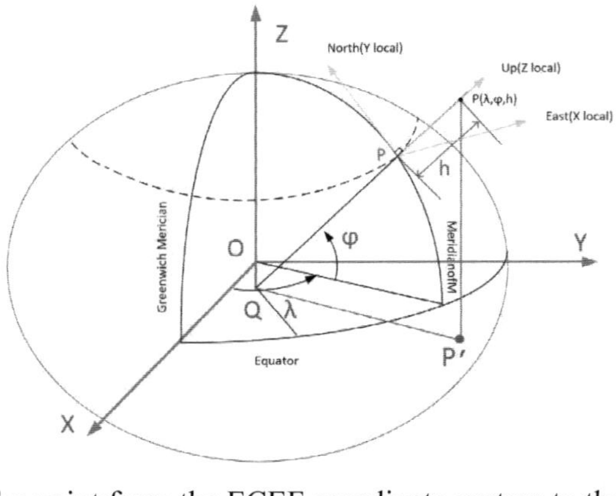

Figure 6. Representing the point from the ECEF coordinate system to the ENU coordinate system.

Firstly, we initiate the rotation operation. ① The direction of the X_{ENU} axis is perpendicular to the meridian where point P is located, and the meridian where point P is located is obtained by counterclockwise rotation of the prime meridian around Z_{ECEF} by λ. Therefore, the ECEF coordinate system needs to be first rotated around Z_{ECEF} until the direction of X_{ECEF} aligns with X_{ENU}, which is a counterclockwise rotation of $\frac{\pi}{2}+\lambda$. Then, we rotate counterclockwise around X_{ECEF} until Z_{ECEF} aligns with Z_{ENU}, which is a counterclockwise rotation of $\frac{\pi}{2}-\varphi$. Finally, the rotation transformation matrix R is obtained and represented by Equation (2).

$$R = R_{xecef}[(\frac{\pi}{2}-\varphi)]R_{zecef}[(\frac{\pi}{2}+\lambda)] = \begin{bmatrix} -\sin\lambda & \cos\lambda & 0 \\ -\cos\lambda\sin\varphi & -\sin\lambda\sin\varphi & \cos\varphi \\ \cos\lambda\cos\varphi & \sin\lambda\cos\varphi & \sin\varphi \end{bmatrix} \quad (2)$$

After the rotation transformation, a coordinate system ENU00 is constructed, with its origin coinciding with the origin of the ECEF coordinate system, as shown in Figure 7.

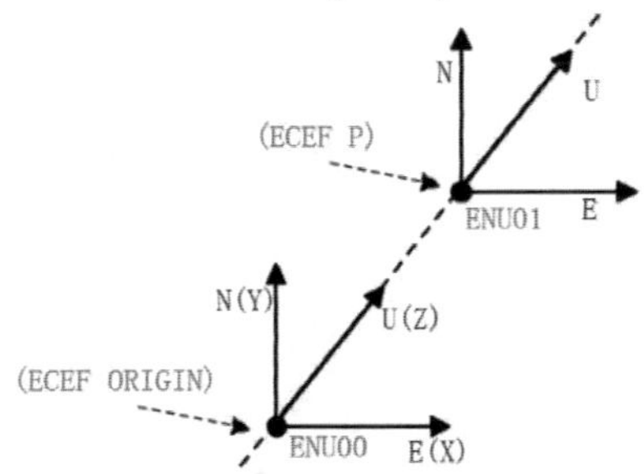

Figure 7. Relationship between the coordinate systems.

Next, we proceed with the translation operation. Now, it is required to construct an ENU coordinate system ENU01 with the origin at point P, as shown in Figure 7. By analysis, it is known that ENU00 and ENU01 have an offset only along the Z-axis, denoted as DistZ01. The coordinates of the origins 0_{ENU00} and 0_{ENU01} of the local coordinate systems ENU00 and ENU01 in the global coordinate system ECEF are given as $0_{ENU00}=(X_{ENU00},Y_{ENU00},Z_{ENU00})$ and $0_{ENU00}=(X_{ENU00},Y_{ENU00},Z_{ENU00})$ respectively. The value of DistZ01 can be represented by Equation (3).

$$\mathrm{DistZ01}=\sqrt{(X_{ENU01}-X_{ENU00})^2+(Y_{ENU01}-Y_{ENU00})^2+(Z_{ENU01}-Z_{ENU00})^2} \tag{3}$$

According to the previous analysis, $0_{ENU00}=(0,0,0)$, and the origin 0_{ENU01} of the ENU01 coordinate system is represented by the position of point P in the ECEF coordinate system. Therefore, $0_{ENU01}=P_{ECEF}=(X_0,Y_0,Z_0)$. In that case, the specific value of DistZ01 is represented as $\mathrm{DistZ01}=\sqrt{(X_0)^2+(Y_0)^2+(Z_0)^2}$. Translating the local coordinate system $ENU00$ along the Z-axis by a distance of DistZ01 will result in the local coordinate system $ENU01$.

3.3. The representation of the global position point set in the ENU coordinate system

Before the drone takes off, a local coordinate system is first established at the take-off point P. Now, the set of ground nodes to be visited is known to be represented in the WGS-84 coordinate system as $S_{GBL}=\{\mathrm{GBLn}(\mathrm{LONn},\mathrm{LATn},\mathrm{ALTn})\,|\,n\le m\}$, where n is the node number, m is the number of nodes, and the global position of the nth node is represented in the WGS-84 coordinate system as LONn, LATn, and ALTn. By calculation, the set of positions of S_{GBL} in the ECEF coordinate system is represented as $S_{ECEF}=\{\mathrm{GBLn}_{ECEF}(\mathrm{Xn},\mathrm{Yn},\mathrm{Zn})\,|\,n\le m\}$. To represent the local coordinate position GBLn_{ENU01} of the nth point in S_{ECEF} in the ENU01 coordinate system, we first need to calculate the displacement Δ_{ECEF} of the point GBLn_{ECEF} in S_{ECEF} relative to the point P_{ECEF} (origin of ENU01). The displacement is denoted as $\Delta_{ECEF}=(\mathrm{Xn},\mathrm{Yn},\mathrm{Zn})^T-(\mathrm{X0},\mathrm{Y0},\mathrm{Z0})^T$.

Next, the position of point GBLn_{ECEF} in the ENU01 coordinate system, GBLn_{ENU01}, is represented using Equation (4).

$$\mathrm{GBLn}_{ENU01}(x_n,y_n,z_n)=\begin{bmatrix} x_n \\ y_n \\ z_n \end{bmatrix}=R\Delta_{ECEF} \tag{4}$$

Finally, the set of positions of ground nodes in the ENU01 coordinate system is obtained and represented as $S_{ENU01}=\{\mathrm{GBLn}_{ENU01}(x_n,y_n,z_n)\,|\,n\le m\}$. In the local coordinate system ENU01, the coordinate origin corresponds to the take-off point of the drone. By utilizing UAV autonomous navigation, the drone can fly to the positions corresponding to the nodes in sequence, thereby completing the aerial-ground cooperative inspection.

When there is a need to repeatedly conduct agricultural monitoring, if the position of the drone take-off point changes, a new local coordinate system will be established, denoted as ENU02. In ENU02, the positions of ground monitoring nodes need to be redefined, as shown in Figure 8, where point P is the coordinate origin of ENU01 and point M is the coordinate origin of ENU02.

In Figure 8, given the positions of the origin of the local coordinate systems ENU01 and ENU02 as P_{ECEF} and M_{ECEF}, respectively, and the set of WGS-84 positions of the ground nodes to be visited, we can calculate the sets of positions of the nodes in ENU01 as $S_{ENU01}=\{\mathrm{GBLn}_{ENU01}(x_n,y_n,z_n)\,|\,n\le m\}$ and in ENU02 as $S_{ENU02}=\{\mathrm{GBLn}_{ENU02}(x_n,y_n,z_n)\,|\,n\le m\}$, respectively.

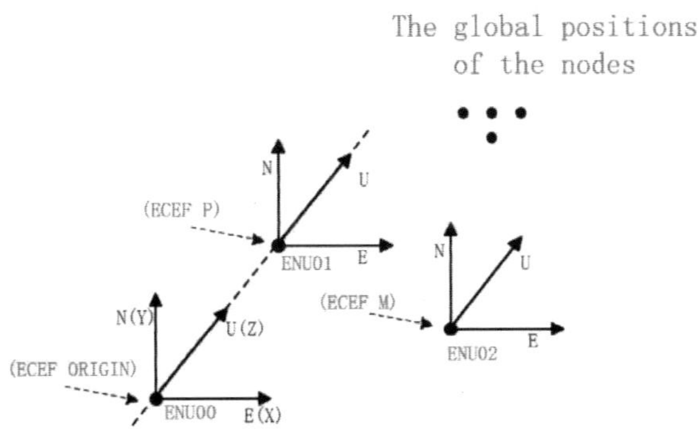

Figure 8. Illustration of the positions of the global set of points in different local ENU coordinate systems.

4. Incremental construction of point cloud maps

4.1. Reason for proposal

In the research on UAV remote sensing for monitoring agricultural information, the spectral and textural characteristics of crop growth and development vary at different stages. In studies involving UAV-based air-ground cooperative monitoring, it is sometimes necessary to obtain data on the continuous growth stages of crops to accurately represent parameters such as the true growth status and indices of the crops. Therefore, throughout the entire growth cycle of crops, multiple monitoring sessions need to be conducted at specific locations to obtain data related to their growth and development.

In this study, a point cloud map of obstacles is constructed using 2D Lidar. Typically, the point cloud map data obtained by 2D Lidar is planar. Obtaining a 3D point cloud map through scanning with 2D Lidar would have a significant cost advantage and effectively reduce the dependency on onboard computer computational power. Chu et al. [13] vertically mounted the 2D Lidar and controlled the scanning device's motion trajectory through an "S-shaped" scanning control strategy to achieve a complete 3D scan of the entire measurement area. Zhu et al. [14] vertically mounted the 2D Lidar and used onboard 2D LiDAR to scan the ground for reconstructing Digital Elevation Models (DEMs). Fang et al. [15] mounted the 2D Lidar on a rotating mechanism composed of a stepper motor to achieve a 3D scanning effect. Bosse et al. [16] proposed a new concept for designing a 3D sensor by combining a 2D laser scanner with a spring. By mounting the sensor on a mobile platform and utilizing the platform's motion to extend the 2D scanning plane, 3D point cloud map data can be obtained.

Building on previous studies [16], a new method is proposed for UAV-mounted 2D Lidar to achieve 3D point cloud mapping. The 2D Lidar is horizontally mounted on the UAV, as shown in Figure 9(a), with the flight controller (PIXHAWK4) combined with high-precision GNSS, providing the UAV's attitude and position information. When the UAV performs start-stop or accelerated flight actions, the UAV will inevitably tilt. With this design, a point cloud map of obstacles can be established near the set flight altitude, as shown in Figure 9(b).

Figure 9(a). 2D Lidar horizontally mounted on UAV.

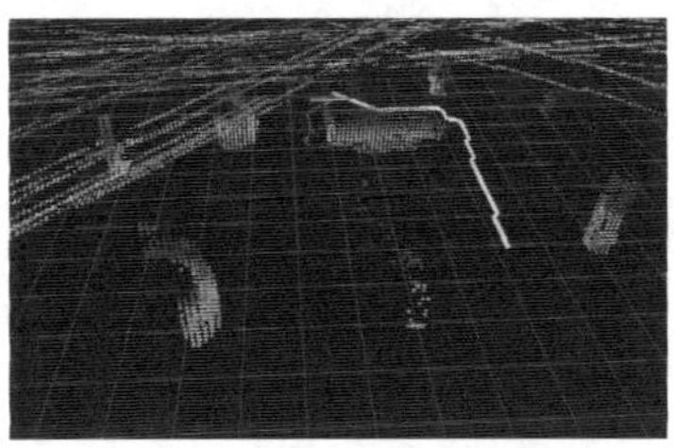

Figure 9(b). Establishment of a point cloud map near the flight altitude.

Considering that if the 3D point cloud map built during the previous flight is not actively saved, the map data will be lost as the monitoring process ends. When starting agricultural monitoring again, a new point cloud map of obstacles must be re-established. By effectively utilizing the previous mapping results and calculated position data, it should be possible to reduce the dependency on onboard computational resources during flight and improve the efficiency of agricultural monitoring. Based on this idea, a method for incrementally constructing a point cloud map of obstacles has been researched and designed.

4.2. Analysis of incremental point cloud map construction

Although the approach mentioned earlier allows for representing the coordinates of ground acquisition nodes in the ENU coordinate system and calculating the coordinates of each point in the obstacle point cloud map, each time a cooperative monitoring task is executed, it requires recalculating the positions of the UAV takeoff point, ground nodes, and regenerating the point cloud map in the ENU coordinate system. This leads to repetitive calculations on the one hand and places a significant computational burden on the UAV onboard computer throughout the entire flight process on the other hand.

Since the positions of ground acquisition nodes and obstacle point cloud maps in the ENU01 and ENU02 coordinate systems are related, as shown in Figure 8, before the UAV takes off for the next mission, it can first determine the relationship between the two ENU coordinate systems by calculating the offset of the origins of these two coordinate systems in the ECEF coordinate system. Once the calculation relationship is established, the computational burden during flight can be reduced by minimizing repetitive calculations.

Let us assume that the coordinates of the origins of ENU01 and ENU02 in the ECEF coordinate system are represented as $0_{ENU01} = P_{ECEF}(X0, Y0, Z0)$ and $0_{ENU02} = M_{ECEF}(X02, Y02, Z02)$, respectively. Therefore, the offset of ENU02 relative to ENU01 can be represented as $\Delta = 0_{ENU02} - 0_{ENU01}$, denoted as $\Delta = (\Delta X, \Delta Y, \Delta Z)$.

The set of all points in the ENU coordinate system is denoted by Equation (5). In Equation (5), $S_{ENU0x} = \{GBLn_{ENU0x}(x_n, y_n, z_n) \mid n \le m\}$ represents the set of positions of all ground nodes, and $PCL_{ENU0x} = \{PCLcn_{ENU0x}(x_n, y_n, z_n) \mid cn \le cm\}$ represents the point cloud map. x represents the ENU coordinate system number, cn represents the point number in the point cloud map, and cm represents the total number of points in the point cloud map.

$$P_{ENU0X} = \{S_{ENU0X}, PCL_{ENU0X}\} \tag{5}$$

Therefore, for all points in ENU01, it is represented as $P_{ENU01} = \{S_{ENU01}, PCL_{ENU01}\}$. For all points in ENU02, it is represented as $P_{ENU02} = \{S_{ENU02}, PCL_{ENU02}\}$.

If the values in the sets P_{ENU01} and P_{ENU02} are calculated independently, the same computational steps are required, as described in Equations (1), (2), and (4), and Figure (9). This independent calculation undoubtedly leads to a waste of computational resources. Utilizing the results of previous calculations can significantly improve the situation. Since the position offset Δ of ENU02 relative to ENU01 has already been calculated, with the knowledge of P_{ENU01}, it is straightforward to compute the values of each element in P_{ENU02} by performing the offset operation, as described in Equation (6). This greatly simplifies the computational workload.

$$\begin{aligned} &P_{ENU02} = \{S_{ENU02}, PCL_{ENU02}\} = \\ &\{\{GBLn_{ENU02} = GBLn_{ENU01} - \Delta \mid n \le m\}, \{PCLcn_{ENU02} = PCLcn_{ENU01} - \Delta \mid cn \le cm\}\} \end{aligned} \tag{6}$$

4.3. Incremental point cloud map construction simulation

We establish a simulation environment in GAZEBO that includes obstacles, as shown in Figure 10. During the initial flight, we assume the global position set of points to be visited is shown in Table 1, consisting of four nodes with global positions. The global coordinates corresponding to the origin of

ENU01 are (8.5455940, 47.3977419, 536.4), and for ENU02, the global coordinates corresponding to its origin are (8.5455671, 47.3977419, 536.3). By calculation, the global positions of points can be determined and represented in both ENU01 and ENU02 coordinate systems.

The positions of the ENU01 and ENU02 coordinate origins in ECEF are represented as $0_{ENU01}(X_0, Y_0, Z_0)$ and $0_{ENU02}(X_{02}, Y_{02}, Z_{02})$, respectively. As mentioned earlier, the offset of ENU02 relative to ENU01 is Δ=(-2, 0, 0). In the ENU02 coordinate system, P_{ENU02} represents the set of positions of all points, obtained through the operation described in Equation (6). Figure 11(a) shows the point cloud map PCL_{ENU01} constructed in ENU01. Figure 11(b) depicts the point cloud map constructed in ENU01 and represented in ENU02, denoted as PCL_{ENU02}. Figure 11(c) shows the new point cloud map constructed in ENU02 by incorporating the point cloud map PCL_{ENU01} from ENU01 in an incremental manner.

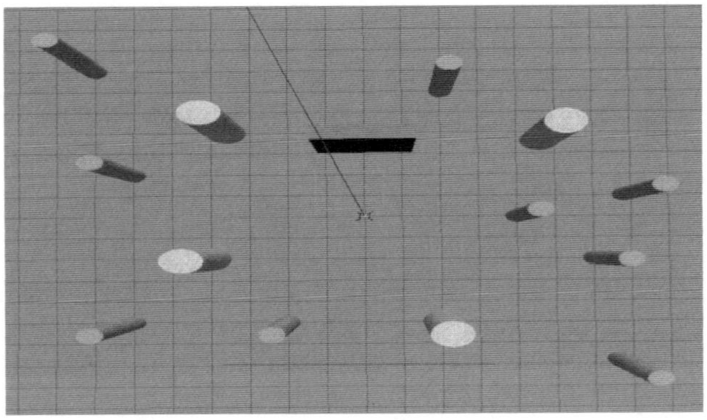

Figure 10. GAZEBO simulation environment with obstacles.

Table 1. Global and local positions of ground nodes.

Point number	Global position (wgs-84)		Local position (enu01)		Local position (enu02)	
	lon	8.5456204	x	1.993	x	4.023
1	lat	47.3977434	y	0.166	y	0.1667
	alt	536.30	z	0.97	z	0.97
	lon	8.5456742	x	6.054	x	8.085
2	lat	47.3977434	y	0.166	y	0.1667
	alt	536.30	z	0.97	z	0.97
	lon	8.5457273	x	10.06	x	12.094
3	lat	47.3977422	y	0.033	y	0.033
	alt	536.30	z	0.97	z	0.97
	lon	8.5458597	x	20.06	x	22.090
4	lat	47.3977417	y	-0.02	y	-0.022
	alt	536.30	z	0.97	z	0.97

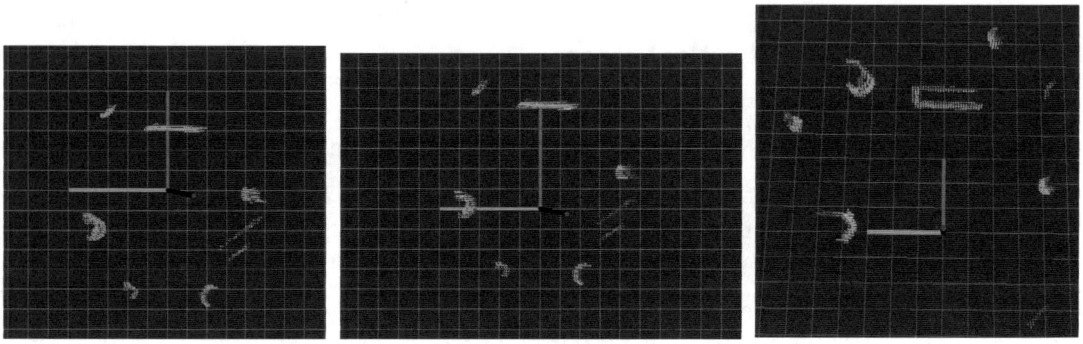

(a) (b) (c)

Figure 11. Mapping process using incremental flights.

5. Conclusion

In this study, a collaborative agricultural monitoring scheme was designed, which, through aerial-ground collaboration, can accurately and comprehensively obtain agricultural information. The representation of global coordinate points and local point clouds in different local coordinate systems was analyzed in detail. By calculating relative offset values, point sets were transformed from one local coordinate representation to another, effectively reducing computational power dependency in a consistent hardware environment. A method was designed to construct a 3D point cloud map by combining 2D LiDAR data with UAV flight attitude changes and flight altitude. Furthermore, the progressive construction of the 3D point cloud map was investigated. The effectiveness of the design approach was validated through a combination of simulation experiments and real-world testing.

Acknowledgments

The authors would like to thank all reviewers for their very helpful comments and suggestions. This work is supported by the Key Research Topic at Kashi University No. 022020097 and the Science and Technology Plan project in Kashgar region No. KS2021016.

References

[1] Radoglou-Grammatikis, P., Sarigiannidis, P., Lagkas, T., & Moscholios, I. (2020) A compilation of UAV applications for precision agriculture. *Computer Networks*, *172*. https://doi.org/10.1016/j.comnet.2020.107148

[2] del Cerro, J., Ulloa, C. C., Barrientos, A., & de León Rivas, J. (2021). Unmanned aerial vehicles in agriculture: A survey. In Agronomy (Vol. 11, Issue 2). https://doi.org/10.3390/agronomy11020203

[3] Gokool, S., Mahomed, M., Kunz, R., Clulow, A., Sibanda, M., Naiken, V., Chetty, K., & Mabhaudhi, T. (2023). Crop Monitoring in Smallholder Farms Using Unmanned Aerial Vehicles to Facilitate Precision Agriculture Practices: A Scoping Review and Bibliometric Analysis. In Sustainability (Switzerland) (Vol. 15, Issue 4). https://doi.org/10.3390/su15043557

[4] Guo, T., Kujirai, T., & Watanabe, T. (2012). MAPPING CROP STATUS FROM AN UNMANNED AERIAL VEHICLE FOR PRECISION AGRICULTURE APPLICATIONS. The International Archives of the Photogrammetry, Remote Sensing and Spatial Information Sciences, XXXIX-B1. https://doi.org/10.5194/isprsarchives-xxxix-b1-485-2012

[5] Huang, Y. B., Thomson, S. J., Hoffmann, W. C., Lan, Y. Bin, & Fritz, B. K. (2013). Development and prospect of unmanned aerial vehicle technologies for agricultural production management. International Journal of Agricultural and Biological Engineering, 6(3). https://doi.org/10.3965/j.ijabe.20130603.001

[6] Liu, Z., & Li, J. (2023). Application of Unmanned Aerial Vehicles in Precision Agriculture. In Agriculture (Switzerland) (Vol. 13, Issue 7). https://doi.org/10.3390/agriculture13071375

[7] Otto, A., Agatz, N., Campbell, J., Golden, B., & Pesch, E. (2018). Optimization approaches for civil applications of unmanned aerial vehicles (UAVs) or aerial drones: A survey. In Networks (Vol. 72, Issue 4). https://doi.org/10.1002/net.21818

[8] Rejeb, A., Abdollahi, A., Rejeb, K., & Treiblmaier, H. (2022). Drones in agriculture: A review and bibliometric analysis. In Computers and Electronics in Agriculture (Vol. 198). Elsevier B.V. https://doi.org/10.1016/j.compag.2022.107017

[9] Singh, A. P., Yerudkar, A., Mariani, V., Iannelli, L., & Glielmo, L. (2022). A Bibliometric Review of the Use of Unmanned Aerial Vehicles in Precision Agriculture and Precision Viticulture for Sensing Applications. In Remote Sensing (Vol. 14, Issue 7). https://doi.org/10.3390/rs14071604

[10] Xiangpeng, F., Jianping, Z., & Yan, X. U. (2021). Research Advances of Monitoring Agricultural Information Using UAV Low-Altitude Remote Sensing. *Journal of Xinjiang University (Natural Science Edition in Chinese and English*, *38*(5). https://doi.org/10.13568/j.cnki.651094.651316

[11] Li, M., Liu, X., & Wang, H. (2023). Completion Time Minimization Considering GNs' Energy for UAV-Assisted Data Collection. *IEEE Wireless Communications Letters*, *12*(12), 2128–2132. https://doi.org/10.1109/LWC.2023.3309625

[12] Tsai, H.-C., Hong, Y.-W. P., & Sheu, J.-P. (2022). Completion Time Minimization for UAV-Enabled Surveillance Over Multiple Restricted Regions. *IEEE Transactions on Mobile Computing*, 1–14. https://doi.org/10.1109/TMC.2022.3200732

[13] Chu, H., Gong, K., Shao, Y., Chang, Z., & Ni, J. (2018). 3D Perception and Reconstruction System Based on 2D Laser Scanner. *2018 Chinese Automation Congress (CAC)*, 1520–1524. https://doi.org/10.1109/CAC.2018.8623391

[14] Zhu, H., Wang, H., & Sun, S. (2018). Research on DEM Reconstruction Based on Airborne 2D Laser Radar. *2018 37th Chinese Control Conference (CCC)*, 8435–8440. https://doi.org/10.23919/ChiCC.2018.8482734

[15] Fang, Z., Zhao, S., Wen, S., & Zhang, Y. (2018). A Real-Time 3D Perception and Reconstruction System Based on a 2D Laser Scanner. *Journal of Sensors*, *2018*, 1–14. https://doi.org/10.1155/2018/2937694

[16] Bosse, M., Zlot, R., & Flick, P. (2012). Zebedee: Design of a Spring-Mounted 3-D Range Sensor with Application to Mobile Mapping. *IEEE Transactions on Robotics*, *28*(5), 1104–1119. https://doi.org/10.1109/TRO.2012.2200990

Dynamic time warping based instantaneous frequency estimation for rotating machineries under non-stationary operating conditions

Xiyue Tao*

Wuhan university of technology, Wuhan, Hubei, 430070, China

*Corresponding author's e-mail: taoxiyue312@whut.edu.cn

Abstract. For tacho-less order tracking, it is crucial to achieve an accurate estimation of instantaneous phase and frequency under non-stationary conditions for fault diagnosis of mechanical systems. Therefore, an instantaneous frequency estimation method based on dynamic time warping for rotating machinery was proposed. Firstly, a mono-component signal is obtained by variational mode decomposition of multi-harmonic vibration signal generated in practical engineering. Then, a stationary reference signal is constructed by the mean frequency of the vibration signal, which is obtained by the signal spectrum. Aligning the vibration signal with the reference signal to obtain the rough phase estimation and warping path. Finally, a polynomial model is constructed to calculate the analytic expression of the smooth warping path to obtain the instantaneous frequency result. The simulation and experiment present that the dynamic time warping-based method can estimate the instantaneous frequency of the signal under non-stationary conditions. Compared with the classical Hilbert transform method, the proposed dynamic time-warping algorithm has stronger noise resistance.

1. Introduction

For rotating machines under non-stationary conditions, order tracking (OT) is usually used to extract the fault order to determine the fault location. OT requires the instantaneous frequency (IF) of a rotating machine as the reference, which can generally obtained from angle encoders or tachometers. However, due to the cost and installation location problems, the practicability of traditional OT is limited. Therefore, vibration-based speed estimate methods have gained more and more attention recently [1].

Nowadays, IF estimation methods of vibration signals are mainly categorized into two groups. The first one is time-frequency distribution, such as wavelet transform [2-3] and short-time Fourier transform [4] (STFT). However, both of them are limited by the Heisenberg uncertainty principle, which makes it challenging to achieve high resolution in both the time and frequency domain. Meanwhile, mode mixing and discontinuous ridges also need to be solved. Another category is phase demodulation using the Hilbert transform [5-6] (HT). However, it is influenced by the end effect, and the speed component needs to be clear enough to be separated from the multi-component signal.

Therefore, an IF estimation algorithm based on dynamic time warping [7] (DTW) is proposed combined with variational mode decomposition (VMD). It is a phase demodulation algorithm, which is not affected by the Heisenberg uncertainty principle. The advantage of the DTW and VMD-based methods is that they can estimate the IF only through the vibration signal captured by the accelerometer and reduce the influence of noise interference. The obtained IF information can be used as a reference for subsequent OT, which lays the data foundation for health monitoring.

2. The principle of DTW to achieve feature alignment

The Euclidean distance is capable of achieving a "one-to-one" mapping for points at the same position in time between two time series, but it cannot quantify the similarity between these two time series and requires that both sequences be of equal length. By allowing the sequences to be stretched or condensed in the time dimension, DTW can achieve feature alignment and gauge the degree of similarity between two time series, which can be generally defined as:

$$\begin{cases} X=[x_1,\ldots,x_n,\ldots,x_N], & n=1,2,\ldots,N \\ Y=[y_1,\ldots,y_m,\ldots,y_M], & m=1,2,\ldots,M \end{cases} \tag{1}$$

The Euclidean distance of the pair (x_n, y_m) can be calculated as:

$$d(n,m)=\sqrt{(x_n-y_m)^2} \tag{2}$$

DTW algorithm can be summarized as finding a time-warping path *P*:

$$P=[p_1,p_2,\ldots,p_k,\ldots,p_K] \tag{3}$$

Here, $p_k=(n,m)$ stands for the *k*-th element of *P*, and (n,m) stands for the point x_n in *X* aligned to y_m in *Y*. Meanwhile, the selection of the warping path *P* should satisfy boundary constraints, subsequence constraints, and monotonicity constraints. There are a number of warping paths that meet the above limitations. However, DTW aims to obtain the optimal path P_{best} with the smallest cumulative Euclidean distance overall points on the path. Therefore, a cumulative cost matrix *C* of size $N\times M$ is constructed to solve this problem, which can be treated as a dynamic programming problem:

$$C[n][m]=d(n,m)+\min\begin{cases} C[n-1][m] \\ C[n][m-1] \\ C[n-1][m-1] \end{cases} \tag{4}$$

where $d(n, m)$ stands for the Euclidean distance between the point x_n in *X* and y_m in *Y*

3. IF estimation algorithm based on VMD and DTW

3.1. Mono-component signal extraction by VMD

In practical engineering applications, due to the influence of load and mechanical components, the vibration signal from the rotating machinery is often multi-components with non-stationary frequency. However, both the traditional HT and the proposed DTW algorithm can only process mono-component signals. Therefore, VMD is introduced here to extract a mono-component signal from a multi-harmonic vibration signal [8].

VMD decomposes the vibration signal into IMF, $\{u_k(t)\}$, with the smallest sum of bandwidths. The constrained variational model can be expressed as:

$$\begin{cases} \min\limits_{\{u_k\},\{\omega_k\}}\left\{\sum_k\left\|\partial_t\left[\left(\delta(t)+\frac{j}{\pi t}\right)\times u_k(t)\right]e^{-j\omega_k t}\right\|_2^2\right\} \\ s.t.\quad \sum_k u_k=f \end{cases} \tag{5}$$

Journal of Physics: Conference Series **2815** (2024) 012044 doi:10.1088/1742-6596/2815/1/012044

where k is the decomposed level, u_k and ω_k are the kth IMF component and its center frequency respectively, ∂_t is the gradient operation of time, and $\delta(t)$ is the unit function with impulse. In order to figure out the constrained variational problem effectively, the penalty parameter α and the Lagrange multiplier $\lambda(t)$ are introduced. Thus, Equation (5) can be rewritten in an updated version:

$$L(\{u_k\},\{\omega_k\},\lambda)=\alpha\sum_k\left\|\partial_t\left[\left(\delta(t)+\frac{j}{\pi t}\right)\times u_k(t)\right]e^{-j\omega_k t}\right\|_2^2 + \left\|f(t)-\sum_k u_k(t)\right\|_2^2+\left\langle\lambda(t),f(t)-\sum_k u_k(t)\right\rangle \tag{6}$$

where <·> stands for the inner product operation.

In the process of solving Equation (5), the iterative update of $u_k^{(n+1)}$, $\omega_k^{(n+1)}$ and $\lambda_k^{(n+1)}$ is carried out by the alternate direction method of multipliers to obtain the optimal solution. Hence, all elements of the IMFs can be restored and reconstructed.

3.2. IF estimation by DTW

The mono-component signal is first extracted from the original multi-components vibration signal by VMD and then normalized to obtain the kth mono-component signal $x_i(t)$:

$$x_i(t)=\cos(2\pi\int f_x(t)dt+\theta) \tag{7}$$

Where $f_x(t)$ and θ are the IF and initial phase of $x_i(t)$ respectively.

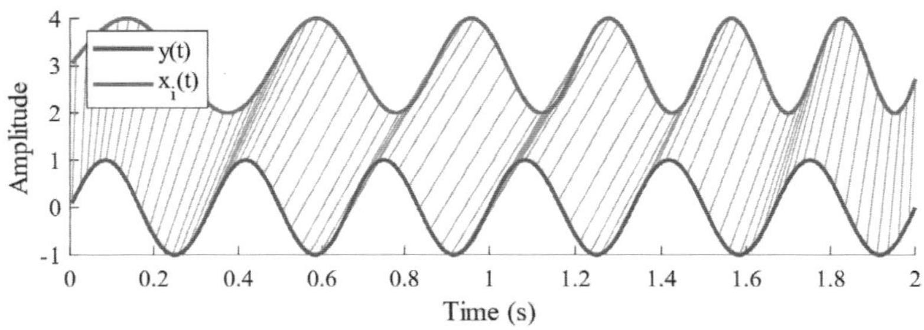

Figure 1. Feature alignment of two signals by DTW.

Since DTW can achieve feature alignment, the phase of the alignment points is considered to be similar for the two signals, as displayed in Figure 1. If the instantaneous phase (IP) of one signal is known, the optimal warping path P_{best} obtained by DTW is able to be used to calculate the IP of the other signal. Construct a reference signal $y(t)$ with a constant frequency f_r:

$$y(t)=\cos(2\pi f_r t+\theta_r) \tag{8}$$

where f_r is the frequency of $y(t)$ which is calculated by the average frequency of $x_i(t)$, and θ_r is the phase of $y(t)$ when t = 0 s. The signal $x_i(t)$ and the reference $y(t)$ are matched by the DTW algorithm over a period of time, and then a rough IP estimation of $x_i(t)$ can be calculated from P_{best}. Finally, $\varphi(t)$, the fitting function expression of P_{best}, is obtained to realize the "one-to-one" mapping of the two signals, reducing the influence of singularities. The IF of $x_i(t)$ can be calculated by the fitting function expression of P_{best} and f_r:

$$f_x(t)=d\varphi(t)/dt\cdot f_r \tag{9}$$

Since the reference signal $y(t)$ is a sinusoidal function, the phase estimate calculated based on the DTW algorithm can reduce the noise effect caused by velocity fluctuations. The process of the DTW and VMD-based method is illustrated in Figure 2.

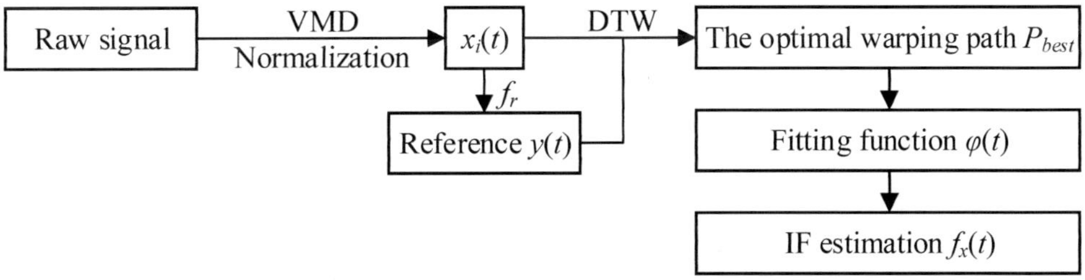

Figure 2. Flowchart of the proposed method.

4. Experiment and analysis

4.1. Simulation analysis

According to the practical application and the signal model proposed in section 3.2, the following mono-component non-stationary simulation signal is constructed:

$$x(t) = \cos\left(2\pi \int f_x(t)dt\right) + n(t) \tag{10}$$

where $n(t)$ stands for Gaussian white noise with a SNR of 0dB. The sampling frequency of the simulation signal is 2000Hz, the duration is 1s, and the signal fundamental frequency $f_x(t)$ is:

$$f_x(t) = 200(t - 0.4)^2 + 150 \tag{11}$$

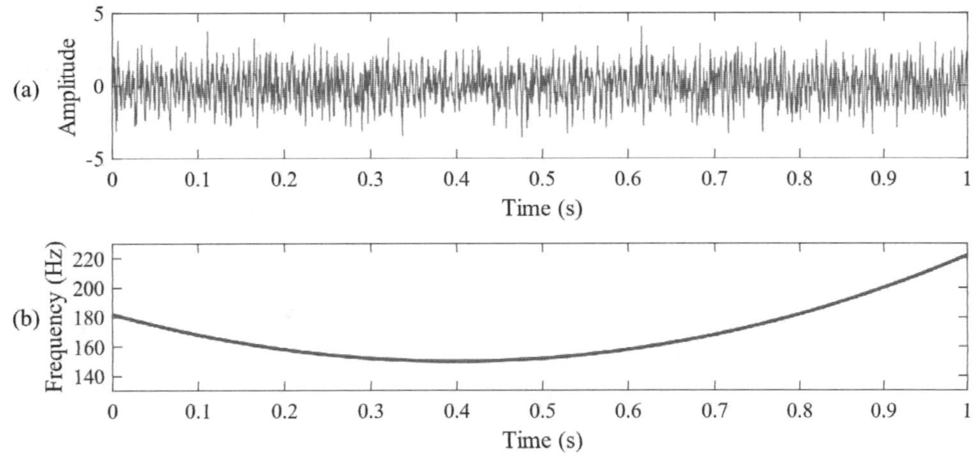

Figure 3. The wave and true IF of the simulated signal.

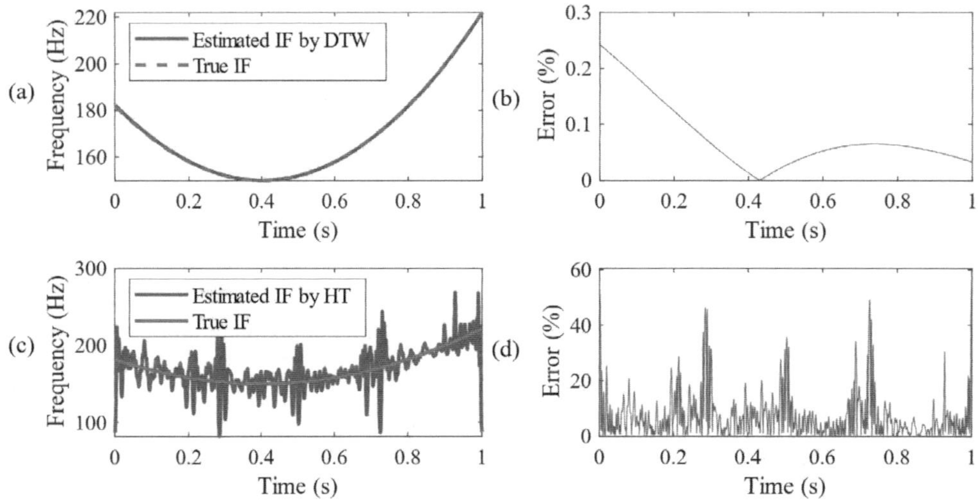

Figure 4. IF estimations and relative errors by DTW and HT.

Figure 3 displays the wave and true time-frequency function of the simulated signal. The IF estimates obtained by HT and DTW are shown in Figure 4. The maximum relative error of IF obtained by HT is 60.52%, and the average relative error is 7.52%. The IF estimated can reflect the trend of the simulation signal, but the local IF cannot be clearly expressed due to the influence of noise. However, the IF estimation relative error obtained by DTW can be less than 0.2420%, and the average relative error is only 0.07742%. Compared with HT, the DTW algorithm can estimate the IF more accurately and has a strong ability to resist noise interference. The main reason why the proposed method is superior to the classical algorithm is that the reference signal constructed is sinusoidal, which has a certain filtering effect and reduces the impact of the noise contained in the target signal on the IF estimation.

4.2. Experimental and engineering applications

In this paper, planetary gearbox data is used to verify the DTW and VMD-based methods. As demonstrated in Figure 5. The experimental system consists of the platform base plate, three-phase variable frequency motor with a rated power of 3kW, rolling bearing rotor system, planetary gearbox, torque loading system, coupling, circulating cooling system, control system, etc. The NI-DAQ acquisition card and accelerometer are employed to acquire the vibration data at a specified frequency sampling rate of 5kHz when the radial pressure of the spindle bearing is 735N, and the motor speed fluctuation range is 600-900r/min.

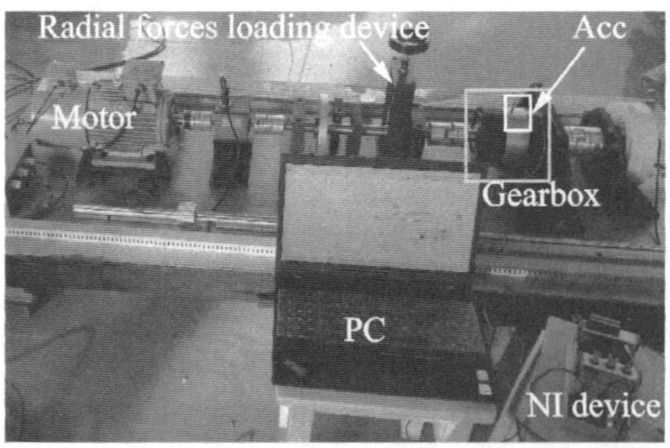

Figure 5. Experimental platform.

Journal of Physics: Conference Series **2815** (2024) 012044 doi:10.1088/1742-6596/2815/1/012044

The wave, Fourier transform, and STFT spectrum of the acquired vibration signal are shown in Figure 6. The mono-component signal with the highest energy with a frequency between 100 and 160Hz is filtered to calculate IF. As shown in figure 7, compared with the classical Hilbert transform, the proposed method achieves more accurate IF estimation. Because the IF set by the control system is a predetermined value, the actual IF may not be exactly equal to the preset speed. However, it is clear that the IF estimated by the proposed method is almost equal to the IF in the STFT spectrum. Therefore, the proposed method is beneficial to the subsequent use of an order-tracking algorithm to locate the fault position.

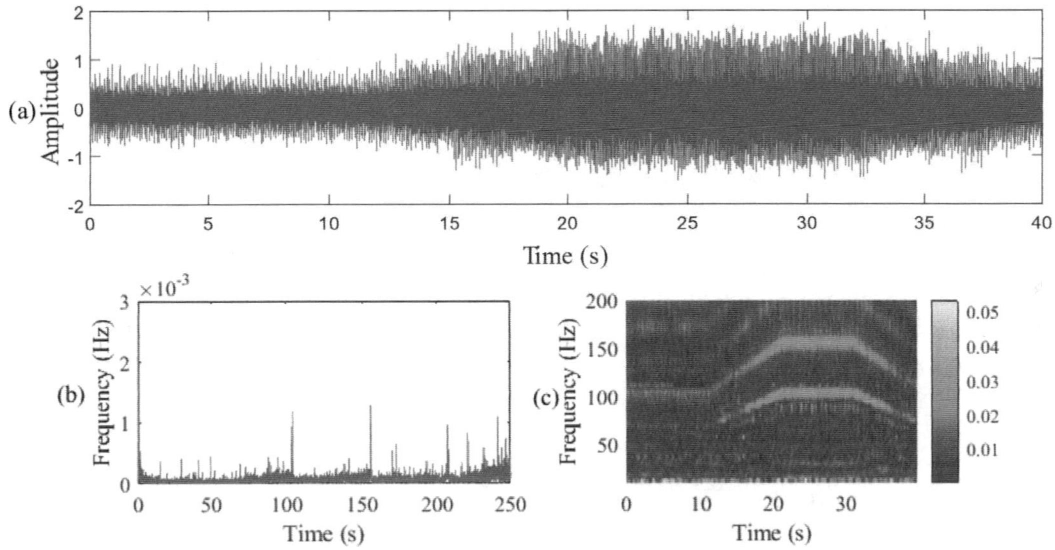

Figure 6. The wave, Fourier transform, and STFT spectrum of the acquired vibration signal.

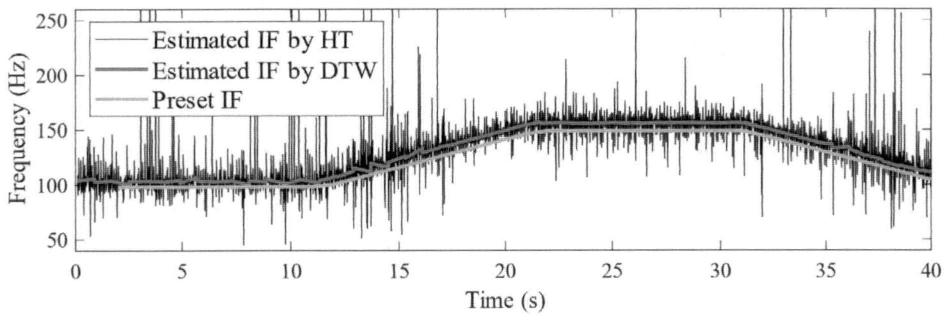

Figure 7. IF estimations by DTW and HT.

5. Conclusion

To address the challenge of diagnosing faults in rotating machinery without a tachometer, VMD is used to extract the mono-component signal from the vibration signal. DTW is introduced to estimate the IF of a non-stationary vibration signal. The obtained IF can be utilized to resample the initial waveform, obtain the order spectrum of the signal, and diagnose the fault. The experimental results show that DTW can extract the IF information from the noisy vibration signal. In mechanical systems that are not suitable for installing tachometers, the DTW and VMD-based methods can be used to analyze the fault of the gearbox and bearing.

References

[1] Choudhury M D, Hong L, Dhupia J S. (2021) A novel tacholess order analysis method for bearings operating under time-varying speed conditions. Measurement, 186: 110-127.
[2] Scheper R A, Teolis A. (2003) Cramer-rao bounds for wavelet transform-based instantaneous frequency estimates. IEEE Transactions on Signal Processing, 51(6): 1593-603.
[3] Wang S, Chen X, Tong C, et al. (2017) Matching synchrosqueezing wavelet transform and application to aeroengine vibration monitoring. IEEE Transactions on Instrumentation and Measurement, 66(2): 360-72.
[4] Guo J, Hao G, Yu J, et al. (2023) A novel solution for improved performance of Time-frequency concentration. Mechanical Systems and Signal Processing, 185: 109784.
[5] Braun S, Feldman M. (2011) Decomposition of non-stationary signals into varying time scales: Some aspects of the EMD and HVD methods. Mechanical Systems and Signal Processing, 25(7): 2608-30.
[6] Feldman M. (2006) Time-varying vibration decomposition and analysis based on the Hilbert transform. Journal of Sound and Vibration, 295(3-5): 518-30.
[7] Rabiner L, Rosenberg A, Levinson S. (1978) Considerations in dynamic time warping algorithms for discrete word recognition. IEEE Transactions on Acoustics, Speech, and Signal Processing, 26(6): 575-82.
[8] Yan X, Jia M. (2019) Application of CSA-VMD and optimal scale morphological slice bispectrum in enhancing outer race fault detection of rolling element bearings. Mechanical Systems and Signal Processing, 122: 56-86.

Journal of Physics: Conference Series **2815** (2024) 012045 doi:10.1088/1742-6596/2815/1/012045

Design of control system for seawater total alkalinity in-situ analyzer

Hongfan Wang[1], Yingying Zhang[1, *], Da Yuan[1], Bingwei Wu[1], Jing Zeng[1] and Maotong Hao[1]

[1] Institute of Oceanographic Instrumentation, Qilu University of Technology (Shandong Academy of Sciences), Shandong Provincial Key Laboratory of Ocean Environmental Monitoring Technology, National Engineering and Technological Research Center of Marine Monitoring Equipment, Qingdao, Shandong, 266061, China.

[*] Corresponding author's e-mail: zhang2020yy@126.com

Abstract: To address the inadequacies of traditional seawater total alkalinity detection methods, which fail to meet the demands for continuous on-site monitoring at sea, a seawater total alkalinity analyzer based on single-point titration and spectrophotometry has been developed. This paper focuses on the design of the analyzer's system architecture, hardware, and control systems. The instrument adopts a modular design concept and utilizes microfluidic technology for detection, with the STM32 serving as the control core. Through the design of the control system for the detection instrument, accurate in-situ automatic measurement of seawater total alkalinity is achieved. Laboratory testing and analysis demonstrate that the instrument's measurement relative error is less than 1%, satisfying the accuracy requirements for seawater total alkalinity measurement. This research provides critical monitoring data support for a timely understanding of marine ecological environments and the marine carbon cycle.

1. Introduction

Human activities have increasingly exacerbated the issue of carbon emissions. As of February 2024, the concentration of atmospheric CO_2 has surged to 421 parts per million (ppm) [1]. Approximately 30% of the carbon dioxide emitted into the atmosphere due to human endeavors is absorbed by the oceans [2], which, upon transforming into carbonic acid within the seawater, induces a reduction in the pH levels of over 0.1 [3]. This decline in seawater pH could significantly alter marine biological processes. For instance, as the oceans grow more acidic, the saturation states of calcium carbonate diminish [4-5], complicating the shell formation process for calcifying organisms [6-7]. The detrimental impacts of ocean acidification on marine calcification might propagate through the food web, thereby transforming the entire ecosystem [8].

Seawater total alkalinity is one of the four key parameters characterizing the marine carbonate system. Owing to its relatively stable variation in the ocean, it allows for precise measurement and analysis [9-10], making it one of the most frequently utilized observational parameters in the study of marine carbonate systems.

To date, several in-situ total alkalinity analyzers have been reported both domestically and internationally. These include the Submersible Autonomous Moored Analyzer for Alkalinity developed by Spaulding et al. [11], which demonstrated accuracy and precision of -2.20±13.10 $\mu mol \cdot kg^{-1}$ (n=86) and ±4.7 $\mu mol \cdot kg^{-1}$ (n=22) respectively in field measurements. The dual pH-AT sensor developed by Briggs et al. [12] utilized Ion-Sensitive Field-Effect Transistor (ISFET) technology to simultaneously measure pH and total alkalinity without reagents, albeit facing challenges in packaging, with an accuracy of 2-10 $\mu mol \cdot kg^{-1}$. The microfluidic total alkalinity analyzer developed by Sonnichsen et al. [13] achieved an accuracy of -0.17±24 $\mu mol \cdot kg^{-1}$ and a precision of 16 $\mu mol \cdot kg^{-1}$. The ISA-TA analyzer developed by Li Qiu et al. [14] realized a precision above 1, though its components are not pressure-resistant, and it is only suitable for surface seawater.

Despite significant advancements in recent years, the performance of in-situ total alkalinity analyzers/sensors still requires further enhancement. In response to the existing challenge that in-situ total alkalinity measuring devices cannot concurrently achieve low power consumption and high accuracy, this study has developed a seawater total alkalinity in-situ analyzer based on a single-drop titration-spectrophotometry method. This paper focuses on the research of the system structure, hardware design, and control design of the seawater total alkalinity in-situ analyzer's detection system, aiming to meet the complex marine conditions faced by oceanographic instruments, reduce the instrument's power consumption, and enhance the instrument's real-time capabilities, stability, and reliability. Additionally, the performance of the instrument was validated through laboratory tests.

2. Principles and methods

2.1 Definition of seawater total alkalinity

Seawater total alkalinity is defined as the moles of hydrogen ions in a kilogram sample where proton acceptors exceed proton donors, with proton acceptors being weak bases with a pKa≤4.5, and proton donors being weak acids with a pKa>4.5. This can be represented by Equation(1):

$$A_T = \left[HCO_3^-\right] + 2\left[CO_3^{2-}\right] + \left[B(OH)_4^-\right] + \left[OH^-\right] - \left[H^+\right] - \left[HSO_4^-\right] - [HF] \quad (1)$$

2.2. Analysis of detection data

The instrument design and development are based on spectrophotometry, which employs the Lambert-Beer law to determine the concentration of a substance. This law states that when a parallel beam of monochromatic light passes vertically through a uniform, non-scattering, light-absorbing material at a specific wavelength, absorption of light occurs. A light source illuminates the sample flow cell with incident light I_0, and if the concentration of the substance being detected in the sample flow cell is higher, or the absorption layer is thicker, the absorption of light is stronger, resulting in weaker transmitted light, meaning the intensity of the emitted light I is lower. This can be represented by Equation (2):

$$A = \lg\left(\frac{I}{I_0}\right) \quad (2)$$

where A represents the absorbance, I_0 is the intensity of the incident light, and I is the intensity of the transmitted light.

For Bromocresol Green (BCG), the maximum absorption wavelength in its acidic state (HI^-'s maximum absorption wavelength) is 444 nm, and in its basic state (I^{2-}'s maximum absorption wavelength) is 616 nm. Based on the dissociation formula of BCG: $\left[HI^-\right] \longleftrightarrow \left[H^+\right] + \left[I^{2-}\right]$, the concentration of hydrogen ions can be calculated from the absorbance of the indicator at these two wavelengths. This calculation is represented by Equation(3).

In the following equation, K_{ind} is the secondary dissociation constant of BCG. Combined with spectrophotometry, the concentration of hydrogen ions can be calculated using the following equation:

$$\left[H^{+}\right]=\frac{K_{ind}\left[HI^{-}\right]}{\left[I^{2-}\right]} \tag{3}$$

$$pH = pK_{ind} + \lg\left(\frac{R-e_1}{e_2 - Re_3}\right) \tag{4}$$

where e_1, e_2, and e_3 represent the ratios of molar absorptivity coefficients, and R is the ratio of the absorbance of the mixed solution at a wavelength of 616 nm to the absorbance of the mixed solution at a wavelength of 444 nm. For BCG, the values of the molar absorptivity coefficient ratios are respectively 0.0013, 2.3148, and 0.1299 [14].

During measurements, since the temperature can affect the molar absorptivity coefficients and electronic components, it is necessary to investigate and characterize the temperature effects. The calculated pH values are corrected using the following formula:

$$pH = (aT+b)\cdot pH \tag{5}$$

3. Overall design

The seawater total alkalinity analyzer designed and developed represents an integrated application of embedded technology, microfluidic technology, and communication technology, thereby achieving accurate measurement of seawater total alkalinity. The overall design structure of the instrument is illustrated in Figure 1. The entire detection system of the instrument can be divided into two parts: the microfluidic sample inlet and outlet system and the signal processing system.

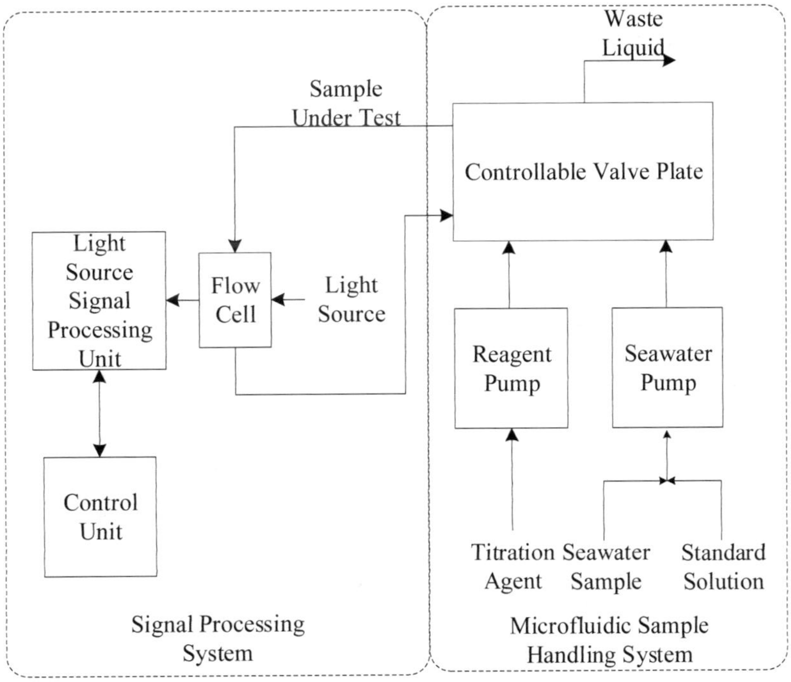

Figure 1. Schematic diagram of the total alkalinity analyzer system for seawater.

The signal processing system comprises various wavelength light sources, a flow cell, a light source signal processing unit, and a control unit. The light source signal processing unit, serving as the core of the signal processing system, includes a photoelectric signal conversion module, a light source signal amplification module, and an AD conversion module. The flow cell employs a cross-shaped design, enabling absorbance detection without disrupting the flow path. Furthermore, a detailed introduction to the control unit is provided in Section 4.

The automatic sample inlet and outlet system consists of a microfluidic peristaltic pump and a controllable valve board. Compared to traditional liquid pumps, the microfluidic peristaltic pump can control the liquid flow rate more precisely, helping to reduce the volume of the entire flow path, thus minimizing the consumption of samples and titrants as well as the detection time. The controllable valve board, when switching flow paths, has a smaller liquid dead volume than traditional flow control systems like the eight-port valve, enabling the instrument to achieve higher precision and accuracy.

During operation, the peristaltic pump drives the seawater sample or standard solution into the flow path of the controllable valve board. Subsequently, the reagent pump propels the titrant into the flow path. Before and after the mixing of the seawater sample and titrant, the light source and light source signal processing unit are activated to detect absorbance. Finally, the control unit analyzes the data to determine total alkalinity.

4. Analysis of hardware design

The core instrument control and detection circuitry adopt a modular design philosophy, employing the STM32F103 microcontroller as the control system MCU. According to functionality, it separately designs and develops various modules: a power control module, a 3-channel LED constant current light source driver module, a 4-channel solenoid valve driver module, a dual-channel three-way valve driver module, a dual-channel liquid pump driver module, a temperature data collection module, a light source signal processing module, a data storage module, and a communication module, with a TF card used as the on-site data storage medium. The overall structure of the control system is depicted in Figure 2. This section will primarily focus on power control and signal processing from light source sources.

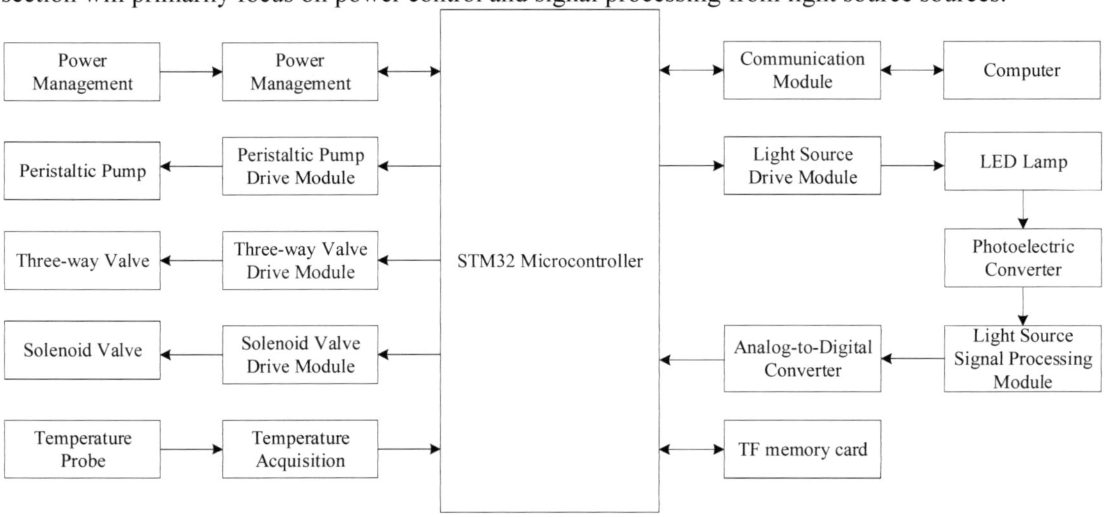

Figure 2. Schematic representation of the control system structure.

4.1. Power control design

As an in-situ automatic detection instrument for the marine environment, the power consumption of the seawater total alkalinity analyzer is a critical factor limiting its continuous monitoring duration. Given the relatively low variation rate of total alkalinity in seawater, the measurement frequency of the analyzer is low, and it remains in standby mode most of the time. Therefore, the standby power consumption of the analyzer plays a decisive role in the working duration of the instrument, with the standby power consumption of various pumps, valves, and LED lights accounting for more than 70% of the total power consumption. To address this issue, a power management module is added to the circuit design, enabling autonomous control over the entire system's +24V, +12V, and +5V power supplies. After the total alkalinity measurement process is completed, the power supply to the high-power-consuming devices is cut off, reducing the system's standby power consumption and achieving

Journal of Physics: Conference Series **2815** (2024) 012045 doi:10.1088/1742-6596/2815/1/012045

a low-power design. Additionally, during standby, the MCU is shut down or put into standby mode to reduce power consumption further. Figure 3 shows the circuit diagram of the power management module. The control end is connected to the microcontroller I/O port; if the control end is set high, the N-channel MOSFET conducts, creating a negative voltage between the gate and source of the P-channel MOSFET that exceeds its conduction threshold. Voltage is output from the drain of the P-channel MOSFET, marking the voltage activation process. Conversely, setting the control end low turns off all MOSFETs, cutting off the power supply to the corresponding power-consuming devices.

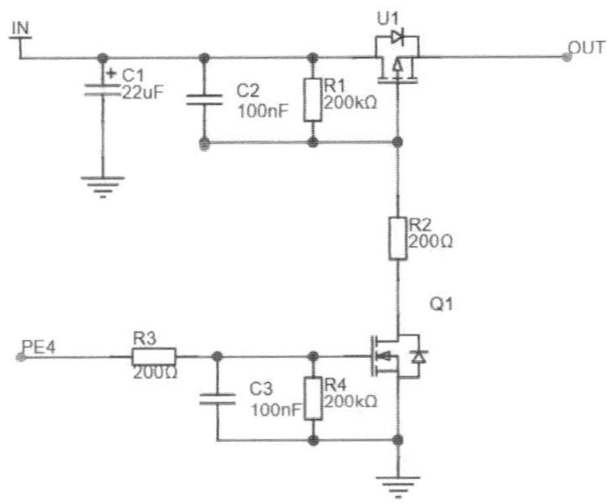

Figure 3. Circuit diagram of the power management module.

4.2. Light source signal processing design

To ensure the stability of the light source, the driving current for the three LED lights was specifically adjusted. This adjustment resulted in a weaker light source signal, necessitating the amplification of the electrical signal converted by the photodiode. To ensure that the amplified light source signal remains undistorted, the light source signal amplification module consists of an inverting amplifier, a second-order active filter, a programmable amplifier, and a voltage follower, ensuring stable detection of the light source signal. To avoid interference in the processing of weak total alkalinity signals within this highly integrated system, a digital-to-analog isolation method was employed during the design process. Figure 4 shows the schematic diagram of the light source signal amplification module.

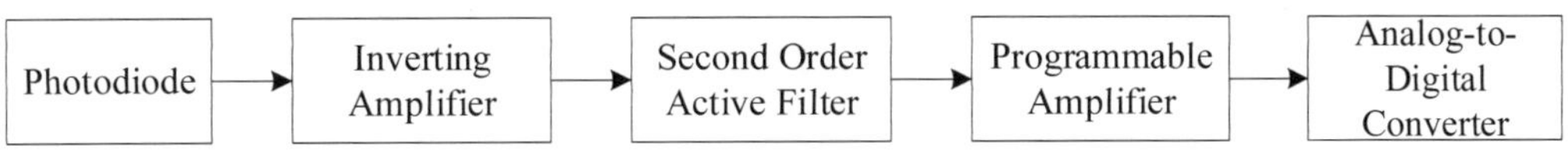

Figure 4. Schematic diagram of the light source signal amplification module.

5. Control software design

The measurement of total alkalinity requires corresponding control logic. When the instrument receives a detection command, the microcontroller controls various devices to execute the detection process, processes the detection data, and saves it. Figure 5 is the flowchart of the program.

Before detection begins, various devices are initialized to ensure normal operation. Then, the detection process is carried out. After obtaining the detection and temperature data, they are processed to obtain total alkalinity data. Upon completion of the detection, the data is uploaded to the host computer and saved for subsequent analysis.

Once the instrument enters the detection process, the first step is to clean the pipelines to eliminate the impact of impurities on the detection. Next, seawater sampling is conducted. After sampling is completed, a period is allowed to ensure the stability of the water pressure inside the pipeline. Then, the photodetection system is activated to measure the absorbance of the seawater sample. The next step involves introducing the titrant for sampling. Once sampling is completed, the seawater sample and the titrant are mixed, followed by photodetection of the mixed solution to obtain its absorbance. Finally, the pipelines are cleaned to prevent interference with subsequent detections.

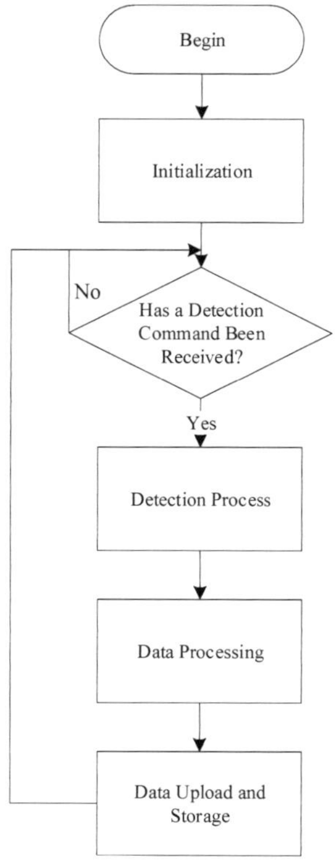

Figure 5. Flowchart of the main control program.

6. Results

6.1. Low power consumption verification

Upon analysis, to ensure the stability of the tested feeble photoelectric conversion signals, the light source signal processing unit remains powered continuously. Through testing, the leakage current of the photodiode used by the control system was found to be 400 μA. The standby current of the light source signal processing unit was 13~14 mA. The microcontroller's power consumption during non-execution phases, entering sleep mode, was approximately 8~9 mA, resulting in a system static power consumption of about 22 mA.

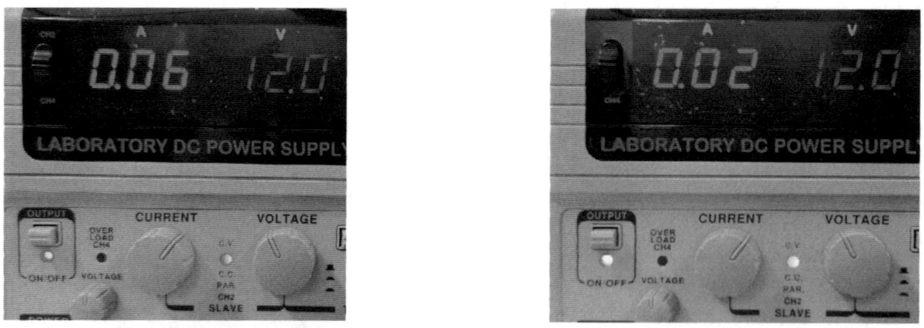

(a) (b)

Figure 6. Instrument standby power consumption comparison.

Relevant tests were conducted in the laboratory to validate the efficacy of the power management module. This sensor is powered by a digital power supply, and comparative experiments were conducted on the system's standby current without and with the design of the power management module. The experimental results are shown in Figure 6.

As indicated by Figure 6, without the power management module, the entire system's standby power consumption was 0.72W. After incorporating the power management module, the system's standby power consumption was reduced to 0.24 W, significantly lowering the power consumption.

6.2. Accuracy verification

Table 1. Experimental data on total alkalinity.

Standard Solution / ($\mu mol \cdot kg^{-1}$)	Measured Data / ($\mu mol \cdot kg^{-1}$)	Absolute Error / ($\mu mol \cdot kg^{-1}$)	Relative Error / %
2000	1990.5	-9.5	-0.48
2100	2097.3	-2.7	-0.13
2250	2266.9	16.9	0.75
2500	2484.6	-15.4	-0.62

In laboratory measurements, four samples with total alkalinity values ranging from 2000 to 2500 $\mu mol \cdot kg^{-1}$ were prepared using a standard solution, including one with the standard solution and others with diluted solutions. The data obtained from the instrument were compared with the standard values to assess the accuracy of the instrument, where the instrument's measured data were rounded to one decimal place. A calibrated seawater total alkalinity analyzer was used to test the sample. Table 1 shows part of the detection data. The selected total alkalinity values cover most of the seawater alkalinity values to meet the requirements of different marine areas. According to the detection results of the laboratory standard samples, the testing instrument's measurement error is <0.1%, meeting the requirements for accurate measurement.

6.3. Precision verification

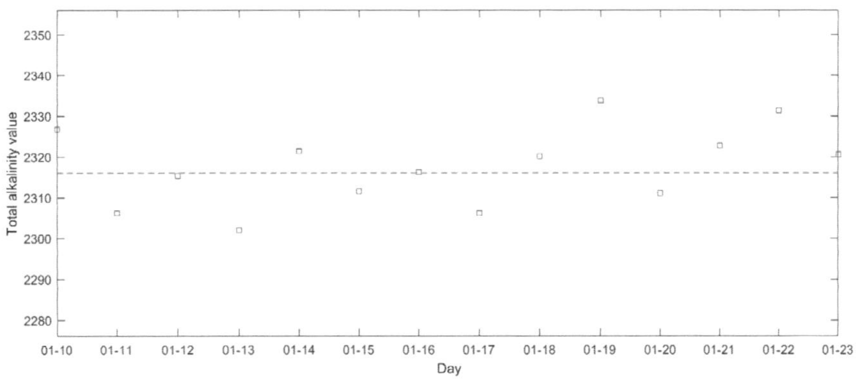

Figure 7. Total alkalinity precision experiment data.

To assess the instrument's precision, a 14-day laboratory test was conducted. During the testing period, the instrument performed a measurement on the sample every 12 hours, using seawater collected and filtered from the Badaguan Pier in Qingdao. Figure 7 presents a portion of the detection data, where the dashed line represents the total alkalinity value measured using the Gran titration method. The experimental data reveal that the instrument's measurement error is $1 \pm 16 \mu mol \cdot kg^{-1}$ (n=28), with a precision of <1%, meeting the requirements for long-term stable measurements.

7. Conclusions

This paper presents the design of a seawater total alkalinity analyzer based on a single-drop titration-spectrophotometry method, featuring its high accuracy and stability. Through the application of key technologies such as modular design and microfluidic technology, the instrument's in-situ detection duration, reliability, and universality have been effectively enhanced, meeting the requirements for seawater total alkalinity detection. Laboratory tests have demonstrated that this sensor detection system possesses low power consumption, high detection accuracy, and excellent stability, fulfilling the needs for long-term in-situ measurement of seawater total alkalinity. Future research on the seawater total alkalinity analyzer will focus on addressing various complex marine environments and further enhancing accuracy.

Acknowledgments

This study was financially supported by the Taishan Scholars Program, i.e., Key R&D Program Projects of Shandong Province (No. 2023CXPT015), National Natural Science Foundation of China (No. U2006222) and Qilu University of Technology (Shandong Academy of Sciences) unveiling project (No. 2022JBZ01-02-02).

References

[1] Global Monitoring Laboratory, (2024). Trends in atmospheric carbon dioxide. https://gml.noaa.gov/ccgg/trends/global.html.

[2] Müller D J, Gruber N, Carter B, et al. (2023) Decadal Trends in the Oceanic Storage of Anthropogenic Carbon From 1994 to 2014. J. AGU Advances., 4(4).

[3] Orr J C, Fabry V J, Aumont O, et al. (2005) Anthropogenic ocean acidification over the twenty-first century and its impact on calcifying organisms. J. Nature., 437: 681–686.

[4] Alvarez C C, Penkman K, Kröger R, et al. (2024) Insights into the response of coral biomineralisation to environmental change from aragonite precipitations in vitro. J. Geochimica et Cosmochimica Acta., 364: 184-194.

[5] Mortel D V H, Delaigue L, Humphreys P M, et al. (2024) Laboratory Observation of the Buffering Effect of Aragonite Dissolution at the Seafloor. J. Journal of Geophysical Research-Biogeosciences., 129(2).

[6] Hoegh-Guldberg O, Poloczanska E S, Skirving W, et al. (2017) Coral Reef Ecosystems under Climate Change and Ocean Acidification. J. Frontiers in Marine Science., 4.

[7] Mcmahon A, Santos I R, Schulz K G, et al. (2018) Determining coral reef calcification and primary production using automated alkalinity, pH and ρCO2 measurements at high temporal resolution. J. Estuarine Coastal and Shelf Science., 209(SEP.30): 80-88.

[8] Thomas W, Mads T S, Julia B K, et al. (2023) Impacts of Climate Change on Marine Foundation Species. J. Annual review of marine science., 16: 247-282.

[9] Wolf-Gladrow D A, Zeebe R E, Klaas C, et al. (2007) Total alkalinity: the explicit conservative expression and its application to biogeochemical processes. J. Marine Chemistry., 106 (1–2): 287–300.

[10] Zeebe R E and Wolf-Gladrow D A. (2001) CO2 in Seawater: Equilibrium, Kinetics, Isotopes. Elsevier Oceanography Series, Amsterdam.

[11] Spaulding R S, Degrandpre M D, Beck J C, et al. (2014) Autonomous in situ measurements of seawater alkalinity. J. Environmental Science & Technology., 48(16): 9573-9581.

[12] Briggs E M, De Carlo E H, Sabine C L, et al. (2020) Autonomous ion-sensitive field effect transistor-based total alkalinity and pH measurements on a barrier reef of Kane'ohe Bay. J. ACS Earth Space Chem., 4: 355–362.

[13] Colin S, Dariia A, Andre H, et al. (2023) An Automated Microfluidic Analyzer for In Situ Monitoring of Total Alkalinity. J. ACS Sensors., 8 (1): 344-352.

[14] Qiu L, Li Q L, Yuan D X, et al. (2023) High-Precision In Situ Total Alkalinity Analyzer Capable of Month-Long Observations in Seawaters. J. ACS sensors., 8(7): 2702-2712.

Study on bird strike analysis of a civil aircraft equipped with satcom antenna

Jun Gao [1,*] **and Yuanzhe Zhu** [1]

[1] Airframe Integration Department, Shanghai Aircraft Design and Research Institute, Shanghai, 201210, China

[*] Corresponding author's e-mail: gaojun1@comac.cc

Abstract: Civil aircraft have specific requirements for the anti-bird strike performance of the structure, so the installation of various types of antennas must carry out bird strike assessment to avoid catastrophic consequences because of bird strikes. In this paper, the anti-bird strike performance of a civil aircraft equipped with an antenna is studied by numerical calculation. It shows that the antenna adapter may be plastically deformed after a bird strike, but it will not be damaged or fall off from the body structure as a whole. Therefore, the satcom antenna will not cause catastrophic consequences after a bird strike and meet the airworthiness requirements.

1. Introduction

During different flight phases, a bird strike collates between a fixed- or rotary-wing aircraft [1].

When humans invented airplanes, bird strikes occurred with them, with increasing frequency and consequences, reflecting the development and global expansion of the aviation industry, resulting in a considerable number of bird strikes. Among the numerous bird strike events, some are well known. In 1905, the Wright brothers finished the first recorded bird strike event [2].

In 1912, Calbraith P. Rodgers recorded another important bird strike event, where he trained with the Wright brothers and conducted his first transcontinental flight over the United States. However, he collided with a group of seagulls and ultimately caused the plane to crash into the sea during an exhibition flight over Long Beach, California, a few months later [3].

On January 15, 2009, American Airlines Flight 1549 was attacked by birds two minutes after take-off, causing both engines to stall and fail. The plane immediately lost all power and had to make an emergency landing in the Hudson River. Fortunately, no one died in the accident. The event was even made into a film and became known as the Miracle on the Hudson [4].

According to incomplete statistics, at least 200 people have died in aircraft crashes caused by bird strikes since 1988. The bird strike incident in the aviation industry is a nightmare for crew members. The aerospace industry spends $2 billion annually on this [1].

Civil aircraft put specific requirements on the anti-bird strike performance of the structure. According to the CS25 promulgated by EASA [5], transport aircraft should be able to land safely after being hit by a 4lb bird. Similar requirements are put forward in the FAR25 [6] promulgated by the FAA in the USA and CCAR25 [7] promulgated by CAAC in China, and the bird body mass struck by the tail bird was increased to 8lb.

According to EASA's policy, the areas affected by the bird strike include the traditional components, such as the nose, wing leading edge, tail leading edge, engine nacelle lip and pylon leading edge, and externally mounted large antennas. This requires us to evaluate the bird strike resistance of the antenna protruding from the outer surface of the aircraft to ensure that it meets the requirements of the airworthiness provisions [8].

Much work has been done on bird strikes worldwide, usually focusing on common areas such as the leading edge of the tail, the leading edge of the wing, the nose, or the engine, usually with analytical or experimental methods. Currently, only Guo et al. [9] have carried out bird strike analysis of radome. However, with the deepening of the study of aircraft bird strikes, industry and reviewers have realized that the antenna structure protrusion on the surface of the aircraft can also be subjected to bird strikes if the antenna is a key piece of equipment affecting the safe takeoff and landing of the aircraft, or the debris from the antenna and auxiliary structures after the bird strike is larger, which may cause further damage.

The purpose of this research is to support the installation of a satcom antenna on civil aircraft, enabling passengers to achieve more convenient communication between the sky and the earth. The paper innovatively demonstrates through numerical calculation that the installed satcom antenna meets the requirements for bird strike. The research plan includes providing bird strike requirements for the antenna, obtaining the dynamic response of the base, obtaining the damage characteristics of the structure after bird strikes through simulation, and ultimately demonstrating airworthiness compliance.

2. Antenna bird strike requirement

2.1. Antenna structure and installation

A satcom antenna is sometimes installed to ensure the comfortable operation of civilian aircraft. The satcom antenna of a certain type of civil aircraft is mounted on the FR1 frame and FR2 frame on the upper fuselage through four connection points. To meet the requirements of the aerodynamic shape of the fuselage, a satcom antenna radome is installed. The radome is divided into two parts: the front and the rear. The radome is installed on the satcom antenna adapter. Figure 1 shows the structure diagram of the satcom antenna, and Figure 2 shows the installation diagram of the satcom antenna.

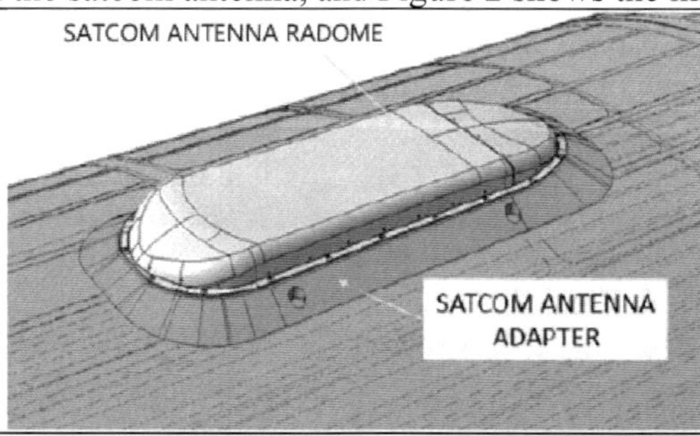

Figure 1. Structure diagram of satcom antenna.

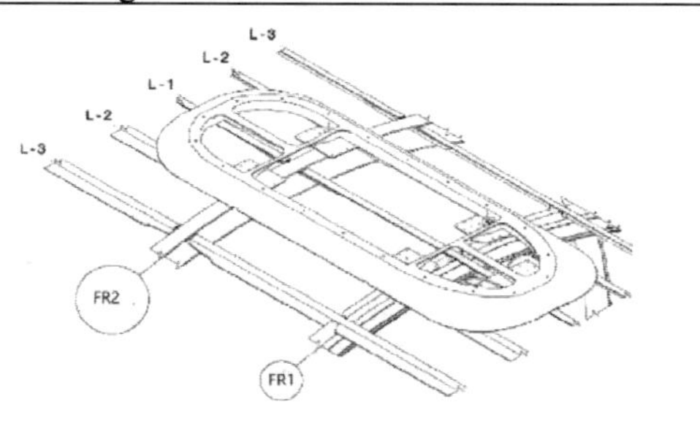

Figure 2. Installation diagram of satcom antenna.

2.2. Bird strike requirements for antennas

Antennas can be divided into two types: large antennas and small antennas. Usually, when an antenna is installed in a skin compartment (within an area of two adjacent frames and two adjacent booms), it is considered small; otherwise, it is called a large antenna. However, even if installed in a skin compartment, if the height of the antenna protrudes beyond the fuselage skin boundary layer, the antenna is considered a large antenna.

According to the civil aviation industry consensus, small antennas and their size and mass are small and usually do not cause serious consequences. Their anti-bird strike performance is generally considered acceptable.

For large antennas, the applicant should demonstrate that any foreseeable circumstances (such as bird impact) will not cause the aircraft to fall off debris or that the fallen debris is insufficient to cause fatal harm to the aircraft.

According to the above requirements, as long as the whole satcom antenna does not fall off, it can ensure the safety of the aircraft. The damage of the bird strike radome and the debris of some components are acceptable, and there is no need to carry out an analysis. Therefore, the object of this study is the antenna adapter, and the criterion is whether it falls off.

3. Numerical calculation of bird strikes

3.1. Bird strike certification method

Generally, there are three methods for studying bird strikes in civil aircraft structures: theoretical analysis, numerical simulation, and experimental study. Among them, the theoretical analysis of bird strikes is cheap and fast. Still, it can only draw qualitative conclusions about the anti-bird strike characteristics of the structure, and it is only used in the early stage of bird strike research. The bird strike test method is the most direct method to study bird strikes, but its test parts manufacturing cost is high, and the test period is long. The numerical calculation method of bird strike has the advantages of rapid and reasonable prediction of bird strike damage, high accuracy, and low cost. It has been rapidly developed and fully applied in the past 20 years. In this paper, the bird strike numerical calculation method is used to evaluate the anti-bird strike performance of the antenna structure to support the installation of civil aircraft antennas.

3.2. Establishment of numerical calculation model

In numerical calculation, the bird body is discrete into SPH units, the shape of which is a cylinder with hemispherical ends, and its structure diagram is shown in Figure 3.

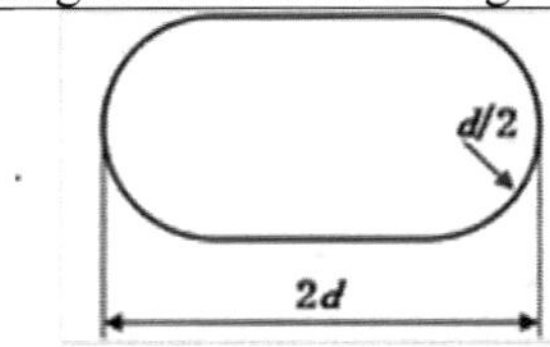

Figure 3. Dimension diagram of bird SPH model.

Monaghan EOS defines the constitutive model of the bird body.

$$P = P_0 + B\left[\left(\frac{\rho}{\rho_0}\right)^{\lambda} - 1\right] \quad (1)$$

where P0 represents the initial pressure, ρ_0 represents the initial density of the bird body, and $\rho_0 = 9.6\times10^{-7}\ kg/mm^3$ is taken in this paper. B is the volume modulus and λ is the state parameter. These two parameters are obtained through analysis and test benchmarking, $B = 0.128GPa$, and $\lambda = 7.98$ in this paper.

To reasonably balance the relationship between calculation accuracy and calculation speed, the element size of the finite element model was selected as 7mm.

Belytschko-Wong-Chiang (BWC) shell elements simulated aluminum alloy structures. The Johnson-Cook constitutive model was adopted for the satcom antenna adapter, regardless of the strain rate and temperature effects of the materials. The constitutive equation of the material model is as follows:

$$\sigma = a + b\varepsilon_p^n \quad (2)$$

In the formula, σ is the stress. ε_p is the plastic strain. a, b and n are the yield stress, hardening coefficient and hardening index of the material respectively, and the failure deletion of the unit is controlled by setting the failure strain.

For the fasteners in the structure, PLNK units are used to simulate and define the transfer of load between them and the structure and their failure. PLINK uses MAT302 material for simulation.

In impacting the bird and the aircraft antennas, the mathematical model must be followed. In the global coordinate system, the mathematical model [10] can be expressed as:

$$[M]\{a\}+[C]\{v\}+[K]\{d\}=\{F^{ex}\} \quad (3)$$

where $[M]$ is the mass matrix, $\{a\}$ is the acceleration vector, $[C]$ is the damping matrix, $\{v\}$ is the velocity vector, $[K]$ is the stiffness matrix, $\{d\}$ is the displacement vector, and $\{F^{ex}\}$ is the external force vector.

We define $[C]\{v\}+[K]\{d\}=\{F^{in}\}$, then:

$$[M]\{a\}=\{F^{ex}\}-\{F^{in}\} \quad (4)$$

In this paper, the asymmetric master-slave contact algorithm defines the contact between the bird body and the structure. An adaptive contact algorithm is used to define the self-contact between structures.

Figure 4 shows the analysis model of the antenna adapter of this civil aircraft satcom antenna. The model includes the antenna adapter and the nearby fuselage structure. The fuselage structure ranges from the front and rear four frames, the left 3 stringer to the right 3 stringer (including the above boundary).

To simulate the support of the fuselage structure for the antenna adapter as much as possible, the boundaries of the front and rear frames are constrained by the partial support of the FR1 and FR2 frames near the left-third and right 3rd stringer. The specific situation is shown in Figure 5, where the red part is a fixed support constraint.

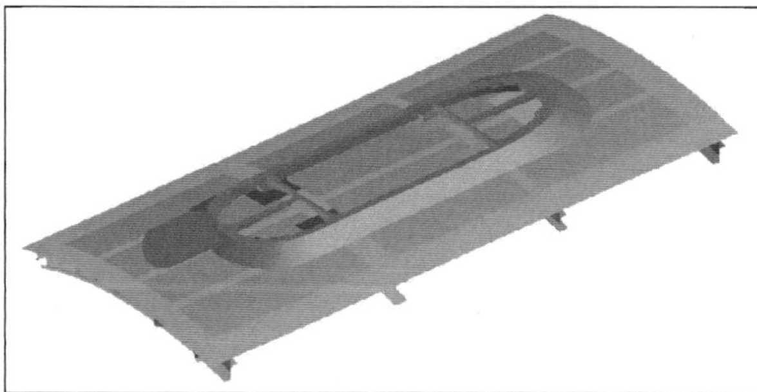

Figure 4. Finite element model of civil aircraft antenna adapter and nearby supporting structure.

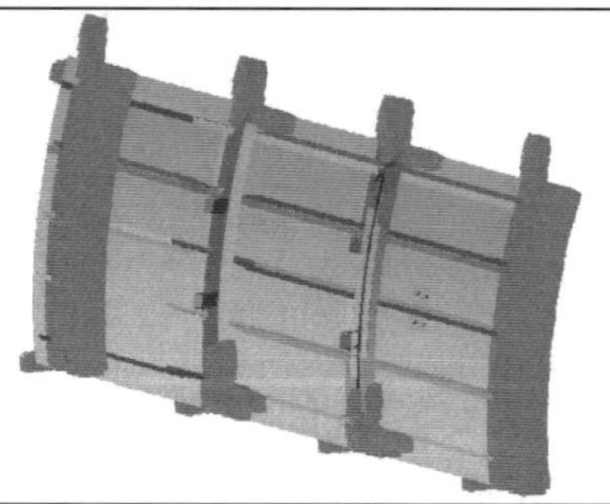

Figure 5. Boundary constraints of structural finite element model.

CCAR25 Airworthiness Standard for Transport Aircraft §25.571(e)-(1) states that an aircraft must be able to safely complete a flight in the event of structural damage after being struck by a 1.80 kg (4 lb) bird. The relative speed of the aircraft and the bird along the flight path of the aircraft is either sea level Vc or 2, 450 m (8, 000 ft) 0.85Vc, whichever is greater. Therefore, the mass of the bird is defined as 1.8kg, and the speed is determined according to the actual aircraft speed.

4. Results of numerical calculation

4.1. Impact cases and qualitative analysis

The bird strike numerical calculation of the satcom antenna adapter and the nearby supporting structure is based on the attitude of the aircraft at 0° angle of attack. The selection of impact cases is mainly the typical bird strike points selected within the possible impact range of the bird body in front of the adapter, as shown in Figure 6.

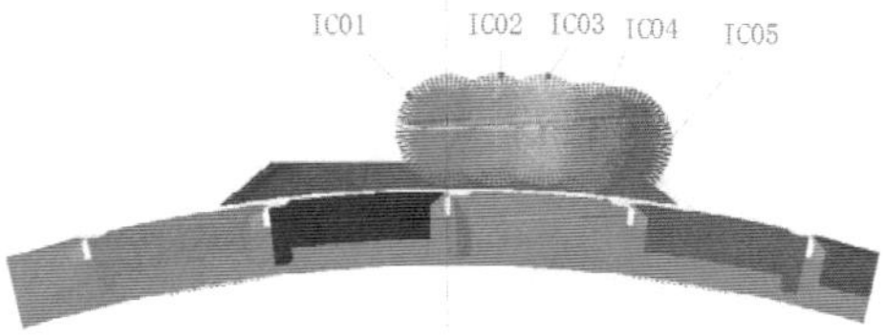

Figure 6. Schematic of the location of the impact cases.

As seen in Figure 6, the height of the antenna adapter structure is small, and the area of the structure colliding with the bird body is small. Only about 1/10 of the bird's body will directly hit the adapter structure, while the rest will slide above the adapter.

4.2. Numerical calculation results

Figure 7 takes IC01 as an example to give the bird strike analysis results of the antenna adapter. According to the results, the bird body dispersed and flew out after striking the structure, and the antenna adapter deformed, but the whole was not damaged or fell off. IC02~IC05 operating condition analysis results are similar, so we will not repeat them.

The plastic deformation values of the structure are given in Figures 8 to 12, which show that the plastic strains of IC01 to IC05 are 0.043, 0.038, 0.044, 0.049, and 0.028, respectively. According to the plastic deformation value, the plastic deformation of IC01 to IC04 impact cases are close, but the plastic deformation of the IC05 impact case is smaller because the shape of the structure near the IC05 impact case is circular, the structural deformation is relatively gentle, and the local plastic deformation value is small.

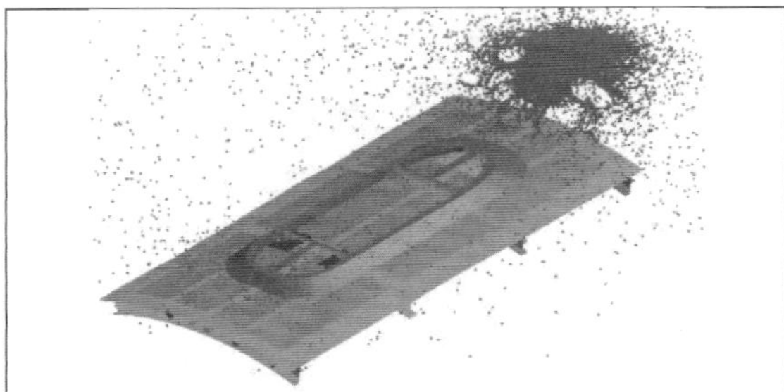

Figure 7. Bird strike results under typical conditions.

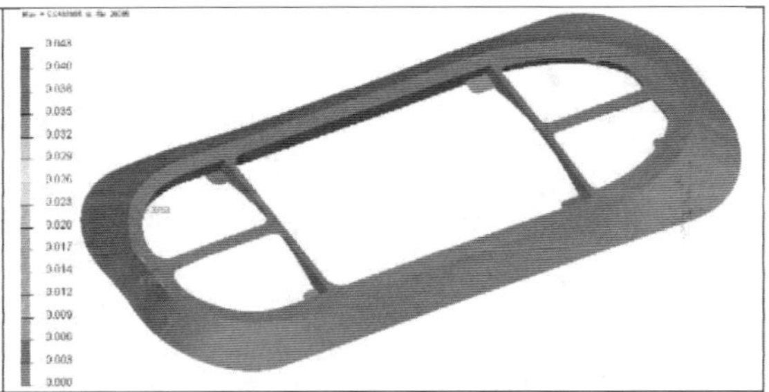

Figure 8. The structural plastic strain of IC01 condition.

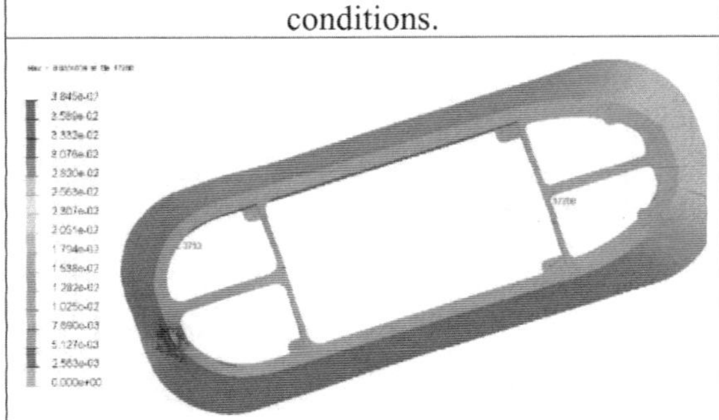

Figure 9. The structural plastic strain of the IC02 condition.

Figure 10. The structural plastic strain of IC03 condition.

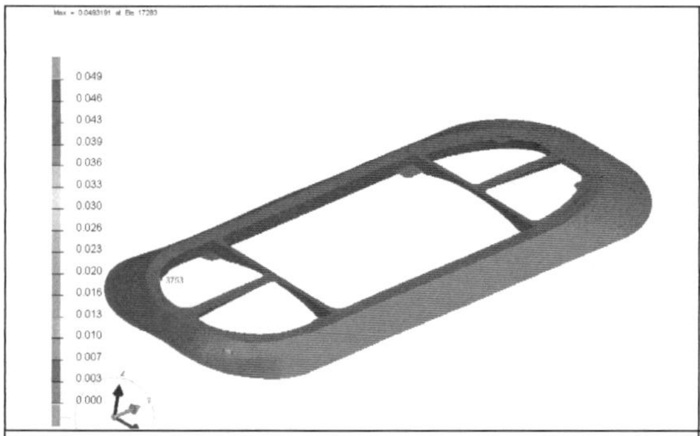

Figure 11. The structural plastic strain of IC04 condition.

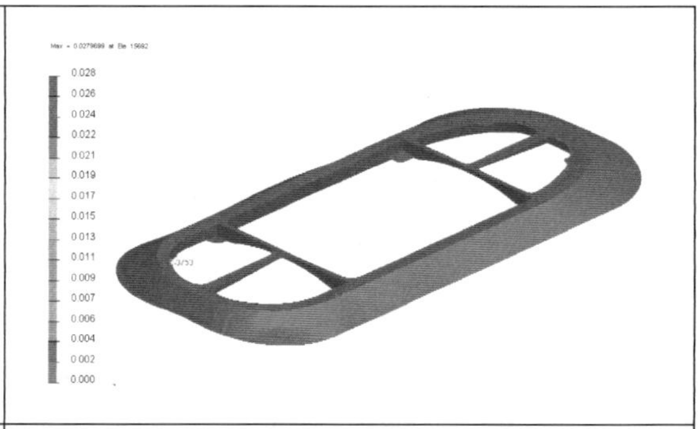

Figure 12. The structural plastic strain of IC05 condition.

Figure 13 shows the bird impact force under different impact cases. It can be seen that the contact force of IC01, IC02, and IC03 gradually decreases, but the pulse width gradually widens. This is because the middle distance is closer to the support column, the local stiffness of the impact area is larger, and the impact angle is larger. When the support column is far away, its support stiffness decreases rapidly, so the contact force decreases, and the pulse width of the impact force increases. However, the contact force and pulse width of IC03, IC04, and IC05 are the same, as they are farther from the middle support column, resulting in a smaller impact area and angle.

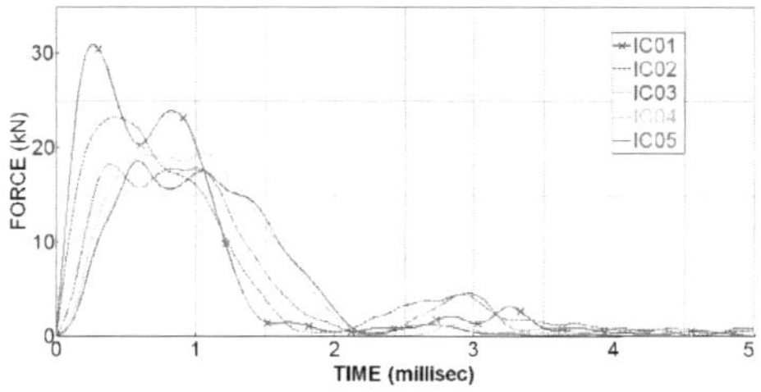

Figure 13. Contact force during bird strike.

5. Conclusion

In this paper, the bird strike numerical calculation of the satcom antenna installed on a certain type of civil aircraft is carried out. The research shows that:

1. The height of the antenna adapter structure analyzed in this paper is small, and the collision area with the bird body is small. Most of the bird body will slide out from the adapter, the antenna adapter structure will not be damaged by bird collision, and the antenna adapter will not partially or completely fall off the fuselage structure.
2. The bird strike angle at the middle position of the antenna adapter is large, and the support stiffness is large, resulting in a large impact force and large plastic deformation of the structure. The arc of the edge structure is also large, and the impact angle is relatively small, resulting in a relatively small impact force and plastic deformation of the structure.
3. When civil aircraft are equipped with antennas, the antenna adapter and other structures should be reasonably designed to reduce the height of the structure, reduce the impact angle, and ensure that the overall fall-off does not occur after the bird strike, to avoid further damage to the aircraft, thus meeting the requirements of the airworthiness clause.

Acknowledgments

The research of this paper has been strongly supported by the relevant colleagues of the Airframe Integration Department and the Fuselage Department of Shanghai Aircraft Design and Research Institute, and I would like to express my thanks here.

References

[1] Ahmed F. El-sayed. Bird Strike in Aviation [M]. 2019.
[2] Hedayati R, Sadighi M. Bird Strike: An Experimental, Theoretical and Numerical Investigation [M]. 2015.
[3] Bilstein R E. Flight in America: from the Wrights to the astronauts [J]. 1994.
[4] Lowe P. Miracle on the Hudson prompts safety review [J]. Aviation International News, 2010, 42.
[5] CS-25, Certification Specifications and Acceptable Means of Compliance for Large Aeroplanes [S]. 2021.11.24
[6] 14 CFR Part 25 - PART 25—AIRWORTHINESS STANDARDS: TRANSPORT CATEGORY AIRPLANES [S]. 2003.11.18
[7] CCAR-25-R4, China Civil Aviation Regulation Part 25 Airworthiness standards: Transport Category Airplanes [S]. Beijing: CAAC, 2011.
[8] EASA CM - S – 001 Issue: 01 Subject Compliance with CS-25 Bird Strike Requirements [Z]. 2012.4.11
[9] Guo Jun, Fu Weichao, Li Xuyang, et al. Simulation analysis of bird strike calculation for civil aircraft radome [J]. Aeronautical Science & Technology, 2021, 32(12): 122-126.
[10] JIA Jian-dong. Bird impact resistance designing and analyzing for aircraft typical structures [D]. Nanjing University of Aeronautics and Astronautics, 2010

Journal of Physics: Conference Series **2815** (2024) 012047 doi:10.1088/1742-6596/2815/1/012047

A self-cleaning glue feeding system based on slit coating

Hong Zhang[1], Guanglei Wang[1], Wei Luo[1], Jie Xu[1], Jiansha Lin[1], Xiaoquan Li [1] and Daiquan He[1,*]

[1]Filter rod forming workshop, Honghe Cigarette Factory of Hongyun Honghe Tobacco, MiLe, Yunan, 652300, China

[*] Corresponding author's e-mail: 36011047@qq.com

Abstract. This paper introduces a self-cleaning glue feeding system. Based on the slit coating process, a mid-line adhesive glue feeding system suitable for KDF3E filter rod forming machine has been developed through innovative design of nozzle structure, improvement of coating process flow, optimization of process parameters, and other methods and measures. After the machine test, we can solve the current glue feeding problem and improve the overall efficiency and quality of mid-line glue.

1. Introduction

The paper mid-line glue feeding system is used to form a paper strip continuously and evenly coated at least one cold glue line, which can realize the bonding of the filter in the rod and the wrapping paper [1] and fix their relative position. This can avoid the position slip of the filter rod material and molding paper in the process of cutting and conveying, causing an unqualified filter rod due to the external force. At present, there are two commonly used ways of gluing in the molding machine (needle-type glue valve spray glue and non-contact atomization spray glue). In the use process, there are problems of pollution equipment, unstable bonding, and poor adaptability of glue materials [2]. This paper presents a kind of self-cleaning through the innovative design of the nozzle structure, improving the coating process and optimization of process parameters, developing a set of mid-line glue systems for the filter molding machine, which greatly improves the coating accuracy and reliability of the system, solve the problem of the current molding machine, and improve the overall efficiency of the mid-line glue [3].

2. Overall scheme design

The scheme design adopts gluing with the slit coating: the glue liquid transported from the glue feed mechanism enters the slit nozzle by the glue pump. When the machine starts, the cut-off valve in the glue loop is opened, and the glue is ejected from the gap and evenly coated on the filter rod, forming paper to complete the glue work. The scheme design is based on the filter rod forming machine named KDF3E. The functional components of the whole machine mainly include a gluing device, a control unit, a circulating water system, and the filling water trolley [4].

The technological process of the filter rod forming machine named KDF3E is shown in Figure 1:

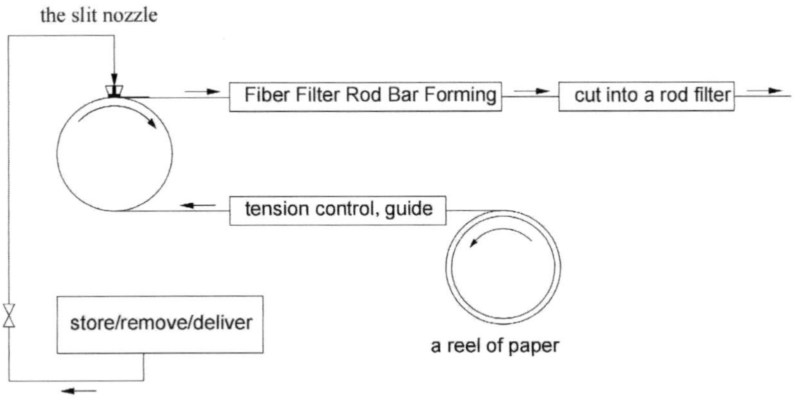

Figure 1. The technological process.

The middle line gluing device adopts an embedded design. In order to keep the overall process of the filter rod forming machine named KDF3E unchanged, the gluing device is installed on the position of the original middle line gluing device, and the paper tape path is basically unchanged.

The overall layout of the equipment is shown in Figure 2:

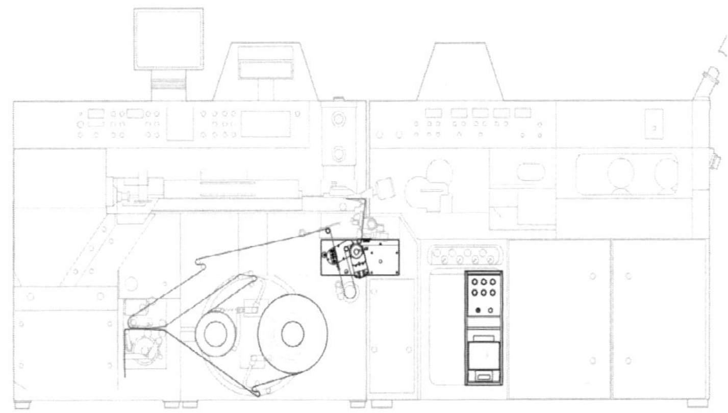

Figure 2. The overall layout.

3. Component design

The cold glue feeding system of filter rod molding machine is composed of glue nozzle, glue box, gear pump and glue pipeline [5]. The gluing device integrates some functions, such as the paper tape guide mechanism, nozzle coating, and cleaning nozzle, and combines the functions of paper tape correction, transmission, gluing, cleaning, and glue breaking; the control components serve as the electric control and gas control center to manually control the machine; the circulating water conveying device is equipped with the water tank for the water cleaning the nozzle, and the clean and used sewage filter is collected back to the water tank; the filling trolley fills the clean water for the circulating water conveying device, emptying the sewage from the sewage tank.

Its layout is shown in Figure 3:

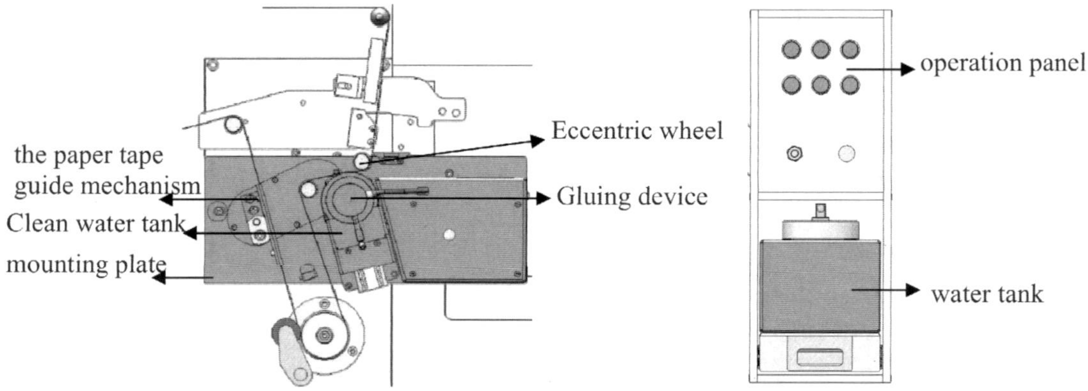

Figure 3. Layout design diagram of the system components.

3.1. Gluing device

For the gluing device, the inner cylinder is set as the power drive rotation to realize the free switch between the working position and the cleaning seal position; the glue nozzle outputs the glue quantitatively and coats the inside of the forming paper to form the required adhesive line. The nozzle is composed of the upper template and the lower template. After the combination of the two templates, a glue buffer cavity and a slit glue opening can be formed. The special design of the nozzle shape ensures that the glue around the glue mouth has a specific flow field to ensure the stability of glue feeding, reduce the leveling time of the glue, and improve the uniformity of the glue. In addition, for the specification requirements of replacing the filter rod molding paper, the glue nozzle is designed as an easily removable and replaceable component, and rapid specification replacement can be realized when needed. At the same time, through the rotation of the nozzle installation seat, the on-off state of the glue feeding can be switched quickly. The through-off structure can block the glue supplied by the pipeline in time and reduce equipment pollution and glue waste.

The structure of the gluing device is shown in the Figure 4:

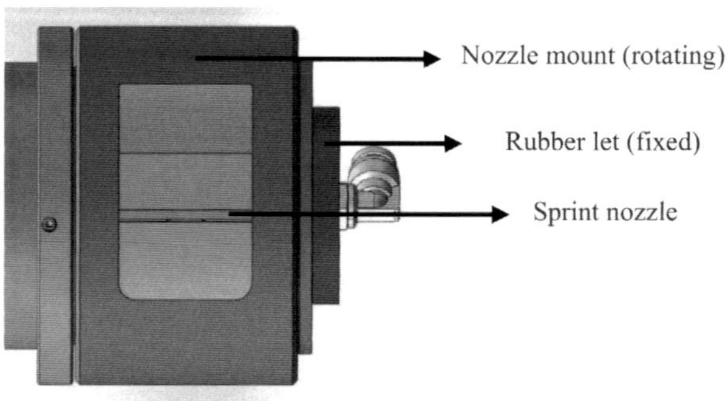

Figure 4. Schematic diagram of the gluing device and the mounting seat.

3.2. Paper belt guide

The paper tape guidance mechanism is a plate guide structure that can adjust the paper tape guide width and center distance to suit different paper tape specifications and different center requirements.

3.3. Eccentric wheel

The adjusting roller is an eccentric structure, and the appropriate eccentric distance is designed to suit the adjustment requirements of the adhesive tape.

Journal of Physics: Conference Series **2815** (2024) 012047 doi:10.1088/1742-6596/2815/1/012047

3.4. Cleaning institutions

The cleaning mechanism is shown in Figure 5:

It is powered by a lifting cylinder, which can drive the cleaning water cylinder up and down and can switch the sealing cleaning position and opening position freely. The cleaning tank is equipped with a nozzle block to spray high-pressure water flushing glue nozzle. There are overflow ports and sealing rings to ensure no water leakage in the cleaning mechanism during work. In addition, the clean water tank is equipped with a dovetail slot for convenient disassembly during maintenance.

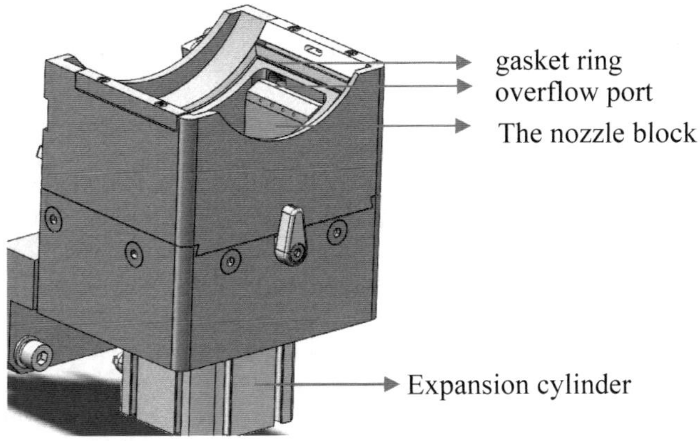

Figure 5. Design drawing of the cleaning mechanism.

3.5. Circulating water conveying device and structure

The circulating water conveying device is shown in Figure 6:

It is equipped with a water tank with a volume of 8 L (available for more than one week) to store the water for cleaning the glue nozzle and collect the sewage from cleaning the glue nozzle. It is equipped with an internal liquid level detector and slide rail at the bottom of the water tank for easy installation and maintenance [6]. The box frame of the circulating water-conveying device is made of aluminum panels. It can be embedded in the lower left corner of the AF-3E.

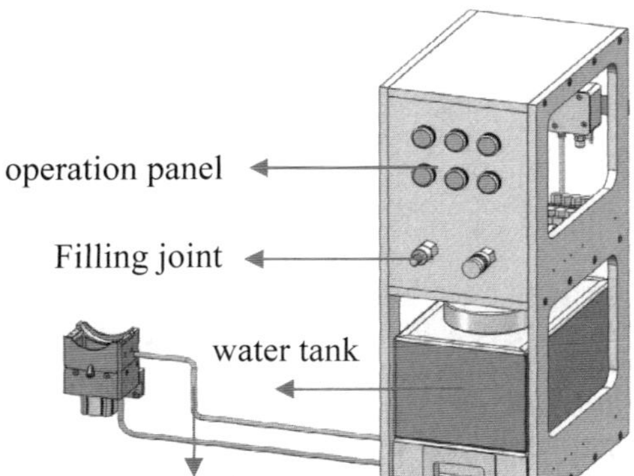

Figure 6. Circulating water delivery device.

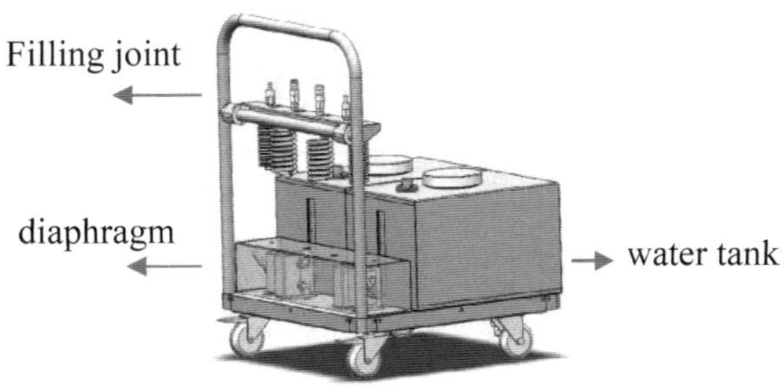

Figure 7. Design drawing of the filling trolley.

3.6. Trolley for adding water

The filling trolley is shown in Figure 7:

It is equipped with a transparent water tank and sewage tank. The two tanks are the same size with a capacity of 46 L. Five pieces of equipment can be centrally filled and discharged at one time. The filling trolley is equipped with a pneumatic pump, and the quick plug connector for docking is designed to prevent misinsertion [7]. When filling the discharge, you can insert the pipe plug into the corresponding plug on the manual operation board and press the corresponding discharge or filling button to start the discharge or filling.

3.7. The glue unit

There is a set of glue supply mechanisms for the filter rod forming machineKDF3E, which uses a servo motor to drive the precision gear pump to realize the precision adjustment of the glue supply. The servo motor synchronously follows the main engine speed to match the glue supply and paper tape speed. We will continue to use this institution without making any modifications to this part.

3.8. Control unit

The control scheme is shown in Figure 8:

Coating itself is a continuous process and the individual elements must be integrated and function at the same speed [8].The control unit uses a Siemens PLC S7-1511 module to control the midline glue system, the encoder collects the speed signal of the host, Lenze8400 TopLine synchronously controls the metering pump motor, and the valve island completes the control of the related cylinder and pump of the gluing device. The system provides an independent power supply and independent logic, and the electrical interface only sends the fault signal to the main machine [9]. When the equipment is running, it is necessary to strictly control the amount of middle-line glue, monitor the effect of middle-line glue at all times, and deal with the faults in the process of middle-line glue application in time [10].

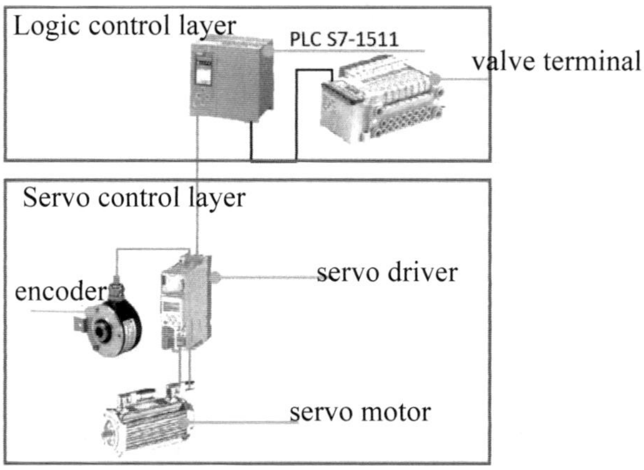

Figure 8. Control scheme.

4. Application effect

The experimental test is shown in Table 1:

After the experimental test, installation, and debugging of the machine, based on the running speed of the molding machine of 5500 pieces per minute, the output and the production quality with the slit coating glue unit are equal to that of the unmodified unit. In the production process, the glue on the middle line of the filter rod was continuously checked, and no glue break was found. The width of the glue line was uniform and continuous, and the quality of the adhesive met the quality requirements. The status of installation and operation is shown in Figure 9. In the production operation, there was no failure of the midline glue device, which led to the shutdown of equipment. In the starting stage, the midline glue was weak, and the glue line was normal in the forming machine. The gluing effect is shown in Figure 10:

Table 1. Run test proceedings and reports.

Production specification: Perimeter 24 mm, weight 0.629 g, suction resistance 310 Pa				
NO.	**Time period**	**Speed (Branch/min)**	**Reason for shutdown**	**Gluing status record**
Condition Record: 0.6 mm nozzle, glue of KDF3, empty discharge of 10.94 g/600 m adhesive amount. Total running time: 295 minutes, no fault time.				
1	18:44-18:58	2852	Adhesive failure	1. The adhesive line is continuous and stable. 2. After running for 3 minutes, a small amount of adhesive buildup is observed. 3. After gluing, there is adhesive imprinting through the wheel glue line. 4. The nozzle is run for 70 minutes to stack 0.8 g of glue/paper ash. 5. There are insufficient or no adhesive lines during the startup phase. 6. After prolonged operation and shutdown, residual glue at the nozzle needs to be treated.
	19:08-19:16	5007	Adhesive failure	
	19:21-20:36	507	Manual operation	
	22:26-23:00	5564	Manual operation	
2	14:15-15:16	5563	Manual operation	
	15:45-16:34	5564	Adhesive failure	
	16:59-17:38	5546	Manual operation	
Condition Record: Replace with the glue of KDF6, and the measured empty discharge amount is 10.45 g/600 m. The adhesive amount is set to 12 g/600 m. Total running time: 377 minutes, no fault time.				
3	11:17-12:05	5563	Manual operation	1. The glue application delay after startup is about 11 seconds. 2. The accumulation of glue at the application site has improved. 3. There is adhesive accumulation on the surface of the paper guide roller. 4. After long-term operation and shutdown, residual glue at the nozzle needs to be treated.
	15:07-15:45	5560	Manual operation	
	15:54-16:17	5563	Manual operation	
4	10:26-11:50	5562	Manual operation	
	13:00-15:02	5800	V-shaped groove adhesive buildup	
	15:10-16:17	5546	Manual operation	
Condition Record: Using the glue of KDF6, the adhesive amount is set to 12 g/600 m, and the measured empty discharge adhesive amount is 12.2 g/600 m. On the afternoon of the 15th, the original glue was replaced, and the measured amount of glue discharged was 13.6 g/600 m. The glue signal on the center line has been changed from a delay of 4 seconds to no delay. Total running time: 451 minutes, no fault time.				
5	10:15-11:22	5800	V-shaped groove adhesive buildup	1. Slight glue leakage. 2. The paper guide roller has accumulated glue, and it needs to

Journal of Physics: Conference Series **2815** (2024) 012047 doi:10.1088/1742-6596/2815/1/012047

	14:30-16:45	5800	Manual operation	be operated for 70 and 100 minutes to handle the accumulated glue. 3. After running for 100 minutes, the nozzle accumulated 1.1 g of adhesive.
6	11:10-12:00	5500	Manual operation	1. The nozzle is sealed overnight without any adhesive discharge, and the adhesive printing during the startup stage is shallow or absent. 2. The accumulation of glue at the application site is faster than before. 3. Glue accumulation on the paper guide roller. 4. After long-term operation and shutdown, residual glue at the nozzle needs to be treated.
	13:51-14:17	5550	Manual operation	
	14:28-16:07	5550	Manual operation	
7	10:20-11:30	5500	Manual operation	
Condition Record: It is replaced with a 0.4 mm wide glue line nozzle. The KDF3 glue is used and the glue amount is set to 12 g/600 m. 11.2 g/600 m is measured. Total running time: 130 minutes, no fault time.				
8	15:30-16:15	5500	Manual operation	1. The startup stage of the offset printing is shallow, and it runs stably in other states 2. After long-term operation and shutdown, residual glue at the nozzle needs to be treated
	10:20	4790	Adhesive failure	
	10:22-11:05	5500	loading machine full	
9	11:07-11:50	5490	Manual operation	

Figure 9. Installation and operation.

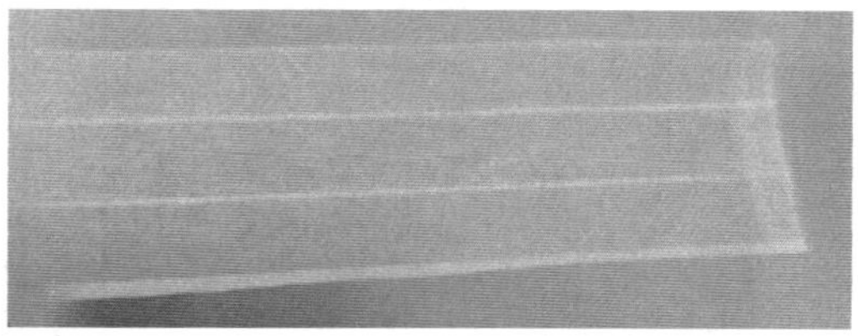

Figure 10. The actual effect after gluing.

5. Conclusions

Through the analysis of the principle of the slit-type coating process, based on the KDF3E filter rod molding machine, combined with the requirements of the midline glue of the molding machine, a set of glue systems with the slit coating process is developed to improve the uniformity and reliability of the midline glue of the molding machine. Quality defects, such as incomplete midline glue caused by clogging of the gluing nozzle, leakage, and accumulation of glue, are eliminated. This can improve the quality of filter rod products, reduce equipment pollution, reduce the frequency and intensity of gluing system maintenance, improve equipment operation efficiency, and achieve the production goals of high quality, high efficiency, and low consumption. The research and development of products provide technical references for the gluing systems of filter rod molding machines and cigarette machines.

Acknowledgments
The research and development of the glue feeding system for the middle line of paper based on the slit coating supported by Hongyun Honghe Tobacco (Group) Co., Ltd. (HYHH2021GY08).

References
[1] Peng X, Liu X and Cai H J (2022) Improvement of the Gluing Auxiliary Device for ZL26B Molding Machine. Manufacture & Upgrading Today, (11):103-105.
[2] Ding X W (2015) Renovation of the centerline adhesive system of KDF-2 filter rod forming machine. China Machinery, 2: 16-17.
[3] Wuhan Yimao Machinery Co., Ltd. A glue supply system for filter rod forming machine: CN201922394692.3. 2020-10-27.
[4] Hongyun Honghe Tobacco (Group) Co., Ltd. A multi-level narrow seam gluing device with self-cleaning centerline glue: CN202210256137.8. 2022-06-28.
[5] Proceedings of the 2014 Annual Conference of Shanghai Tobacco Society. Design and application of multi-function glue nozzle of new filter rod molding machine. 2015:1-11.
[6] Xiangyang Zedong Chemical Group Co., Ltd. A circulating water conveying device: CN201920011967.8. 2019-11-05.
[7] Xinxiang Xingxing Technology Co., Ltd. A lubricating grease filling car: CN202022833472.9. 2021-10-22.
[8] Gutoff E B, Cohen E D (2016) Water-and solvent-based coating technology[M]//Multilayer flexible packaging. William Andrew Publishing,2016: 205-234.
[9] Ding X S and Zhang Y (2014) Improvement method for independent drive of the central line adhesive in the molding machine. New Technologies & New Processes, 12: 9-11.DOI: 10.3969/j.issn.1003-5311.2014.12.04.
[10] Wei L, Li X K and Shen T Q. Design of a Constant Pressure Control Device for Line Glue in ZL29 Filter rod forming machine. China Equipment Engineering,8:102-104.

Journal of Physics: Conference Series **2815** (2024) 012048 doi:10.1088/1742-6596/2815/1/012048

A Markov method for a mechanical system reliability assessment using discrete degradation data

Hang Shen[1], Zheng Nie[1,*], Jie Xia[2] and Qibin Wang[2,*]

[1] Xi'an Institute of Electromechanical Information Technology, Xi'an 710065, China.
[2] School of Mechano-Electronic Engineering, Xidian University, Xi'an 710071, China.

[*] Corresponding author: Zheng Nie (e-mail: niezheng@126.com), Qibin Wang (e-mail: qbwangpap@126.com).

Abstract. A homogenous continuous-time Markov model (HCTMM) has the advantage of high accuracy and is usually used in reliability analysis. However, the transition intensities (rates) between any two states cannot be accurately obtained due to the discrete data obtained from a mechanical system and little expert knowledge. This paper proposes a method in which the transition intensity matrix of an HCTMM can be obtained indirectly using the actual discrete degradation data of a mechanical system. Firstly, the one-step transition probability matrix can be calculated by a hidden Markov model using the discrete degradation data. Secondly, the unit-time transition intensity matrix can be calculated from the one-step transition probability matrix. At last, an HCTMM is established to assess the reliability of a mechanical system based on the unit-time transition intensity matrix. Moreover, considering the maintenance activities cannot restore a mechanical system as a fire-new one, the random transition of a mechanical system after maintenance is expressed by a quasi-renewal process in this paper. Finally, three cases are studied to validate the proposed method.

1. Introduction

Reliability assessment of a mechanical system is essential in improving operational efficiency and running security ability. In the practical project application, mechanical systems cannot run with perfect performance levels (rate) for a long time. With increasing time, the performance of a mechanical system becomes lower and lower until failure. The performance degradation data of mechanical systems, such as precision data, wear data, and so on, is usually obtained discretely. Therefore, using discrete degradation data to assess the reliability of a mechanical system is necessary.

The mechanical system usually shows multiple states ranging from perfect functioning, through degradation to complete failure. This type of system is called a multi-state system which can run with many different performance levels [1]. With a no-memory assumption that the next state of a mechanical system depends only on its current state, the random behavior of a mechanical system can be described through a Markov process [2]. There are a great number of published papers that studied reliability analysis by the Markov process. Zhu et al. [3] proposed a method for assessing long-term fatigue reliability using Markov models, simulating a monopile baseline turbine that accounted for environmental factors and fatigue load uncertainty. Zhang and Cao [4] developed models to assess the reliability of heat exchange systems in deep-sea submarines using Markov analysis, focusing on common-cause failure state transition models. Jiang et al. [5] established a DBN for sensor networks using Markov chains to capture the reliability degradation of measurement sensors over time. Liu et al. [6] constructed a Bayesian structure, combining prior and observation data, to obtain the posterior distribution of transition intensity, and calculated point and interval estimates through Markov chain Monte Carlo simulation. Reznícek et al. [7] researched a method to solve the failure distribution function of non-homogeneous Markov chains, utilizing differential formulas to calculate time slices with adjustable parameters. Furtadoa and Ribeiro [8] used a semi-Markov continuous-time process to predict the degradation of operating railway bridges.

HMM is a statistical model based on the discontinuous-time Markov chain (MC). This model consists of two stochastic processes. One is a hidden description of state transition with an MC having a finite number of states, and the other is an emission description of the stochastic connection between each state and observed variables. Many researchers are focusing on studies using HMMs for reliability assessment. Murthy et al. [9] studied the reliability assessment of phasor measurement units by HMM. Fort et al. [10] investigated a common reliability analysis method, in which the system availability was obtained using HMM. Feng et al. [11] employed the Gaussian Mixture Hidden Markov Model for reliability estimation to mitigate reliance on historical data in mechanical system lifetime analysis. Liu et al. [12] investigated the operational performance reliability for a CNC machine using HMM and the hidden semi-Markov model, respectively. Kouadri et al. [13] took advantage of the HMM and PCA model based on machine learning (ML) to propose a new FDD method to improve the performance of WEC converter (WECC) in different availability, reliability, and required security under the conditions. Han et al. [14] developed a data-driven tool wear rate estimation method based on a hierarchical Dirichlet process HMM. Deep A et al. [15] introduced an HMM-integrated model that incorporates an absorbing state representing failure occurrence for continuous condition monitoring.

For repairable mechanical systems, maintenance activities must be considered in reliability modeling. Maintenance can be classified into categories such as perfect, imperfect, minimal, worse, and worst [16]. Generally, imperfect maintenance, falling between perfect and minimal, is more realistic for characterizing maintenance effects. Wang and Pham [16] developed the quasi-renewal process, which is widely used for imperfect repair modeling. Wang et al. [17] introduced a model that updates the failure rate function and studied dynamic imperfect preventive maintenance decision-making for wind turbines. Furthermore, Wang et al. [18] suggested two degradation models suitable for multivariate systems with time-dependent covariates and imperfect maintenance activities.

The above studies have proved that the Markov model is a very effective method for reliability assessment. It is worth noting that all of the studies have mainly concerned the different Markov models and different forms of transition intensities. Almost all of the research assumes that the transition intensities are obtained from specialists or simulations. There is seldom a paper that studies the estimation of transition intensities using actual observation data and the existing studies have stringent requirements for the degradation data. In fact, the degradation data is various and could not meet the stringent requirements of the previous methods. Therefore, this paper introduces a general method for the transition intensities estimation from the actual discrete performance degradation statistics of a mechanical system. Firstly, the one-step transition probabilities can be obtained by an HMM using the discrete degradation data. Then, the transition intensities can be calculated from the one-step transition probabilities. An HCTMM is established based on this estimation. Finally, the reliability of a mechanical system can be obtained. Moreover, the maintenance activities of a mechanical system are considered in the reliability analysis.

This paper is structured into four main sections. The review of the published research on reliability evaluation by the Markov process is summarized in Section 1. The HMM is reviewed in Section 2.1 which discretely characterizes the degradation process of a mechanical system. In Section 2.2, the unit-time transition intensity matrix of a mechanical system is obtained via the one-step transition probability matrix, which is updated via the hidden Markov model. The reliability function and state probability functions of a mechanical system are evaluated based on the unit-time transition intensity matrix in Section 2.3. The quasi-renewal process of a mechanical system is presented in Section 2.4.

Then, three example cases are provided to demonstrate the application of the suggested method in Section 3. And conclusions are summarized in Section 4.

2. Theory

2.1 Hidden Markov model for calculating the one-step transition probability matrix for a mechanical system

Due to the complexity of a mechanical system, it is difficult to discover a one-to-one correspondence between the performance observations and the mechanical system state. The HMM is a statistics model based on MC that can be used to perceive the hidden state of a mechanical system using performance observations.

A general HMM is shown in Figure 1 and it is completely defined as follows [9]:

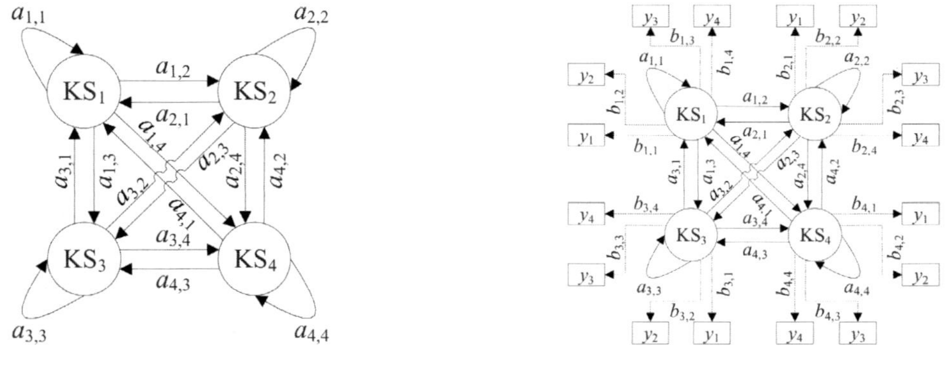

(a) (b)

Figure 1. (a) State-transition diagram of the four states MC. (b) Diagram of the four hidden states and four observations HMM.

1) *N* is the number of hidden states of MC, and the states are represented as KS_1, KS_2, ..., and KS_N.

2) *M* is the number of observation symbols, and they are denoted by the set $Y=\{y_1, y_2, \cdots, y_M\}$, where y_m represents the mth observation symbol in the alphabet.

3) The state transition probability matrix $\boldsymbol{A}=[a_{i,j}]_{N\times N}$ is expressed as:

$$a_{i,j}=P\{S_{t+1}=\mathrm{KS}_j \mid S_t=\mathrm{KS}_i\},\ 1\le i, j\le N \tag{1}$$

where S_t denotes the state at time *t*.

$a_{i,j}$ should satisfy the condition $a_{i,j}\ge 0$ and $\sum_{j=1}^{N} a_{i,j}=1$. The assumption is that the probabilities of state transitions do not depend on the specific time at which the transitions take place. This can be expressed for any t_1 and t_2 as follows:

$$P\{S_{t_1+1}=\mathrm{KS}_j \mid S_{t_1}=\mathrm{KS}_i\}=P\{S_{t_2+1}=\mathrm{KS}_j \mid S_{t_2}=\mathrm{KS}_i\} \tag{2}$$

4) The observation probability matrix $\boldsymbol{B}=[b_{j,m}]_{N\times M}$ is expressed as:

$$b_{j,m}=P\{o_t=y_m \mid S_t=\mathrm{KS}_j\},\ 1\le j\le N, 1\le m\le M \tag{3}$$

where o_t expresses the observation symbol at time *t*.

Following stochastic constraints, the observation probabilities should satisfy the condition $b_{j,m}\ge 0$ and $\sum_{m=1}^{M} b_{j,m}=1$. The key assumption is that the present observation is statistically independent from the prior observations. This can be expressed mathematically for a sequence of observations:

$$O=o_1,o_2,\cdots,o_T \tag{4}$$

$$P\{O \mid S_1,S_2,\cdots S_T,\lambda\}=\prod_{t=1}^{T}P(o_t \mid S_t,\lambda) \tag{5}$$

5) The initial state probability, i.e., $\boldsymbol{\Pi}=\{\pi_i\}$, can be expressed:

$$\pi_i=P\{S_1=\mathrm{KS}_i\},\ 1\le i\le N \tag{6}$$

where π_i should satisfy the condition $\pi_i \ge 0$ and $\sum_{i=1}^{N}\pi_i=1$.

Therefore, an HMM λ can be completely defined with five elements: N, M, $\boldsymbol{A}$, $\boldsymbol{B}$, $\boldsymbol{\Pi}$. It can be compactly denoted as follows:

$$\lambda=(\boldsymbol{A},\boldsymbol{B},\boldsymbol{\Pi}) \tag{7}$$

The following are the three primary problems associated with HMM [10]:

- Evaluation problem: For a given HMM and sequence, the probability that such an HMM generates such a sequence of observations is calculated and derived using FBA.
- Decoding problem: For a given HMM and observations, the sequence of hidden states most likely to be created by those observations is computed and obtained by the Viterbi algorithm.
- Learning problem: The parameters and general structure of an HMM model fitting a given set of training observations are computed and obtained by BWA.

The explanation of the above algorithms in detail is given in [9]. This paper concentrates on the computation of one-step transition probability matrix A for a complex system, which can be accomplished through the BWA, which is essentially a variant of the Expectation-Maximization algorithm (Figure 2).

For updating the model parameters, two variables are defined as follows:

$$\xi_t(i,j)=P\left(S_t=\mathrm{KS}_i,S_{t+1}=\mathrm{KS}_j \mid O,\lambda\right) \tag{8}$$

$$\gamma_t(i)=P\left(S_t=\mathrm{KS}_i \mid O,\lambda\right) \tag{9}$$

These two variables can be calculated as follows:

$$\xi_t(i,j)=\frac{\alpha_t(i)a_{ij}\beta_{t+1}(j)b_{jo_{t+1}}}{P(O\mid\lambda)}=\frac{\alpha_t(i)a_{ij}\beta_{t+1}(j)b_{jo_{t+1}}}{\sum_{i=1}^{N}\sum_{j=1}^{N}\alpha_t(i)a_{ij}\beta_{t+1}(j)b_{jo_{t+1}}} \tag{10}$$

$$\gamma_t(i)=\frac{\alpha_t(i)\beta_t(i)}{P(O\mid\lambda)}=\frac{\alpha_t(i)\beta_t(i)}{\sum_{i=1}^{N}\alpha_t(i)\beta_t(i)} \tag{11}$$

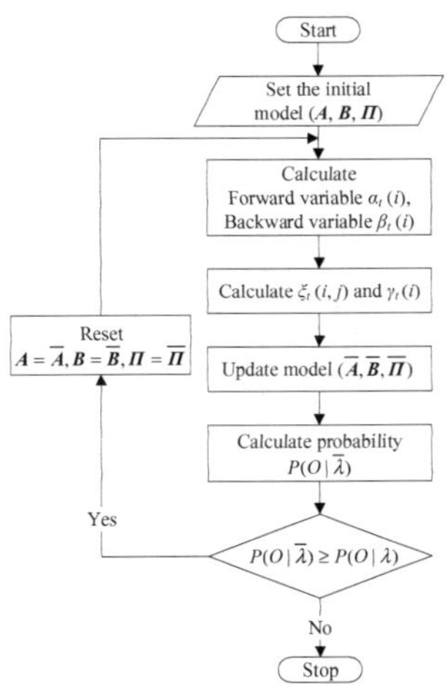

Figure 2. Flowchart for updating model parameters by the BWA.

In the figure, the Forward variable $\alpha_t(i)$ is the probability in state KS_i at time t for a given HMM. Backward variable $\beta_{t+1}(j)$ is the probability describing part of observation sequence after time $t+1$ for a given HMM and the hidden state KS_j at time $t+1$. Both forward and backward variables can be obtained by the FBA.

The updated one-step transition probability matrix $\overline{A} = \left[\overline{a_{i,j}}\right]_{N\times N}$ can be calculated using the BWA until the maximum value of $P\ (O\ |\ \lambda)$ is got as follows:

$$\overline{a_{i,j}} = \frac{\text{the excepted number of jumps from state } KS_i \text{ to } KS_j}{\text{the excepted number of leaving state } KS_i} = \frac{\sum_{t=1}^{T-1}\xi_t(i,j)}{\sum_{t=1}^{T-1}\gamma_t(i)} \tag{12}$$

2.2 Calculation of the unit-time transition intensity matrix for a mechanical system

The stochastic process can be defined as an HCTMM if the process contains two natures as follows:

1 Markov nature:

$$P\big(X(s+t)=j\,|\,X(s)=i,\,X(u),\,0\le u\le s\big)=P\big(X(s+t)=j\,|\,X(s)=i\big) \tag{13}$$

This indicates that all previous states can be disregarded. Then, $\{X(t), t\ge 0\}$ is referred to as a continuous-time MC.

2 Homogeneous time nature:

$$P\big(X(s+t)=j\,|\,X(s)=i\big)=P\big(X(t)=j\,|\,X(0)=i\big) \tag{14}$$

If a mechanical system is observed to be in state i during the last inspection, the probability of the system being in state j ($j\in\{1, 2, \cdots, N\}$) after Δt is denoted as $p_{i,j}(\Delta t)$. Without considering maintenance activities, the state transition probability matrix $P(\Delta t)$ of a mechanical system can be denoted as:

$$\boldsymbol{P}(\Delta t)=\begin{bmatrix} p_{1,1}(\Delta t) & p_{1,2}(\Delta t) & \cdots & p_{1,N}(\Delta t) \\ 0 & p_{2,2}(\Delta t) & \cdots & p_{2,N}(\Delta t) \\ \vdots & \vdots & \ddots & \vdots \\ 0 & 0 & \cdots & p_{N,N}(\Delta t) \end{bmatrix} \tag{15}$$

Similar in [6], the instantaneous transition intensity of an HCTMM from state i to any other state j is assumed to be a constant and is expressed by $\lambda_{i,j}$, where $i\le j$ and $i\in\{1,2,\cdots,N-1,N\}$. When $i=j$, $\lambda_{i,i}$ is the instantaneous transition intensity of leaving state i, and $\lambda_{i,i}=-\sum_{j=1\ and\ j\ne i}^{N}\lambda_{i,j}$. The HCTMM without considering maintenance activity can be expressed with the instantaneous transition intensity matrix Λ:

$$\boldsymbol{\Lambda}=\begin{bmatrix} \lambda_{1,1} & \lambda_{1,2} & \cdots & \lambda_{1,N} \\ 0 & \lambda_{2,2} & \cdots & \lambda_{2,N} \\ \vdots & \vdots & \ddots & \vdots \\ 0 & 0 & 0 & \lambda_{N,N} \end{bmatrix} \tag{16}$$

where $\lambda_{i,j} = \lim_{\Delta t\to 0}\frac{\Pr\{X(t+\Delta t)=j\,|\,X(t)=i\}}{\Delta t}$ is the first derivative of the transition probability $p_{i,j}(\Delta t)$ versus time interval Δt and it is as follows:

$$\lambda_{i,j} = \lim_{\Delta t\to 0}\frac{p_{i,j}(\Delta t)-p_{i,j}(0)}{\Delta t} \tag{17}$$

where $p_{i,j}(0)=0$ when $i \neq j$ or $p_{i,j}(0)=1$ when $i=j$. Then, the transition probability $p_{i,j}(\Delta t)$ and the transition probability matrix $P(\Delta t)$ can be denoted by the instantaneous transition intensity $\lambda_{i,j}$ and the time interval Δt as follows:

$$p_{i,j}(\Delta t)=\begin{cases}\lambda_{i,j}\cdot\Delta t, & i\neq j\\ 1+\lambda_{i,j}\cdot\Delta t, & i=j\end{cases} \tag{18}$$

$$\boldsymbol{P}(\Delta t)=\begin{bmatrix}1+\lambda_{1,1}\cdot\Delta t & \lambda_{1,2}\cdot\Delta t & \cdots & \lambda_{1,N}\cdot\Delta t\\ 0 & 1+\lambda_{2,2}\cdot\Delta t & \cdots & \lambda_{2,N}\cdot\Delta t\\ \vdots & \vdots & \ddots & \vdots\\ 0 & 0 & 0 & 1+\lambda_{N,N}\cdot\Delta t\end{bmatrix} \tag{19}$$

Setting $\Delta t=1$, similar to $\overline{A}$ matrix proposed in HMM, the $P(1)$ matrix means the transition probability matrix per unit of time, which can be denoted as follows:

$$\boldsymbol{P}(1)=\begin{bmatrix}1+\lambda_{1,1} & \lambda_{1,2} & \cdots & \lambda_{1,N}\\ 0 & 1+\lambda_{2,2} & \cdots & \lambda_{2,N}\\ \vdots & \vdots & \ddots & \vdots\\ 0 & 0 & 0 & 1+\lambda_{N,N}\end{bmatrix} \tag{20}$$

Making a comparison of the transition intensity matrix Λ (Formula (16)) with the unit-time transition probability matrix $P(1)$ (Formula (20)), it is obvious to discover that the elements of those two matrices are uniform when $i \neq j$. The elements of the unit-time transition probability matrix $P(1)$ are 1 bigger than the elements of Λ matrix when $i=j$. In other words, the transition intensity matrix Λ can be expressed with the $P(1)$ matrix and the identity matrix E. Therefore, the unit time transition intensity matrix $\overline{\Lambda}$ is as follows:

$$\overline{\boldsymbol{\Lambda}}=\boldsymbol{P}(1)-\boldsymbol{E}=\overline{\boldsymbol{A}}-\boldsymbol{E} \tag{21}$$

where $\boldsymbol{E}=\begin{bmatrix}1 & 0 & \cdots & 0\\ 0 & 1 & \cdots & 0\\ \vdots & \vdots & \ddots & \vdots\\ 0 & 0 & 0 & 1\end{bmatrix}$, and the size of E is the same as $P(1)$.

2.3 Markov model for reliability assessment of a mechanical system

With an HCTMM assumption, $p_i(t)$ of a mechanical system can be calculated as follows:

$$\frac{\mathrm{d}p_i(t)}{\mathrm{d}t}=\sum_{j=1}^{i-1}p_j(t)\lambda_{j,i}-p_i(t)\sum_{j=i+1}^{N}\lambda_{i,j} \tag{22}$$

By solving Formula (22) under the initial conditions $p_1(0)=1$ and $p_i(0)=0,(i\neq 1)$, one can obtain the state probabilities $p_i(t)$ of a mechanical system as time functions. At time t, the dependability of a mechanical system is expressed as the probability of the system's capability not being less than the user's needs at that time [6]. It means the probability that the system's state value will not surpass the user's desired state value (state 1 expresses the perfect performance and state N expresses the complete failure). If the user demand is state u ($1 \leq u \leq N$), the reliability function is formulated as follows:

$$R(t,u)=\sum_{i=1}^{N}p_i(t)\cdot 1(i-u\leq 0) \tag{23}$$

where $1(\cdot)$ is a unity function, i.e., 1(TRUE)=1, and 1(FALSE)=0.

Generally, the last one state value N (complete failure state) is denoted as the only one state value more than the user demand. Then the reliability function is written as follows:

$$R(t) = 1 - p_N(t) \tag{24}$$

2.4 Quasi-renewal process of a mechanical system

It is assumed that a multi-state mechanical system is restored to its perfect state (State 1) after maintenance. However, even if this system is recovered to its perfect performance after maintenance, it cannot be expected as a fire-new one. Imperfect maintenance is more actual to characterize the maintenance effect which falls in between perfect maintenance and minimal maintenance. Similar to the original quasi-renewal process, the extended quasi-renewal process also captures the decreasing trend in the lifetime and illustrates the increasing trend in transition intensities from the best state to lower states following maintenance activities.

A sequence of nonnegative random variables $\{X_i^1, X_i^2, \ldots, X_i^n\}$ is defined, where X_i^n ($i \in \{1, 2, \ldots, N\}$, $n \in \{1, 2, \ldots\}$) denotes the random duration of the mechanical system to remain in the ith state during the nth maintenance cycle. If $X_i^1 = Z_i^1$, $X_i^2 = \alpha Z_i^2$, $X_i^3 = \alpha^2 Z_i^3$, … where rand variables Z_i^n, $n \in \{1, 2, \ldots\}$ are i. i. d., and $0 < \alpha < 1$. The counting process $\{N(t), t \geq 0\}$, which represents the mechanical system's transition time from best to worst state, is reduced after maintenance. This phenomenon is known as a quasi-renewal process with decreasing parameter α, in which the intensity of the transition between any two states increases proportionally after each maintenance activity. Thus, $p_i^n(t)$ in the nth maintenance cycle is denoted as follows:

$$p_i^n(t) = p_i\left(\alpha^{1-n} t\right) \tag{25}$$

Especially, if α=1, the quasi-renewal process will be transformed into the ordinary renewal process, and the system that has been repaired is considered a fire-new system in the following cycle. Then, the reliability function in the nth maintenance cycle $Rn(t)$ can be calculated via Formula (24) as follows:

$$R^n(t) = 1 - p_N\left(\alpha^{1-n} t\right) \tag{26}$$

The flowchart of a mechanical system reliability assessment using the proposed method is shown in Figure 3. The mechanical system is divided into N different performance levels and the observation values are divided into M different observation levels. The initial AN×N matrix, BN×M matrix, and Π are set for an initial HMM. The performance index levels of a mechanical system are quantified according to the divided observation levels and collected periodically. By training the HMM using the quantified sequences of performance indexes, one can get the updated $\overline{A}$ matrix of a mechanical system. The unit-time transition intensity matrix of the mechanical system can be obtained by Formula (21). By solving the Kolmogorov differential formulas, the state probabilities can be calculated. The reliability of mechanical systems can be calculated via Formula (24).

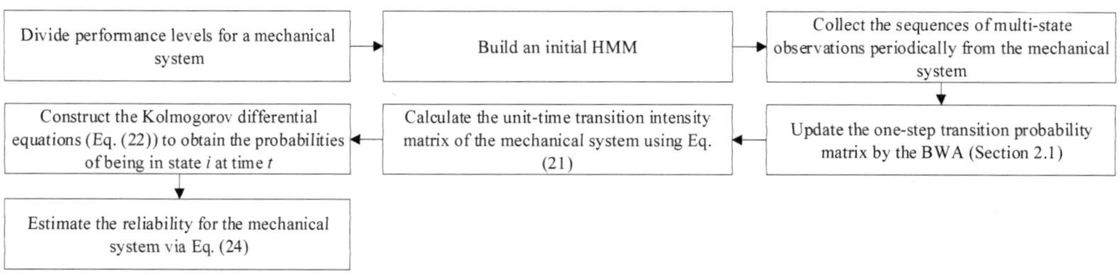

Figure 3. Flowchart of a mechanical system reliability assessment using the proposed method.

3. Results and discussion

Three example cases are provided to demonstrate the efficacy of the suggested method. A Markov process is simulated by three different matrices to compare the effectiveness of the three methods in the first case. In the second case, reliability evaluation for CNC equipment will be demonstrated. Reliability analysis for a mechanical system considering imperfect maintenance will be illustrated in the third case.

3.1 Method validation

In this case, three different matrices $(\boldsymbol{\Lambda}, \boldsymbol{P}(1), \overline{\boldsymbol{\Lambda}})$ are used to calculate $p_i(t)$ and $R(t)$ of a system.

A. Transition intensity matrix Λ

It is assumed that a system follows a four-state HCTMM with a transition intensity matrix Λ given as follows:

$$\boldsymbol{\Lambda} = \begin{bmatrix} -0.006 & 0.002 & 0.002 & 0.002 \\ & -0.004 & 0.002 & 0.002 \\ & & -0.002 & 0.002 \\ & & & 0 \end{bmatrix} \tag{27}$$

The state probabilities $p_i(t)$, i=1, 2, 3, 4 of the system will be calculated under the initial conditions $p_1(0)=1$ and $p_i(0)=0,(i \neq 1)$ as follows:

$$\begin{aligned} \frac{\mathrm{d}p_1(t)}{\mathrm{d}(t)} &= -0.006p_1(t) \\ \frac{\mathrm{d}p_2(t)}{\mathrm{d}(t)} &= 0.002p_1(t) - 0.004p_2(t) \\ \frac{\mathrm{d}p_3(t)}{\mathrm{d}(t)} &= 0.002p_1(t) + 0.002p_2(t) - 0.002p_3(t) \\ \frac{\mathrm{d}p_4(t)}{\mathrm{d}(t)} &= 0.002p_1(t) + 0.002p_2(t) + 0.002p_3(t) \end{aligned} \tag{28}$$

The system reliability will be calculated via Formula (24) and the result is shown in Figure 4. The reliability shows a decreasing trend with the increasing time.

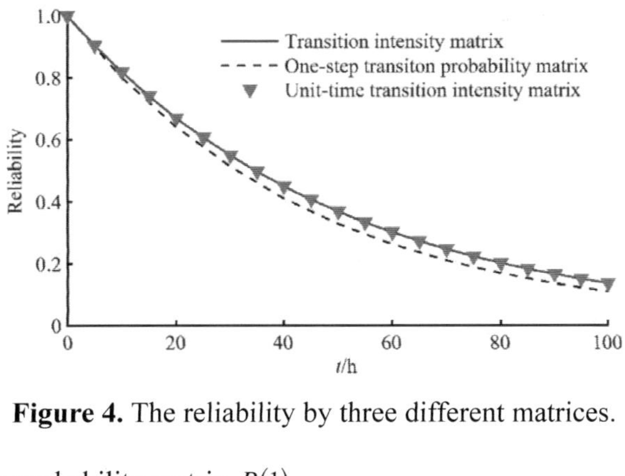

Figure 4. The reliability by three different matrices.

B. One-step transition probability matrix $P(1)$

The $P(1)$ matrix is used to describe a special Markov process, in which the time and state space are both discrete. The time step is set as 10 ($\Delta t = 10$) and $p_{ij}(1)$ can be obtained via the transition intensity matrix (Formula (27)) and time step Δt. The $P(1)$ matrix is:

$$\boldsymbol{P}(1)=\begin{bmatrix}0.94 & 0.02 & 0.02 & 0.02\\ & 0.96 & 0.02 & 0.02\\ & & 0.98 & 0.02\\ & & & 1\end{bmatrix} \tag{29}$$

The n-step transition probability matrix $P(n)$ is:

$$\boldsymbol{P}(n)=\left[\boldsymbol{P}(1)\right]^n=\begin{bmatrix}0.94 & 0.02 & 0.02 & 0.02\\ & 0.96 & 0.02 & 0.02\\ & & 0.98 & 0.02\\ & & & 1\end{bmatrix}^n \tag{30}$$

Then, the state probability Πn after n steps transition will be calculated using the initial state probability Π and the $P(1)$ matrix as follows:

$$\boldsymbol{\Pi}_n=\boldsymbol{\Pi}\boldsymbol{P}(n)=\boldsymbol{\Pi}\left[\boldsymbol{P}(1)\right]^n \tag{31}$$

The result under the same initial state probabilities in Example A is shown in Figure 4.

C. Proposed unit-time transition intensity matrix

Similar to Example B, the time step is set to 10 and the $P(1)$ matrix is the same as Formula (29). The unit-time transition intensity matrix $\overline{\boldsymbol{\Lambda}}$ can be calculated via Formula (21):

$$\overline{\boldsymbol{\Lambda}}=\boldsymbol{P}(1)-\boldsymbol{E}=\begin{bmatrix}-0.06 & 0.02 & 0.02 & 0.02\\ & -0.04 & 0.02 & 0.02\\ & & -0.02 & 0.02\\ & & & 0\end{bmatrix} \tag{32}$$

The state probabilities $p_i(t)$, i=1, 2, 3, 4 will be calculated as follows:

$$\begin{aligned}\frac{\mathrm{d}p_1(t)}{\mathrm{d}(t)}&=-0.06p_1(t)\\ \frac{\mathrm{d}p_2(t)}{\mathrm{d}(t)}&=0.02p_1(t)-0.04p_2(t)\\ \frac{\mathrm{d}p_3(t)}{\mathrm{d}(t)}&=0.02p_1(t)+0.02p_2(t)-0.02p_3(t)\\ \frac{\mathrm{d}p_4(t)}{\mathrm{d}(t)}&=0.02p_1(t)+0.02p_2(t)+0.02p_3(t)\end{aligned} \tag{33}$$

The result is also shown in Figure 4. In the Figure, the reliability results by the three different matrices show the same tendency and the reliability values decrease with the increasing time. It is pretty clear that the unit-time transition intensity result is exactly identical to the transition intensity matrix result. There is a little error between the $P(1)$ matrix result and the Λ matrix result. This is because $P(1)$ the matrix describes a state transition after per unit time Δt (one step) and it assumes that no state transition activity during the unit time Δt. The state probabilities calculated using $P(1)$ matrix cannot continually depict the state transition. By the proposed method, $P(1)$ matrix can be transformed into a continuous $\overline{\Lambda}$ matrix, which can more accurately describe the transition and $p_{ij}(t)$ can be easily obtained using the HMM proposed in Section 2.1. The comparisons of the three matrices are shown in Table 1. The method of obtaining the transition strength matrix through specialist knowledge or simulation is the most accurate method, whose calculation precision is high, it is used as a standard to calculate the average error of other methods. The result of the $P(1)$ matrix has a certain average error with the Λ matrix and low accuracy. The unit-time transition intensity matrix has the same calculation precision as the transition intensity matrix and can be easily obtained using the actual degradation data.

Table 1. Comparisons of three matrices.

Matrix	Acquisition	Precision
Transition intensity matrix	Specialist knowledge or simulation	High
One-step transition probability matrix	Actual discrete degradation data	Low
Unit-time transition intensity matrix	Actual discrete degradation data	High

3.2 Mechanical system

In the practical project application, as the operation time increases, the mechanical system will run from perfect to degraded to complete failure. Degradation of mechanical systems is manifested in changes in performance indicators. By collecting and analyzing the performance index degradation data of the mechanical system, the reliability can be evaluated [12] (Figure 3).

An example shown in [19] is considered, where it is appropriate to assess the reliability of a Computer Numerical Control (CNC) machine with the one-step transition probability matrix. There are many factors influencing the performance of a CNC machine and all of them focus on the movement precision. In this experiment, the spindle executes circular motion on the XOY plate, driven by CNC programming. Circular motion indicators are sampled every 20 hours. 24 sets of circular trajectory data were collected in total. The interpolation method is used to expand the data sets to improve the calculation accuracy. The discrete data statistics of performance indexes are shown in Table A1 in Appendix A. The CNC machine circular motion accuracy failure standard is shown in Table A2 in Appendix A, according to which the proposed CNC machine is divided into four performance levels named normal (KS1), light degradation (KS2), severe degradation (KS3), and failure (KS4). The CNC performance failure standard matrix H can be obtained as follows:

$$\boldsymbol{H} = \begin{bmatrix} 0.019 & -0.015 & 0.025 & 0.030 \\ 0.024 & -0.020 & 0.030 & 0.035 \\ 0.029 & -0.030 & 0.035 & 0.040 \end{bmatrix} \tag{34}$$

The initial A matrix and B matrix are set as Formulas (35) and (36) respectively.

$$\boldsymbol{A} = \begin{bmatrix} 0.3853 & 0.2959 & 0.2063 & 0.1125 \\ & 0.2198 & 0.4064 & 0.3738 \\ & & 0.4989 & 0.5011 \\ & & & 1 \end{bmatrix} \tag{35}$$

$$\boldsymbol{B} = \begin{bmatrix} 0.2343 & 0.1668 & 0.1296 & 0.4693 \\ 0.2575 & 0.3889 & 0.1219 & 0.2317 \\ 0.1961 & 0.2752 & 0.0932 & 0.4355 \\ 0.2710 & 0.2810 & 0.2115 & 0.2365 \end{bmatrix} \tag{36}$$

Then, it can identify the transition probability matrix from the state structure that likely generated the observed historical data sequence, using BWA. The updated $\overline{A}$ matrix can be obtained.

$$\overline{\boldsymbol{A}} = \begin{bmatrix} 0.9443 & 0.0455 & 0.0102 & 0 \\ & 0.6142 & 0.3857 & 0.0001 \\ & & 0.9665 & 0.0335 \\ & & & 1 \end{bmatrix} \tag{37}$$

According to the proposed method, the $\overline{A}$ matrix can be transformed into a unit-time transition intensity matrix via Formula (21), which can describe the continued transition more accurately. The transformed matrix is given as follows:

$$\overline{\Lambda} = \overline{A} - E = \begin{bmatrix} -0.0557 & 0.0455 & 0.0102 & 0 \\ & -0.3858 & 0.3857 & 0.0001 \\ & & -0.0335 & 0.0335 \\ & & & 0 \end{bmatrix} \tag{38}$$

Then, the CNC machine state probabilities will be obtained. The state probability is plotted in Figure 5. The reliability of the CNC machine can be calculated via Formula (24) and plotted in Figure 6.

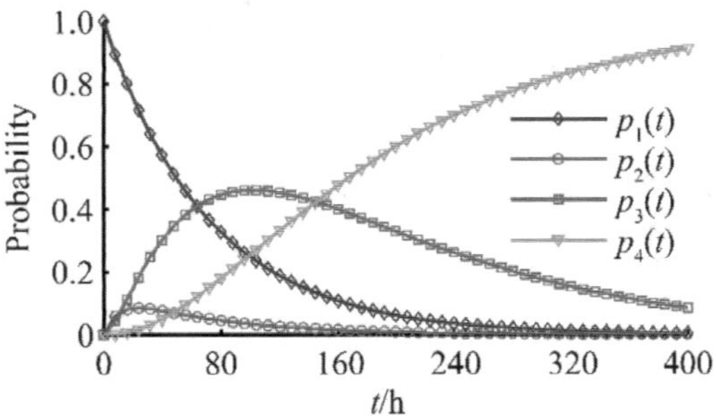

Figure 5. CNC machine state probability $p_i\,(t)$.

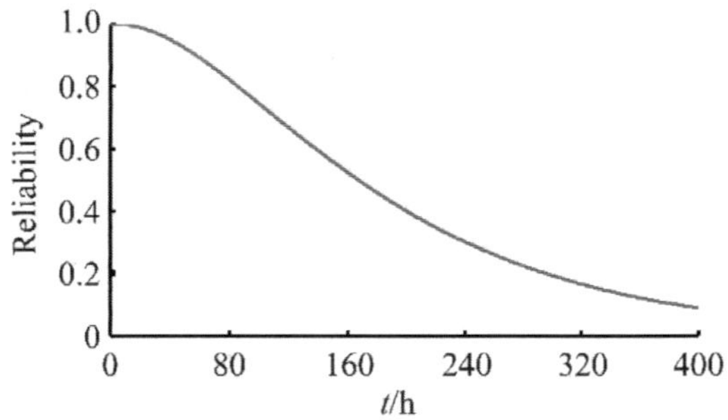

Figure 6. Reliability of the CNC machine.

As shown in Figure 5, the probability in state KS1 decreases and the probability in state KS4 increases. The probability in states KS2 and KS3 increases firstly and decreases with increasing time. The CNC machine is in different states at different time intervals. The CNC machine is in the state KS1 in the first 72 hours, then in the state KS3. About 160 hours later, the CNC machine translates into the state KS4. The reliability of CNC machine tools decreases with time, which reflects the accuracy level of CNC machines. As shown in Table A1, there is a rising tendency of the movement precision data. So, the assessment result is effective in reflecting the reliability of the CNC machine.

3.3 Reliability considering imperfect maintenance

Generally, maintenance behaviors cannot restore a mechanical system to a fire-new one for a repairable mechanical system. Imperfect maintenance is more actual to characterize the maintenance effect, which falls in between perfect maintenance and minimal maintenance. In this paper, the CNC machine is considered as imperfect maintenance. So, it is assumed that α=0.8. The state probabilities and reliability of a CNC machine can be calculated via Formulas (25) and (26) based on the above work. The state probabilities and reliability in each maintenance cycle are shown in Figures 7 and 8. The failure rate ($p_4^n(t)$) becomes higher and the reliability becomes lower with the increasing maintenance times. The results show that the condition of the CNC machine in each maintenance cycle is worse than that in the previous maintenance cycle, and the maintenance activities cannot restore the CNC machine to a new condition. Only after repair can a faulty CNC machine be restored to its best condition. But it will degrade faster during the next maintenance cycle.

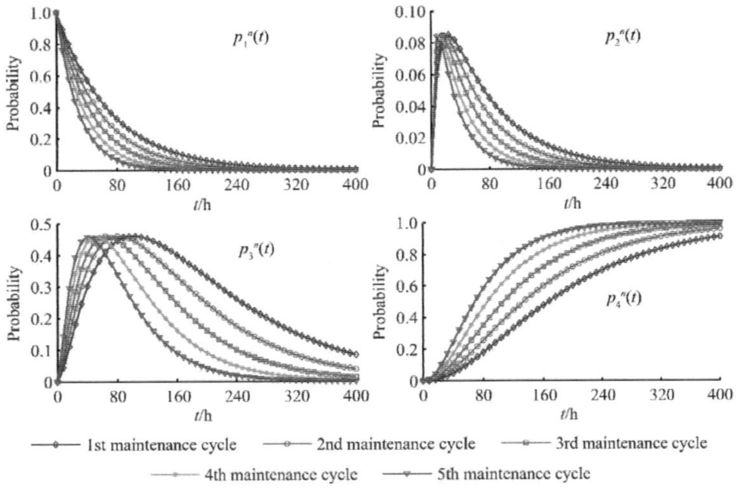

Figure 7. CNC machine state probabilities in each maintenance cycle.

Above all, this paper proposed a general method of a mechanical system reliability assessment, which can get the unit-time transition intensity matrix using the common discrete performance degradation data of a mechanical system and have the same calculation precision compared with the transition intensity matrix. The reliability and state change of a CNC machine are obtained by the proposed method. Based on the above study, the reliability of a CNC machine is analyzed considering imperfect maintenance. The results reveal that the reliability becomes lower with the increasing maintenance times.

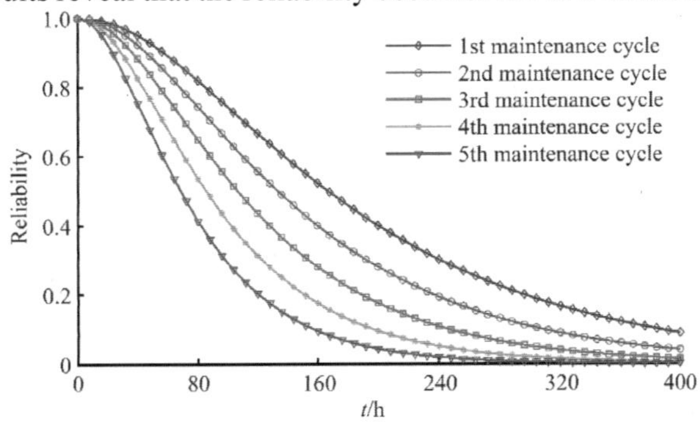

Figure 8. Reliability of the CNC machine in each maintenance cycle.

4. Conclusions

The reliability assessment of a mechanical system using discrete degradation data is studied in the paper. The conclusions may be summarized as follows:

(1) A general method for assessing mechanical system reliability is proposed in this paper. The $\overline{\Lambda}$ matrix can be obtained by the $P(1)$ matrix, which can be calculated using the HMM and the discrete degradation data of a mechanical system. An HCTMM is established based on the $\overline{\Lambda}$ matrix. The reliability of a mechanical system can be obtained by the HCTMM.

(2) Three different matrix methods are used to evaluate the reliability of a system. The results show that there is little error between the $P(1)$ matrix result and the Λ matrix result. The $\overline{\Lambda}$ matrix result is exactly identical to the Λ matrix result. What's more, the $\overline{\Lambda}$ matrix can be obtained easily using the discrete degradation data.

(3) With a CNC machine as an example, the reliability of a mechanical system is accurately evaluated using the discrete degradation data by the proposed method. The results show that the reliability of the CNC machine decreases with the increasing time and it reflects the precision level of the CNC machine. Moreover, maintenance activities are considered in the reliability of a mechanical system. The results show that the failure rate becomes higher and the reliability becomes lower with the increasing maintenance times at any use time.

Acknowledgement
This project was supported by the National Natural Science Foundation of China (52375120)

Appendix A

Table A1 The circular motion degradation data of the CNC machine.

No.	Radial deviation $F_{\min}$ (mm)	Radial deviation $F_{\max}$ (mm)	Circular deviation G (mm)	Circular hysteresis H (mm)
1	-0.01245	0.01665	0.02910	0.02290
2	-0.01235	0.01700	0.02935	0.02270
3	-0.01245	0.01725	0.02970	0.02210
4	-0.01225	0.01725	0.02975	0.02300
5	-0.01240	0.01830	0.03070	0.02320
6	-0.01275	0.01830	0.03160	0.02290
7	-0.01435	0.01665	0.03100	0.01940
8	-0.01430	0.01790	0.03220	0.02095
9	-0.00920	0.01850	0.03220	0.02170
10	-0.01385	0.01900	0.03290	0.02260
11	-0.01360	0.01925	0.03280	0.02310
12	-0.01345	0.01885	0.03235	0.02385
13	-0.03215	0.02010	0.03285	0.02170
14	-0.02960	0.01795	0.03435	0.02280
15	-0.02960	0.01845	0.03470	0.02080
16	-0.02935	0.01825	0.03445	0.02140
17	-0.02900	0.01795	0.03455	0.02220
18	-0.03845	0.02510	0.04080	0.03810
19	-0.04625	0.03045	0.04155	0.03500
20	-0.05149	0.03550	0.04525	0.03560
21	-0.04660	0.03100	0.04555	0.03470
22	-0.04630	0.03100	0.04595	0.03500
23	-0.04630	0.03105	0.04570	0.03500
24	-0.04660	0.03130	0.04615	0.03560

Table A2 CNC machine circular motion accuracy failure standard.

Performance levels	$F_{\min}$ (mm)	$F_{\max}$ (mm)	G (mm)	H (mm)
KS1	(-0.015, 0)	(0, 0.019)	(0, 0.03)	(0, 0.025)
KS2	(-0.02,-0.015)	(0.019, 0.024)	(0.03, 0.035)	(0.025, 0.03)
KS3	(-0.03, -0.02)	(0.024, 0.029)	(0.035, 0.04)	(0.03, 0.035)
KS4	(-∞, -0.03)	(0.029, ∞)	(0.04, ∞)	(0.035, ∞)

References

[1] Kuo W and Zuo M J 2003 *Optimal reliability modeling: principles and applications* (John Wiley & Sons)

[2] Frenkel I, Khvatskin L and Lisnianski A 2017 On importance assessment of aging multi-state system *AIP 11th International Conference on Mathematical Problems in Engineering Aerospace and Sciences (ICNPAA)* (La Rochelle: Univ La Rochelle) **1798(1)** 020056

[3] Zhu D P, Ding Z X, Huang X G and Li X 2023 Probabilistic modeling for long-term fatigue reliability of wind turbines based on Markov model and subset simulation *International Journal of Fatigue.* 173

[4] Zhang K and Cao H B 2021 Reliability analysis of heat exchanging system of deep-sea manned submersibles using Markov model *Quality Engineering.* **33(3)** 487-496

[5] Jiang Q, Gao D-c, Zhong L, Guo S, and Xiao A 2020 Quantitative sensitivity and reliability analysis of sensor networks for well kick detection based on dynamic Bayesian networks and Markov chain *Journal of Loss Prevention in the Process Industries*. 66

[6] Liu Y, Lin P, LI Y and Huang H 2015 Bayesian Reliability and Performance Assessment for Multi-State Systems *IEEE Transaction on Reliability.* **64(1)** 394-409

[7] Reznícek J, Kohlík M and Kubátová H 2020 Non-Homogeneous Continuous Time Markov Chains Calculations *The 23rd Euromicro Conference on Digital System Design (DSD)*

[8] Furtado F and Ribeiro D.2023 Railway Bridge Management System Based on Visual Inspections with Semi-Markov Continuous Time Process *Ksce Journal of Civil Engineering.* **27(1)** 233-250

[9] Murthy C, Mishra A, Ghosh D, Roy D S and Mohanta D K 2014 Reliability analysis of phasor measurement unit using hidden Markov model *IEEE Systems Journal.* **8(4)** 1293-1301

[10] Fort A, Mugnaini M and Vignoli V 2015 Hidden Markov models approach used for life parameters estimations *Reliability Engineering & System Safety.* **136(8)** 85-91

[11] Feng Y, Li W J, Zhang K, Li X L, Cai W F and Liu R N 2022 Morphological Component Analysis-Based Hidden Markov Model for Few-Shot Reliability Assessment of Bearing *Machines.* **10(6)**

[12] Liu F, Zhu H and Liu B 2016 Analysis of operational performance reliability for CNC equipment *IEEE Proceedings of 2016 Sixth International Conference on Instrumentation & Measurement, Computer, Communication and Control* (Harbin) 155-8

[13] Kouadri A, Hajji M, Harkat M F, Abodayeh K, Mansouri M, Nounou H and Nounou M 2020 Hidden Markov model based principal component analysis for intelligent fault diagnosis of wind energy converter systems *Renewable Energy.* **150** 598-606

[14] Han D Y, Yu J S and Tang D Y 2021 An HDP-HMM based approach for tool wear estimation and tool life prediction *Quality Engineering.* **33(2)** 208-220

[15] Deep A, Zhou S Y, Veeramani D and Chen Y 2023 HMM-Based Joint Modeling of Condition Monitoring Signals and Failure Event Data for Prognosis *Ieee Transactions on Reliability.* **72(3)** 878-888

[16] Wang H and Pham H 1996 A quasi renewal process and its applications in imperfect maintenance *International Journal of Systems Science.* **27(4)** 1055-62

[17] Wang J H, Zhang X H, Zeng J C and Zhang, Y Z 2020 Optimal dynamic imperfect preventive maintenance of wind turbines based on general renewal processes *International Journal of Production Research.* **58(22)** 6791-6810

[18] Wang X L, Gaudoin O, Doyen L, Bérenguer C and Xie M 2021 Modeling multivariate degradation processes with time-variant covariates and imperfect maintenance effects *Applied Stochastic Models in Business and Industry.* **37(3)** 592-611

[19] Wu J 2008 Methods and applications of operational reliability estimation for CNC equipment based on performance parameters *Ph.D. dissertation* Huazhong University of Science and Technology

GNSS location error reduction method for microtremor survey system based on EMD-CNN-LSTM

Wei Wu and Xingang Mou*

School of Mechanical and Electrical Engineering, Wuhan University of Technology, Wuhan, Hubei, 430070, China

*Corresponding author's e-mail: sunnymou@whut.edu.cn

Abstract. Large errors exist when the microtremor survey system uses the global navigation satellite system (GNSS) for static localization. Aiming at the problem that the existing methods cannot effectively weaken the random error and multipath error, an error weakening method based on Empirical Mode Decomposition (EMD), Convolutional Neural Networks (CNN), and Long Short-term Memory Networks (LSTM) is proposed. The model first uses EMD to decompose the high-frequency random error, then reconstructs the low-frequency component and extracts the local features using CNN, and finally learns the change rule of multipath error using LSTM and weakens it. The model can remove random errors in the early stage while reducing the interference of noise on the neural network in the later stage and then improve the accuracy of localization. The experimental results show that the model can effectively improve the localization accuracy in the case of short-time measurements so that the localization accuracy in the E, N, and U directions can be improved by 74.57%, 74.76%, and 71.86%, respectively, which is more than 10% higher than the localization accuracy improvement rate of the existing CNN-LSTM model.

1. Introduction

The microtremor survey system is more and more widely used in the construction of underground facilities because of its advantages, such as strong site adaptability and high operational efficiency [1]. In microtremor surveys, the positional accuracy between detectors has high requirements, and GNSS is a commonly used localization method in the field of microtremor surveys, but there are many factors affecting the localization accuracy in GNSS localization. Although most of the errors are completely eliminated or greatly weakened after differential technology processing, multipath error and random error amplified by differential become the main errors in short baseline positioning [2].

In recent years, deep learning techniques have achieved some results in weakening GNSS errors with their excellent nonlinear model capability. Tao uses the CNN-LSTM method to mine the multipath features in the GNSS coordinate sequence and perform weakening [3]. Min uses signal-to-noise ratio and satellite geometric parameters as the features of the multipath model based on the DNN technique to perform correction [4]. Liu uses the CNN-GRU fusion model to mine the deep change law of multipath error and perform real-time prediction and correction [5]. Tong proposed a Conv-GRU-MHM model to predict multipath error in real-time using the repetition period of satellites [6]. However, the current deep learning methods for weakening GNSS positioning errors often focus only on multipath errors and do not take random errors into account. In addition to affecting the GNSS positioning accuracy, random errors also affect the learning effect of deep learning methods on multipath errors.

The deep learning models about time series are the Transformer class and RNN class. The Transformer class model is more complex and requires a large amount of training data [7], which cannot be met by the microtremor survey system GNSS localization data. The RNN class model is simple and requires a small amount of data, and it can be applied to the microtremor survey system GNSS localization. Therefore, to address the problem that the GNSS static positioning error of the microtremor survey system cannot be effectively weakened by the existing deep learning methods, this paper proposes an EMD-CNN-LSTM model based on the RNN class model to simultaneously weaken the random error and multipath error.

2. Structure and workflow of the microtremor survey system

In this paper, the microtremor survey system is developed independently by the laboratory, as shown in Figure 1. The system consists of several physically independent probes and a control center, and the wireless communication between the control center and the probes is carried out through Wi-Fi. The system uses the carrier phase differential positioning technique in GNSS positioning; the reference detector sends its own GNSS data to each mobile detector, and the mobile detector combines its own GNSS data to solve its own position information.

Figure 1. Structure of the system.

Figure 2. Workflow of the Microtremor survey.

The operation flow of the microtremor survey is shown in Figure 2, which first analyzes the detection site and then selects a suitable microtremor survey algorithm in combination with the detection target. Commonly used algorithms are Frequency-Wavenumber [8] (FK) and Spatial Auto-Correlation [9] (SPAC). If the FK method is used for the microtremor survey, the irregular table array is firstly laid out according to the terrain environment, and then the relative position between each detector is measured. This measurement of detector position information when the detector is in a stationary state is called static localization.

3. EMD-CNN-LSTM error weakening method

3.1 Overall framework of the model

There are many modal components in GNSS coordinate sequences, and the use of the empirical mode decomposition tightly coupled framework to excavate the variation rules of each modal component will lead to the superposition of errors of each component.

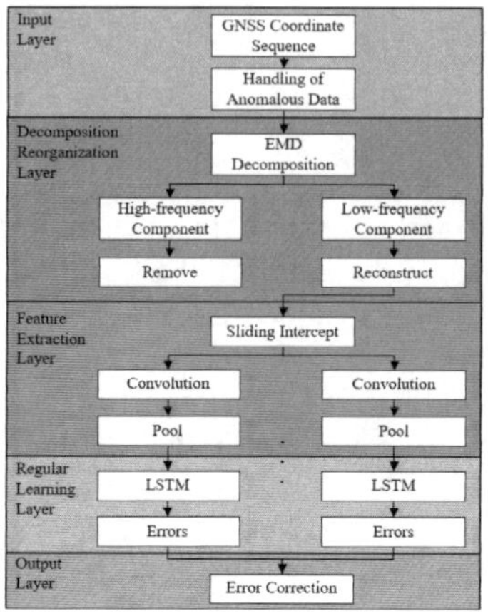

Figure 3. EMD-CNN-LSTM loosely coupled modeling framework.

Therefore, a loosely coupled framework based on EMD-CNN-LSTM is designed to reduce the GNSS static positioning errors of the microtremor survey system, and the overall framework of the model is shown in Figure 3. The overall model is divided into an input layer, a decomposition and reconstruction layer, a feature extraction layer, a law learning layer, and an output layer. The input layer preprocesses the data and removes the outliers; the decomposition and reconstruction layer uses EMD to extract the high-frequency random errors in the GNSS coordinate sequences; the feature extraction layer uses CNN to extract the local features of the GNSS coordinate sequences; the regular learning layer uses LSTM to dig the deep rule of change of multipath errors; the output layer weakens the learned multipath error and output the final localization result.

3.2 Decomposing the reconfiguration layer

There are measurement noises in the GNSS coordinate sequences caused by equipment and external environments, such as thermal noise of electronic devices, signal quantization errors, and software calculation errors. In addition to affecting the positioning accuracy, it will also cause interference to the subsequent neural network model and reduce the learning ability of the neural network on the change rule of multipath error. GNSS data are more random and have strong nonstationary and nonlinear characteristics due to the existence of multipath errors, and the high-frequency random errors can be effectively extracted for weakening using the EMD method. The specific steps are as follows: (1) find all the local maxima and minima of the GNSS coordinate sequence $x(t)$; (2) fit the upper envelope $U(t)$ and lower envelope $L(t)$ of $x(t)$ according to the cubic spline interpolation method; (3) compute the mean value $M(t)$ of the upper envelope $U(t)$ and the lower envelope $L(t)$; (4) subtract the mean value of envelope $M(t)$ from the original sequence $x(t)$ to obtain the candidate IMF sequence $y(t)$. To obtain the candidate IMF sequence y(t); (5) output an IMF component of the original sequence if $y(t)$ satisfies the constraints, and replaces $x(t)$ with the remaining residual term $r(t)$, or replace $x(t)$ with $y(t)$ if it does not satisfy them; (6) repeat Steps 1-5 until the decomposition is completed when the residual term $r(t)$ is a monotonic function or a constant. The EMD decomposition results in the original sequence as:

$$x(t) = \sum IMF_i + r(t) \tag{1}$$

3.3 Feature extraction layer

In this paper, in order to deeply excavate the law of GNSS coordinate sequence, the model pays attention to both local and global change rules. Firstly, CNN is used to fully extract the local features of the GNSS coordinate sequence so as to improve the learning efficiency of the subsequent steps on

Journal of Physics: Conference Series **2815** (2024) 012049 doi:10.1088/1742-6596/2815/1/012049

the GNSS change rule. The CNN intercepts the time sequence segment x through a sliding window; the convolutional layer weights and biases are W and b, $\otimes$ denote the convolution operation, and are nonlinearly mapped by the activation function σ. The local feature q can be represented as:

$$q = \sigma(b + x \otimes W) \tag{2}$$

The structure of the feature extraction layer is shown in Figure 4, where a sliding intercept of the GNSS coordinate sequence is first performed using a window of length 100; then, three one-dimensional convolution kernels of length 11 are used for the convolution operation, using a maximum pooling operation with a step size of three; and finally three one-dimensional convolution kernels of length 11 are again used for the convolution operation, using a maximum pooling operation with a step size of three.

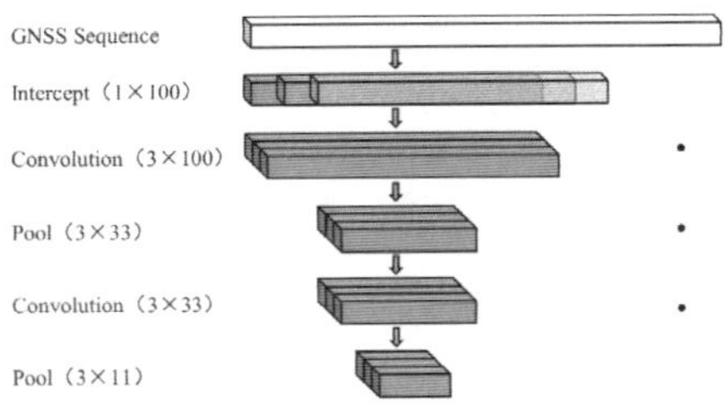

Figure 4. Structure of feature extraction layer.

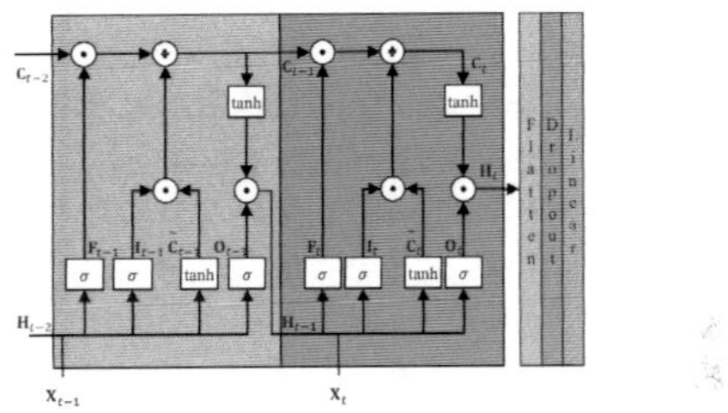

Figure 5. Structure of regulation learning layer.

3.4 Regulation learning layer

There are multipath errors in GNSS coordinate sequences in addition to high-frequency irregular random errors. The multipath error is the error generated by the satellite signal being reflected by the surrounding environment [10]. The multipath error varies with the spatial position of the satellite when both the GNSS receiver and the surrounding environment remain stable.

The regular learning layer of this model uses LSTM, which can learn and memorize long-time sequence information and can extract useful information not only from a single data point but also from the whole time series [11]. The local features extracted by CNN are used to train it, which can effectively learn the change rule of multipath error and then predict the error. The regular learning layer is shown in Figure 5. First, the local features extracted by the feature extraction layer are sequentially input into the LSTM recurrent unit to obtain the hidden state, where the size of the hidden layer is 12. then the data is expanded using the Flatten layer. To prevent overfitting, a dropout layer regularization with a scale of 0.1 is used, and finally, the error value of the prediction is output using the fully connected linear layer.

4 Experimental results and analysis

In order to verify the validity of the EMD-CNN-LSTM loosely coupled model in this paper, GNSS static positioning data was collected first to make a dataset, and then related experiments were designed for analysis.

4.1 Data acquisition and processing

The data in this paper were collected on October 18, 2023, on the roof of a university building, as shown in Figure 6, and 30,000 sampling points were intercepted as the experimental data in this paper after removing the abnormal data. The mean value of the long-time measurement is used as the reference value, where E denotes the due east direction, N denotes the due north direction, and U denotes the elevation direction. The first 80% of the dataset is used as the training set, and the second 20% as the test set.

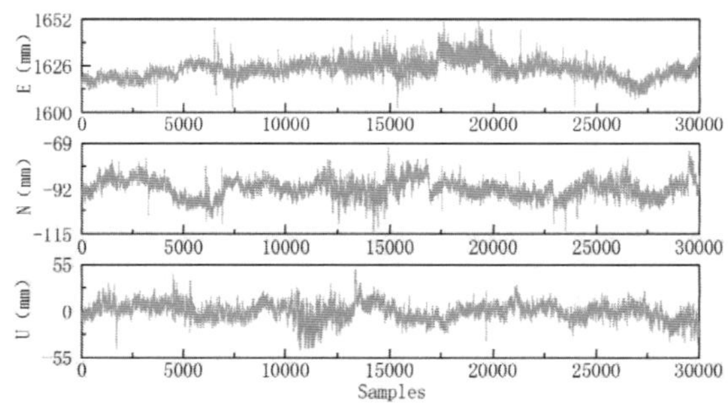

Figure 6. Sequence of GNSS coordinates in ENU direction.

In this paper, we use mean percentage error (MAPE) and mean absolute error (MAE) to evaluate the model effect of different models; we use root mean square error (RMSE) to measure the improvement of localization accuracy and the improvement rate in this paper refers to the percentage of RMSE reduction of the localization results after error weakening relative to the localization results before weakening.

4.2 Random error weakening analysis

In this experiment, in order to study the effect of removing different degrees of high-frequency components on the positioning accuracy, the E-direction data is taken as an example for experiment and analysis, and the multi-scale decomposition of the GNSS coordinate sequence is carried out by using the EMD method to obtain 12 modal components and one residual component as shown in Figure 7. It can be seen that as the degree of decomposition increases, the frequency of the components decreases, and the volatility and randomness are also decreasing. Therefore, the GNSS coordinate sequence becomes smooth after removing the high-frequency components, which is favorable to the regularity mining effect of CNN and LSTM.

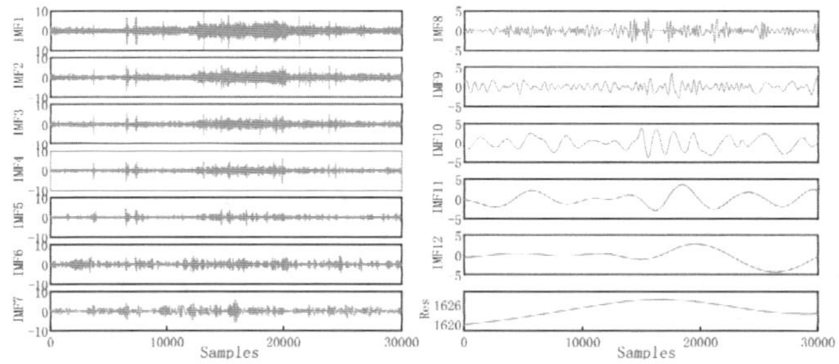

Figure 7. EMD decomposition results for E-direction data.

The high-frequency components in front are removed, and then the remaining components are reconstructed and fed into the CNN-LSTM network for learning. As shown in Figure 8, the horizontal coordinates represent all the components before removing the IMFx component. The results show that the whole model works best when only the highest frequency IMF1 component is removed.

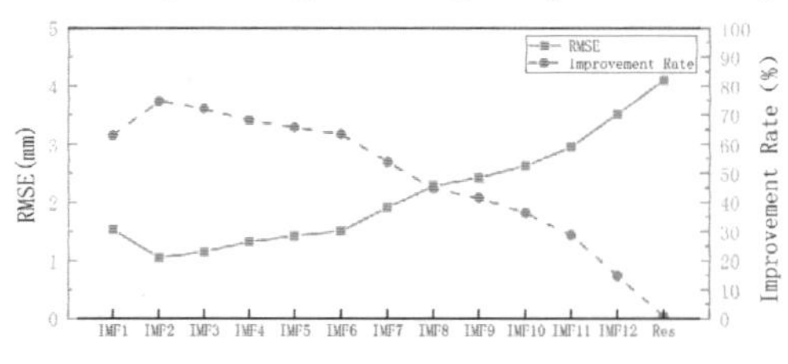

Figure 8 Reconstruction results of different low-frequency components in the E direction.

4.3 Model comparison
Both LSTM and Gated Recurrent Unit (GRU) can process time series, but there are differences in performance in different tasks. In this experiment, we investigate the effect of different models on the improvement rate of GNSS positioning accuracy by constructing EMD-CNN-GRU and EMD-CNN-LSTM models and comparing experiments with existing CNN-GRU and CNN-LSTM models. As shown in Table 1, the EMD-CNN-LSTM model is the best in all three directions. It can be seen that the EMD method can improve the improvement rate by more than 10%, which indicates that the EMD method can not only eliminate the random error but also facilitate the learning of the change rule of multipath error by the neural network.

Table 1. Comparison of improvement rate of positioning accuracy.

Direction	CNN-GRU	CNN-LSTM	EMD-CNN-GRU	EMD-CNN-
E	60.95%	62.97%	73.99%	74.57%
N	59.40%	61.17%	73.79%	74.76%
U	56.51%	58.52%	71.39%	71.86%

Table 2. Comparison of model effects.

Direction	CNN-GRU		CNN-LSTM		EMD-CNN-GRU		EMD-CNN-LSTM	
	MAPE	MAE	MAPE	MAE	MAPE	MAE	MAPE	MAE
E	3.35	1.25	2.82	1.18	2.33	0.84	2.27	0.82
N	1.73	1.46	1.33	1.40	1.08	0.96	1.07	0.93
U	2.38	3.16	2.39	3.00	1.85	2.11	1.59	2.08

The modeling effect of each model was measured using MAPE and MAE, and the results are shown in Table 2; the EMD-CNN-LSTM model has the smallest value of each evaluation index, which indicates that the model has the best modeling effect, in which the model using LSTM is better than the model using GRU.

4.4 Weakening analysis of positioning error
This experiment is to investigate the EMD-CNN-LSTM model for weakening the localization error in different directions. The ENU direction data with the high-frequency component IMF1 removed is put into the CNN-LSTM network for training and testing, respectively. Among them, the E, N, and U direction accuracies reach 1.05 mm, 1.17 mm, and 2.61 mm, with improvement rates of 74.57%, 74.76%, and 71.86%, respectively, and the improvement rates of localization accuracy are all stabilized at more than 70%. As shown in Figure 9, it can be seen that the GNSS coordinate sequence becomes smooth, and both random error and multipath error are significantly weakened.

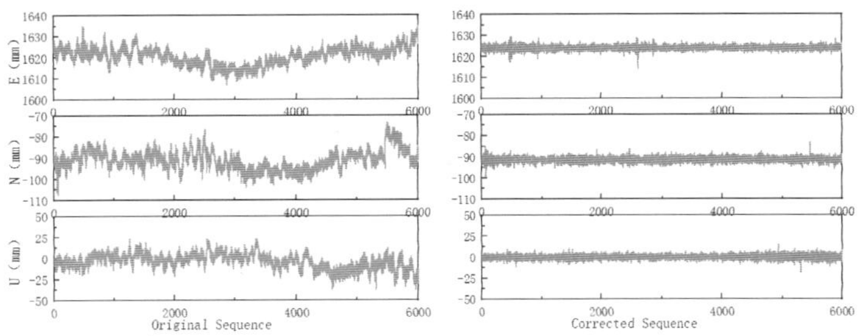

Figure 9 Original and corrected sequences of GNSS coordinates.

5. Conclusion
Existing deep learning methods do not consider the random error in GNSS static localization, resulting in a limited error weakening effect. In this paper, we propose an EMD-CNN-LSTM loosely coupled model to weaken the error in GNSS static localization of microtremor survey system, and the results show that the model can effectively weaken the random error and multipath error at the same time, which improves the localization accuracy to the millimeter level.

Journal of Physics: Conference Series **2815** (2024) 012049 doi:10.1088/1742-6596/2815/1/012049

Reference

[1] Yan Y, Sun C, Lin T, et al. Surface wave exploration technology for the tunnel overlying strata with vehicle vibration source [J]. Tunnelling and Underground Space Technology, 2023, 134: 105014. [CrossRef]

[2] Kato H, Meguro J. Improvement of Differential-GNSS Positioning by Estimating Code Double-Difference-Error Using Machine Learning [J]. IEICE TRANSACTIONS on Information and Systems, 2023, 106(12): 2069-2077. [CrossRef]

[3] Tao Y, Liu C, Chen T, et al. Real-time multipath mitigation in multi-GNSS short baseline positioning via CNN-LSTM method [J]. Mathematical Problems in Engineering, 2021, 2021: 1-12. [CrossRef]

[4] Min D, Kim M, Lee J, et al. DNN-Based Approach to Mitigate Multipath Errors of Differential GNSS Reference Stations [J]. IEEE Transactions on Intelligent Transportation Systems, 2022, 23(12): 25047-25053. [CrossRef]

[5] Liu C Y, Liu C, Tao Y, et al. Real-time GNSS short baseline multipath error reduction method incorporating CNN-GRU [J]. Science of Surveying and Mapping, 2022, 47(11): 40-47+72. [CrossRef]

[6] Tong R, Liu C, Tao Y, et al. ConvGRU-MHM: a CNN GRU-enhanced MHM for mitigating GNSS multipath [J]. Measurement Science and Technology, 2024, 35(4): 045007. [CrossRef]

[7] Geng S, Zhu Z, Wang Z, et al. LW-ViT: The Lightweight Vision Transformer Model Applied in Offline Handwritten Chinese Character Recognition [J]. Electronics, 2023, 12(7): 1693. [CrossRef]

[8] Akbari A, Ashourifar H, Pour G B, et al. A frequency-wavenumber imaging algorithm for detection of subsurface targets in layered mediums [J]. Arabian Journal of Geosciences, 2023, 16(3): 163. [CrossRef]

[9] Du Q, Liu Z. Application of the two-station spatial autocorrelation method in engineering exploration [J]. Arabian Journal of Geosciences, 2022, 15(13): 1181. [CrossRef]

[10] Weng D, Hou Z, Meng Y, et al. Characterization and mitigation of urban GNSS multipath effects on smartphones [J]. Measurement, 2023, 223: 113766. [CrossRef]

[11] Bareth R, Yadav A, Gupta S, et al. Daily average load demand forecasting using the LSTM model based on historical load trends [J]. IET Generation, Transmission & Distribution, 2024, 18(5): 952-962. [CrossRef]

Journal of Physics: Conference Series **2815** (2024) 012050 doi:10.1088/1742-6596/2815/1/012050

Theoretical and experimental study on a novel gear drive with high transmission efficiency

Chao Tian [1], Luhe Zhang [1,*], Dongyu Wang [1], Liang Cheng [1], Yangyong Ni [1] and Bingkui Chen [1]

[1] Chongqing University, Chongqing 400044, China

[*] Corresponding author's e-mail: lh.zhang@cqu.edu.cn

Abstract: This paper introduces a novel gear drive characterized by high transmission efficiency. The gear incorporates cubic curves with corresponding tangents defining the normal section tooth profile. The generation principle of this new gear drive is thoroughly established. Utilizing gear meshing theory, key parameters such as the relative motion velocity, normal vector, and meshing equation are derived. Subsequently, the fundamental principle of the conjugate curve is formulated. The methodology for generating the tooth profile is also explored. An efficiency test was conducted on the gear, and the results demonstrate a superior transmission efficiency relative to that of traditional involute gears.

1. Introduction

Gear transmission is considered a key basic part of equipment manufacturing and defense industry, and it is also regarded as a symbol of industry. Gears are widely used as industrial basic components in automotive, marine, and aerospace applications. At present, the more widely used gears are involute. Although involute gears have the advantages of being simple to manufacture and insensitive to changes in center distance, they have a low load-carrying capacity and poor lubrication. Conventional gears are generally designed according to the principle of conjugate surfaces, and the geometric properties of conjugate surfaces greatly affect the transmission performance of gears. Therefore, many scholars have studied the mathematical principles of conjugate surfaces and tooth profiles.

Litvin and Fuentes [1] have systematically studied the principle of gear meshing from a geometric point of view. Raikhman [2] proposed the principle of formation of conjugate surfaces by generating lines to provide a new perspective on gear machining. Kapelevich [3] proposed a method for studying and designing asymmetric toothed gears that increases load capacity, reduces weight, and decreases size. Chen [4] proposed a helical gearing system characterized by pure rolling and a variable helix angle and conducted a detailed investigation of its meshing and mechanical properties. Luo et al. [5] obtained a gear with good load-bearing capacity by using the cosine curve as the tooth profile and established the relevant theory through mathematical derivation. Komori et al. [6] studied a novel tooth profile that gives spur gears zero relative curvature at the point of meshing contact by means of a concave-convex contact pattern. Robison et al. [7] studied the performance of different types of cycloid profiles and proposed an optimization method that considers multiple performance objectives, which is able to screen out suitable types of cycloid curves under different working conditions. Zschippang et al. [8] proposed a method for generating face gears that incorporates helix and axial intersection angles to mathematically represent their geometry accurately.

Based on the previous research and facing the great demand for high-performance gear transmission

Journal of Physics: Conference Series **2815** (2024) 012050 doi:10.1088/1742-6596/2815/1/012050

in modern industry, this paper proposes a novel gear with high transmission efficiency. This paper introduces the general principle of the conjugate curve, utilizes differential geometry and gear geometry, establishes a spatial coordinate system, and deduces the coordinate transformation matrix between gears. The relative motion velocity and normal vector are solved, and the meshing equation is established to derive the conjugate curve. Mathematical models of normal tooth profile and tooth surface are established. A test bench is established to test the efficiency of the designed prototype, which sets up an experimental foundation for the engineering application of this gear. Finally, the research content of this paper is summarized. The related research content will have a positive impact on enriching gear meshing theory and meeting the demand for high-performance gear transmission, which has a broad engineering application prospect.

2. The basic principle of conjugate curve gear meshing

2.1. Coordinate system and coordinate transformation

Conjugate curves refer to two continuous curves that maintain consistent point contact in the process of mutual motion. As shown in Figure 1, $S_n(O_n\text{-}x_n, y_n, z_n)$ and $S_m(O_m\text{-}x_m, y_m, z_m)$ represent fixed coordinates in space. $S_1(O_1\text{-}x_1, y_1, z_1)$ and $S_2(O_2\text{-}x_2, y_2, z_2)$ are the movable coordinates fixed on Pinion 1 and Gear 2, respectively. The gears rotate at fixed angular speeds ω_1 and ω_2 along the z_n and z_1 axes, respectively, in opposite directions. Point P is the tangent contact point of the curves Γ_1 and Γ_2. The z axis coincides with z_1, the z_m axis coincides with z_2, and the z_n axis is parallel to z_m. The center distance between Pinion 1 and Gear 2 is a.

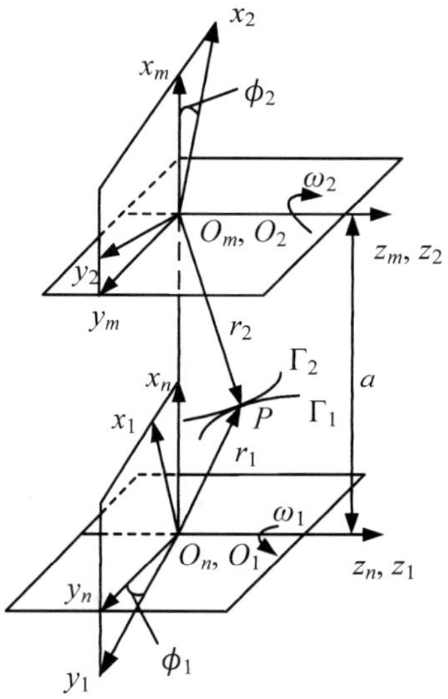

Figure 1. Coordinate system of conjugate curve gear.

According to the established spatial coordinate system and parameter definitions, the relationship between the coordinate systems can be expressed by using the transformation matrix as follows:

$$\boldsymbol{M}_{21}=\begin{bmatrix}\cos(\phi_1+\phi_2) & -\sin(\phi_1+\phi_2) & 0 & -a\cos\phi_2\\ \sin(\phi_1+\phi_2) & \cos(\phi_1+\phi_2) & 0 & -a\sin\phi_2\\ 0 & 0 & 1 & 0\\ 0 & 0 & 0 & 1\end{bmatrix} \tag{1}$$

2.2. Relative motion velocity and normal vector

In the process of establishing the meshing principle, to ensure that the two gears mesh correctly, neither interfering nor disengaging from each other, and to ensure continuous contact and reduce frictional wear, it is necessary to solve for the relative speed of motion at the point of contact. As shown in Figure 1, in the movable coordinate system S_1, the relative velocity is denoted by:

$$\boldsymbol{v}_1^{(12)} = \left[-y_1\left(\omega_1+\omega_2\right)-a\omega_2\sin\phi_1\right]\boldsymbol{i}_1+\left[x_1\left(\omega_1+\omega_2\right)-a\omega_2\cos\phi_1\right]\boldsymbol{j}_1 \tag{2}$$

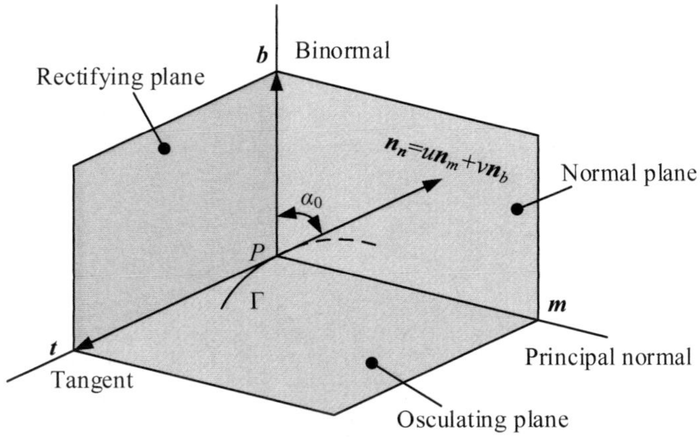

Figure 2. The basic trihedron of the curve.

According to gear geometry theory and application [1], the basic trihedron of the curve Γ_1 at Point P is established, as shown in Figure 2. At this point, three mutually perpendicular unit vectors are identified, namely the vector $\boldsymbol{t}$, the vector $\boldsymbol{m}$, and the vector $\boldsymbol{b}$. The vector $\boldsymbol{b}$ can be computed by $\boldsymbol{b}=\boldsymbol{t}\times\boldsymbol{m}$. The normal vector $\boldsymbol{n}_n$ of the curve Γ_1 at Point P can be represented by the vector $\boldsymbol{m}$ and the vector $\boldsymbol{b}$. The contact α_0 angle is expressed as the angle between vector $\boldsymbol{b}$ and vector $\boldsymbol{n}_n$.

$$\boldsymbol{n}_n = u\boldsymbol{n}_m + v\boldsymbol{n}_b \tag{3}$$

The general space curve Γ_1 in the coordinate system S_1 can be expressed in the parametric equation as:

$$\boldsymbol{r}_1(\varphi) = x_1(\varphi)\boldsymbol{i}_1 + y_1(\varphi)\boldsymbol{j}_1 + z_1(\varphi)\boldsymbol{k}_1 \tag{4}$$

where φ is the curve parameter.

Therefore, the tangent vector $\boldsymbol{n}_t$ of the curve can be derived as:

$$\boldsymbol{n}_t(\varphi) = \boldsymbol{r}_1' / \left|\boldsymbol{r}_1'\right| \tag{5}$$

Similarly, the principal normal vector $\boldsymbol{n}_m$ can be described as:

$$\boldsymbol{n}_m(\varphi) = \frac{\left(\boldsymbol{r}_1'\cdot\boldsymbol{r}_1'\right)\boldsymbol{r}_1'' - \left(\boldsymbol{r}_1'\cdot\boldsymbol{r}_1''\right)\boldsymbol{r}_1'}{\left|\boldsymbol{r}_1'\right|\left|\boldsymbol{r}_1'\times\boldsymbol{r}_1''\right|} \tag{6}$$

2.3. Meshing equations and conjugate curves

To ensure that the conjugate curves Γ_1 and Γ_2 maintain continuous tangential contact without interference and disengagement, the relative velocity of motion is perpendicular to the common normal. The conjugate curve meshing equation can be described as:

$$\boldsymbol{n}_n \cdot v_1^{(12)} = 0 \tag{7}$$

Combined with Equations (2), (3) and (7), we can derive the complete expression of the meshing equation as follows:

$$-ai_{21}(um_{x_1}+vb_{x_1})\sin\phi_1 - ai_{21}(um_{y_1}+vb_{y_1})\cos\phi_1 = (1+i_{21})[y_1(um_{x_1}+vb_{x_1}) - x_1(um_{y_1}+vb_{y_1})] \tag{8}$$

where the relationship is $i_{21}=\omega_2/\omega_1$.

This equation simplifies and can be described as:

$$U\sin\phi_1 - V\cos\phi_1 = W \tag{9}$$

There exists a relationship between the parameter φ of the curve and the motion parameter ϕ_1. The relationship is determined by Equation (9). When two curves are conjugate to each other and make conjugate contact motion and when the equation $\boldsymbol{r}_1$ of one of the curves Γ_1 is given, the equation $\boldsymbol{r}_2$ of the other curve Γ_2 can be deduced from the relation $\boldsymbol{r}_2=M_{21}\boldsymbol{r}_1$ and the meshing equation, which can be represented as:

$$\begin{cases} x_2 = \cos[(1+i_{21})\phi_1]x_1 - \sin[(1+i_{21})\phi_1]y_1 - a\cos(i_{21}\phi_1) \\ y_2 = \sin[(1+i_{21})\phi_1]x_1 + \cos[(1+i_{21})\phi_1]y_1 - a\sin(i_{21}\phi_1) \\ z_2 = z_1 \\ U\sin\phi_1 - V\cos\phi_1 = W \end{cases} \tag{10}$$

2.4. Methods of tooth surface generation

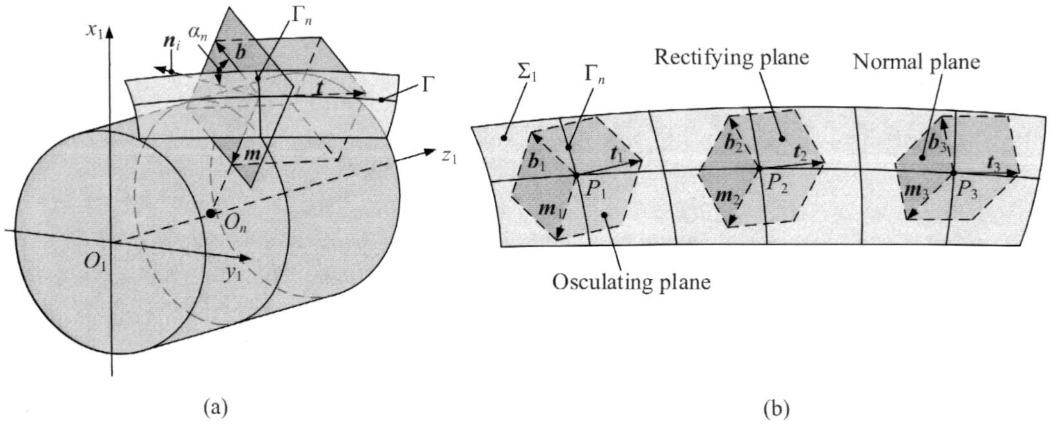

Figure 3. Graph of tooth construction.

Although a pair of conjugate curves can achieve tangential contact motion, the curves cannot be directly loaded. Therefore, tooth surfaces need to be constructed so that they can be used in a transmission system. The normal face tooth profile motion method is proposed to construct the tooth surface, as shown in Figure 3. When a conjugate curve Γ is given, the vector $\boldsymbol{t}$, vector $\boldsymbol{m}$, and vector $\boldsymbol{b}$ of Point P on it can be derived to construct the basic trihedron, as shown in Figure 3 (a). We construct the tooth profile curve Γ_n in the normal plane, which passes through Point P. When the tooth profile curve Γ_n moves continuously along the curve Γ, a continuous tooth surface Σ_1 is constructed. The coordinate system S_σ is established with the unit vector $\boldsymbol{m}_1$, the unit vector $\boldsymbol{b}_1$, and the unit vector $\boldsymbol{t}_1$ at Point P, as shown in Figure 3 (b). Given the normal cross-section tooth profile equation $\boldsymbol{r}_n$, the tooth surface equation $\boldsymbol{r}_{\Sigma_1}$ can be solved by a coordinate transformation $\boldsymbol{M}_{1\sigma}$, which is expressed as:

$$\boldsymbol{r}_{\Sigma_1} = \boldsymbol{M}_{1\sigma}\boldsymbol{r}_n = \begin{bmatrix} \boldsymbol{i}_1\cdot\boldsymbol{t}_1 & \boldsymbol{i}_1\cdot\boldsymbol{m}_1 & \boldsymbol{i}_1\cdot\boldsymbol{b}_1 & \\ \boldsymbol{j}_1\cdot\boldsymbol{t}_1 & \boldsymbol{j}_1\cdot\boldsymbol{m}_1 & \boldsymbol{j}_1\cdot\boldsymbol{b}_1 & \boldsymbol{r}_\Gamma \\ \boldsymbol{k}_1\cdot\boldsymbol{t}_1 & \boldsymbol{k}_1\cdot\boldsymbol{m}_1 & \boldsymbol{k}_1\cdot\boldsymbol{b}_1 & \\ 0 & 0 & 0 & \end{bmatrix}\boldsymbol{r}_n \tag{11}$$

3. Performance experiment

3.1 Manufacturing and testing plan

In this experimental study, which utilizes a limited number of gears for scientific research, milling was chosen as the gear machining method due to its cost-effectiveness and rapid production capability. The relevant parameters for the pinion and gear are detailed in Table 1. As illustrated in Figure 4 (a), the machining was performed on a DMU75 series machining center, which is equipped with the Siemens

840D control system. This system is noted for its excellent operability, high machining precision, and high production efficiency. The machine structure offers robust rigidity, with a positioning accuracy of 5 micrometers. Figures 4 (b) and 4 (c) display the pinion and gear, respectively, during the machining process.

(a) (b) (c)

Figure 4. Pinion and gear machining processes.

Table 1. Design parameters of pinion and gear.

Parameters	Value
Normal module m_n (mm)	3
Normal pressure angle α_n (°)	30
Number of teeth of the pinion and gear Z	13, 65
Center distance a (mm)	130
Helical angle β (°)	25.8419
Pinion and gear tooth width b (mm)	50

This experiment uses an electric power closed experimental bench. Compared with other types of experimental benches, it has a simple mechanical structure, which is easy to build and maintain and has fewer energy conversion links, low energy consumption, and better economy. The load and the drive are controlled by frequency conversion, fast response time, high control accuracy, and convenient control. The layout of the test system is shown in Figure 5, with spring couplings connecting the drive motor, load motor, and torque speed sensors I and II. The temperature, torque, and speed data are transmitted to the computer through a data collector. Meanwhile, the computer is connected to the motor control cabinet, which is utilized to control the output parameters of the motor, facilitating real-time observation and change of system speed and load torque. To ensure the safety of the test, a non-contact torque speed sensor is used, which is highly accurate and has a detachable design for easy and flexible installation.

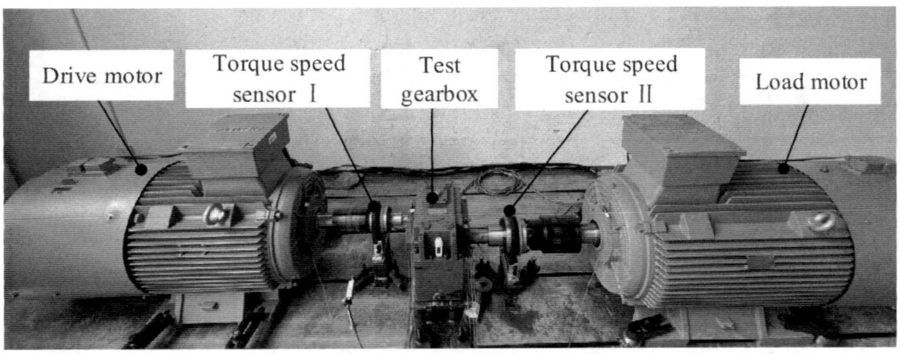

Figure 5. The test site.

After the pinion and gear have been manufactured, they are assembled into the reducer case. Because the newly machined tooth surfaces will undergo a slight deformation at the contact position under the action of the load moment, resulting in the expansion of the contact area, this process is not a real working state. To optimize the contact condition of the tooth surfaces and to collect real efficiency data, it is necessary to run the gears together before starting the experiment.

3.2 Efficiency tests

The gearing efficiency can be expressed by $\eta = n_o T_o / n_i T_i$. The prototype was designed for a power of 30 kW and a rated speed of 1, 500 r/min. The program sets the input speed from 500 r/min to 1, 500 r/min, and the load torque from 300 Nm to 1, 500 Nm.

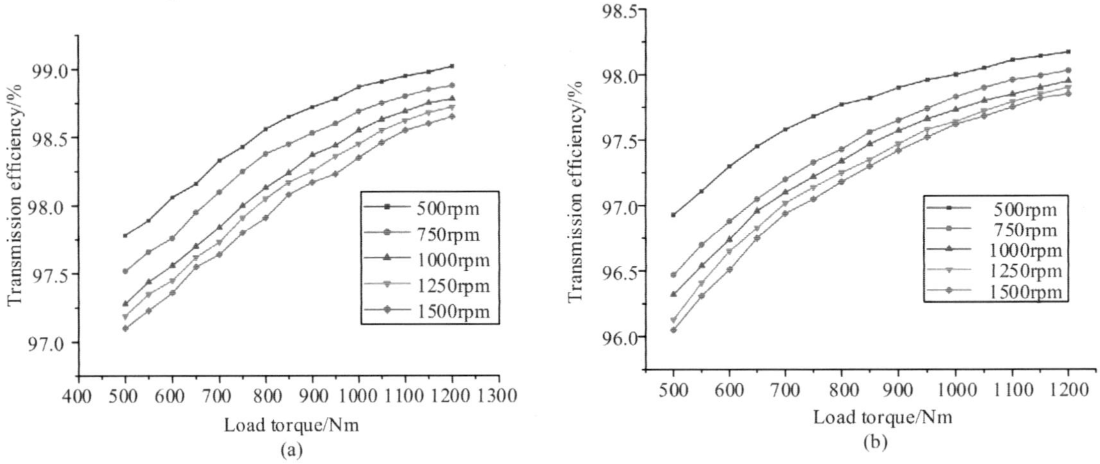

Figure 6. Transmission efficiency.

The designed gear transmission efficiency is shown in Figure 6 (a). When the input speed is 500 r/min and the load torque is 1, 200 Nm, the transmission efficiency of the prototype reaches 99.02%. When the input speed is 1, 500 r/min and the load torque is 500 Nm, the minimum transmission efficiency of the prototype is 97.1%. Therefore, the range of transmission efficiency of the prototype under test conditions is 97.1%-99.02%. When the speed is maintained at a certain value, the load torque is increased from 500 Nm to 1, 200 Nm, and the transmission efficiency is gradually improved. As the load torque is consistently maintained, there is a noticeable decline in transmission efficiency as speed increases.

Efficiency tests are conducted on involute gears with the same drive parameters, and their test conditions are consistent with the design prototype. The transmission efficiency curve is shown in Figure 6 (b). With the input speed of 500 r/min and load torque of 1, 200 Nm, the involute gear transmission efficiency reaches the maximum value of 98.07%. Involute gearing efficiency ranges from 96.05% to 98.07% under test conditions. At an input speed of 1, 500 r/min and a load torque of 500 Nm, the minimum efficiency of the involute gearing is 96.05%. When the input speed is constant, the efficiency of the gear

rises with the increase of the load torque. When the load torque is a certain value, the gear efficiency declines with the increase in speed.

When the rotational speed is kept constant, the efficiency increases with the increase of load torque. This is because in the case of low load, the system's friction loss and churning loss power account for a large proportion. With the increase in the load, the part of the loss of power accounted for more and more small, and the system's transmission efficiency can be improved. When the load torque is maintained at a certain level, the transmission efficiency of the system decreases with the increase of the rotational speed, because the churning loss, friction loss, and wind resistance loss of the system increase with the increase of the rotational speed.

The transmission efficiency of the novel type of gear transmission is 97.1%-99.02%, while the transmission efficiency of the involute gear transmission is 96.05%-98.07%. Therefore, the new gear transmission designed in this paper has higher transmission efficiency and has a broad engineering application prospect.

4. Conclusions

This paper proposes a novel gear transmission theory with high transmission efficiency and conducts experimental research on this gear. The research results of the paper are of great significance to enrich the gear meshing theory and meet the demand for high-performance gear transmission. The main conclusions are as follows:

(1) This paper introduces the general principle of the conjugate curve, utilizes differential geometry and gear geometry, establishes a spatial coordinate system, and deduces the coordinate transformation matrix between gears. The relative motion velocity and normal vector are solved, and the meshing equation is established to derive the conjugate curve. Mathematical models of normal tooth profile and tooth surface are established.

(2) A gear efficiency test platform was constructed to evaluate the performance of the designed gear. Test results indicate that this gear exhibits higher transmission efficiency compared to traditional involute gears. The meshing theory of the conjugate curve gear drive introduced in this paper enhances the existing gear drive theory and offers significant potential for engineering applications.

Acknowledgment

The research work in this article was supported by the Major Projects in Aviation Engines and Gas Turbines (2019-1V0001-0068).

References

[1] Litvin F. L. and Fuentes A. (2004). Gear geometry and applied theory. Cambridge University Press, New York.

[2] Raikhman G. N. (2002). Theory of conjugate surface formation by generating lines. J. Mechanism and Machine Theory, 37: 999-1, 005.

[3] Kapelevich A. L. (2000). Geometry and design of involute spur gears with asymmetric teeth. J. Mechanism and Machine Theory, 35: 117-130.

[4] Chen Z. (2020). Computerized design, simulation of meshing, and stress analysis of pure rolling cylindrical helical gear drives with variable helix angle. J. Mechanism and Machine Theory, 153: 103, 962.

[5] Luo S. M., Wu Y., and Wang J. (2008). The generation principle and mathematical models of a novel cosine gear drive. J. Mechanism and Machine Theory, 43: 1, 543-1, 556.

[6] Komori T., Ariga Y., and Nagata S. (1990). New gears profile having zero relative curvature at many contact points (Logix tooth profile). J. Journal of Mechanisms, Transmissions, and Automation in Design, 112: 430-436.

[7] Andrew J. Robison, Andrea Vacca. (2021). Performance comparison of epitrochoidal, hypotrochoidal, and cycloidal gerotor gear profiles. Mechanism and Machine Theory, 158: 104, 228.

[8] H.A. Zschippang, S. Weikert, K.A. Küçük, K. Wegener. (2019). Face-gear drive: Geometry generation and tooth contact analysis. Mechanism and Machine Theory, 142: 103,576.

Journal of Physics: Conference Series **2815** (2024) 012051 doi:10.1088/1742-6596/2815/1/012051

A damping method of grid strapdown inertial navigation system for polar region

Kefei Yuan [1], Xiaohui Yang [2,*], Guopeng Tian [3] and Yueyang Ben [4]

[1] Wuhan Second Ship Design Institute, Wuhan, Hubei, 430010, China
[2] School of Intelligent Science and Engineering, Harbin Engineering University, Harbin University, Heilongjiang, 150001, China
[3] School of Intelligent Science and Engineering, Harbin Engineering University, Harbin University, Heilongjiang, 150001, China
[4] School of Intelligent Science and Engineering, Harbin Engineering University, Harbin University, Heilongjiang, 150001, China

[*]Corresponding author's e-mail: 17325365583@163.com

Abstract: The conventional north-pointing inertial navigation system (INS) always points to the geographic north. However, when the system crosses the pole in the polar region, the geographic north quickly changes with 180 degrees, and the rate of change of the geographic north direction will be infinite. Secondly, when the latitude approaches to 90 degrees, which will lead to singularities in math for the angular velocity caused by vehicle motion. With its special definition and mechanization, the grid coordinate system can solve the above problems. Nevertheless, the error analysis of the grid strapdown INS shows that there are Schuler periodic and Foucault periodic oscillations, which badly degrade the accuracy of the strap-down INS. To address the problem, the external horizontal damping technology is adopted in this paper, wherein the external velocity is provided by the Doppler velocity log (DVL). Finally, the feasibility and the effectiveness of the external damping technology were verified by simulation.

1. Introduction

The polar regions have received widespread attention from countries worldwide due to their rich oil and gas reserves, strategic position, and advantages in shortening navigation distances [1]. Global warming has accelerated the melting of glaciers in the polar regions, which makes it feasible for ships to navigate in the polar regions. Nowadays, inertial navigation and satellite navigation are the most popular navigation systems used in polar regions. Still, due to the low distribution of satellites over the polar regions and the frequent influence of solar storms, magnetic storms, and ionospheric storms, satellite signals are prone to be interfered with or even work abnormally [2]. The INS does not need to receive external signals nor radiate signals to complete the navigation work, which is a completely autonomous navigation equipment and is also an important way for modern ships to realize navigation capability in the polar region [3]. The INS has a variety of mechanics choreography, wherein the north-pointing inertial navigation mechanization has a simple concept and clear physical meaning. However, when the north-pointing inertial navigation mechanical choreography passes the pole in high latitudes, the coordinate axis will rotate 180 degrees, and a singularity will appear in the control angular rate, causing the control angular rate to be infinite and there will also be a calculation overflow problem in the vertical component. The mechanization of the grid INS can solve the positioning and orientation problem in high latitude areas[4]. For ocean vehicle which adopt the INSs in undamped operating conditions works for a long time, the system will produce oscillation errors that accumulate over time. Therefore, the oscillation error must be attenuated [5]. In this paper, a representative external horizontally damped inertial guidance network is designed for Schuler period oscillation and Foucault period oscillation [6], which can not only solve the problem of navigation in high latitude areas, but also suppress the navigation error of Schuler period oscillation, and at the same time, the Foucault period oscillation error is eliminated [7].

2. Inertial navigation algorithms for grid coordinate systems

2.1. Coordinate system definition

i is a geocentric inertial coordinate system originating at Earth's center of mass. The z-axis aligns with Earth's polar axis, while the x-axis and y-axis lie in the equatorial plane, towards a fixed star.

e is an Earth-centered Earth-fixed (ECEF) coordinate system that remains fixed to the Earth. It rotates around the z-axis at the rate ω_{ie} of Earth's rotation. The z-axis aligns with the Earth's polar axis, while the x-axis is formed by the intersection of the Greenwich meridian plane with the Earth's equatorial plane. Additionally, the y-axis lies within the equatorial plane and forms the right-handed system with the x-axis and y-axis [8].

b is a body coordinate system and originates at the vehicle's center of mass. The x-axis is the right along the horizontal axis of the vehicle, the y-axis is the front along the vehicle, and the z-axis is the top along the vertical axis of the vehicle. xyz constitutes a right-handed and right-angled coordinate system.

t is a geographical coordinate system with the center of mass of the vehicle as the origin, the x-axis and y-axis in the geographic plane, the x-axis pointing east, the y-axis pointing north, and the z-axis points vertically upward, and the xyz forms a right-handed system.

The grid coordinate system G uses a plane parallel to the Greenwich plane at point p as the grid plane, and the local horizontal plane at the vehicle's location as the tangent plane. The intersection of the grid plane and the tangent plane defines the north direction of the grid. The north axis of the grid G_N is formed by the intersection of the grid plane and the meridian tangent plane. The grid up axis G_U aligns with the up axis of the geographic coordinate system, passing through the point p. The grid's east axis G_E lies in the local horizontal plane and is perpendicular to the gird north axis. The angle between G_N and the geographic north angle is referred to as the grid azimuth σ. The following geometric relationships are shown in Figure 1:

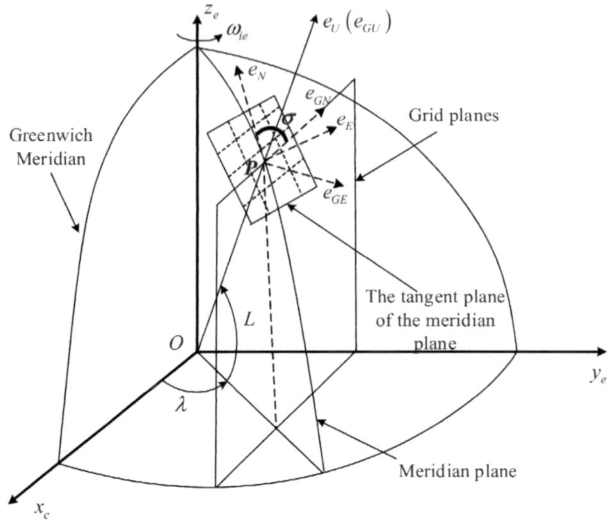

Figure 1. Grid Coordinate System.

(e_{ex}, e_{ey}, e_{ez}) is the set of unit vectors in each axis of the earth frame, and (e_{GE}, e_{GN}, e_{GU}) is the set of unit vectors in each axis of the grid coordinate system. The vectors e_{ey} and e_{GN} are perpendicular to each other and, therefore, satisfy the relation that the inner product is zero:

$$e_{ey} \cdot e_{GN} = 0 \tag{1}$$

where e_{ey} is the earth frame north direction and e_{GN} is the grid coordinate system north direction. The specific expressions for the vectors of e_{ey} and e_{GN} are as follows, respectively:

$$e_{ey} = -\sin\varphi\sin\lambda e_y + \cos\varphi\sin\lambda e_z + \cos\lambda e_x \tag{2}$$

$$e_{Gy} = \cos\sigma e_y - \sin\sigma e_x \tag{3}$$

where (e_x, e_y, e_z) is the set of unit vectors in each axis of the geographic coordinate system, φ and λ represent the latitude information and longitude information of the vehicle's location. The sine and cosine of the grid azimuth angle σ can be described as:

$$\sin\sigma = \frac{\sin\lambda\sin\varphi}{\sqrt{1-\cos^2\varphi\sin^2\lambda}} \tag{4}$$

$$\cos\sigma = \frac{\cos\lambda}{\sqrt{1-\cos^2\varphi\sin^2\lambda}} \tag{5}$$

It is known that there is a transformation matrix from the earth frame to the geographic coordinate system in the INS:

$$\boldsymbol{C}_e^t = \begin{bmatrix} -\sin\lambda & \cos\lambda & 0 \\ -\sin\varphi\cos\lambda & -\sin\varphi\sin\lambda & \cos\varphi \\ \cos\varphi\cos\lambda & \cos\varphi\sin\lambda & \sin\varphi \end{bmatrix} \tag{6}$$

According to the coordinate system rotation relationship, the following relationship exists between the geographic frame and the grid frame:

$$C_t^G = \begin{bmatrix} \cos\sigma & -\sin\sigma & 0 \\ \sin\sigma & \cos\sigma & 0 \\ 0 & 0 & 1 \end{bmatrix} \tag{7}$$

Therefore, the relation $\mathbf{C}_e^G$ between the Earth system and the grid system can be obtained.

2.2. Mechanization of the grid coordinate system

With the grid coordinate system G selected as the navigation frame, the attitude update differential equation can be expressed as:

$$\dot{C}_b^G = C_b^G \Omega_{Gb}^b \tag{8}$$

where:

$$\omega_{Gb}^b = \omega_{ib}^b - \omega_{iG}^b = \omega_{ib}^b - C_G^b \omega_{iG}^G \tag{9}$$

$$C_b^G = C_t^G C_b^t \tag{10}$$

where $\mathbf{C}_b^t$ is the attitude matrix of the conventional north-pointing INS, $\boldsymbol{\omega}_{ib}^b$ is the gyro output, $\mathbf{C}_b^G$ is the attitude matrix of the grid INS, and $\boldsymbol{\omega}_{Gb}^b$ is the projection of the rotation angular velocity of the vehicle coordinate system relative to the grid coordinate system on the vehicle coordinate system.

The control angular velocity of the grid INS, namely $\boldsymbol{\omega}_{iG}^G$, is expressed as:

$$\omega_{iG}^G = \omega_{ie}^G + \omega_{eG}^G \tag{11}$$

The specific expressions of $\boldsymbol{\omega}_{ie}^G$ and $\boldsymbol{\omega}_{eG}^G$ can be written as follows:

$$\omega_{ie}^G = C_t^G \omega_{ie}^t = \begin{bmatrix} -\Omega\cos\varphi\sin\sigma \\ \Omega\cos\varphi\cos\sigma \\ \Omega\sin\varphi \end{bmatrix} \tag{12}$$

$$\omega_{eG}^G = \begin{bmatrix} -\frac{V_{GN}}{R} & \frac{V_{GE}}{R} & -\frac{\sin\sigma\cos\varphi V_{GN}}{R\sin\varphi} \end{bmatrix}^T \tag{13}$$

The velocity differential equation of the grid INS is:

$$\dot{v}^G = C_b^G f^b - (2\omega_{ie}^G + \omega_{eG}^G) \times v^G + g^G \tag{14}$$

For ocean INS, the vertical channel is usually not considered, so the position differential equation is expressed as:

$$\dot{\varphi} = \frac{V_{GN}\cos\sigma - V_{GE}\sin\sigma}{R} \tag{15}$$

$$\dot{\lambda} = \frac{V_{GE}\cos\sigma + V_{GN}\sin\sigma}{R\cos\varphi} \tag{16}$$

where $\mathbf{V}^G = [V_{GE} \quad V_{GN} \quad V_{GU}]^T$ is the grid velocity. Unlike the conventional north-pointing INS, the grid system replaces the north-pointing coordinate system as the navigational coordinate system, and the existence of the angle σ between the geographic system and the grid system results in a different mechanization.

3. Grid system inertial navigation error analysis

To analyze the performance of the grid INS, it is necessary to explore and study the error equations of the grid INS and carry out the error analysis. The error equation of the grid INS can be obtained through the previous derivation [9].

For convenience, the velocity error equation is directly given as:

$$\begin{aligned} \delta\dot{V}_{Gx} &= 2\Omega\sin\varphi\delta V_{GN} - \phi_{GN} g + \delta f_{GE} \\ \delta\dot{V}_{Gy} &= -2\Omega\sin\varphi\delta V_{GE} + \phi_{GE} g + \delta f_{GN} \end{aligned} \tag{17}$$

where ϕ_{GE} and ϕ_{GN} represent the grid east and grid north misalignment angles, while δf_{GE} and δf_{GN} indicate the grid east and north accelerometer zero bias. Determining the accuracy value of the grid east and grid north misalignment angles can be challenging in polar regions. Therefore, to make the error equations better applied within the polar region, the rotation error angle $\delta\theta_G$ of the computational system relative to the ideal platform system is used as the position error when deriving the positional error equations.

$$\begin{aligned} \delta\dot{\theta}_{GE} &= -\frac{1}{R}\delta V_{GN} \\ \delta\dot{\theta}_{GN} &= \frac{1}{R}\delta V_{GE} \end{aligned} \tag{18}$$

The attitude error equation is expressed as:

$$\begin{bmatrix} \dot{\phi}_{GE} \\ \dot{\phi}_{GN} \\ \dot{\phi}_{GU} \end{bmatrix} = \boldsymbol{A}\begin{bmatrix} \phi_{GE} \\ \phi_{GN} \\ \phi_{GU} \end{bmatrix} + \boldsymbol{B}\begin{bmatrix} \delta V_{GE} \\ \delta V_{GN} \\ \delta\theta_{GE} \\ \delta\theta_{GN} \end{bmatrix} + \begin{bmatrix} \varepsilon_{GE} \\ \varepsilon_{GN} \\ \varepsilon_{GU} \end{bmatrix} \tag{19}$$

where:

$$\boldsymbol{A} = \begin{bmatrix} 0 & \Omega\sin\varphi & -\Omega\cos\varphi\cos\sigma \\ -\Omega\sin\varphi & 0 & -\Omega\cos\varphi\sin\sigma \\ \Omega\cos\varphi\cos\sigma & \Omega\cos\varphi\sin\sigma & 0 \end{bmatrix} \tag{20}$$

$$\boldsymbol{B} = \begin{bmatrix} 0 & -\frac{1}{R} & \frac{\Omega\sin\sigma\cos\sigma\cos^2\varphi}{\sin\varphi} & -\Omega\sin\varphi \\ \frac{1}{R} & 0 & \frac{\Omega(1-\cos^2\sigma\cos^2\varphi)}{\sin\varphi} & 0 \\ 0 & -\frac{\cot\varphi\sin\sigma}{R} & -\Omega\cos\sigma\cos\varphi & -\Omega\sin\sigma\cos\varphi \end{bmatrix} \tag{21}$$

ε_{GE}, ε_{GN}, and ε_{GU} are the gyro drift of the grid system. The velocity, position, and attitude errors are chosen as state variables and form a seven-dimensional matrix, where the characteristic equations are derived as:

$$\begin{aligned} \Delta(s) &= |sI - F| \\ &= s(s^2+\omega_s^2)\left[s^4 + 2s^2(\omega_s^2 + 2\omega_f^2) + \omega_s^4\right] \\ &\approx s(s^2+\omega_s^2)\left[s^2 + (\omega_s+\omega_f)^2\right]\left[s^2 + (\omega_s-\omega_f)^2\right] \end{aligned} \tag{22}$$

where ω_s represents the Schuler frequency and ω_f represents the Foucault frequency. Consequently, the grid INS exhibits both Schuler period oscillation and Foucault period oscillation. The damping technology is integrated to suppress these oscillations and enhance navigation accuracy.

4. The damping design of the grid inertial navigation system

The Schuler period oscillation error is mainly manifested in the horizontal channel, and when the Schuler period oscillation error is damped, it is called horizontal damping. Since the vehicle usually sails with acceleration, to overcome the error generated by the vehicle maneuvering, the external horizontally damped grid INS is commonly used in engineering applications [10].

4.1. External horizontal damping

External horizontal damping working state: The characteristic of external horizontal damping is introducing external speed information. When the speed of the carrier is constant, the introduced external information has no impact on the system. When the carrier maneuvers, the external speed information adds a forward channel to the horizontal loop and compensates. This can attenuate the Schuler oscillation error component and compensate for the carrier acceleration interference error caused by a horizontal damping network. Of course, the error of the external speed will also bring new errors to the system, so the external speed itself is required to be highly accurate and the error change to be relatively small.

Considering that, in reality, ships or aircraft and other equipment have external speed measurement equipment, such as the Electronic speed log [11], the Doppler velocity log, or the Doppler radar, etc., this paper uses a high-precision Doppler log as the external speed measurement device. External measured speed information can be introduced into the damped INS to form external speed compensation. For the damped INS, taking the external horizontal damping of the grid north channel as an example [12], a horizontal damping network is added after the first integrator and before the second integrator, which is shown as follows:

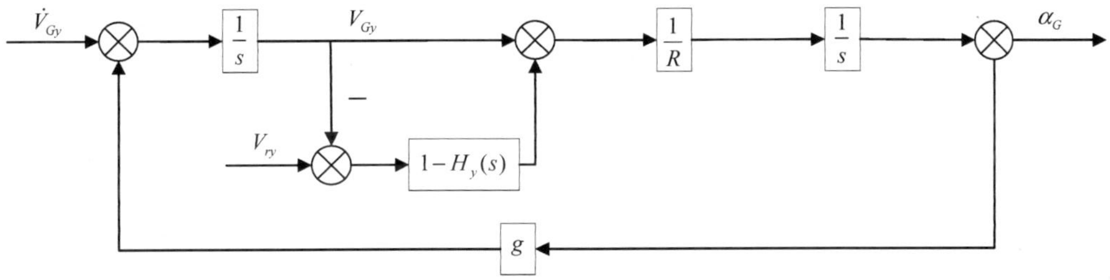

Figure 2. External horizontal damping block diagram of grid north channel.

As depicted in Figure 2, it is evident that the external speed is added to the system through the compensate channel, namely $1-H_y(s)$. When the value of the damping network $H_y(s)$ is chosen to be 1, it implies that the system is not damped at all, that is, the system is an undamped system. The system meets Schuler's adjustment conditions. When the value of the damping network $H_y(s)$ is not 1, the external velocity information is added to the system. If an appropriate damping network is selected, the effects of the external and velocity channels on the horizontal attitude error can be offset. This is the purpose of introducing external speed.

The parameter selection of the damping network generally needs to satisfy: The gain of the damping network is close to 0 at ω_s, and the damping coefficient is close to 0.5 so that the gain of $H(s)$ is 1 in the steady state, which ensures the stability of the system [13].

To unify the work of the mathematical model, some coefficients and parameters that are unified in the equations are given. The horizontal damping network is expressed as:

$$H_x(s) = \frac{A(s+\omega_1)(s+\omega_4)}{(s+\omega_2)(s+\omega_3)} \tag{23}$$

$$H_y(s) = \frac{B(s+\omega_5)(s+\omega_8)}{(s+\omega_6)(s+\omega_7)} \tag{24}$$

where A, B, ω_1, ω_2, ω_3,..., and ω_8 are the parameters in the damping network. Given the external speed state parameter K_1, the gyro drift compensation amount is U_x, U_y, and U_z. The platform motion equation can be obtained:

$$\left.\begin{aligned} \dot{\alpha} &= \omega_{cx} - \omega_x - \gamma\omega_y + \beta\omega_z + \varepsilon_x \quad \alpha(0) = \alpha_0 \\ \dot{\beta} &= \omega_{cy} - \omega_y - \alpha\omega_z + \gamma\omega_x + \varepsilon_y \quad \beta(0) = \beta_0 \\ \dot{\gamma} &= \omega_{cz} - \omega_z - \beta\omega_z + \alpha\omega_y + \varepsilon_z \quad \gamma(0) = \gamma_0 \end{aligned}\right\} \tag{25}$$

where ω_x, ω_y, and ω_z are the control angular velocity of the grid INS in grid east, grid north, and grid up, respectively, and their specific expressions are shown as follows:

$$\left.\begin{aligned} \omega_x &= -\Omega\cos\varphi\sin\sigma - \frac{V_{GN}}{R} \\ \omega_y &= \Omega\cos\varphi\cos\sigma + \frac{V_{GE}}{R} \\ \omega_z &= \Omega\sin\varphi - \frac{\sin\sigma\cos\varphi V_{GN}}{R\sin\varphi} \end{aligned}\right\} \tag{26}$$

The accelerometer output equation is given by:

$$\left.\begin{aligned} A_{px} &= A_x + \gamma A_y - \beta A_z + \Delta A_x \\ A_{py} &= A_y + \alpha A_z - \gamma A_x + \Delta A_y \end{aligned}\right\} \tag{27}$$

For convenience, the cross-coupling terms γA_y and γA_x are neglected; and the term $A_z = g$ is substituted into Equation (27). The simplified accelerometer output equation is obtained as follows:

$$\left.\begin{aligned} A_{px} &= A_x - \beta g + \Delta A_x \\ A_{py} &= A_y + \alpha g + \Delta A_y \end{aligned}\right\} \tag{28}$$

Then, the unified control equations of the computer are given by:

$$\left.\begin{aligned} \dot{V}_{cx} &= A_{px} + (2\Omega\sin\varphi_c + \frac{V_{cx}}{R_N} tg\varphi_c)V_{cy} \\ \dot{V}_{cy} &= A_{py} - (2\Omega\sin\varphi_c + \frac{V_{cx}}{R_N} tg\varphi_c)V_{cx} \end{aligned}\right\} \tag{29}$$

$$\left.\begin{aligned} \omega_{cx} &= -\frac{1}{R_M}\left[\left(V_{cy} - K_1V_{ry}\right)H_y + K_1V_{ry}\right] + U_x \\ \omega_{cy} &= \Omega\cos\varphi_c + \frac{1}{R_N}\left[\left(V_{cx} - K_1V_{rx}\right)H_x + K_1V_{rx}\right] \\ &\quad + DW\left(V_{cy} - K_1V_{ry}\right)H_y/R_M + U_y \\ \omega_{cz} &= \Omega\sin\varphi_c + \frac{1}{R_N}\left[\left(V_{cx} - K_1V_{rx}\right)H_x + K_1V_{rx}\right]tg\varphi_c \\ &\quad - EW\left(V_{cy} - K_1V_{ry}\right)H_y/R_M + U_z \end{aligned}\right\} \tag{30}$$

$$\left.\begin{array}{l}\dot{\varphi}_c=-\dfrac{1}{R_M}\left[\left(V_{cy}-K_1V_{ry}\right)H_y+K_1V_{ry}\right] \quad \varphi_c(0)=\varphi_0 \\ \dot{\lambda}_c=\dfrac{1}{R_N}\left[\left(V_{cx}-K_1V_{rx}\right)H_x+K_1V_{rx}\right]\sec\varphi_c \quad \lambda_c(0)=\lambda_0\end{array}\right\} \tag{31}$$

When the grid INS operates in the external horizontal damping working state, by assigning reasonable values to the damping parameters, setting the unified mathematical model parameters as $A=B=1$, $\omega_1=\omega_5=8.5\times10^{-4}$, $\omega_2=\omega_6=8.0\times10^{-3}$, $\omega_3=\omega_7=9.412\times10^{-2}$, $\omega_4=\omega_8=1.0\times10^{-2}$, and the external speed state parameter $K_1=1$, the horizontal damping network can be obtained as:

$$H_x(s)=H_y(s)=\frac{(s+8.5\times10^{-4})(s+1.0\times10^{-2})}{(s+8.0\times10^{-3})(s+9.412\times10^{-2})} \tag{32}$$

4.2. Simulation experiments

With an initial acceleration of zero, an initial velocity of zero in the east direction, an initial velocity of $2m/s$ in the north direction, an initial latitude of $85°$, a longitude of $120°$, a heading of $45°$, a gyro drift of $0.01°/h$, and an accelerometer zero bias of $1\times10^{-5}g$, the simulation time is 72 hours, and the simulation is shown below:

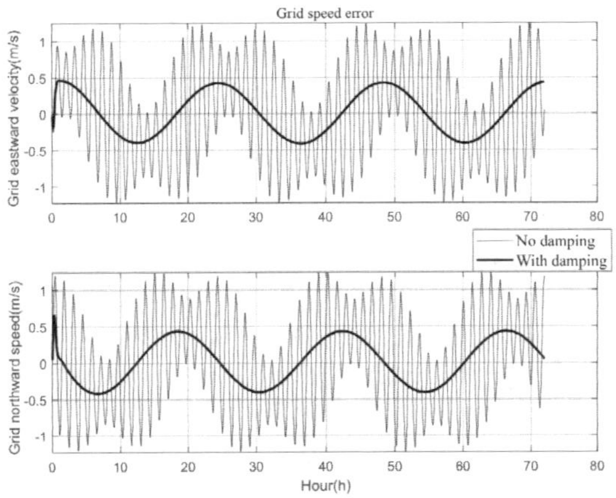

Figure 3. External horizontal damping velocity error.

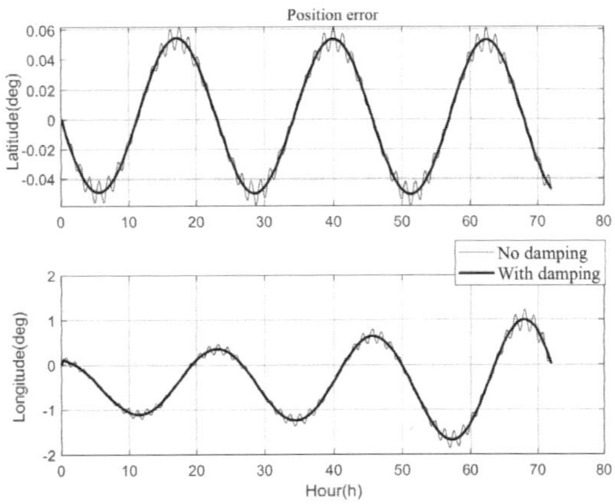

Figure 4. External horizontal damping position error.

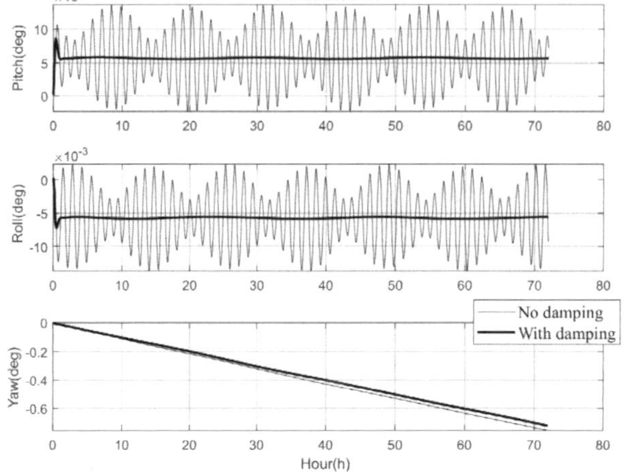

Figure 5. External horizontal damping attitude error.

From the above simulation results shown in Figures 3-5, it can be seen that after the incorporation of the external horizontal damping network, Schuler period oscillation and Foucault period oscillation in the velocity error, position error, and attitude error of the grid INS are well suppressed, which improve the accuracy of the grid INS. The simulation results prove the effectiveness of the external velocity level damping network designed in the grid inertial guidance system.

5. Conclusion

This paper uses navigation in high-latitude areas of the polar regions as the background to derive the mechanical arrangement based on the grid coordinate system and the error equation of the grid coordinate system. In addition, error analysis was conducted using classical control theory based on the error equation of the grid INS. The results show that there are Schuler and Foucault periodic oscillations with a period of 84.4 minutes in the grid INS. An external horizontal damping system based on the grid coordinate system was designed to restrain periodic oscillation errors in horizontal channels. The error curve of the INS in high latitude areas was obtained by simulation of the external horizontal damping system of the grid frame. The external horizontal damping system can suppress Schuler's periodic oscillation and Foucault's periodic oscillation without affecting the steady-state error of the system. The research in this paper has certain practical application value for polar navigation.

Acknowledgments

Natural Science Foundation of Heilongjiang Province of China under Grant YQ2021E011, in part by the National Natural Science Foundation of China under Grant 52371368.

References

[1] Zhao, Y.; Ben, Y. and Li, Q. Inertial navigation systems for polar marine navigation: a survey. Navig. Position. Timing 2020, 7, 1-10. [CrossRef]
[2] Ming, T. Review of polar integrated navigation algorithm. In Proceedings of the Journal of Physics: Conference Series, 2019; p. 032017. [CrossRef]
[3] Hongyang, H.; Bao, L.; Feng, Z. and Hongqiong, T. A fast intelligent algorithm-based positioning error suppression method for strap-down inertial navigation system. IEEE Sensors Journal 2023, 23, 5278-5287. [CrossRef]

[4] Ben, Y.; Cui, W. and Li, Q. An Improved Damping Method for Grid Inertial Navigation System in Polar Region. IEEE Transactions on Instrumentation and Measurement 2022, 71, 1-13. [CrossRef]
[5] Liu, C.; Wu, Q.; Hu, P. and Zhang, R. Design of variable damping INS for ships based on the variation of reference velocity error. Journal of Sensors 2021, 2021, 1-15. [CrossRef]
[6] Liu, Y.; Sun, Y.; Li, B.; Wang, X. and Yang, L. Experimental Analysis of Deep-Sea AUV Based on Multi-Sensor Integrated Navigation and Positioning. Remote Sensing 2024, 16, 199. [CrossRef]
[7] Yang, S.; Li, Q.; Jiang, S. and Liu, Y. A damping method for polar navigation based on grid mechanization. Navig. Positioning Timing 2016, 3, 27-32. [CrossRef]
[8] Ben, Y.; Zang, X.; Li, Q.; Liu, X. and Chen, H. System reset for underwater strapdown inertial navigation system. Ocean Engineering 2019, 182, 552-562. [CrossRef]
[9] Yang, W. and Feng, X. Research and implementation of underwater SINS/DVL integrated navigation and positioning algorithm. In Proceedings of the Fourth International Conference on Artificial Intelligence and Electromechanical Automation (AIEA 2023), 2023; pp. 893-901. [CrossRef]
[10] Kui, L.; Fang, L. and Yefeng, X. Research on internal damping algorithm of marine inertial navigation system. In Proceedings of the Intelligent Computing and Information Science: International Conference, ICICIS 2011, Chongqing, China, January 8-9, 2011. Proceedings, Part II, 2011; pp. 256-261. [CrossRef]
[11] Weiquan, H.; Yanling, H.; Jianhua, C. and Gang, L.; Jiangang, F.; Xiaobing, B. Research of the inertial navigation system with variable damping coefficients horizontal damping networks. In Proceedings of the Oceans' 04 MTS/IEEE Techno-Ocean'04 (IEEE Cat. No. 04CH37600), 2004; pp. 1272-1276. [CrossRef]
[12] Zhou, Z.; Liang, X.; Shi, Z.; Wang, G.; Lang, J. and Liu, L. A New Method to Restrain Schuler Periodic Oscillation in Inertial Navigation System Based on Normal Time–Frequency Transform. IEEE Transactions on Instrumentation and Measurement 2022, 71, 1-15. [CrossRef]
[13] Huang, R.; Wang, Q.; Xiang, Z.; Nie, X.; Zhou, J. and Luo, H. A water track laser Doppler velocimeter for use in underwater navigation. Measurement Science and Technology 2024, 35, 056301. [CrossRef]

Journal of Physics: Conference Series **2815** (2024) 012052 doi:10.1088/1742-6596/2815/1/012052

Optimization of milling parameters based on GA-BP neural network

Liqin Miao[1,2,*], Chaoneng Liao[2], Dashun Zhang[1], Huaidan Liang[2] and Di Gao[1]

[1]Department of numerical design&manufacture, Changchun Research Institute of Equipment and Technology, Changchun, Jilin, 130012, China
[2]College of Mechanical and Electrical Engineering, Changchun University of Science and Technology, Changchun, Jilin, 130012, China

[*]Corresponding author's e-mail: miaolq_1980@163.com

Abstract. Since the selection of machining parameters usually relies on the experience of workers or traditional calculation formulas, the traditional prediction method has the disadvantages of a complicated arithmetic process, large prediction deviation, and high consumption cost, making it challenging to meet the increasing demand of production and processing. Therefore, this paper proposes a machining quality prediction model based on the GA-BP neural network. Through experiments, it verifies the data-fitting ability of the prediction model and then takes the prediction model as the optimization objective, culminating in a multi-objective optimization model for process parameters based on the NSGA-II algorithm. Experiments demonstrate that the cutting force and surface roughness obtained by the optimization model are 3.6% and 10.0% lower than those obtained by the empirical parameters, respectively, leading to reductions of 3.6% and 10.6%, which verified the optimization effect of the model.

1. Introduction

The rational selection of cutting parameters is crucial for the guidance of production processing. In recent years, many researchers have made unremitting efforts in the field of cutting parameter optimization. Khan et al. [1] proposed a multi-objective milling parameter optimization model that takes into account the improvement of energy efficiency, surface quality, and material removal rate, and they solved the model by using the NSGA-II algorithm, grey correlation analysis, and Taguchi's method. Their results showed that the proposed algorithm has a certain degree of reliability. Zhao et al. [2] established a mathematical model for the multi-objective optimal design of cylindrical gear transmission with maximum overlap, minimum volume, and the same bending strength as the optimization objects and solved the model by NSGA-II and obtained the Pareto optimal solution. Wang et al. [3] took the minimum cost, energy consumption, and surface roughness as the optimization objects and solved the model by NSGA-II algorithm, resulting in the optimal solution. Zhang et al. [4] optimized the machining parameters during the laser cladding process with a multi-objective optimization algorithm with the optimization objectives of reducing the surface roughness and improving the hardness of the workpiece. By comparing the optimization effects of several algorithms, it was concluded that NSGA-II can seek the optimal parameter set in a better and faster way.

Wang et al. [5] proposed a multi-objective parameter optimization model of face milling with the objectives of minimizing cost and energy consumption, and they obtained the optimal solution by using

Journal of Physics: Conference Series **2815** (2024) 012052 doi:10.1088/1742-6596/2815/1/012052

the NSGA-II algorithm. Multi-objective parameter optimization model and solved the optimization model by evolutionary strategy based optimization method. Leo Kumar et al. [6] analyzed the effect of cutting conditions on the optimization objective in micro-cutting machining by taking surface roughness and machining efficiency as the optimization objective and solved the optimization by GA.

In this study, a processing quality prediction model based on GA-BP neural network is proposed. The data estimation ability and prediction accuracy of the prediction model and empirical formulas are compared on the basis of simulation experimental data, and the results show that the prediction model is obviously superior to the empirical formulas. Then, a multi-objective optimization model for process parameters based on the NSGA-II algorithm is proposed using the prediction model as an objective function, and the optimization effect of the model is validated to be better than that of the empirical parameter values through experiments.

2. Establishment of processing quality prediction model based on GA-BP neural network

2.1 Introduction to neural networks

Back Propagation neural network is one of the most widely used artificial neural networks[7]. Combined with the practical problems of this study, a three-layer BP neural network structure combining cutting parameters with cutting force and cutting temperature is constructed, as shown in Figure. 1.

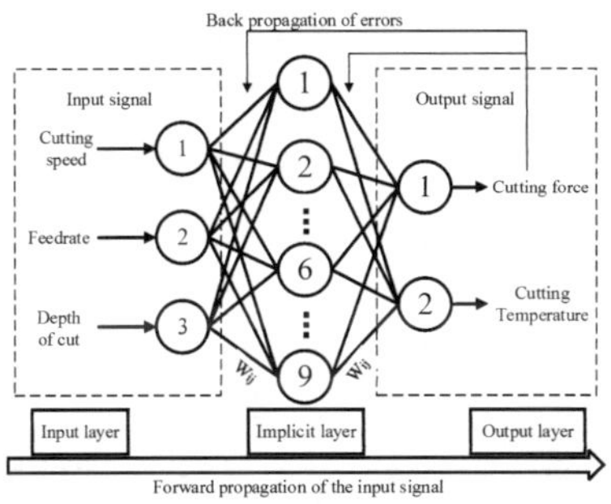

Figure 1. Structure of three-layer BP neural network.

2.2 Construction of neural networks

1) Construct the topology

The input layer is three neurons involving three factors: milling speed v_c, feed per tooth f_z and back draft a_p. The output layer has two neurons: cutting force and surface roughness. The number of nodes in the hidden layer is calculated to be nine by using the empirical formula (1) and the repeated experiment method[8], which are shown in the following equations, respectively.

$$N_{hid} = 2N_{in} + 1,\ N_{hid} = \sqrt{N_{in}\left(N_{out}+1\right)+1}, N_{hid} < \sqrt{N_{in}+N_{out}} + random\left(1\sim 10\right) \quad (1)$$

where: N_{in} - number of input layer neurons; N_{hid} - number of hidden layer neurons;

N_{out} - number of neurons in the output layer;

In summary, a 3×9×2 BP neural network structure is established in this study.

2) Define the activation function

Currently commonly used neural network activation functions include Sigmoid, Tanh and ReLU functions[8], and their specific expressions, respectively, as shown in equation (2).

$$f(x) = \frac{1}{1+e^{-x}},\ f(x) = \frac{e^{x}-e^{-x}}{e^{x}+e^{-x}},\ f(x) = \begin{cases} x, & x \ge 0 \\ 0, & x < 0 \end{cases} \quad (2)$$

The prediction error is minimized when the Tanh function is used in both hidden and output layers. Therefore, after comprehensive consideration, the Tanh function is chosen as the activation function of the neural network in this study.

3) Determine the training function and initial parameters

Currently, the commonly used training functions for neural networks are Bayesian regularization, Levenberg-Marquardt (LM)[9], and Quantized Conjugate Gradient[10], and their counterparts are trainbr, trainlm, and trainscg. Combined with the specifics of this study, it was decided to select the trainlm as the training function.

4)Construct the BP neural network structure

After completing the above topology, activation function, and specific parameter settings, the structure of the neural network is built by programming in MATLAB software, as shown in equation (3).

$$net = newff\left(W, S, P, PF, BFT, BLF, DDF, WF, OWF, IWF\right) \tag{3}$$

where: W - matrix of input data; S - network implicit layer; P - matrix of mapped value data;

BFT - training function; PF - activation function, here tansig function is used;

The rest of the parameters are set as per default.

2.3 GA optimized neural network

Genetic algorithm belongs to a kind of bionic algorithm, which originates from the imitation of the way of biological genetic information transmission, and was proposed by John Holland in the 1960s[11]. As the traditional BP neural network in dealing with the problem, there are shortcomings such as slow convergence speed, training with randomness, and susceptibility to falling into the local optimal solution, which can easily affect the accuracy of the solution results.

So in this paper, GA is chosen to optimize the neural network in order to improve the estimation ability of the prediction model to the data and the prediction accuracy. The GA-BP optimization process is shown in Figure. 2.

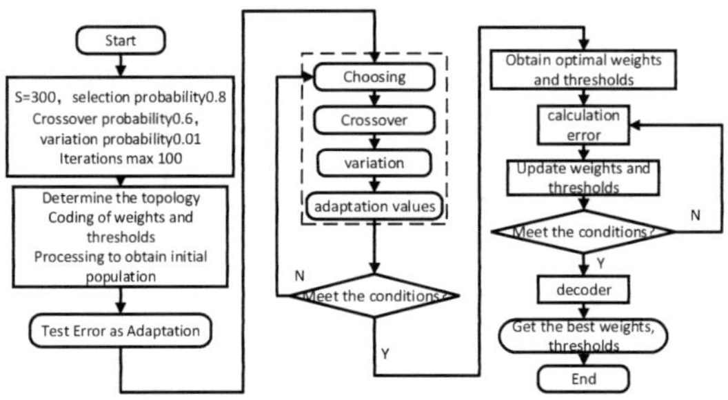

Figure 2. GA-BP flow chart.

2.4 Utility validation of predictive models

Based on the simulation experimental data, the relative errors of the prediction model and the empirical formula about the experimental data were found, respectively, which judged the fitting effect of the prediction model and the empirical formula. The relative error comparison is shown in Figure. 3.

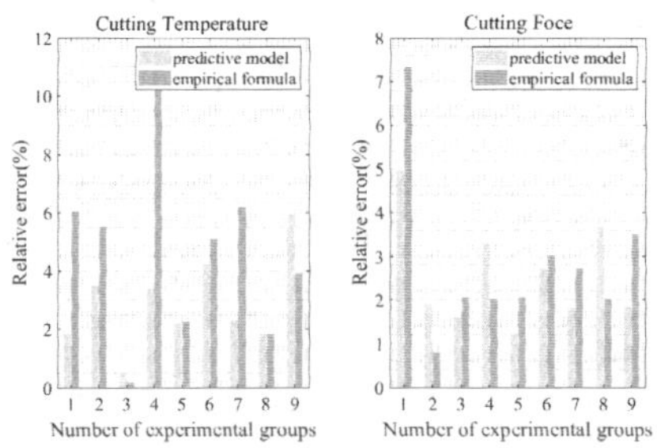

Figure 3. Comparison of the errors of the two methods.

For the relative error of cutting temperature, the prediction model errors in the remaining experimental groups are less than or equal to the empirical formulas, except for the data in Groups 3 and 9.

For the relative error of cutting force, except for the 2nd, 4th, and 8th data groups, the prediction model error is greater than the empirical formula, and the prediction model error in the remaining experimental groups is less than the empirical formula.

It is concluded that the predictive model fits the data better than the empirical formula.

Finally, based on the latter three groups of data, after the comparison experiments, the final relative error comparisons were obtained, as shown in Tables 1 and 2.

Table 1. Comparison of relative error of cutting force.

Number	Actual value	BP projected value	BP error	Formulas projected value	formula error
Test 25	149.7648	158.2987	5.70%	158.3128	5.71%
Test 26	203.3780	208.3997	2.47%	207.3248	1.94%
Test 27	276.0729	268.1066	-2.89%	258.4387	-6.39%

Table 2. Comparison of Relative Error of Cutting Temperature.

Number	Actual value	BP projected value	BP error	Formulas projected value	formula error
Test 25	385.1350	385.6668	0.14%	399.3898	3.70%
Test 26	434.5771	420.5241	3.23%	436.9018	5.3%
Test 27	455.6562	462.0646	1.41%	470.1522	3.18%

The final obtained error comparison plots of the prediction model and the empirical formulae regarding the actual values, respectively, are shown in Figure. 4.

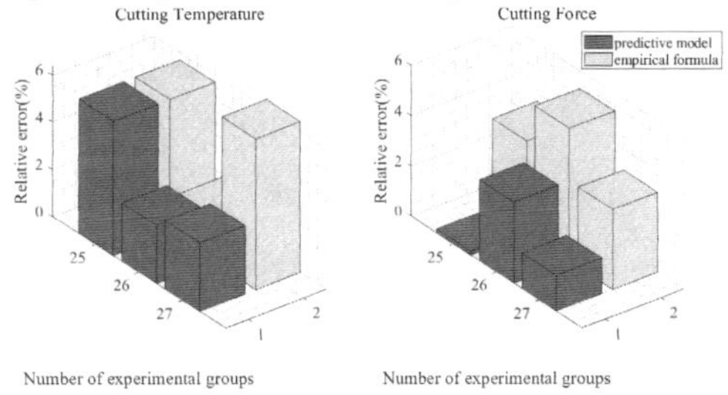

Figure 4. Comparison of the error validation of the two methods.

In the case of cutting temperature comparison, the prediction errors of the remaining experimental data are less than the empirical formulas except for data set 26.

In the case of cutting force comparison, the prediction errors of all data are smaller than the empirical formula.

This shows that the prediction model has better estimation ability and prediction accuracy of the data than the empirical formula.

3. Establishment of a multi-objective optimization model based on NSGA-II

3.1 Principles of the NSGA-I algorithm

The second generation non-dominated sorted genetic algorithm is one of the classical evolutionary algorithms, which is an improved version of the traditional genetic algorithm[12], finding extensive utilizations in solving multi-objective optimization algorithms problems in engineering due to the advantages of this algorithm, such as good flexibility, fast convergence, and so on. The flowchart of the NSGA-II optimization algorithm is shown below Figure. 5.

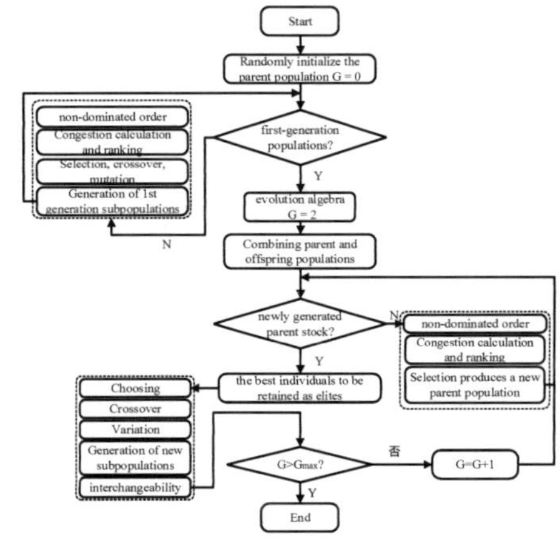

Figure 5. NSGA-II optimization algorithm process.

3.2 Initial parameter setting

1) Population size and the maximum number of evolutionary generations

Through the experiment, the population size was set at 200, and the maximum number of evolutionary generations was set to 80 after comprehensive consideration.

2) Selection of crossover and mutation operators

Considering the characteristics of the optimization problem in this study and the complexity of the search space and other factors, the crossover frequency was set to 0.5, and the variation operator was set to 0.05.

3.3 Determination of objective function

The processing quality prediction model in the above paper is used as the objective function, and considering the different degrees of importance of the optimization objectives, the objective function of NSGA-II is established here through the weight coefficients. So, the final objective function is shown in equation (4).

$$F\left(X\right)=W_1\cdot sim(net_1,X)+W_2\cdot sim\left(net_2,X\right) \tag{4}$$

where: W - weighting coefficients and $W_1+W_2=1$; $Sim()$ - function for neural network to call; X -The cutting parameter matrix to be optimized;

The specific expression of net_i, as shown in equation (3), represents the GA-BP machining quality prediction model.

3.4 Setting of constraints

1) The constraint range of the input parameters, as shown in equation (5).

$$X_{min}<X_i<X_{max} \tag{5}$$

Where X_{min} denotes the lower limit of the cutting parameter and X_{max} denotes the upper limit of the cutting parameter.

2) Cutting speed constraints and constraints on the feed amount and depth of cut constraint

Considering the range of machining capacity of the machine tool, the value of cutting speed here is taken in the range of $\mathrm{v}\in\frac{[60,150]\mathrm{m}}{\mathrm{min}}$, the value of the feed here is taken in the range of $f_z\in[0.08,0.16]mm/r$, and the depth of cut here takes the value range of $a_p\in[0.4,0.8]mm$.

3.5 Utility validation of multi-objective optimization models

Based on the simulation experimental data, the minimum cutting force and minimum surface roughness are taken as the optimization objectives, which are solved iteratively by the multi-objective optimization algorithm constructed above, and the experimental data are shown in Table 3.

Table 3. Experimental data table.

Number	v (m/min)	f_z (mm/r)	a_p (mm)	F (N)	R_a (μm)
Test 1	30	0.1	0.5	144.3575	3.4437
Test 2	30	0.15	0.5	170.0997	3.0374
Test 3	30	0.2	0.5	200.4913	2.5135
Test 4	40	0.1	0.4	118.7513	3.5881
Test 5	40	0.15	0.4	142.3147	3.4828
Test 6	40	0.2	0.4	153.3793	3.2369
Test 7	50	0.1	0.6	185.9356	3.1997
Test 8	50	0.15	0.6	235.1828	2.7315
Test 9	50	0.2	0.6	276.0729	2.1961

The corresponding Pareto optimization iteration process of the optimization algorithm obtained after the iterative optimization search is shown in Figure 6.

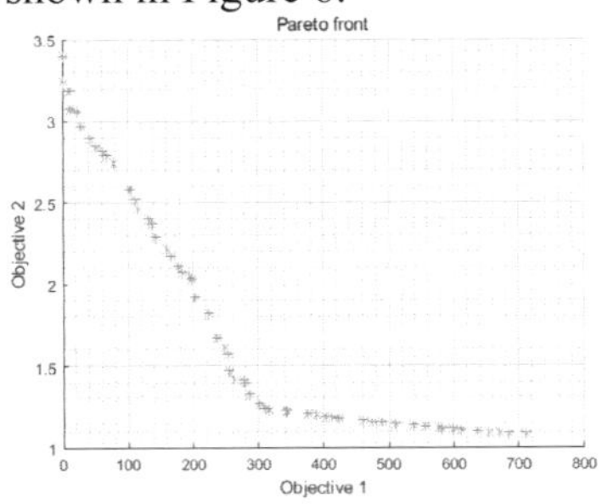

Figure 6. Diagram of Pareto iterative process.

After an iterative optimization search through the NSGA-II optimization algorithm, the final results of 60 groups of optimized process parameters are obtained. Combined with the constraints of data screening $R_a < 2.0\mu m$, the optimal five groups of data are obtained, the information of which is shown in Table 4.

Table 4. Table of optimal five sets of parameters.

Number	v (m/min)	f_z (mm/r)	a_p (mm)	F (N)	R_a (μm)
Test 1	97.1134	0.1037	0.5063	126.2100	1.4204
Test 2	85.2234	0.1048	0.4059	198.5694	1.6548
Test 3	77.0662	0.1021	0.3000	205.3982	1.5062
Test 4	104.6564	0.1015	0.3164	301.0415	1.5680
Test 5	91.4218	0.1048	0.1883	217.5160	1.3106
Average	91.0962	0.1034	0.2634	229.1300	1.4920

The average values of cutting force and surface roughness for these five sets of data were found and then compared with the data obtained from the empirical machining parameters, and the comparison is shown in Table 5.

Table 5. Comparison results of optimization results with empirical parameters.

	v (m/min)	f_z (mm/r)	a_p (mm)	F (N)	R_a (μm)
Empirical parameter	75	0.11	0.60	221.1680	1.3490
Optimization results	91.0962	0.1034	0.2634	229.1300	1.4920
Enhancement effect				3.6%	10.6%

The process parameters optimized by the algorithm showed a reduction of 3.6% in cutting force and 10.6% in surface roughness compared to the empirical parameters.

4. Conclusions

1) The relative error between the prediction model and the actual value ranges from a minimum of 2.47% to a maximum of 5.7%, and the relative error between the empirical formula and the actual value ranges from a minimum of 1.94% to a maximum of 6.39%. For the cutting temperature, the relative error between the prediction model and the actual value ranges from a minimum of 0.14% to a maximum of 3.23%, while the relative error between the empirical formula and the actual value ranges from a minimum of 3.18% to a maximum of 5.3%. It is concluded that the predictive model generally fits the data better than the empirical formula.

2) With the machining quality prediction model as the optimization objective, a multi-objective optimization model for process parameters based on the NSGA-II algorithm is proposed, and it is concluded that the cutting force and surface roughness obtained from the model optimization have been reduced by 3.6% and 10.6%, respectively, compared with the empirical machining parameters, which indicates that the multi-objective process parameter optimization model is reliable.

References

[1] Khan A M, Jamil M, Salonitis K, et al. Multi-objective optimization of energy consumption and surface quality in nanofluid SQCL assisted face milling[J]. Energies, 2019, 12(4): 710.

[2] ZHAO Ning and YANG Jie. Multi-objective optimization design of high overlap cylindrical gear transmission[J]. Mechanical Drives, 2012, 36(07): 43-46.

[3] Wang Q, Liu F and Wang X. Multi-objective Optimization of Machining parameters considering energy consumption[J]. The International Journal of Advanced Manufacturing Technology, 2014, 71(5-8): 1133-1142.

[4] ZHANG Shengjiang, WANG Mingdi, NI Chao, et al. Multi-objective optimization of laser cladding process parameters on QT800-2 ductile iron surface[J]. Surface technology, 2021, 50(07): 74-82.

[5] Wang Y C, Kim D W, Katayama H, et al. Optimization of machining economics and energy consumption in face milling operations[J]. The International Journal of Advanced Manufacturing Technology, 2018, 99(9-12): 2093-2100.

[6] Leo Kumar S P. Measurement and Uncertainty Analysis of Surface Roughness and Material Removal Rate in Micro Turning Operation and Process Parameters Optimization[J]. Measurement, 2019, 140: 538-547.

[7] Liang Shuting. Research on the theory and technology of high-precision molding of polyetheretherketone denture chair-side[D]. Jilin: Jilin University, 2023.

[8] Kuri-Morales Angel. Closed determination of the number of neurons in the hidden layer of a multi-layered perceptron network[J]. Soft Computing, 2017, 21(3): 597-609.

[9] SL Shen, N Zhang, A Zhou, et al. Enhancement of neural networks with an alternative activation function tanhLU[J]. Expert Systems With Applications, 2022, Vol.199: 117181.

[10] Zhou H, Zhang YL, Zhang DH, et al. An anisotropic magnetostriction model based on BP neural network combining Levenberg-Marquardt algorithm and particle swarm optimization[J]. International Journal of Applied Electromagnetics and Mechanics, 2017, 55, 193-S20.

[11] Liu Yuan and Pang Baojun. Damage pattern recognition based on Bayesian regularized BP neural network for hypervelocity impact damage of aluminum flat plate[J]. Vibration and Shock, 2016, 35(12): 22-27.

[12] Hao-Kai Zhu. Research on multi-objective feature selection method based on improved NSGA-II[D]. Jiangsu: Nanjing University of Information Engineering, 2022.

Comprehensive simulation analysis of wave-induced dynamics and mesh adaptivity on submarine-launched missiles

Yue Hong [1], Xiaoqun Ju [1], Haoyu Sun [2], Changli Yu [1] and Xingming Zhang [3,*]

[1] School of Ocean Engineering, Harbin Institute of Technology at Weihai, Weihai, Shandong, 264200, China
[2] Department of Electrical and Electronic Engineering, University of Manchester, Manchester, M139PL, United Kingdom
[3] Shandong Institute of Shipbuilding Technology, Weihai, Shandong, 264200, China
Xiaoqun Ju and Yue Hong are the first authors of this paper.

*Corresponding author's e-mail: zhxm@hit.edu.cn

Abstract: Using the ANSYS simulation platform, we constructed a model of the underwater launching environment and submarine-launched vehicle (SLV) motion based on its launching environment and motion characteristics. We conducted numerical simulations using dynamic mesh technology to study the effects of waves on the attitude and pressure distribution of SLV during the launch process and to analyze the changes in its displacement, angle of rotation, force, as well as the perturbation rules of waves on SLV's attitude during motion. This will provide a theoretical basis for the in-depth study of the impacts of underwater environmental factors on SLV launching.

1. Introduction

Submarine-launched weapons generally rely on underwater vehicles as launch platforms, characterized by good concealment, a wide range of mobility, and high survivability [1]. When a submarine-launched vehicle (SLV) is out of the water and exposed to waves and other marine environments, the fluid force it is subjected to will change abruptly, affecting its attitude. Since waves are the most important form of disturbance in surface seawater, their impact has an important influence on the attitude disturbance of SLV, so studying the impact of waves on SLV out of the water is an important research content in underwater launching.

The wave disturbance on SLV belongs to the fluid-solid coupling problem, and the mathematical models for solving it mainly include the Navier-Stokes (N-S) model and the ideal potential flow model. For the numerical simulation of underwater launching, domestic and foreign scholars have carried out relevant investigations. Li et al. [2] used the mixture multiphase flow model to solve the hydrodynamic force of the horizontal launching process. Lindau et al. [3] established an implicit preconditioned algorithm that was applied to the multiphase flow problem. Weiland et al. [4] studied the effects of launch water depth and horizontal incoming currents on the air curtain launch of SLV. Shen [5] used FLUENT software, dynamic grid technology, and a multiphase flow model to perform axisymmetric and three-dimensional calculations of SLV underwater and out-of-water process, analyzing its dynamic characteristics and the change rule of the surrounding flow field. Bai [6] took the SLV underwater and out-of-water process as the research object and analyzed the influence of boat speed, waves, currents, sea wind, etc., on SLV underwater and out-of-water ballistic. This paper uses

Journal of Physics: Conference Series **2815** (2024) 012053 doi:10.1088/1742-6596/2815/1/012053

ANSYS software to realize the motion of SLV based on the six-degrees-of-freedom motion model and dynamic mesh technology [8]. In contrast, the VOF model and the 5th-order Stokes wave theory are used to build waves [9] to study the motion response of the SLV out of water under the waves [10] and to analyze the motion rule of this SLV to provide references for the research of launching of related types of SLVs [11].

This study explores the initial disturbances caused by wave impacts during missile surfacing. It delves into the adaptive mesh refinement techniques for capturing the complex fluid-structure interactions during this transient phase. The innovative aspects of this research include using dynamic meshing to accurately model the changing interface between air and water and applying the Volume of Fluid (VOF) method for improved phase distinction. Our main research content involves a detailed parametric study of mesh adaptivity effects on the missile's exit dynamics from water, presenting a comprehensive plan to explore the influences of environmental factors like wave pattern variability and missile speed on launch dynamics. This extensive simulation work aims to optimize SLV design for enhanced stability and performance in marine launch scenarios.

2. Theory and methodology

This paper uses a VOF (Volume of Fluids) multiphase flow model to simulate the phase interactions during the submerged ejecta discharge process. The phase distribution at the interface and the location of the interface are described by the field of phase volume fraction α_i. The volume fraction of the phase i is defined as shown in Equation (1), and the sum of the volume fractions of all the phases in the grid cell must be 1.

$$\alpha_i = \frac{V_i}{V} \tag{1}$$

$$\sum_{i=1}^{N} \alpha_i = 1 \tag{2}$$

The volume fraction transfer equation for the VOF model is as follows:

$$\frac{\partial}{\partial_t}\int \alpha_i dV + \oint \alpha_i v \cdot da = \int (S_{\alpha_i} - \frac{\alpha_i}{\rho_i}\frac{D_{\rho_i}}{D_t})dV - \int \frac{1}{\rho_i}\nabla \cdot (\alpha_i \rho_i v_{d_{r,i}})dV + \int \frac{\mu_t}{\sigma_t \rho}\nabla_{\alpha_i} \cdot da \tag{3}$$

In Equation (3), a is the surface area vector. $v_{d_{r,i}}$ is the diffusion velocity. S_{α_i} is the source term of phase i. μ_t is the turbulent kinetic viscosity. σ_t is the turbulent Schmidt number. $\frac{D_{\rho_i}}{D_t}$ is the material or Lagrangian derivative of the phase density, ρ_i.

3. Motion modeling and mesh segmentation

According to the theoretical research method used by Zhong et al. [7], the bare projectile is used as the basis for equal scale reduction. A cube area of $20\text{m} \times 10\text{m} \times 5\text{m}$ is established to simulate the water around the projectile exit position. The water depth is 5m, while the SLV is placed 1m underwater. The wave is set to 1 meter in height, 7 meters in wavelength, and 2.1 seconds in period. To analyze the effect of wave force on the missile out of the water, the model is set as a fixed, rigid body for the first 9 seconds without considering its free motion.

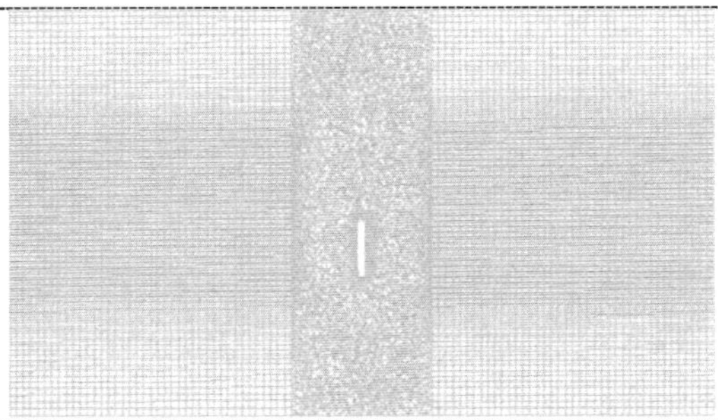

Figure 1. Mesh profile of the computational area.

As Figure 1 shown, only a strip mesh is refined around the submarine and the free surface to reduce the computational volume. The motion process involves the use of dynamic grid technology. In this model, the speed of the launching platform is ignored. The submarine is pushed out of the water by the fixed thrust in the stationary state [13], so the tet meshing area is appropriately reduced in the horizontal direction to reduce the number of meshes. The structured grid is utilized on meshing regions and refined in the middle region to ensure a good waveform [14]. The minimum element size in the computational domain is $3.27\times10^{-7}\,m^3$, the maximum element size is $7.44\times10^{-3}\,m^3$, and the total number of meshes is 630924.

Figure 2. Calculation of the initial phase diagram of the region.

The flow field is bounded by a velocity inlet on the left [12], a free outflow on the right, and wall surfaces on the top and bottom, as depicted in Figure 2. The front and back boundaries perpendicular to the Z axis are symmetry surfaces. The SLV is treated as a wall boundary. It has a mass of 35kg, a thrust of 400N, and an initial velocity of zero.

4. Equations of motion of a projectile

The six-degree-of-freedom equation of motion for the projectile is:

$$\frac{\partial G}{\partial t}+\overline{\omega}\times G=F \tag{4}$$

$$\frac{\partial L}{\partial t}+\overline{\omega}\times L+v\times G=M \tag{5}$$

In Equations (4) and (5), G is the total momentum of the projectile and the surrounding fluid. L is the total momentum moment relative to the origin of the projectile system. The angular velocity of the projectile is $\overline{\omega}=\left(\overline{\omega_x},\overline{\omega_y},\overline{\omega_z}\right)$, and the translational velocity of the projectile is $v=\left(v_x,v_y,v_z\right)$. F and M are the external force and moment of the projectile, respectively. G and L can be expressed as the dot product of the inertia tensor and velocity vector of the system:

Journal of Physics: Conference Series **2815** (2024) 012053 doi:10.1088/1742-6596/2815/1/012053

$$(G,L)^T = (A+J)\cdot U \tag{6}$$

In Equation (6), the transposed column vector of the six-dimensional velocity vector is $U=(v_x,v_y,v_z)^T$. A is the matrix of additional fluid mass coefficients, and J is the matrix of inertia coefficients of the projectile.

5. Submarine out-of-water simulation results and analysis

5.1. Full flow field analysis of the out-of-water process

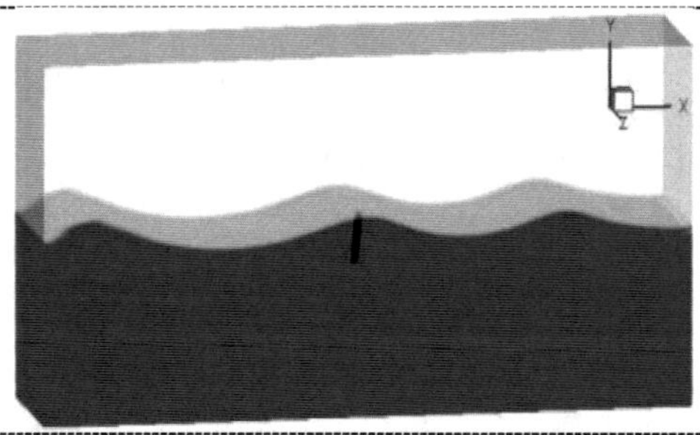

(a) t=0.354 s

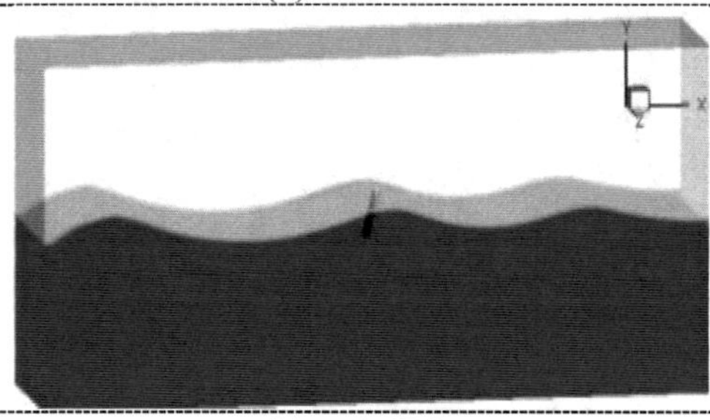

(b) t=0.504 s

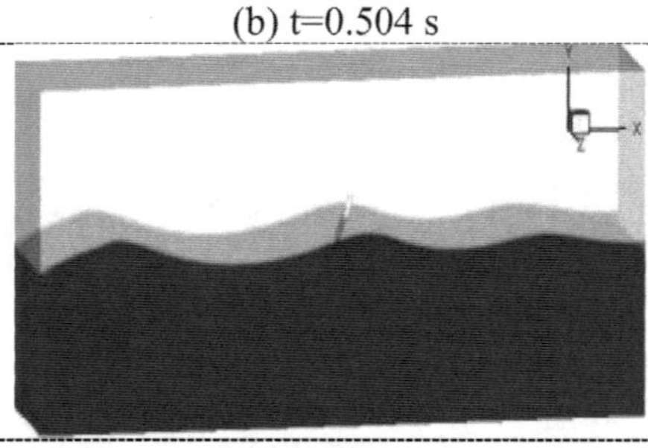

(c) t=0.634 s

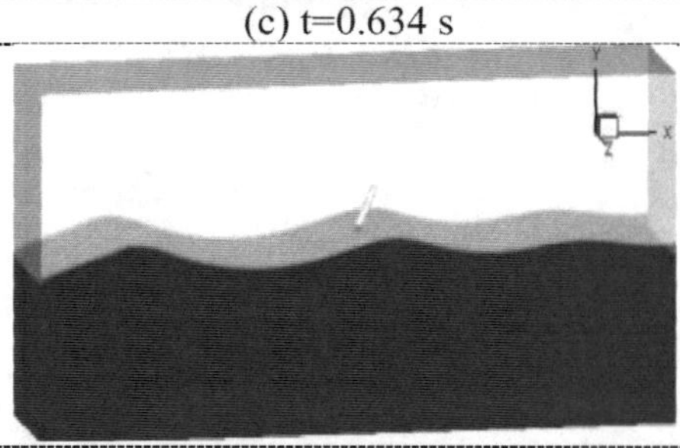

(d) t=0.734 s

Figure 3. Missile out of the water process.

Journal of Physics: Conference Series **2815** (2024) 012053 doi:10.1088/1742-6596/2815/1/012053

Figure 3 displays the time diagram of the submarine-launched vehicle from launch to surfacing. It demonstrates that during the surfacing process, wave force causes a certain angle of deflection. The deflection angle and direction at the moment of emergence are closely related to the phase of the waves. As wave motion is a periodic process, the waveforms of different periodic moments have varying effects on the SLV. The second-order STOKES wave model indicates a clear tendency for motion in the positive direction of the X-axis because of the presence of a drift velocity.

5.2. Projectile pressure analysis

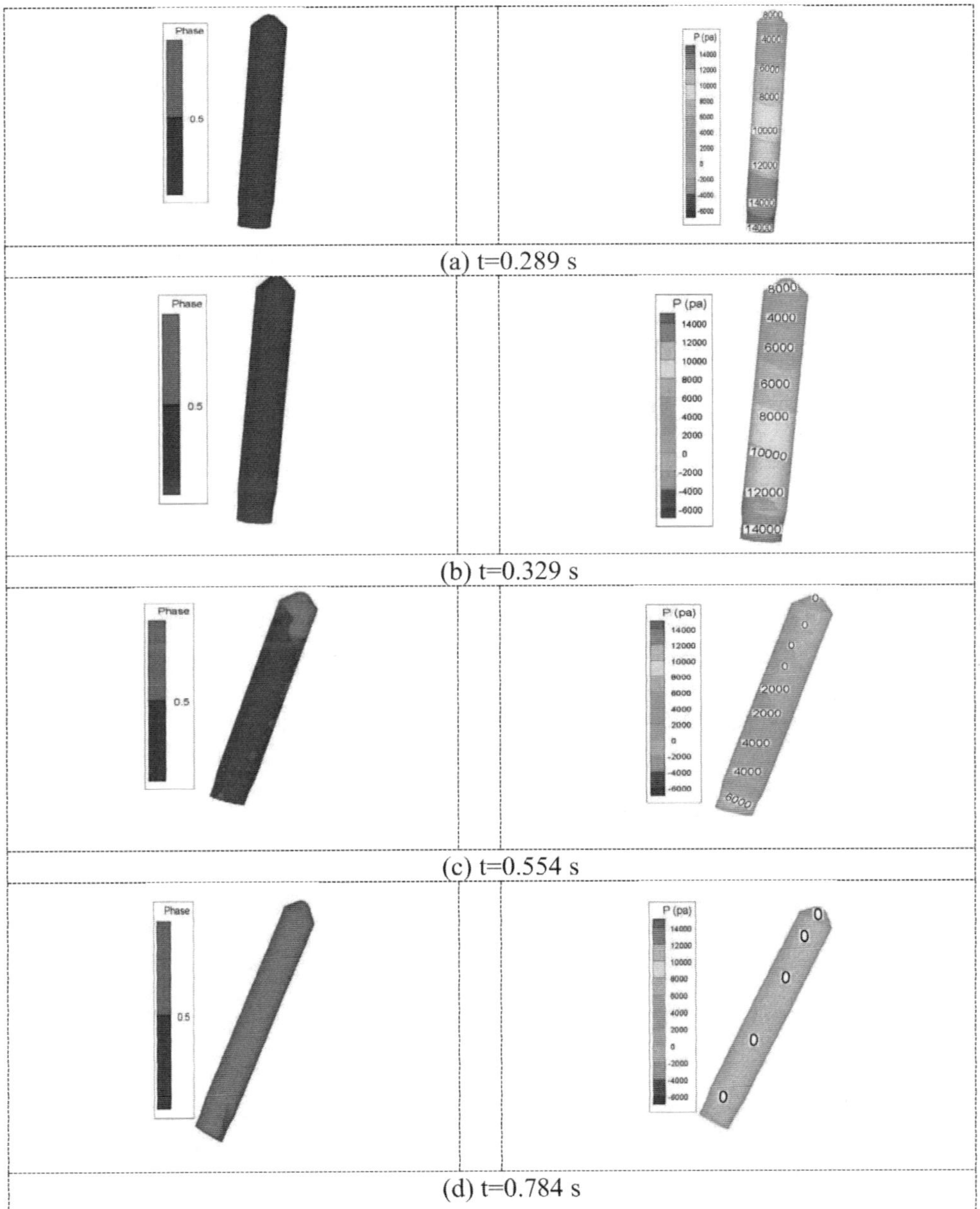

(a) t=0.289 s

(b) t=0.329 s

(c) t=0.554 s

(d) t=0.784 s

Figure 4. Missile's phase and pressure diagram.

Figure 4 displays phase and pressure diagrams for the SLV at four different moments during its emergence from the water. It is noted that the pressures are relative pressures based on atmospheric pressure. In Figure 4(a), the submarine is completely submerged. Its pressure increases from top to bottom along the body, with the high-pressure region at the head mainly due to the drag pressure caused by the relative motion of the fluid as it moves. As the submarine moves forward, the pressure in its front region gradually decreases [15]. Figure 4(c) indicates that when the submarine is out of the water, the pressure in the upper part is reduced to atmospheric pressure. Due to the small air resistance, the pressure change in the head is not obvious. When it is completely out of the water, its surface pressure equals the atmospheric pressure, as shown in Figure 4(d).

5.3. Analysis of submarine kinematic parameters

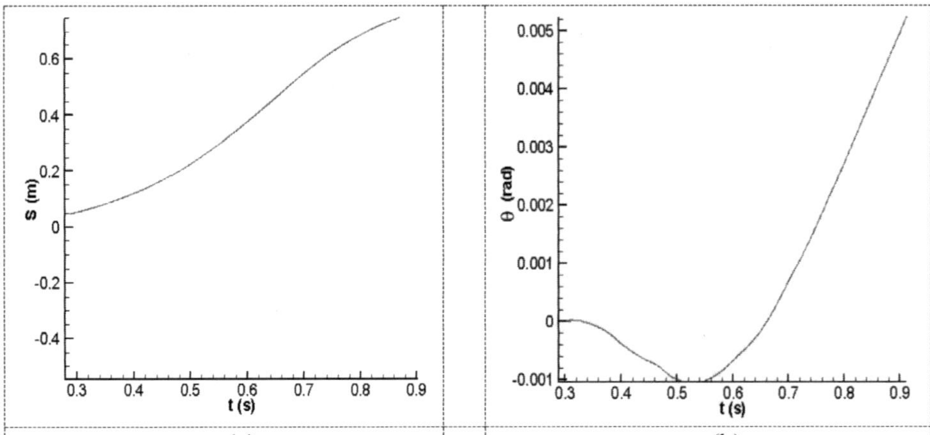

(a) (b)

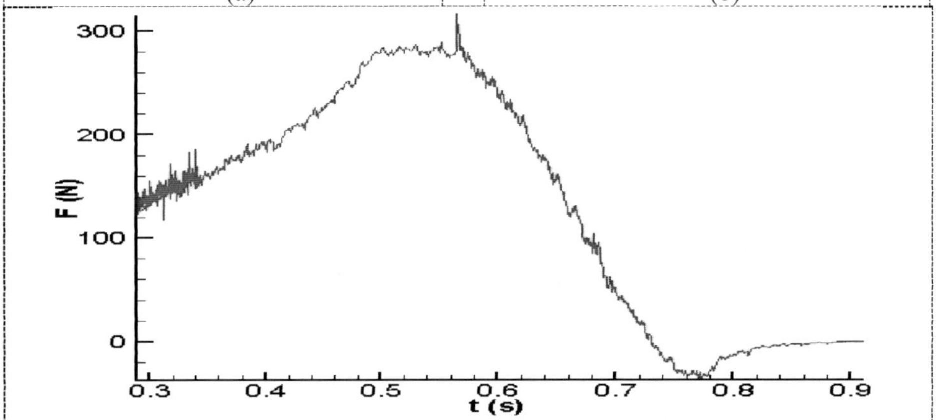

(c)

Figure 5. X-axis displacement, rotation angle, and force.

Figure 5 shows the parameters of the submarine's motion in the X-axis direction [16]. The wave force causes displacement of the submarine's center of mass in the X-direction. The fluid force in the horizontal direction reaches its maximum between 0.5 and 0.6 seconds, and the velocity of the center of mass moving in the X direction is also maximum during this period. The slope of the curve is highest between 0.5 and 0.6 seconds, as shown in Figure 5(a).

Journal of Physics: Conference Series **2815** (2024) 012053 doi:10.1088/1742-6596/2815/1/012053

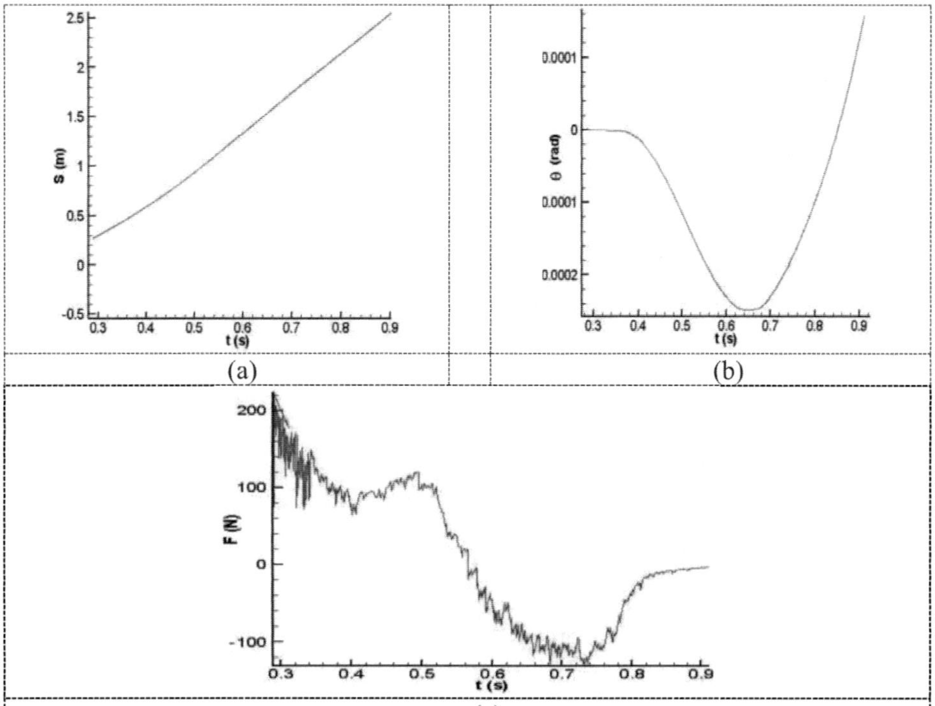

(a) (b)

(c)

Figure 6. Y-axis displacement, rotation angle, and force.

Figure 6 shows the kinematic parameters of the submarine in the Y-axis direction. The center of mass is significantly displaced in the Y-direction due to thrust. The buoyancy force on the submarine ejector gradually decreases while exiting the water, and thus, the fluid force on the Y-direction gradually decreases, as shown in Figure 6(c). There is a strong oscillation of the curve during this process, mainly caused by the free surface wave motion. The buoyancy force always maintains a consistent direction and change pattern. However, the wave force acting on an SLV varies due to location, speed, size, and direction differences. This results in a complex coupling between the wave force and buoyancy force, leading to the fluid force oscillation.

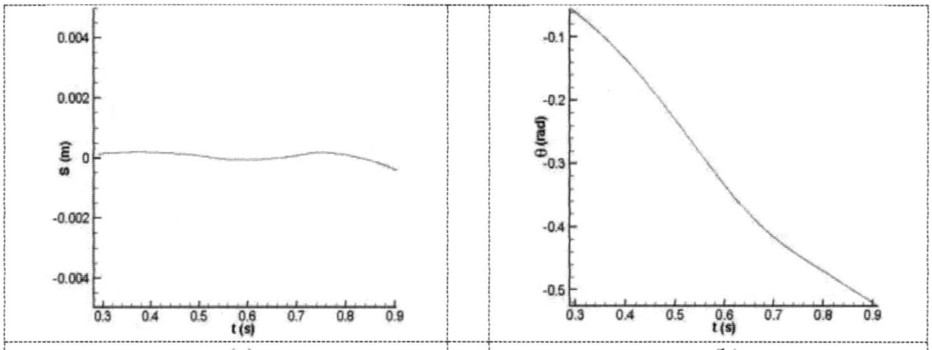

(a) (b)

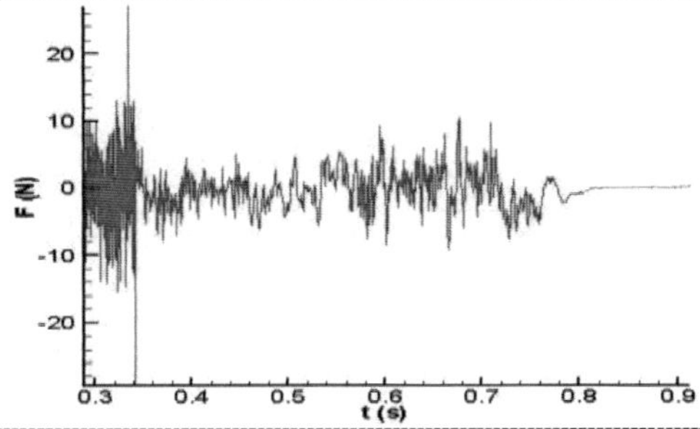

(c)

Figure 7. Z-axis displacement, rotation angle, and force.

Figure 7(b) shows that the submarine experiences a large deflection around the Z-axis. The wave force hits the head of the submarine projectile in the process of emerging from the water, which leads to a deflection moment around its center of mass. With the increase of the water emerging part, the position of the wave force gradually moves down, the distance between the action point and the center of mass is shortened, and the rotation tendency is slowed down.

6. Conclusion

In conclusion, this study presents a comprehensive analysis of the impact of wave effects on submarine-launched vehicles (SLVs). Using the ANSYS simulation platform, it was observed that wave action significantly affects the SLV's attitude and pressure distribution during the launch process. The findings highlight the complex interplay between wave forces and SLV dynamics, emphasizing the importance of considering environmental factors in the design and operation of such systems. The study contributes valuable insights for future research and development in underwater vehicle technology, particularly in understanding and mitigating the effects of marine disturbances on SLV performance. This research advances our knowledge in the field and lays a foundation for more sophisticated and reliable SLV designs in challenging marine environments.

Acknowledgment

This work is based on Project Shandong Provincial Key Research and Development Plan (No. 2022CXGC020410) and Project Shandong Provincial Key Research and Development Plan (No. 2021CXGC010702).

References

[1] Takamure, K., Uchiyama, T. (2023) Energy transition in air of a sphere launched vertically upward in water. Ocean Engineering, 207: 32. Doi: 10.1016/j.oceaneng.2020.107426.

[2] Li, D., Sankaran, V., Lindau, J.W., et al. (2005) A unified computational formulation for multi-component and multi-phase flows. Doi: 10.2514/6.2005-1391.

[3] Lindau, J.W., Kunz, R.F., Boger, D.A., et al. (2002) High reynolds number, unsteady, multiphase CFD modeling of cavitating flows. Journal of Fluids Engineering: Transactions of the ASME, 2002(3): 124.

[4] Weiland, C.J., Vlachos, P.P., Yagla, J. J. (2010) Concept analysis and laboratory observations on a water piercing missile launcher. Ocean Engineering, 37(11-12): 959-965. Doi: 10.1016/j.oceaneng.2010.03.009.

[5] Shen, L.H. (2009) Study on fluid characteristics of underwater and exiting vehicles. Harbin: Harbin Institute of Technology.

[6] Bai, Y.L. (2013) Study on dynamics and nonlinear control method of submarine-launched missile in muti-medium environment. Harbin: Harbin Institute of Technology.

[7] Zhong, X.F., Lin, L.M., Wu, Y.X., et al. (2012) New Method in Suppressing Vortex-Induced Vibration of Marine Riser. Applied Mechanics and Materials, 226-228: 9-12. Doi: 10.4028/www.scientific.net/AMM.226-228.9.

[8] Meng Jiacheng, Wei Tingting, Wang Yufei, Bai Zhaofeng, Wu Junxia, Gao Duorui & Xie Xiaoping. (2023). Two-stage frequency compensation for Doppler shift on BPSK transceiver with LDPC codes for free-space optical communication systems. Frontiers in Physics. Doi: 10.3389/FPHY.2023.1099867.

[9] Li Yue, Zhang Xueting & Shen Zhangyi. (2022). YOLO-Submarine Cable: An Improved YOLO-V3 Network for Object Detection on Submarine Cable Images. Journal of Marine Science and Engineering, 10(8). Doi: 10.3390/JMSE10081143.

[10] Glowacki Oskar. (2022). Distinguishing subaerial and submarine calving with underwater noise. Journal of Glaciology, 272. Doi: 10.1017/JOG.2022.32.

[11] Cao Chen, Ge Yongqiang, Ren Xueyu, Peng Xiaoqing, Chen Jiawang, Lu Zhifei & Zheng Xinlong. (2022). Experimental research on submarine cable monitoring method based on MEMS sensor. Micro and Nano Engineering. Doi: 10.1016/J.MNE.2022.100130.

[12] Sun Jinhuan, Zhang Chengchun, Sun Xiaowei, Du Tianyu, Shen Chun & Fan Shuqun. (2022). Weakening wave impact on submarine-launched vehicles by imitating humpback whale belly grooves. Ocean Engineering. Doi: 10.1016/J.OCEANENG.2021.110451.

[13] Shi Yao, Ren Jinyi, Gao Shan & Pan Guang. (2021). Numerical Investigation on Air Film Fusion of Pressure-Equalizing Exhaust around Shoulder Ventilation of Submarine-Launched Vehicle. Journal of Marine Science and Engineering, 10(1). Doi: 10.3390/JMSE10010039.

[14] (2020). Engineering - Marine Science and Engineering; Cyprus University of Technology Researchers Describe Recent Advances in Marine Science and Engineering (An Analysis Tool for the Installation of Submarine Cables in an S-Lay Configuration Including "In and Out of Water" Cable Segments). Computer Weekly News.

[15] (2020). Energy; Researchers from Harbin University of Science and Technology Report Findings in Energy (Design of Partial Discharge Test Environment for Oil-Filled Submarine Cable Terminals and Ultrasonic Monitoring). Energy & Ecology.

[16] Hai Long Huang, Wei Cao & Cong Wang. (2014). Numerical Simulation of the Launching of Underwater Vehicle. Applied Mechanics and Materials, 556-562. Doi: 10.4028/www.scientific.net/AMM.556-562.4404.

AUTHOR INDEX